Nonlinear and Mixed-Integer Optimization

TOPICS IN CHEMICAL ENGINEERING

A Series of Textbooks and Monographs

SERIES TITLES

Receptors: Models for Binding, Trafficking, and Signaling
D. Laufenburger and J. Linderman

Process Dynamics, Modeling, and Control
B. Ogunnaike and W. H. Ray

An Introduction to Theoretical and Computational Fluid Dynamics
C. Pozrikidis

Optical Rheometry of Complex Fluids
G. Fuller

Nonlinear and Mixed-Integer Optimization: Fundamentals and Applications
C. Floudas

Nonlinear and Mixed-Integer Optimization

Fundamentals and Applications

CHRISTODOULOS A. FLOUDAS

New York Oxford
OXFORD UNIVERSITY PRESS
1995

Oxford University Press

Oxford New York
Athens Auckland Bangkok Bombay
Calcutta Cape Town Dar es Salaam Delhi
Florence Hong Kong Istanbul Karachi
Kuala Lumpur Madras Madrid Melbourne
Mexico City Nairobi Paris Singapore
Taipei Tokyo Toronto

and associated companies in
Berlin Ibadan

Published by Oxford University Press, Inc.,
198 Madison Avenue, New York, New York 10016

Oxford is a registered trademark of Oxford University Press

Library of Congress Cataloging-in-Publication Data
Floudas, Christodoulos A.
Nonlinear and mixed-integer optimization : fundamentals and applications / Christodoulos A. Floudas.
p. cm. — (Topics in chemical engineering)
Includes bibliographical references and index.
ISBN 0-19-510056-5
1. Mathematical optimization. 2. Nonlinear theories. 3. Integer programming.
4. Chemical engineering—Mathematical models.
I. Title. II. Series: Topics in chemical engineering (Oxford University Press)
QA402.5.F587 1995
660'.28'015193—dc20 95-10917

9 8 7 6 5 4 3 2 1

Printed in the United States of America
on acid-free paper

To my wife, Fotini

Preface

Nonlinear and Mixed-Integer Optimization addresses the problem of optimizing an objective function subject to equality and inequality constraints in the presence of continuous and integer variables. These optimization models have many applications in engineering and applied science problems and this is the primary motivation for the plethora of theoretical and algorithmic developments that we have been experiencing during the last two decades.

This book aims at presenting the fundamentals of nonlinear and mixed-integer optimization, and their applications in the important area of process synthesis and chemical engineering. The first chapter introduces the reader to the generic formulations of this class of optimization problems and presents a number of illustrative applications. For the remaining chapters, the book contains the following three main parts:

- Part 1: Fundamentals of Convex Analysis and Nonlinear Optimization
- Part 2: Fundamentals of Mixed-Integer Optimization
- Part 3: Applications in Process Synthesis

Part 1, comprised of three chapters, focuses on the fundamentals of convex analysis and nonlinear optimization. Chapter 2 discusses the key elements of convex analysis (i.e., convex sets, convex and concave functions, and generalizations of convex and concave functions), which are very important in the study of nonlinear optimization problems. Chapter 3 presents the first and second order optimality conditions for unconstrained and constrained nonlinear optimization. Chapter 4 introduces the basics of duality theory (i.e., the primal problem, the perturbation function, and the dual problem) and presents the weak and strong duality theorem along with the duality gap. Part 1 outlines the basic notions of nonlinear optimization and prepares the reader for Part 2.

Part 2, comprised of two chapters, addresses the fundamentals and algorithms for mixed-integer linear and nonlinear optimization models. Chapter 5 provides the basic ideas in mixed-integer linear optimization, outlines the different methods, and discusses the key elements of branch and bound approaches. Chapter 6 introduces the reader to the theoretical and algorithmic developments in mixed-integer nonlinear optimization. After a brief description of the motivation and the formulation of such models, the reader is introduced to (i) decomposition-based approaches (e.g., Generalized Benders Decomposition, Generalized Gross Decomposition), (ii) linearization-based methods (e.g., Outer Approximation and its variants with Equality Relaxation and Augmented Penalty, and Generalized Outer Approximation), and (iii) comparison between decomposition- and linearization-based methods.

Part 3, consisting of four chapters, deals with important application areas in chemical engineering. Chapter 7 discusses the components of a chemical process system, defines the objectives in the area of process synthesis, and presents the different approaches in process synthesis. Subsequently, the reader is introduced to modeling issues in mixed-integer nonlinear optimization problems of process synthesis. Chapter 8 presents the application area of heat exchanger network synthesis. The reader is introduced to optimization models that correspond to (i) targeting methods for minimum utility cost and minimum number of matches, (ii) decomposition-based methods, and (iii) simultaneous optimization approaches for the synthesis of heat recovery networks. Chapter 9 presents applications of mixed-integer nonlinear optimization in the area of separations. In particular, the synthesis of sharp heat-integrated distillation columns and the synthesis of non-sharp separation columns are addressed. Chapter 10 discusses the application of mixed-integer nonlinear optimization methods in the synthesis of reactor networks with complex reactions and in the synthesis of prototype chemical processes consisting of reactor-separator-recycle systems.

The main objectives in the preparation of this book are (i) to acquaint the reader with the basics of convex analysis and nonlinear optimization without presenting the proofs of the theoretical results and the algorithmic details, which can be found in several other textbooks, (ii) to introduce the reader to the elementary notions of mixed-integer linear optimization first and to the theory and methods of mixed-integer nonlinear optimization next, which are not discussed in other textbooks, and (iii) to consider several key application areas of chemical engineering process synthesis and design in which the mixed-integer nonlinear optimization models and methods apply naturally. Special efforts have been made so as to make this book self-contained, and establish only the needed fundamentals in Part 1 to be used in Part 2. The modeling issues and application areas in Part 3 have been selected on the basis of the most frequently studied in the area of process synthesis in chemical engineering. All chapters have several illustrations and geometrical interpretations of the theoretical results presented; they include a list of recommended books and articles for further reading in each discussed topic, and the majority of the chapters contain suggested problems for the reader. Furthermore, in Part 3 the examples considered in each of the application areas describe the resulting mathematical models fully with the key objective to familiarize the reader with the modeling aspects in addition to the algorithmic ones.

This book has been prepared keeping in mind that it can be used as a textbook and as a reference. It can be used as a textbook on the fundamentals of nonlinear and mixed-integer optimization and as a reference for special topics in the mixed-integer nonlinear optimization part and the presented application areas. Material in this book has been used in graduate level courses in Optimization and Process synthesis at Princeton University, while Parts 1 and 2 were presented in a graduate level course at ETH. Selected material, namely chapters 3, 5, 7, and 8, has been used in the undergraduate design course at Princeton University as an introduction to optimization and process synthesis.

A number of individuals and institutions deserve acknowledgment for different kinds of help. First, I thank my doctoral students, postdoctoral associates, colleagues, and in particular, the chairman, Professor William B. Russel, at Princeton University for their support in this effort. Second, I express my gratitude to my colleagues in the Centre for Process Systems Engineering at Imperial College and in the Technical Chemistry at ETH for the stimulating environment and support they provided during my sabbatical leave. Special thanks go to Professors John Perkins, Roger W. H. Sargent, and David W. T. Rippin for their instrumental role in a productive and enjoyable sabbatical. Third, I am indebted to several colleagues and students who have provided inspiration, encouragement, extensive feedback, and helped me to complete this book. The thoughtful comments and constructive criticism of Professors Roger W. H. Sargent, Roy Jackson, Manfred Morari, Panos M. Pardalos, Amy R. Ciric, and Dr. Efstratios N. Pistikopoulos have helped enormously to improve the book. Claire Adjiman, Costas D. Maranas, Conor M. McDonald, and Vishy Visweswaran critically read several manuscript drafts and suggested helpful improvements. The preparation of the camera-ready copy of this book required a significant amount of work. Special thanks are reserved for Costas D. Maranas, Conor M. McDonald and Vishy Visweswaran for their time, LaTex expertise, and tremendous help in the preparation of this book. Without their assistance the preparation of this book would have taken much longer time. I am also thankful for the excellent professional assistance of the staff at Oxford University Press, especially Karen Boyd, who provided detailed editorial comments, and senior editor Robert L. Rogers. Finally and most importantly, I am very grateful to my wife, Fotini, and daughter, Ismini, for their support, encouragement, and forbearance of this seemingly never ending task.

C.A.F.

Princeton, New Jersey
March 1995

Contents

Nonlinear and Mixed-Integer Optimization

Chapter 1 Introduction

This chapter introduces the reader to elementary concepts of modeling, generic formulations for nonlinear and mixed integer optimization models, and provides some illustrative applications. Section 1.1 presents the definition and key elements of mathematical models and discusses the characteristics of optimization models. Section 1.2 outlines the mathematical structure of nonlinear and mixed integer optimization problems which represent the primary focus in this book. Section 1.3 illustrates applications of nonlinear and mixed integer optimization that arise in chemical process design of separation systems, batch process operations, and facility location/allocation problems of operations research. Finally, section 1.4 provides an outline of the three main parts of this book.

1.1 Mathematical and Optimization Models

A plethora of applications in all areas of science and engineering employ mathematical models. A mathematical model of a system is a set of mathematical relationships (e.g., equalities, inequalities, logical conditions) which represent an abstraction of the real world system under consideration. Mathematical models can be developed using (i) fundamental approaches, (ii) empirical methods, and (iii) methods based on analogy. In (i), accepted theories of sciences are used to derive the equations (e.g., Newton's Law). In (ii), input–output data are employed in tandem with statistical analysis principles so as to generate empirical or "black box" models. In (iii), analogy is employed in determining the essential features of the system of interest by studying a similar, well understood system.

A mathematical model of a system consists of four key elements:

(i) Variables,

(ii) Parameters,

(iii) Constraints, and

(iv) Mathematical relationships.

The variables can take different values and their specifications define different states of the system. They can be continuous, integer, or a mixed set of continuous and integer. The parameters are fixed to one or multiple specific values, and each fixation defines a different model. The constants are fixed quantities by the model statement.

The mathematical model relations can be classified as equalities, inequalities, and logical conditions. The model equalities are usually composed of mass balances, energy balances, equilibrium relations, physical property calculations, and engineering design relations which describe the physical phenomena of the system. The model inequalities often consist of allowable operating regimes, specifications on qualities, feasibility of heat and mass transfer, performance requirements, and bounds on availabilities and demands. The logical conditions provide the connection between the continuous and integer variables.

The mathematical relationships can be algebraic, differential, integrodifferential, or a mixed set of algebraic and differential constraints, and can be linear or nonlinear.

An optimization problem is a mathematical model which in addition to the aforementioned elements contains one or multiple performance criteria. The performance criterion is denoted as objective function, and it can be the minimization of cost, the maximization of profit or yield of a process for instance. If we have multiple performance criteria then the problem is classified as multi–objective optimization problem. A well defined optimization problem features a number of variables greater than the number of equality constraints, which implies that there exist degrees of freedom upon which we optimize. If the number of variables equals the number of equality constraints, then the optimization problem reduces to a solution of nonlinear systems of equations with additional inequality constraints.

1.2 Structure of Nonlinear and Mixed-Integer Optimization Models

In this book we will focus our studies on nonlinear and mixed integer optimization models and present the fundamental theoretical aspects, the algorithmic issues, and their applications in the area of *Process Synthesis* in chemical engineering. Furthermore, we will restrict our attention to algebraic models with a single objective. The structure of such nonlinear and mixed integer optimization models takes the following form:

$$
\begin{aligned}
\min_{\boldsymbol{x},\boldsymbol{y}} \quad & f(\boldsymbol{x},\boldsymbol{y}) \\
s.t. \quad & \boldsymbol{h}(\boldsymbol{x},\boldsymbol{y}) = \boldsymbol{0} \\
& \boldsymbol{g}(\boldsymbol{x},\boldsymbol{y}) \leq \boldsymbol{0} \\
& \boldsymbol{x} \in \boldsymbol{X} \subseteq \Re^n \\
& \boldsymbol{y} \in \boldsymbol{Y} \text{ integer}
\end{aligned}
\tag{1.1}
$$

where $\boldsymbol{x}$ is a vector of n continuous variables, $\boldsymbol{y}$ is a vector of integer variables, $\boldsymbol{h}(\boldsymbol{x},\boldsymbol{y}) = \boldsymbol{0}$ are m equality constraints, $\boldsymbol{g}(\boldsymbol{x},\boldsymbol{y}) \leq \boldsymbol{0}$ are p inequality constraints, and $f(\boldsymbol{x},\boldsymbol{y})$ is the objective function.

Formulation (1.1) contains a number of classes of optimization problems, by appropriate consideration or elimination of its elements. If the set of integer variables is empty, and the objective function and constraints are linear, then (1.1) becomes a linear programming LP problem. If the set of integer variables is empty, and there exist nonlinear terms in the objective function and/or constraints, then (1.1) becomes a nonlinear programming NLP problem. The fundamentals of nonlinear optimization are discussed in Part 1 of this book. If the set of integer variables is nonempty, the integer variables participate linearly and separably from the continuous, and the objective function and constraints are linear, then (1.1) becomes a mixed-integer linear programming MILP problem. The basics of mixed-integer linear optimization are discussed in Part 2, Chapter 5, of this book. If the set of integer variables is nonempty, and there exist nonlinear terms in the objective function and constraints, then (1.1) is a mixed-integer nonlinear programming MINLP problem. The fundamentals of MINLP optimization are discussed in Chapter 6. The last class of MINLP problems features many applications in engineering and applied science, and a sample of these are discussed in Part 3 of this book. It should also be mentioned that (1.1) includes the pure integer linear and nonlinear optimization problems which are not the subject of study of this book. The interested reader in pure integer optimization problems is referred to the books by Nemhauser and Wolsey (1988), Parker and Rardin (1988), and Schrijver (1986).

1.3 Illustrative Applications

Mixed-integer nonlinear optimization problems of the form (1.1) are encountered in a variety of applications in all branches of engineering, applied mathematics, and operations research. These represent currently very important and active research areas, and a partial list includes:

(i) **Process Synthesis**

  Heat Exchanger Networks
  Distillation Sequencing
  Mass Exchange Networks
  Reactor–based Systems
  Utility Systems
  Total Process Systems

(ii) **Design, Scheduling, and Planning of Batch Processes**

  Design and Retrofit of Multiproduct Plants
  Design and Scheduling of Multipurpose Plants

(iii) **Interaction of Design and Control**

(iv) **Molecular Product Design**

(v) **Facility Location and Allocation**

(vi) **Facility Planning and Scheduling**

(vii) **Topology of Transportation Networks**

Part 3 of this book presents a number of major developments and applications of MINLP approaches in the area of *Process Synthesis*. The illustrative examples for MINLP applications, presented next in this section, will focus on different aspects than those described in Part 3. In particular, we will consider: the binary distillation design of a single column, the retrofit design of multiproduct batch plants, and the multicommodity facility location/allocation problem.

1.3.1 Binary Distillation Design

This illustrative example is taken from the recent work on interaction of design and control by Luyben and Floudas (1994a) and considers the design of a binary distillation column which separates a saturated liquid feed mixture into distillate and bottoms products of specified purity. The objectives are the determination of the number of trays, reflux ratio, flow rates, and compositions in the distillation column that minimize the total annual cost. Figure (1.1) shows a superstructure for the binary distillation column.

Formulation of the mathematical model here adopts the usual assumptions of equimolar overflow, constant relative volatility, total condenser, and partial reboiler. Binary variables q_i denote the existence of trays in the column, and their sum is the number of trays N. Continuous variables represent the liquid flow rates L_i and compositions x_i, vapor flow rates V_i and compositions y_i, the reflux R_i and vapor boilup VB_i, and the column diameter D_i. The equations governing the model include material and component balances around each tray, thermodynamic relations between vapor and liquid phase compositions, and the column diameter calculation based on vapor flow rate. Additional logical constraints ensure that reflux and vapor boilup enter only on one tray and that the trays are arranged sequentially (so trays cannot be skipped). Also included are the product specifications. Under the assumptions made in this example, neither the temperature nor the pressure is an explicit variable, although they could easily be included if energy balances are required. A minimum and maximum number of trays can also be imposed on the problem.

For convenient control of equation domains, let $TR = \{1, \ldots, N\}$ denote the set of trays from the reboiler to the top tray and let $\{N_f\}$ be the feed tray location. Then $AF = \{N_f + 1, \ldots, N\}$ is the set of trays in the rectifying section and $BF = \{2, \ldots, N_f - 1\}$ is the set of trays in the stripping section. The following equations describe the MINLP model.

a. **Overall material and component balance**

$$\begin{aligned} D + B - F &= 0 \\ Dx_D + Bx_B - Fz &= 0 \end{aligned}$$

b. **Total condenser**

$$\begin{aligned} V_N - \sum_{i \in AF} R_i - D &= 0 \\ y_N - x_D &= 0 \end{aligned}$$

c. **Partial reboiler**

$$
\begin{aligned}
B + \sum_{i \in BF} VB_i - L_2 &= 0 \\
Bx_B + \left(\sum_{i \in BF} VB_i \right) y_B - L_2 x_2 &= 0
\end{aligned}
$$

d. **Phase equilibrium**

$$
\begin{aligned}
\alpha x_i - y_i[1 + x_i(\alpha - 1)] &= 0 \quad i = 2, \ldots, N \\
\alpha x_B - y_B[1 + x_B(\alpha - 1)] &= 0
\end{aligned}
$$

e. **Component balances**

$$
\begin{aligned}
L_i x_i + V_i y_i - L_{i+1} x_{i+1} - V_{i-1} y_{i-1} - R_i x_D &= 0 \quad i \in AF \\
L_i x_i + V_i y_i - L_{i+1} x_{i+1} - V_{i-1} y_{i-1} - Fz &= 0 \quad i = N_f \\
L_i x_i + V_i y_i - L_{i+1} x_{i+1} - V_{i-1} y_{i-1} - VB_i y_B &= 0 \quad i \in BF
\end{aligned}
$$

f. **Equimolar overflow**

$$
\begin{aligned}
V_i - V_{i-1} &= 0 \quad i \in AF \\
L_i - L_{i+1} &= 0 \quad i \in BF
\end{aligned}
$$

g. **Diameter**

$$
\begin{aligned}
v - k_v \cdot ff \cdot \sqrt{(\rho_L - \rho_V)/\rho_V} &= 0 \\
Di^2 - \frac{4 \cdot V_{N_f} \cdot MW}{\pi \cdot v \cdot \rho_V} &= 0
\end{aligned}
$$

h. **Reflux and boilup constraints**

$$
\begin{aligned}
R_i - F_{max}(q_i - q_{i+1}) &\leq 0 \quad i \in AF \\
VB_i - F_{max}(q_i - q_{i-1}) &\leq 0 \quad i \in BF
\end{aligned}
$$

i. **Product specifications**

$$
\begin{aligned}
x_D^{spec} - x_D &\leq 0 \\
x_B - x_B^{spec} &\leq 0
\end{aligned}
$$

j. **Sequential tray constraints**

$$
\begin{aligned}
q_i - q_{i-1} &\leq 0 \quad i \in AF \\
q_i - q_{i+1} &\leq 0 \quad i \in BF
\end{aligned}
$$

The economic objective function to be minimized is the cost, which combines the capital costs associated with building the column and the utility costs associated with operating the column. The form for the capital cost of the column depends upon the vapor boilup, the number of trays, and the column diameter

$$\begin{aligned} \text{Cost} &= \beta_{tax}(C_{LPS}\Delta H_{vap} + C_{CW}\Delta H_{cond})V + f(N, Di)/\beta_{pay} \ , \\ f(N, Di) &= 12.3[615 + 324Di^2 + 486(6 + 0.76N)Di] + 245N(0.7 + 1.5Di^2) \ , \end{aligned}$$

where the parameters include the tax factor β_{tax}, the payback period β_{pay}, the latent heats of vaporization ΔH_{vap} and condensation ΔH_{cond}, and the utility cost coefficients c_{LPS}, c_{CW}.

The model includes parameters for relative volatility α, vapor velocity v, tray spacing flow constant k_v, flooding factor ff, vapor ρ_V and liquid ρ_L densities, molecular weight MW, and some known upper bound on column flow rates F_{max}.

The essence of this particular formulation is the control of tray existence (governed by q_i) and the consequences for the continuous variables. In the rectifying section, all trays above the tray on which the reflux enters have no liquid flows, which eliminates any mass transfer on these trays where $q_i = 0$. The vapor composition does not change above this tray even though vapor flows remain constant. Similarly, in the stripping section, all trays below the tray on which the vapor boilup enters have no vapor flows and the liquid composition does not change below this tray even though liquid flows remain constant. The reflux and boilup constraints ensure that the reflux and boilup enter on only one tray.

It is worth noting that the above formulation of the binary distillation design features the binary variables q_i separably and linearly in the set of constraints. The objective function, however, has products of the diameter D_i and the number of trays N_i which are treated as integer variables.

1.3.2 Retrofit Design of Multiproduct Batch Plants

This illustrative example is taken from Fletcher *et al.* (1991) and corresponds to a retrofit design of multiproduct batch plants. Multiproduct batch systems make a number of related products employing the same equipment operating in the same sequence. The plant operates in stages and during each stage, taking a few days or weeks, a product is made. Since the products are different, each product features a different relationship between the volume at each stage and the final batch size, and as a result the limiting stage and batch size may be different for each product. Furthermore, the processing times at each stage may differ, as well as the limiting stage and cycle time for each product.

In preparation for the problem formulation, we define the products by the index i, and the total number manufactured by the fixed parameter N. One of the objectives is to determine the batch size, B_i, which is the quantity of product i produced in any one batch. The batch stages are denoted by the index j, and the total number of stages in the batch plant is the fixed parameter M. Each batch stage is assumed to consist of a number of units which are identical and operate in parallel. The number of units in a batch stage j of an existing plant is denoted by N_j^{old} and the units in the stage are denoted by the index m. The size of a unit in a batch stage of an existing plant is denoted by $(V_j^{old})_m$.

In a retrofit batch design, we optimize the batch plant profitability defined as the total production value minus the cost of any new equipment. The objective is to obtain a modified batch plant structure, an operating strategy, the equipment sizes, and the batch processing parameters. Discrete decisions correspond to the selection of new units to add to each stage of the plant and their type of operation. Continuous decisions are represented by the volume of each new unit and the batch processing variables which are allowed to vary within certain bounds.

New units may be added to any stage j in parallel to existing units. These new units at stage j are denoted by the index k, and binary variables y_{jk} are introduced so as to denote whether a new unit k is added at stage j. Upper bounds on the number of units that can be added at stage j and to the plant are indicated by Z_j and Z^U, respectively.

The operating strategy for each new unit involves discrete decisions since it allows for the options of

Option B_m: *operate in phase with existing unit m to increase its capacity*

Option C: *operate in sequence with the existing units to decrease the stage cycle time*

These are denoted by the binary variables $(y^B_{ijk})_m$ and y^C_{ijk}, respectively, and take the value of 1 if product i is produced via operating options B_m or C for the new unit k in stage j. The volume of the k_{th} new unit in stage j is denoted by V_{jk}, and the processing volume of product i required is indicated by $(V^B_{ijk})_m$ or V^C_{ijk} depending on the operating alternative.

The MINLP model for the retrofit design of a multiproduct batch plant takes the following form:

Objective function

$$MAX \sum_{i=1}^{N} p_i n_i B_i - \sum_{j=1}^{M} \sum_{k=1}^{Z_j} (K_j y_{jk} + c_j V_{jk})$$

a. Production targets

$$n_i B_i \leq Q_i, \quad i = 1, \ldots, N$$

b. Limiting cycle time constraints

$$N_j^{old} + \sum_{k=1}^{Z_j} y^C_{ijk} \geq \frac{t_{ij}}{T_{Li}}, \quad i = 1, \ldots, N; \; j = 1, \ldots, M$$

c. Operating time period constraint

$$\sum_{i=1}^{N} n_i T_{Li} \leq H$$

d. Upper bound on total new units

$$\sum_{j=1}^{M} \sum_{k=1}^{Z_j} y_{jk} \leq Z^U$$

e. Lower bound constraints for new units

$$V_{jk} \geq V_{jk}^{L} y_{jk}, \quad j = 1, \ldots, M; \; k = 1, \ldots, Z_j$$

f. Operation in phase or in sequence

$$y_{jk} = \sum_{m=1}^{N_j^{old}} (y_{ijk}^{B})_m + y_{ijk}^{C}, \quad i = 1, \ldots, N; \; j = 1, \ldots, M; \; k = 1, \ldots, Z_j$$

g. Volume requirement for option B_m

$$\sum_{k=1}^{Z_j} \left(V_{ijk}^{B}\right)_m + \left(V_j^{old}\right)_m \geq S_{ij} B_i, \quad i = 1, \ldots, N; \; j = 1, \ldots, M; \; m = 1, \ldots, Z_j^{old}$$

h. Volume requirement for option C

$$U\left(1 - y_{ijk}^{C}\right) + V_{ijk}^{C} \geq S_{ij} B_i, \quad i = 1, \ldots, N; \; j = 1, \ldots, M; \; m = 1, \ldots, N_j$$

i. Processing volume restrictions of new units

$$\left.\begin{array}{rcl} (V_{ijk}^{B})_m & \leq & V_{jk} \\ (V_{ijk}^{B})_m & \leq & U(y_{ijk}^{B})_m \end{array}\right\} \quad \begin{array}{l} i = 1, \ldots, N \\ j = 1, \ldots, M \\ k = 1, \ldots, Z_j \\ m = 1, \ldots, N_j^{old} \end{array}$$

$$\left.\begin{array}{rcl} V_{ijk}^{C} & \leq & V_{jk} \\ V_{ijk}^{C} & \leq & U y_{ijk}^{C} \end{array}\right\} \quad \begin{array}{l} i = 1, \ldots, N \\ j = 1, \ldots, M \\ k = 1, \ldots, Z_j \end{array}$$

j. Distinct arrangements of new units

$$y_{jk} \geq y_{j,k+1}, \quad i = 1, \ldots, N; \; k = 1, \ldots, Z_j$$

The above formulation is a mixed-integer nonlinear programming MINLP model and has the following characteristics. The binary variables appear linearly and separably from the continuous variables in both the objective and constraints, by defining a new set of variables $w_i = t_{ij}/T_{Li}$ and including the bilinear constraints $w_i T_{Li} = t_{ij}$. The continuous variables n_i, B_i, T_{Li}, w_i appear nonlinearly. In particular, we have bilinear terms of $n_i B_i$ in the objective and constraints, bilinear terms of $n_i T_{Li}$ and $w_i T_{Li}$ in the constraints. The rest of the continuous variables V_j, $(V_{ijk}^{B})_m$, V_{ijk}^{C} appear linearly in the objective function and constraints.

1.3.3 Multicommodity Facility Location–Allocation

The multicommodity capacity facility location–allocation problem is of primary importance in transportation of shipments from the original facilities to intermediate stations and then to the destinations. In this illustrative example we will consider such a problem which involves I plants, J distribution centers, K customers, and P products. The commodity flow of product p which is shipped from plant i, through distribution center j to customer k will be denoted by the continuous variable x_{ijkp}. It is assumed that each customer k is served by only one distribution center j. Data are provided for the total demand by customer k for commodity p, D_{kp}, the supply of commodity p at plant i denoted as S_{ip}, as well as the lower and upper bounds on the available throughput in a distribution center j denoted by V_j^L and V_j^U, respectively.

The objective is to minimize the total cost which includes shipping costs, setup costs, and throughput costs. The shipping costs are denoted via linear coefficients c_{ijkp} multiplying the commodity flows x_{ijkp}. The setup costs are denoted by f_j for establishing each distribution center j. The throughput costs for distribution center j consists of a constant, v_j multiplying a nonlinear functionality of the flow through the distribution center.

The set of constraints ensure that supply and demand requirements are met, provide the logical connection between the existence of a distribution, the assignment of customers to distribution centers, and the demand for commodities, and make certain that only one distribution center is assigned to a customer.

The binary variables correspond to the existence of a distribution center j, and the assignment of a customer k to a distribution center j. These are denoted as z_j and y_{jk}, respectively. The continuous variables are represented by the commodity flows x_{ijkp}. The mathematical formulation of this problem becomes

Objective function

$$\min \sum_{i\in I}\sum_{j\in J}\sum_{k\in K}\sum_{p\in P} c_{ijkp}x_{ijkp} + \sum_{j\in J}\left[f_i z_j + v_j\left(\sum_{j\in J}\sum_{k\in K}\sum_{p\in P} x_{ijkp}\right)^{2.5}\right]$$

a. Supply requirements

$$\sum_{j\in J}\sum_{k\in K} x_{ijkp} \leq S_{ip}, \quad i\in I,\ p\in P$$

b. Demand constraints

$$\sum_{i\in I} x_{ijkp} = D_{kp}y_{jk}, \quad j\in J,\ k\in K,\ p\in P$$

c. Logical constraints

$$\sum_{k\in K}\sum_{p\in P} D_{kp}y_{jk} \leq V_j^U z_j, \quad j\in J$$

$$\sum_{k\in K}\sum_{p\in P} D_{kp}y_{jk} \geq V_j^L z_j, \quad j\in J$$

d. Assignment constraints

$$\sum_{j \in J} y_{jk} = 1, \quad k \in K$$

e. Nonnegativity and integrality conditions

$$\begin{aligned} x_{ijkp} &\geq 0, \quad i \in I, \; j \in J, \; k \in K, \; p \in P \\ y_{jk}, \; z_j &= 0 - 1 \; j \in J, \; k \in K \end{aligned}$$

Note that in the above formulation the binary variables y_{jk}, z_j appear linearly and separably in the objective function and constraints. Note also that the continuous variables x_{ijkp} appear linearly in the constraints while we have a nonlinear contribution of such terms in the objective function.

1.4 Scope of the Book

The remaining chapters of this book form three parts. Part 1 presents the fundamental notions of convex analysis, the basic theory of nonlinear unconstrained and constrained optimization, and the basics of duality theory. Part 1 acquaints the reader with the important aspects of convex analysis and nonlinear optimization without presenting the proofs of the theoretical results and the algorithmic details which can be found in several other textbooks. The main objective of Part 1 is to prepare the reader for Part 2. Part 2 introduces first the elementary notions of mixed-integer linear optimization and focuses subsequently on the theoretical and algorithmic developments in mixed-integer nonlinear optimization. Part 3 introduces first the generic problems in the area of *Process Synthesis*, discusses key ideas in the mathematical modeling of process systems, and concentrates on the important application areas of heat exchanger networks, separation system synthesis, and reactor-based system synthesis.

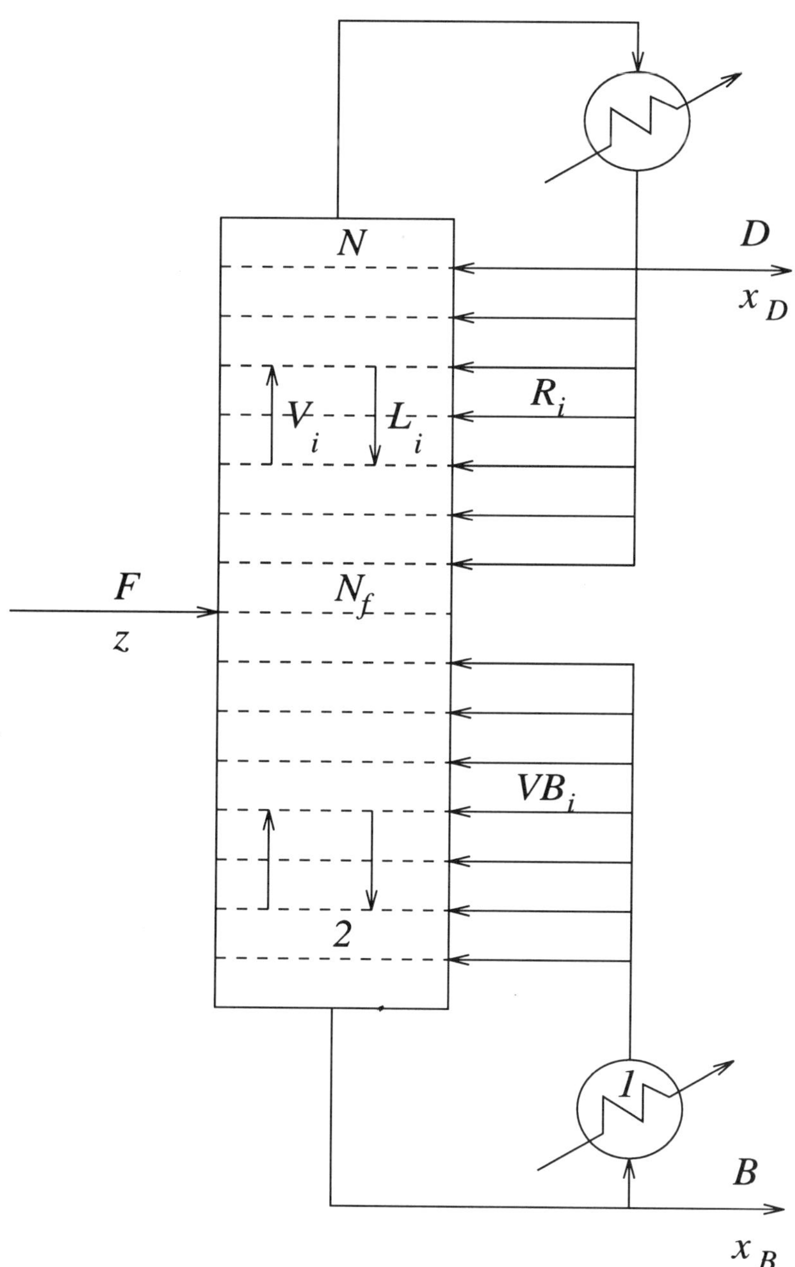

Figure 1.1: Superstructure for distillation column

Part 1

Fundamentals of Convex Analysis and Nonlinear Optimization

Chapter 2 Convex Analysis

This chapter discusses the elements of convex analysis which are very important in the study of optimization problems. In section 2.1 the fundamentals of convex sets are discussed. In section 2.2 the subject of convex and concave functions is presented, while in section 2.3 generalizations of convex and concave functions are outlined.

2.1 Convex Sets

This section introduces the fundamental concept of convex sets, describes their basic properties, and presents theoretical results on the separation and support of convex sets.

2.1.1 Basic Definitions

Definition 2.1.1 (Line) Let the vectors $\boldsymbol{x}_1$, $\boldsymbol{x}_2 \in \Re^n$. The *line* through $\boldsymbol{x}_1$ and $\boldsymbol{x}_2$ is defined as the set:

$$\{\boldsymbol{x} \mid \boldsymbol{x} = (1-\lambda)\boldsymbol{x}_1 + \lambda \boldsymbol{x}_2 , \lambda \in \Re\}.$$

Definition 2.1.2 (Closed line segment) Let the vectors $\boldsymbol{x}_1$, $\boldsymbol{x}_2 \in \Re^n$. The *closed line segment* through $\boldsymbol{x}_1$ and $\boldsymbol{x}_2$ is defined as the set:

$$\{\boldsymbol{x} \mid \boldsymbol{x} = (1-\lambda)\boldsymbol{x}_1 + \lambda \boldsymbol{x}_2,\ 0 \le \lambda \le 1\}.$$

The *open*, *closed-open*, and *open-closed* line segments can be defined similarly by modifying the inequalities for λ.

Illustration 2.1.1 Consider the line in $\Re^2$ which passes through the two points $x_1 = (1, 1)$ and $x_2 = (2, 3)$. The equation of this line is

$$2x - y = 1.$$

that is any point (x, y) satisfying the above equation lies on the line passing through $(1,1)$ and $(2,3)$. From definition 2.1.1, we can express any point x as

$$x = (1-\lambda)x_1 + \lambda x_2, \quad \text{or}$$

$$\begin{bmatrix} x \\ y \end{bmatrix} = (1-\lambda)\begin{bmatrix} 1 \\ 1 \end{bmatrix} + \lambda \begin{bmatrix} 2 \\ 3 \end{bmatrix} = \begin{bmatrix} 1+\lambda \\ 1+2\lambda \end{bmatrix}.$$

For $\lambda = 0.5$, we obtain $(x, y) = (1.5, 2)$, which lies on the line segment between $(1,1)$ and $(2,3)$. For $\lambda = 2$, we obtain $(x, y) = (3, 5)$, which lies on the line but not on the line segment between $(1,1)$ and $(2,3)$.

Definition 2.1.3 (Half-space) Let the vector $c \in \Re^n$, $c \neq 0$, and the scalar $z \in \Re$. The *open half-space* in $\Re^n$ is defined as the set:

$$\{x \mid c^t x < z, x \in \Re^n\}.$$

The *closed half-space* in $\Re^n$ is defined as the set:

$$\{x \mid c^t x \leq z, x \in \Re^n\}.$$

Definition 2.1.4 (Hyperplane) The *hyperplane* in $\Re^n$ is defined as the set:

$$\{x \mid c^t x = z, x \in \Re^n\}.$$

Illustration 2.1.2 The hyperplane in $\Re^2$

$$H = \{x \mid x_1 - x_2 = -1\}$$

divides $\Re^2$ into the half-spaces H_1 and H_2 as shown in Figure 2.1.

Definition 2.1.5 (Polytope and polyhedron) The intersection of a finite number of *closed half-spaces* in $\Re^n$ is defined as a *polytope*. A bounded polytope is called a *polyhedron*.

Definition 2.1.6 (Convex set) A set $S \in \Re^n$ is said to be *convex* if the *closed line segment* joining any two points x_1 and x_2 of the set S, that is, $(1 - \lambda)\, x_1 + \lambda\, x_2$, belongs to the set S for each λ such that $0 \leq \lambda \leq 1$.

Illustration 2.1.3 (Examples of convex sets) The following are some examples of convex sets:

(i) Line

(ii) Open and closed half-space

(iii) Polytope, polyhedron

(iv) All points inside or on the circle

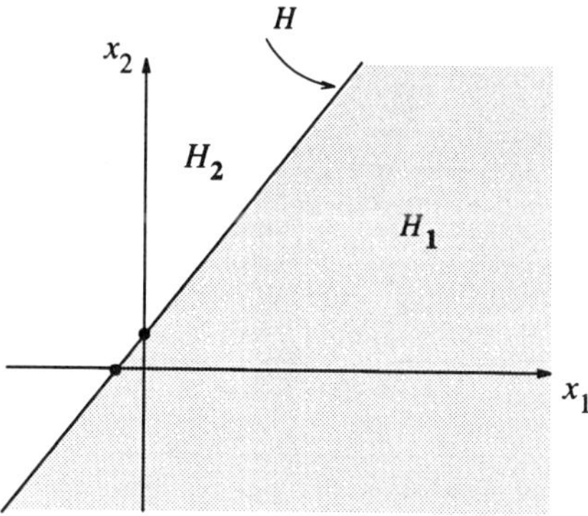

Figure 2.1: Half-spaces

(v) All points inside or on a polygon

Figure 2.2 illustrates convex and nonconvex sets.

Lemma 2.1.1 (Properties of convex sets) Let S_1 and S_2 be convex sets in $\Re^n$. Then,

(i) The intersection $S_1 \cap S_2$ is a convex set.

(ii) The sum $S_1 + S_2$ of two convex sets is a convex set.

(iii) The product θS_1 of the real number θ and the set S_1 is a convex set.

Definition 2.1.7 (Extreme point (vertex)) Let S be a convex set in $\Re^n$. The point $x \in S$ for which there exist no two distinct $x_1, x_2 \in S$ different from x such that $x \in [x_1, x_2]$, is called a *vertex* or *extreme point* of S.

Remark 1 A convex set may have no vertices (e.g., a line, an open ball), a finite number of vertices (e.g., a polygon), or an infinite number of vertices (e.g., all points on a closed ball).

Theorem 2.1.1 (Characterization of extreme points)
Let the polyhedron $S = \{x \mid Ax = b, x \geq 0\}$, where A is an $m \times n$ matrix of rank m, and b is an m vector. A point x is an extreme point of S if and only if A can be decomposed into $A = [B, N]$ such that:

$$x^t = [x_B, x_N] = [B^{-1}b, 0]$$

where B is an $m \times m$ invertible matrix satisfying $B^{-1} b \geq 0$, N is an $m \times (n - m)$ matrix, and x_B, x_N are the vectors corresponding to B, N.

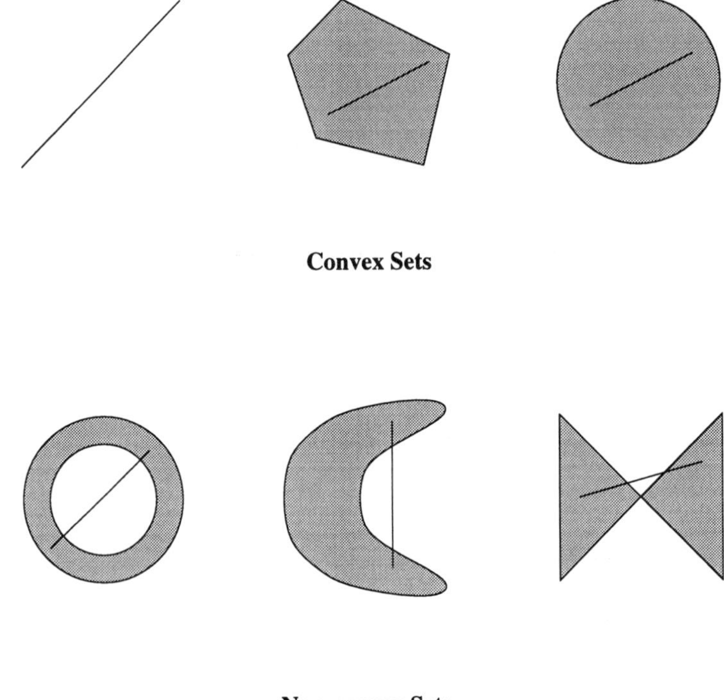

Figure 2.2: Convex and nonconvex sets

Remark 2 The number of extreme points of S is less than or equal to the maximum number of possible ways to select m columns of $\boldsymbol{A}$ to form $\boldsymbol{B}$, which is

$$\frac{n!}{m!(n-m)!} \cdot$$

Thus, S has a finite number of extreme points.

2.1.2 Convex Combination and Convex Hull

Definition 2.1.8 (Convex combination) Let $\{\boldsymbol{x}_1, \ldots, \boldsymbol{x}_r\}$ be any finite set of points in $\Re^n$. A convex combination of this set is a point of the form:

$$\begin{aligned} &\lambda_1 \boldsymbol{x}_1 + \ldots + \lambda_r \boldsymbol{x}_r \ , \\ &\lambda_1 + \ldots + \lambda_r = 1 \ , \\ &\qquad \lambda_1, \ldots, \lambda_r \geq 0. \end{aligned}$$

Remark 1 A convex combination of two points is in the closed interval of these two points.

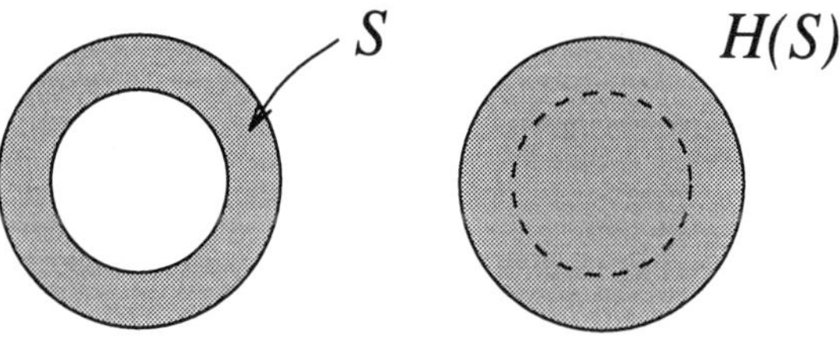

Figure 2.3: Convex hull

Definition 2.1.9 (Simplex) Let $\{x_0, \ldots, x_r\}$ be $r+1$ distinct points in $\Re^n$, $(r \leq n)$, and the vectors $x_1 - x_0, \ldots, x_r - x_0$ be linearly independent. An r-simplex in $\Re^n$ is defined as the set of all convex combinations of $\{x_0, \ldots, x_r\}$.

Remark 2 A 0-simplex (i.e., $r = 0$) is a point, a 1-simplex (i.e., $r = 1$) is a closed line segment, a 2-simplex (i.e., $r = 2$) is a triangle, and a 3-simplex (i.e., $r = 3$) is a tetrahedron.

Definition 2.1.10 (Convex hull) Let S be a set (convex or nonconvex) in $\Re^n$. The convex hull, $H(S)$, of S is defined as the intersection of all convex sets in $\Re^n$ which contain S as a subset.

Illustration 2.1.4 Figure 2.3 shows a nonconvex set S and its convex hull $H(S)$. The dotted lines in $H(S)$ represent the portion of the boundary of S which is not on the boundary of $H(S)$.

Theorem 2.1.2
The convex hull, $H(S)$, of S is defined as the set of all convex combinations of S. Then $x \in H(S)$ if and only if x can be represented as

$$x = \sum_{i=1}^{r} \lambda_i x_i \ ,$$
$$\sum_{i=1}^{r} \lambda_i = 1 \ ,$$
$$\lambda_i \geq 0 \, , \, i = 1, \ldots, r \ ,$$
$$x_i \in S \, , \, i = 1, \ldots, r \ ,$$

where r is a positive integer.

Remark 3 Any point x in the convex hull of a set S in $\Re^n$ can be written as a convex combination of at most $n+1$ points in S as demonstrated by the following theorem.

Theorem 2.1.3 (Caratheodory)
Let S be a set (convex or nonconvex) in $\Re^n$. If $x \in H(S)$, then it can be expressed as

$$x = \sum_{i=1}^{n+1} \lambda_i x_i \ ,$$

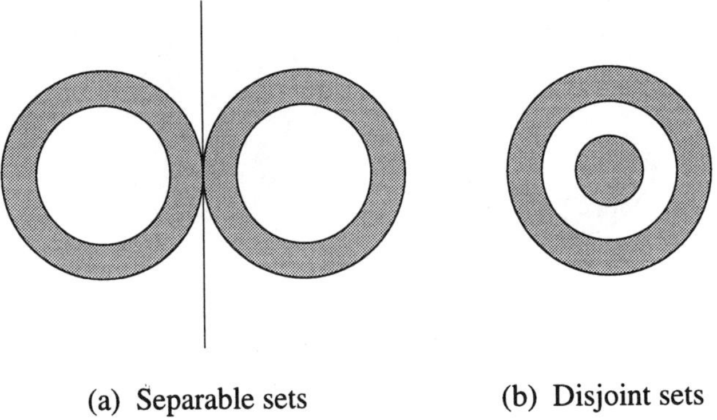

(a) Separable sets (b) Disjoint sets

Figure 2.4: Separating hyperplanes and disjoint sets

$$\sum_{i=1}^{n+1} \lambda_i = 1 \ ,$$
$$\lambda_i \geq 0 \, , \, i = 1, \ldots, n+1 \ ,$$
$$x_i \in S \, , \, i = 1, \ldots, n+1 \ .$$

2.1.3 Separation of Convex Sets

Definition 2.1.11 (Separating hyperplane) Let S_1 and S_2 be nonempty sets in $\Re^n$. The hyperplane

$$\{x \mid c^t x = z \ , \ x \in \Re^n\}$$

is said to separate (strictly separate) S_1 and S_2 if

$$x \in S_1 \Longrightarrow c^t x \leq z \ (c^t x < z) \ ,$$
$$x \in S_2 \Longrightarrow c^t x \geq z \ (c^t x > z).$$

Illustration 2.1.5 Figure 2.4(a) illustrates two sets which are separable, but which are neither disjoint or convex. It should be noted that separability does not imply that the sets are disjoint. Also, two disjoint sets are not in general separable as shown in Figure 2.4(b).

Theorem 2.1.4 (Separation of a convex set and a point)
Let S be a nonempty closed convex set in $\Re^n$, and a vector y which does not belong to the set S. Then there exist a nonzero vector c and a scalar z such that

$$c^t y > z$$

and

$$x \in S \Longrightarrow c^t x \leq z \ .$$

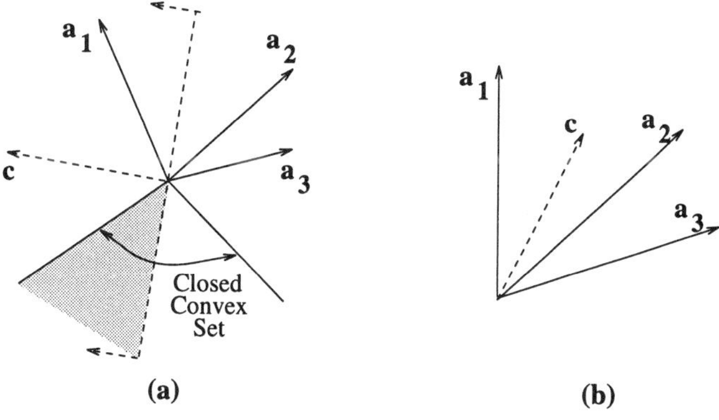

Figure 2.5: Illustration of Farkas' theorem

Theorem 2.1.5 (Farkas)
Let $\boldsymbol{A}$ be an $m \times n$ matrix and $\mathbf{c}$ be an n vector. Exactly one of the following two systems has a solution:

- *System 1:* $\boldsymbol{Ax} \leq 0$ *and* $\mathbf{c}^t\boldsymbol{x} > 0$ *for some* $\boldsymbol{x} \in \Re^n$.
- *System 2:* $\boldsymbol{A}^t\boldsymbol{y} = \boldsymbol{c}$ *and* $\boldsymbol{y} \geq 0$ *for some* $\boldsymbol{y} \in \Re^m$.

Illustration 2.1.6 Consider the cases shown in Figure 2.5(a,b). Let us denote the columns of A^t as $\boldsymbol{a}_1$, $\boldsymbol{a}_2$, and $\boldsymbol{a}_3$. System 1 has a solution if the closed convex cone defined by $\boldsymbol{Ax} \leq 0$ and the open half-space defined by $\mathbf{c}^t\boldsymbol{x} > 0$ have a nonempty intersection. System 2 has a solution if $\mathbf{c}$ lies within the convex cone generated by $\boldsymbol{a}_1$, $\boldsymbol{a}_2$, and $\boldsymbol{a}_3$.

Remark 1 *Farkas'* theorem has been used extensively in the development of optimality conditions for linear and nonlinear optimization problems.

Theorem 2.1.6 (Separation of two convex sets)
Let S_1 and S_2 be nonempty disjoint convex sets in $\Re^n$. Then, there exists a hyperplane

$$\{\boldsymbol{x} \mid \mathbf{c}^t\boldsymbol{x} = z \ , \ z \neq 0 \ , \ \boldsymbol{x} \in \Re^n\},$$

which separates S_1 and S_2; that is

$$\boldsymbol{x} \in S_1 \Longrightarrow \mathbf{c}^t\boldsymbol{x} \leq z \ ,$$
$$\boldsymbol{x} \in S_2 \Longrightarrow \mathbf{c}^t\boldsymbol{x} \geq z \ .$$

Theorem 2.1.7 (Gordan)
Let $\boldsymbol{A}$ be an $m \times n$ matrix and $\boldsymbol{y}$ be an m vector. Exactly one of the following two systems has a solution:

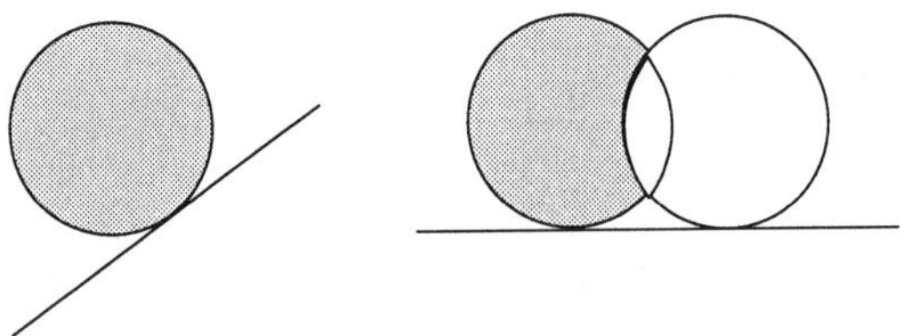

Figure 2.6: Supporting hyperplanes

- *System 1:* $\mathbf{A}\mathbf{x} < 0$ *for some* $\mathbf{x} \in \Re^n$.
- *System 2:* $\mathbf{A}^t\mathbf{y} = 0$ *and* $\mathbf{y} \geq 0$ *for some* $\mathbf{y} \in \Re^m$.

Remark 2 *Gordan's* theorem has been frequently used in the derivation of optimality conditions of nonlinearly constrained problems.

2.1.4 Support of Convex Sets

Definition 2.1.12 (Supporting hyperplane) Let S be a nonempty set in $\Re^n$, and $\bar{x}$ be in the boundary of S. The supporting hyperplane of S at $\bar{x}$ is defined as the hyperplane:

$$\{x \mid c^t(x - \bar{x}) = 0\}$$

that passes through $\bar{x}$ and has the property that all of S is contained in one of the two closed half-spaces:

$$\{x \mid c^t(x - \bar{x}) \leq 0\}$$

or

$$\{x \mid c^t(x - \bar{x}) \geq 0\}$$

produced by the hyperplane.

Illustration 2.1.7 Figure 2.6 provides a few examples of supporting hyperplanes for convex and nonconvex sets.

2.2 Convex and Concave Functions

This section presents (i) the definitions and properties of convex and concave functions, (ii) the definitions of continuity, semicontinuity and subgradients, (iii) the definitions and properties of differentiable convex and concave functions, and (iv) the definitions and properties of local and global extremum points.

2.2.1 Basic Definitions

Definition 2.2.1 (Convex function) Let S be a convex subset of $\Re^n$, and f(x) be a real valued function defined on S. The function f(x) is said to be convex if for any $x_1, x_2 \in S$, and $0 \leq \lambda \leq 1$, we have

$$f[(1-\lambda)x_1 + \lambda x_2] \leq (1-\lambda)f(x_1) + \lambda f(x_2).$$

This inequality is called *Jensen's inequality* after the Danish mathematician who first introduced it.

Definition 2.2.2 (Strictly convex function) Let S be a convex subset of $\Re^n$, and f(x) be a real valued function defined on S. The function f(x) is said to be strictly convex if for any $x_1, x_2 \in S$, and $0 < \lambda < 1$, we have

$$f[(1-\lambda)x_1 + \lambda x_2] < (1-\lambda)f(x_1) + \lambda f(x_2).$$

Remark 1 A strictly convex function on a subset S of $\Re^n$ is convex on S. The converse, however, is not true. For instance, a linear function is convex but not strictly convex.

Definition 2.2.3 (Concave function) Let S be a convex subset of $\Re^n$, and f(x) be a real valued function defined on S. The function f(x) is said to be concave if for any $x_1, x_2 \in S$, and $0 \leq \lambda \leq 1$, we have

$$f[(1-\lambda)x_1 + \lambda x_2] \geq (1-\lambda)f(x_1) + \lambda f(x_2).$$

Remark 2 The function $f(x)$ is concave on S if and only if $-f(x)$ is convex on S. Then, the results obtained for convex functions can be modified into results for concave functions by multiplication by -1 and vice versa.

Definition 2.2.4 (Strictly concave function) Let S be a convex subset of $\Re^n$, and f(x) be a real valued function defined on S. The function f(x) is said to be strictly concave if for any $x_1, x_2 \in S$, and $0 < \lambda < 1$, we have

$$f[(1-\lambda)x_1 + \lambda x_2] > (1-\lambda)f(x_1) + \lambda f(x_2).$$

Illustration 2.2.1 Figure 2.7 provides an illustration of convex, concave, and nonconvex functions in $\Re^1$.

2.2.2 Properties of Convex and Concave Functions

Convex functions can be combined in a number of ways to produce new convex functions as illustrated by the following:

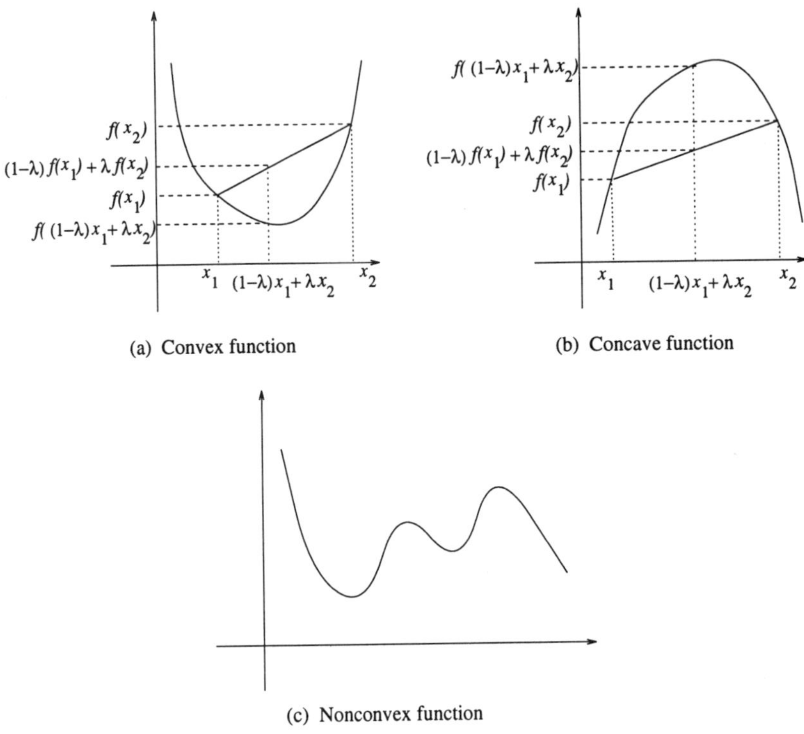

(a) Convex function (b) Concave function

(c) Nonconvex function

Figure 2.7: Convex, concave and nonconvex functions

(i) Let $f_1(\boldsymbol{x}), \ldots, f_n(\boldsymbol{x})$ be convex functions on a convex subset S of $\Re^n$. Then, their summation

$$f_1(\boldsymbol{x}) + \ldots + f_n(\boldsymbol{x})$$

is convex. Furthermore, if at least one $f_i(\boldsymbol{x})$ is strictly convex on S, then their summation is strictly convex.

(ii) Let f($\boldsymbol{x}$) be convex (strictly convex) on a convex subset S of $\Re^n$, and λ is a positive number. Then, λ f($\boldsymbol{x}$) is convex (strictly convex).

(iii) Let f($\boldsymbol{x}$) be convex (strictly convex) on a convex subset S of $\Re^n$, and g($\boldsymbol{y}$) be an increasing convex function defined on the range of f($\boldsymbol{x}$) in $\Re$. Then, the composite function g[f($\boldsymbol{x}$)] is convex (strictly convex) on S.

(iv) Let $f_1(\boldsymbol{x}), \ldots, f_n(\boldsymbol{x})$ be convex functions and bounded from above on a convex subset S of $\Re^n$. Then, the pointwise supremum function

$$f(\boldsymbol{x}) = \max\{f_1(\boldsymbol{x}), \ldots, f_n(\boldsymbol{x})\}$$

is a convex function on S.

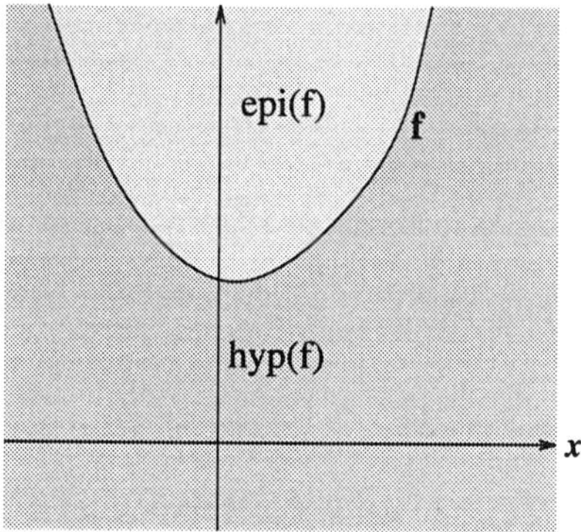

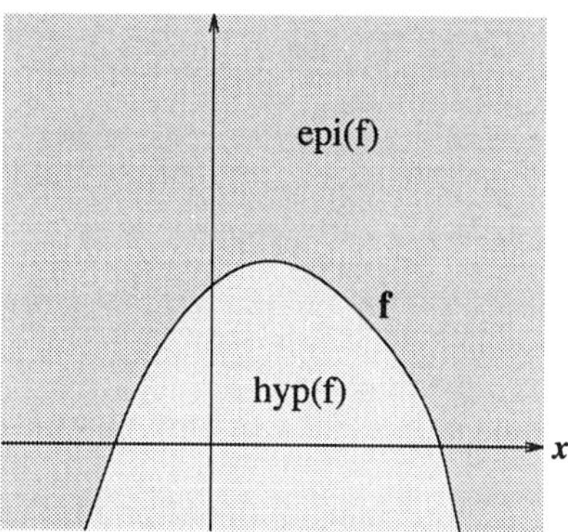

Figure 2.8: Epigraph and hypograph of a function

(v) Let $f_1(x), \ldots, f_n(x)$ be concave functions and bounded from below on a convex subset S of $\Re^n$. Then, the pointwise infimum function

$$f(x) = \min\{f_1(x), \ldots, f_n(x)\}$$

is a concave function on S.

Definition 2.2.5 (Epigraph of a function) Let S be a nonempty set in $\Re^n$. The epigraph of a function $f(x)$, denoted by $epi(f)$, is a subset of $\Re^{n+1}$ defined as the set of $(n+1)$ vectors (x, y):

$$\{\,(x, y) : f(x) \leq y, x \in S, y \in \Re\}.$$

Definition 2.2.6 (Hypograph of a function) The hypograph of $f(x)$, denoted by $hyp(f)$, is a subset of $\Re^{n+1}$ defined as the set of $(n+1)$ vectors (x, y):

$$\{\,(x, y) : f(x) \geq y, x \in S, y \in \Re\}.$$

Illustration 2.2.2 Figure 2.8 shows the epigraph and hypograph of a convex and concave function.

Theorem 2.2.1
Let S be a nonempty set in $\Re^n$. The function $f(x)$ is convex if and only if $epi(f)$ is a convex set.

Remark 1 The epigraph of a convex function and the hypograph of a concave function are convex sets.

2.2.3 Continuity and Semicontinuity

Definition 2.2.7 (Continuous function) Let S be a subset of $\Re^n$, $x^0 \in S$, and $f(x)$ a real valued function defined on S. $f(x)$ is continuous at x^0 if either of the following equivalent conditions hold:

Condition 1: For each $\epsilon_1 > 0$, there exists an $\epsilon_2 > 0$:
$\| \boldsymbol{x} - \boldsymbol{x}^o \| < \epsilon_2$, $\boldsymbol{x} \in S$ implies that

$$-\epsilon_1 < f(\boldsymbol{x}) - f(\boldsymbol{x}^0) < \epsilon_1 .$$

Condition 2: For each sequence $\boldsymbol{x}^1, \boldsymbol{x}^2, .. \boldsymbol{x}^n$ ($\boldsymbol{x} \in S$) converging to $\boldsymbol{x}^0$,

$$\lim_{n \to \infty} f(\boldsymbol{x}^n) = f(\lim_{n \to \infty} \boldsymbol{x}^n) = f(\boldsymbol{x}^0) ,$$

$f(\boldsymbol{x})$ is continuous on S if it is continuous at each $\boldsymbol{x}^0 \in S$.

Definition 2.2.8 (Lower semicontinuous function) $f(\boldsymbol{x})$ is lower semicontinuous at $\boldsymbol{x}^0$ if either of the following equivalent conditions hold:

Condition 1: For each $\epsilon_1 > 0$, there exists an $\epsilon_2 > 0$:
$\| \boldsymbol{x} - \boldsymbol{x}^o \| < \epsilon_2$, $\boldsymbol{x} \in S$ implies that

$$-\epsilon_1 < f(\boldsymbol{x}) - f(\boldsymbol{x}^0) .$$

Condition 2: For each sequence $\boldsymbol{x}^1, \boldsymbol{x}^2, \ldots, \boldsymbol{x}^n$ ($\boldsymbol{x} \in S$) converging to $\boldsymbol{x}^0$,

$$\lim_{n \to \infty} \inf f(\boldsymbol{x}^n) \geq f(\lim_{n \to \infty} \boldsymbol{x}^n) = f(\boldsymbol{x}^0) ,$$

where $\lim_{n \to \infty} \inf f(\boldsymbol{x}^n)$ is the infimum of the limit points of the sequence $f(\boldsymbol{x}^1), f(\boldsymbol{x}^2), \ldots, f(\boldsymbol{x}^n)$.

$f(\boldsymbol{x})$ is lower semicontinuous on S if it is lower semicontinuous at each $\boldsymbol{x}^0 \in S$.

Definition 2.2.9 (Upper semicontinuous function) $f(\boldsymbol{x})$ is upper semicontinuous at $\boldsymbol{x}^0$ if either of the following equivalent conditions hold:

Condition 1: For each $\epsilon_1 > 0$, there exists an $\epsilon_2 > 0$:
$\| \boldsymbol{x} - \boldsymbol{x}^o \| < \epsilon_2$, $\boldsymbol{x} \in S$ implies that

$$f(\boldsymbol{x}) - f(\boldsymbol{x}^0) < \epsilon_1 .$$

Condition 2: For each sequence $\boldsymbol{x}^1, \boldsymbol{x}^2, \ldots, \boldsymbol{x}^n$ ($\boldsymbol{x} \in S$) converging to $\boldsymbol{x}^0$,

$$\lim_{n \to \infty} \sup f(\boldsymbol{x}^n) \leq f(\lim_{n \to \infty} \boldsymbol{x}^n) = f(\boldsymbol{x}^0) ,$$

where $\lim_{n \to \infty} \sup f(\boldsymbol{x}^n)$ is the supremum of the limit points of the sequence $f(\boldsymbol{x}^1), f(\boldsymbol{x}^2), \ldots, f(\boldsymbol{x}^n)$.

$f(\boldsymbol{x})$ is upper semicontinuous on S if it is upper semicontinuous at each $\boldsymbol{x}^0 \in S$.

Remark 1 $f(\boldsymbol{x})$ is lower semicontinuous at $\boldsymbol{x}^0 \in S$ if and only if $-f(\boldsymbol{x})$ is upper semicontinuous at $\boldsymbol{x}^0 \in S$.

Remark 2 $f(\boldsymbol{x})$ is continuous at $\boldsymbol{x}^0 \in S$ if and only if it is both lower and upper semicontinuous at $\boldsymbol{x}^0 \in S$.

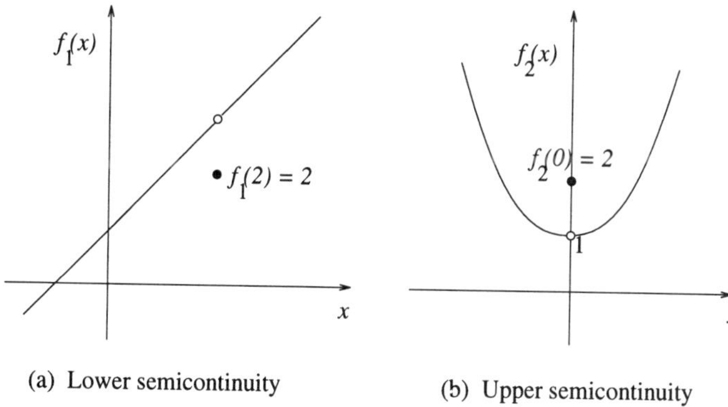

(a) Lower semicontinuity

(b) Upper semicontinuity

Figure 2.9: Lower and upper semicontinuous functions

Illustration 2.2.3 Consider the functions

$$f_1(x) = \begin{cases} x+1 & for\ x \neq 2\ , \\ 2 & for\ x = 2\ , \end{cases}$$

$$f_2(x) = \begin{cases} x^2+1 & for\ x \neq 0\ , \\ 2 & for\ x = 0\ , \end{cases}$$

which are shown in Figure 2.9. $f_1(x)$ is lower semicontinuous at $x = 2$, while $f_2(x)$ is upper semicontinuous at $x = 2$. Hence, $f_1(x)$ is lower semicontinuous and $f_2(x)$ is upper semicontinuous.

Theorem 2.2.2
Let S be a nonempty convex set in $\Re^n$ and $f(x)$ be a convex function. Then, $f(x)$ is continuous on the interior of S.

Remark 3 Convex and concave functions may not be continuous everywhere but the points of discontinuity have to be on the boundary of S.

Theorem 2.2.3
Let $f_i(x)$ be a family of lower (upper) semicontinuous functions on S. Then

(i) Its least upper bound (greatest lower bound)

$$f(x) = \sup_i f_i(x) \quad [f'(x) = \inf_i f_i(x)]$$

is lower (upper) semicontinuous on S.

(ii) If the family $f_i(x)$ is finite, its greatest lower bound (least upper bound)

$$f'(x) = \inf_i f_i(x) \quad [f(x) = \sup_i f_i(x)]$$

is lower (upper) semicontinuous on S.

2.2.4 Directional Derivative and Subgradients

Definition 2.2.10 (Directional derivative) Let S be a nonempty convex set in $\Re^n$, $\boldsymbol{x}^0 \in S$, and $\boldsymbol{y}$ be a nonzero vector such that $(\boldsymbol{x}^0 + \lambda \boldsymbol{y}) \in S$ for sufficiently small and strictly positive λ. The directional derivative of $f(\boldsymbol{x})$ at the point $\boldsymbol{x}^0$, along the direction $\boldsymbol{y}$, denoted as $f'(\boldsymbol{x}^0, \boldsymbol{y})$, is defined as the limit ($\pm\infty$ included) of

$$f'(\boldsymbol{x}^0, \boldsymbol{y}) = \lim_{\lambda \to 0+} \frac{f(\boldsymbol{x}^0 + \lambda \boldsymbol{y}) - f(\boldsymbol{x}^0)}{\lambda}.$$

Definition 2.2.11 (Subgradient of convex function) Let S be a nonempty convex set in $\Re^n$ and $f(\boldsymbol{x})$ be a convex function. The subgradient of $f(\boldsymbol{x})$ at $\boldsymbol{x}^0 \in S$, denoted by the vector $\boldsymbol{d}$, is defined as

$$f(\boldsymbol{x}) \geq f(\boldsymbol{x}^0) + \boldsymbol{d}^T(\boldsymbol{x} - \boldsymbol{x}^0) \quad \forall\, \boldsymbol{x} \in S. \tag{2.1}$$

Remark 1 The right-hand side of the above inequality (2.1) is a linear function in $\boldsymbol{x}$ and represents the first-order Taylor expansion of $f(\boldsymbol{x})$ around $\boldsymbol{x}^0$ using the vector $\boldsymbol{d}$ instead of the gradient vector of $f(\boldsymbol{x})$ at $\boldsymbol{x}^0$. Hence, $\boldsymbol{d}$ is a subgradient of $f(\boldsymbol{x})$ at $\boldsymbol{x}^0$ if and only if the first-order Taylor approximation always provides an underestimation of $f(\boldsymbol{x})$ for all $\boldsymbol{x}$.

Illustration 2.2.4 Consider the convex function

$$f(\boldsymbol{x}) = x^2 + 1,$$

the set $S = \{x \mid -2 \leq x \leq 2\}$ and the point $\boldsymbol{x}^0 = 1$. Let us assume that $d = 2$. The right-hand side of (2.1) is

$$f(\boldsymbol{x}^0) + d(x - \boldsymbol{x}^0) = 2 + 2(x - 1) = 2x.$$

Note that (2.1) holds for $d = 2$, and hence $d = 2$ is a subgradient for $f(\boldsymbol{x})$ at $\boldsymbol{x}^0 = 1$ (see also Figure 2.10).

Definition 2.2.12 (Subgradient of a concave function) Let S be a nonempty convex set in $\Re^n$ and $f(\boldsymbol{x})$ be a concave function. The subgradient of $f(\boldsymbol{x})$ at $\boldsymbol{x}^0 \in S$, denoted by the vector $\boldsymbol{d}$, is defined as:

$$f(\boldsymbol{x}) \leq f(\boldsymbol{x}^0) + \boldsymbol{d}^T(\boldsymbol{x} - \boldsymbol{x}^0) \quad \forall\, \boldsymbol{x} \in S. \tag{2.2}$$

Definition 2.2.13 (Subdifferential) The set of all subgradients of a function $f(\boldsymbol{x})$ at $\boldsymbol{x}^0 \in S$, denoted by $\partial f(\boldsymbol{x}^0)$, is the subdifferential of $f(\boldsymbol{x})$ at $\boldsymbol{x}^0$.

Theorem 2.2.4
Let S be a nonempty convex set in $\Re^n$. If, for all points $\boldsymbol{x}^0 \in int\ S$ there exists a subgradient vector $\boldsymbol{d}$:

$$f(\boldsymbol{x}) \geq f(\boldsymbol{x}^0) + \boldsymbol{d}^T(\boldsymbol{x} - \boldsymbol{x}^0) \quad \forall\, \boldsymbol{x} \in S\,,$$

then, $f(\boldsymbol{x})$ is convex on $int\ S$.

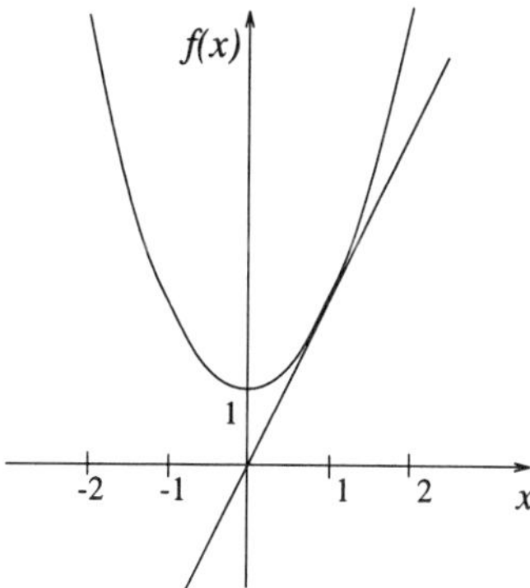

Figure 2.10: Subgradient of a function

2.2.5 Differentiable Convex and Concave Functions

Definition 2.2.14 (Differentiable function) Let S be a nonempty open set in $\Re^n$, $f(x)$ be a function defined on S, and $x^0 \in S$. Then, $f(x)$ is differentiable at x^0 if for all $(\Delta x = x - x^0) \in \Re^n$ such that $(x^0 + \Delta x) \in S$ we have

$$f(x^0 + \Delta x) = f(x^0) + \nabla f(x^0)\Delta x + \alpha(x^0, \Delta x) \parallel \Delta x \parallel$$

with

$$\lim_{\Delta x \to 0} \alpha(x^0, \Delta x) = 0 \ ,$$

where $\alpha(x^0, \Delta x)$ is a function of Δx, and $\nabla f(x^0)$ is the gradient of $f(x)$ evaluated at x^0 (the n-dimensional vector of the partial derivatives of $f(x)$ with respect to $x_1, x_2, \ldots, x_n$); that is,

$$\nabla f(x^0) = \left[\frac{\partial f(x^0)}{\partial x_1}, \ldots, \frac{\partial f(x^0)}{\partial x_n} \right] .$$

$f(x)$ is differentiable on S if it is differentiable at each $x^0 \in S$.

Remark 1 If $f(x)$ is differentiable at x^0, then $f(x)$ is continuous at x^0 and $\nabla f(x^0)$ exists. However, the converse is not true.

Remark 2 If $\nabla f(x^0)$ exists and $\nabla f(x)$ is continuous at x^0, then $f(x)$ is differentiable at x^0.

Definition 2.2.15 (Twice differentiable function) Let S be a nonempty open set in $\Re^n$, $f(x)$ be a function defined on S, and $x^0 \in S$. $f(x)$ is twice differentiable at x^0 if for all $(\Delta x = x - x^0) \in \Re^n$ such that $(x^0 + \Delta x) \in S$ we have

$$f(x^0 + \Delta x) = f(x^0) + \nabla f(x^0)\Delta x + \frac{\Delta x^T \nabla^2 f(x^0) \Delta x}{2} + \beta(x^0, \Delta x) \parallel \Delta x \parallel^2 \ ,$$

with

$$\lim_{\Delta x \to 0} \beta(x^0, \Delta x) = 0 \ ,$$

where $\beta(x^0, \Delta x)$ is a function of Δx, and $\nabla^2 f(x^0)$ is the Hessian of $f(x)$ evaluated at x^0, that is, an $n \times n$ matrix whose ij^{th} element is $\frac{\partial^2 f(x^0)}{\partial x_i \partial x_j}$.

Remark 3 If $\nabla f(x)$ is differentiable at x^0 (i.e., it has continuous partial derivatives at x^0), then $f(x)$ is twice differentiable at x^0.

Remark 4 If $\nabla^2 f(x)$ is continuous at x^0, then

$$\frac{\partial^2 f(x^0)}{\partial x_i \partial x_j} = \frac{\partial}{\partial x_i}\left(\frac{\partial f(x^0)}{\partial x_j}\right), \quad \text{and}$$

$\nabla^2 f(x^0)$ is symmetric $((\nabla^2 f(x))_{ij} = (\nabla^2 f(x))_{ji})$.

Theorem 2.2.5
Let S be a nonempty open set in $\Re^n$, and $f(x)$ a differentiable function at $x^0 \in S$.

(i) If $f(x)$ is convex at $x^0 \in S$, then

$$f(x) - f(x^0) \geq \nabla f(x^0)(x - x^0) \quad \forall\, x \in S.$$

(ii) If $f(x)$ is concave at $x^0 \in S$, then

$$f(x) - f(x^0) \leq \nabla f(x^0)(x - x^0) \quad \forall\, x \in S.$$

Theorem 2.2.6
Let S be a nonempty open set in $\Re^n$, and $f(x)$ a differentiable function on S.

(i) $f(x)$ is convex on S if and only if

$$f(x^2) - f(x^1) \geq \nabla f(x^1)(x^2 - x^1) \quad \forall\, x^1, x^2 \in S.$$

(ii) $f(x)$ is concave on S if and only if

$$f(x^2) - f(x^1) \leq \nabla f(x^1)(x^2 - x^1) \quad \forall\, x^1, x^2 \in S.$$

Remark 5 The above two theorems can be directly extended to strictly convex and strictly concave functions by replacing the inequalities $\geq$ and $\leq$ with strict inequalities $>$ and $<$.

Illustration 2.2.5 Figure 2.11 shows a differentiable convex and concave function, as well as their linearizations around a point x^1. Note that the linearization always underestimates the convex function and always overestimates the concave function.

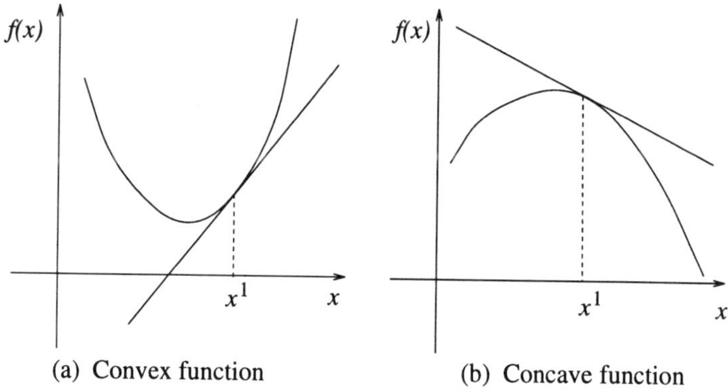

(a) Convex function (b) Concave function

Figure 2.11: Differentiable functions and linearizations

Theorem 2.2.7
Let S be a nonempty open set in $\Re^n$, and $f(\boldsymbol{x})$ be a twice differentiable function at $\boldsymbol{x}^0 \in S$.

(i) If $f(\boldsymbol{x})$ is convex at $\boldsymbol{x}^0$, then

$$\nabla^2 f(\boldsymbol{x}^0) \quad \text{is positive semidefinite (i.e. } z^T \nabla^2 f(\boldsymbol{x}^0) \cdot z \geq 0, \ \forall\, z \in \Re^n\text{).}$$

(ii) If $f(\boldsymbol{x})$ is concave at $\boldsymbol{x}^0$, then

$$\nabla^2 f(\boldsymbol{x}^0) \quad \text{is negative semidefinite (i.e. } z^T \nabla^2 f(\boldsymbol{x}^0) \cdot z \leq 0, \ \forall\, z \in \Re^n\text{).}$$

Theorem 2.2.8
Let S be a nonempty open set in $\Re^n$, and $f(\boldsymbol{x})$ be a twice differentiable function at $\boldsymbol{x}^0 \in S$.

(i) $f(\boldsymbol{x})$ is convex on S if and only if

$$\nabla^2 f(\boldsymbol{x}) \quad \text{is positive semidefinite on } S \text{ for all } \boldsymbol{x} \in S.$$

(ii) $f(\boldsymbol{x})$ is concave on S if and only if

$$\nabla^2 f(\boldsymbol{x}) \quad \text{is negative semidefinite on } S \text{ for all } \boldsymbol{x} \in S.$$

Remark 6 (i) If $f(\boldsymbol{x})$ is strictly convex at $\boldsymbol{x}^0$, then $\nabla^2 f(\boldsymbol{x}^0)$ is positive semidefinite (not necessarily positive definite).

(ii) If $\nabla^2 f(\boldsymbol{x}^0)$ is positive definite, then $f(\boldsymbol{x})$ is strictly convex at $\boldsymbol{x}^0$.

(iii) If $f(\boldsymbol{x})$ is strictly concave at $\boldsymbol{x}^0$, then $\nabla^2 f(\boldsymbol{x}^0)$ is negative semidefinite (not necessarily negative definite).

(iv) If $\nabla^2 f(\boldsymbol{x}^0)$ is negative definite, then $f(\boldsymbol{x})$ is strictly concave at $\boldsymbol{x}^0$.

Remark 7 Theorem 2.2.8 provides the conditions for checking the convexity or concavity of a function $f(x)$. These conditions correspond to positive semidefinite (P.S.D.) or negative semidefinite (N.S.D.) Hessian of $f(x)$ for all $x \in S$, respectively. One test of PSD or NSD Hessian of $f(x)$ is based on the sign of eigenvalues of the Hessian. If all eigenvalues are greater than or equal to zero for all $x \in S$, then the Hessian is PSD and hence the function $f(x)$ is convex. If all eigenvalues are less or equal than zero for all $x \in S$ then the Hessian is NSD and therefore the function $f(x)$ is concave.

Illustration 2.2.6 Consider the function

$$f(x_1, x_2, x_3) = x_1^2 + x_2^2 + x_3^2 + 2x_1x_2.$$

The Hessian of $f(x_1, x_2, x_3)$ is

$$H = \begin{pmatrix} 2 & 2 & 0 \\ 2 & 2 & 0 \\ 0 & 0 & 2 \end{pmatrix}$$

The eigenvalues of the Hessian of $f(x)$ are calculated from

$$|H - \lambda I| = 0$$

which becomes

$$\begin{vmatrix} 2-\lambda & 2 & 0 \\ 2 & 2-\lambda & 0 \\ 0 & 0 & 2-\lambda \end{vmatrix} = 0.$$

After algebraic manipulation the determinant becomes

$$(2-\lambda)(-\lambda)(4-\lambda) = 0$$

which implies that

$$\lambda = 2, 0, 4 \geq 0$$

Therefore, the function $f(x_1, x_2, x_3)$ is convex.

2.2.6 Minimum (Infimum) and Maximum (Supremum)

Definition 2.2.16 (Minimum) Let $f(x)$ be a function defined on the set S. If there exists $x^\star \in S$:

$$x \in S \quad \text{implies} \quad f(x) \geq f(x^\star),$$

then $f(x^\star)$ is called the minimum of $f(x)$ on S, denoted by

$$f(x^\star) = \min_{x \in S} f(x).$$

Definition 2.2.17 (Infimum) Let $f(x)$ be a function defined on the set S. If there exists a number α:

(i)

$$x \in S \quad \text{implies} \quad f(x) \geq \alpha, \text{ and}$$

(ii) for sufficiently small $\epsilon > 0$ there exists $x \in S$:

$$\alpha + \epsilon > f(x) \ ,$$

then α is the infimum of $f(x)$ on S, denoted by

$$\alpha = \inf_{x \in S} f(x) \ .$$

Definition 2.2.18 (Maximum) Let $f(x)$ be a function defined on the set S. If there exists $x^\star \in S$:

$$x \in S \quad \text{implies} \quad f(x) \leq f(x^\star),$$

then $f(x^\star)$ is called the maximum of $f(x)$ on S, denoted by

$$f(x^\star) = \max_{x \in S} f(x).$$

Definition 2.2.19 (Supremum) Let $f(x)$ be a function defined on the set S. If there exists a number β:

(i)

$$x \in S \quad \text{implies} \quad f(x) \leq \beta, \text{ and}$$

(ii) for sufficiently small $\epsilon > 0$ there exists $x \in S$:

$$\beta - \epsilon < f(x) \ ,$$

then β is the supremum of $f(x)$ on S, denoted by

$$\beta = \sup_{x \in S} f(x) \ .$$

Remark 1 If we admit the points $\pm\infty$, then every function $f(x)$ has a supremum and infimum on the set S.

Remark 2 The minimum (maximum) of a function $f(x)$, if it exists, must be finite, and is an attained infimum (supremum); that is,

$$f(x^\star) = \min_{x \in S} f(x) = \inf_{x \in S} f(x) \quad [\, f(x^\star) = \max_{x \in S} f(x) = \sup_{x \in S} f(x) \,].$$

Remark 3 Not all functions have a minimum (maximum). For instance, e^x has no maximum on $\Re$, and e^{-x} has no minimum on $\Re$. However, e^x has a supremum of $+\infty$ on $\Re$, and e^{-x} has an infimum of 0 on $\Re$.

Theorem 2.2.9 (Existence of minimum (maximum))
A function $f(x)$ defined on a set S in $\Re^n$ exhibits a minimum (maximum) $x^\star \in S$ if

(i) $f(x)$ is lower (upper) semicontinuous on S, and

(ii) S is closed and bounded.

Illustration 2.2.7 Consider the function

$$f(x) = \begin{cases} x^2+1 & \text{for } x \neq 0 \\ 2 & \text{for } x = 0 \end{cases}$$

in the closed and bounded set $-1 \leq x \leq 1$.

$\left\{ \inf_{-1 \leq x \leq 1} f(x) = 1 \right\}$, but no minimum exists since $f(x)$ is not lower semicontinuous.

Illustration 2.2.8 Consider the function

$$f(x) = x + 1$$

in the open and bounded set $-1 < x < 1, x \in \Re$.

$\left\{ \inf_{-1 < x < 1} f(x) = 0 \right\}$, but no minimum exists because the set is not closed.

Illustration 2.2.9 Consider the function

$$f(x) = e^x, \quad x \in \Re.$$

$\left\{ \sup_{x \in \Re} f(x) = +\infty \right\}$, but no maximum exists since the set is unbounded.

2.2.7 Feasible Solution, Local and Global Minimum

Consider the problem of minimizing $f(x)$ subject to $x \in S$.

Definition 2.2.20 (Feasible solution) A point $x \in S$ is a feasible solution to this problem.

Definition 2.2.21 (Local minimum) Suppose that $x^\star \in S$ and that there exists an $\epsilon > 0$ such that

$$f(x) \geq f(x^\star) \qquad \forall\, x \in S : \| x - x^\star \| < \epsilon ,$$

then $x^\star$ is a local minimum.

Definition 2.2.22 (Global minimum) Suppose that $x^\star \in S$ and

$$f(x) \geq f(x^\star) \qquad \forall\, x \in S ,$$

then $x^\star$ is a global minimum.

Definition 2.2.23 (Unique global minimum) Suppose that $\boldsymbol{x}^\star \in S$ and

$$f(\boldsymbol{x}) > f(\boldsymbol{x}^\star) \qquad \forall\, \boldsymbol{x} \in S\,,$$

then $\boldsymbol{x}^\star$ is the unique global minimum.

Theorem 2.2.10
Let S be a nonempty convex set in $\Re^n$ and $\boldsymbol{x}^\star \in S$ be a local minimum.

(i) If $f(\boldsymbol{x})$ is convex, then $\boldsymbol{x}^\star$ is a global minimum.

(ii) If $f(\boldsymbol{x})$ is strictly convex, then $\boldsymbol{x}^\star$ is the unique global minimum.

2.3 Generalizations of Convex and Concave Functions

This section presents the definitions, properties and relationships of quasi-convex, quasi-concave, pseudo-convex and pseudo-concave functions.

2.3.1 Quasi-convex and Quasi-concave Functions

Let S be a nonempty convex set in $\Re^n$.

Definition 2.3.1 (Quasi-convex function) $f(\boldsymbol{x})$ is quasi-convex if

$$f[(1-\lambda)\boldsymbol{x}_1 + \lambda \boldsymbol{x}_2] \le \max[\, f(\boldsymbol{x}_1), f(\boldsymbol{x}_2)\,]$$

$$\text{for all } \lambda \in [0,1] \quad \text{and all } \boldsymbol{x}_1, \boldsymbol{x}_2 \in S.$$

Definition 2.3.2 (Quasi-concave function) $f(\boldsymbol{x})$ is quasi-concave if

$$f[(1-\lambda)\boldsymbol{x}_1 + \lambda \boldsymbol{x}_2] \ge \min[\, f(\boldsymbol{x}_1), f(\boldsymbol{x}_2)\,]$$

$$\text{for all } \lambda \in [0,1] \quad \text{and all } \boldsymbol{x}_1, \boldsymbol{x}_2 \in S.$$

Note that $f(\boldsymbol{x})$ is quasi-concave if $-f(\boldsymbol{x})$ is quasi-convex.

Illustration 2.3.1 Figure 2.12 shows a quasi-convex and quasi-concave function.

Remark 1 A convex function is also quasi-convex since

$$f[(1-\lambda)\boldsymbol{x}_1 + \lambda \boldsymbol{x}_2] \le (1-\lambda) f(\boldsymbol{x}_1) + \lambda f(\boldsymbol{x}_2) \le \max[f(\boldsymbol{x}_1), f(\boldsymbol{x}_2)]$$

Similarly a concave function is also quasi-concave since

$$f[(1-\lambda)\boldsymbol{x}_1 + \lambda \boldsymbol{x}_2] \ge (1-\lambda) f(\boldsymbol{x}_1) + \lambda f(\boldsymbol{x}_2) \ge \min[f(\boldsymbol{x}_1), f(\boldsymbol{x}_2)]$$

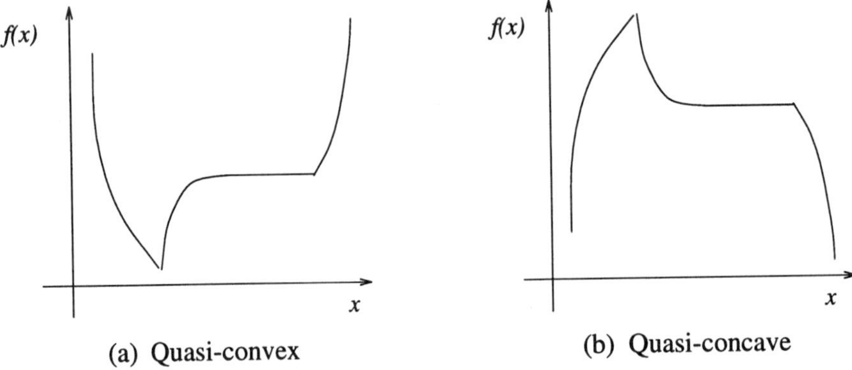

(a) Quasi-convex (b) Quasi-concave

Figure 2.12: Quasi-convex and quasi-concave functions

Theorem 2.3.1
Consider the function $f(x)$ on a convex set $S \in \Re^n$, and

$$\begin{aligned} S_\alpha &= \{x \mid x \in S,\ f(x) \leq \alpha\}, \quad \text{and} \\ S_\beta &= \{x \mid x \in S,\ f(x) \geq \beta\}. \end{aligned}$$

Then,

(i) $f(x)$ is quasi-convex on S if and only if S_α is convex for each $\alpha \in \Re$.

(ii) $f(x)$ is quasi-concave on S if and only if S_β is convex for each $\beta \in \Re$.

Definition 2.3.3 (Strictly quasi-convex function) $f(x)$ is strictly quasi-convex if

$$f[(1-\lambda)x_1 + \lambda x_2] < \max[\,f(x_1), f(x_2)\,]$$

for all $\lambda \in (0,1)$ and all $x_1, x_2 \in S$, $f(x_1) \neq f(x_2)$.

Definition 2.3.4 (Strictly quasi-concave function) $f(x)$ is strictly quasi-concave if

$$f[(1-\lambda)x_1 + \lambda x_2] > \min[\,f(x_1), f(x_2)\,]$$

for all $\lambda \in (0,1)$ and all $x_1, x_2 \in S$, $f(x_1) \neq f(x_2)$.

Note that $f(x)$ is strictly quasi-concave if $-f(x)$ is strictly quasi-convex.

Illustration 2.3.2 Figure 2.13 shows a strictly quasi-convex and strictly quasi-concave function.

Theorem 2.3.2
Let $f(x)$ be a lower (upper) semicontinuous function on the convex set S in $\Re^n$. If $f(x)$ is strictly quasi-convex (strictly quasi-concave) on S, then

$f(x)$ is quasi-convex (quasi-concave) on S, but the converse is not true.

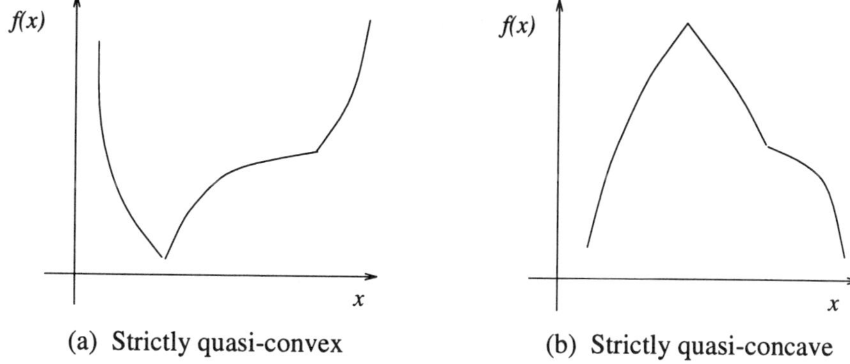

(a) Strictly quasi-convex (b) Strictly quasi-concave

Figure 2.13: Strictly quasi-convex and quasi-concave functions

Theorem 2.3.3
Let $f(x)$ be a function on the convex set S in $\Re^n$, and let $x^\star \in S$ be a local minimum (maximum) of $f(x)$. If $f(x)$ is strictly quasi-convex (strictly quasi-concave) on S, then

$f(x^\star)$ is a global minimum (maximum) of $f(x)$ on S.

2.3.2 Properties of Quasi-convex and Quasi-concave Functions

Quasi-convex and quasi-concave functions satisfy the following properties:

(i) Let $f(x)$ be a quasi-convex (quasi-concave) function on a subset S of $\Re^n$ and $g(y)$ be a nondecreasing function defined on the range of $f(x)$ in $\Re$. Then the composite function $g(f(x))$ is quasi-convex (quasi-concave) on S.

(ii) If $f(x)$ is either a positive or negative quasi-concave function on a subset S of $\Re^n$, then $\frac{1}{f(x)}$ is quasi-convex on S.

(iii) If $f(x)$ is either a positive or negative quasi-convex function on a subset S of $\Re^n$, then $\frac{1}{f(x)}$ is quasi-concave on S.

Remark 1 Note that the summation of quasi-convex functions is not necessarily a quasi-convex function as in convex functions. Also, note that the summation of convex and quasi-convex functions is not necessarily a convex or quasi-convex function.

Remark 2 Convex and concave functions do not satisfy properties (ii) and (iii) of the quasi-convex and quasi-concave functions. For instance, it is true that the reciprocal of a positive concave function is convex, but the reverse does not hold. As an example consider the function $f(x) = e^x$ which is convex and whose reciprocal is also convex.

2.3.3 Differentiable Quasi-convex, Quasi-concave Functions

Theorem 2.3.4
Let $f(x)$ be differentiable on a nonempty open convex set S in $\Re^n$. Then, $f(x)$ is quasi-convex if and only if for every $x_1, x_2 \in S$

$$f(x_1) \leq f(x_2) \text{ implies that } (x_1 - x_2)^T \nabla f(x_2) \leq 0.$$

Similarly, f(x) is quasi-concave if and only if for every $x_1, x_2 \in S$

$$f(x_1) \geq f(x_2) \text{ implies that } (x_1 - x_2)^T \nabla f(x_2) \geq 0.$$

For twice differentiable quasi-concave functions $f(x)$ on the open, nonempty convex set S in $\Re^n$, a direction z orthogonal to ∇f exhibits the following interesting properties:

(i) If $\bar{x} \in S, z \in \Re^n$ and $z^T \nabla f(\bar{x}) = 0$, then

$$z^T \nabla^2 f(\bar{x}) z \leq 0.$$

(ii) The Hessian of $f(x)$ has at most one positive eigenvalue at every $x \in S$.

Remark 1 From property (ii) we observe that the generalization of concavity to quasi-concavity is equivalent to allowing the existence of at most one positive eigenvalue of the Hessian.

2.3.4 Pseudo-convex and Pseudo-concave Functions

Let S be a nonempty open set in $\Re^n$ and let $f(x)$ be a differentiable function on S.

Definition 2.3.5 (Pseudo-convex function) $f(x)$ is pseudo-convex if for every $x_1, x_2 \in S$,

$$f(x_1) < f(x_2)$$

implies that $(x_1 - x_2)^T \nabla f(x_2) < 0$.

Definition 2.3.6 (Pseudo-concave function) $f(x)$ is pseudo-concave if for every $x_1, x_2 \in S$,

$$f(x_1) > f(x_2)$$

implies that $(x_1 - x_2)^T \nabla f(x_2) > 0$.

Note that $f(x)$ is pseudo-concave if $-f(x)$ is pseudo-convex.

2.3.5 Properties of Pseudo-convex and Pseudo-concave Functions

Pseudo-convex and pseudo-concave functions exhibit the following properties:

(i) Let $f(x)$ be a pseudo-convex (pseudo-concave) function on a subset S of $\Re^n$ and $g(y)$ be a differentiable function defined on the range of $f(x)$ in $\Re$ and which satisfies $g'(y) > 0$. Then the composite function $g[f(x)]$ is pseudo-convex (pseudo-concave) on S.

(ii) If $f(x)$ is a positive or negative pseudo-concave function on a subset S of $\Re^n$, then $\frac{1}{f(x)}$ is pseudo-convex on S.

2.3.6 Relationships among Convex, Quasi-convex and Pseudo-convex Functions

The relationships among convex, quasi-convex and pseudo-convex functions are summarized in the following:

(i) A convex differentiable function is pseudo-convex,

(ii) A convex function is strictly quasi-convex,

(iii) A convex function is quasi-convex,

(iv) A pseudo-convex function is strictly quasi-convex, and

(v) A strictly quasi-convex function which is lower semicontinuous is quasi-convex.

Summary and Further Reading

In this chapter, the basic elements of convex analysis are introduced. Section 2.1 presents the definitions and properties of convex sets, the definitions of convex combination and convex hull along with the important theorem of Caratheodory, and key results on the separation and support of convex sets. Further reading on the subject of convex sets is in the excellent books of Avriel (1976), Bazaraa *et al.* (1993), Mangasarian (1969), and Rockefellar (1970).

Section 2.2 discusses the definitions and properties of convex and concave functions, the definitions of continuity, lower and upper semicontinuity of functions, the definitions of subgradients of convex and concave functions, the definitions and properties of differentiable convex and concave functions, the conditions of convexity and concavity along with their associated tests, and the definitions of extremum points. For further reading, refer to Avriel (1976), Mangasarian (1969), and Rockefellar (1970).

Section 2.3 focuses on the generalizations of convex and concave functions and treats the quasi-convex, quasi-concave, pseudo-convex and pseudo-concave functions, and their properties. Further reading in this subject is the excellent book of Avriel *et al.* (1988).

Problems on Convex Analysis

1. Show that the interior of a convex set is convex.

2. Show that an open and closed ball around a point $x \in \Re^n$ is a convex set.

3. Show that the function
$$f(x_1, x_2, x_3) = e^{x_1^2+x_2^2+x_3^2}$$
is strictly convex.

4. Show that the function
$$f(x) = \sum_{i=1}^{n} c_i e^{\alpha_i x},$$
where α_i represents fixed vectors in $\Re^n$ and c_i are positive real numbers, is convex.

5. Show that the function
$$f(x_1, x_2) = x_1^2 - 4x_1x_2 + 4x_2^2 - \ln(x_1x_2)$$
is strictly convex for x_1, x_2 strictly positive.

6. Determine whether the function
$$f(x_1, x_2) = 5e^{x_1} - e^{x_2}$$
is convex, concave, or neither.

7. Prove property (iii), (iv), and (v) of convex and concave functions.

8. Show that the function
$$f(x_1, x_2, x_3) = x_1 \ln\left(\frac{x_1}{x_1 + x_2 + x_3}\right)$$
with $x_1, x_2, x_3 > 0$ and $x_1 + x_2 + x_3 = \alpha$, $\alpha > 0$ is convex.

9. Show that the function
$$f(x_1, x_2) = \sum_i w_i \sqrt{(x_1 - \alpha_i)^2 + (x_2 - \beta_i)^2}$$
with fixed values of the parameters $w_i \geq 0, \alpha_i, \beta_i$ is convex.

10. Determine whether the function
$$f(x_1, x_2) = \sqrt{(x_1 - \alpha_1)^2 + (x_2 - \beta_1)^2} - \sqrt{(x_1 - \alpha_2)^2 + (x_2 - \beta_2)^2}$$
with $\alpha_1, \beta_1, \alpha_2, \beta_2$ fixed values of parameters is convex, concave, or neither.

11. Determine whether the function

$$f(x) = \left\{ \begin{array}{ll} x^2 & \text{if } x < 0 \\ x^2 + 0.75 & \text{if } x \geq 0 \end{array} \right\}$$

is quasi-concave.

12. Show that the function

$$f(x_1, x_2) = \frac{\alpha_1 x_1 + \alpha_2 x_2 + \alpha_3}{\beta_1 x_1 + \beta_2 x_2 + \beta_3}$$

with $\beta_1 x_1 + \beta_2 x_2 + \beta_3 > 0$ is both pseudo-convex and pseudo-concave.

13. Show that the function

$$f(x_1, x_2) = -x_1 x_2$$

with $x_1 \geq 0, x_2 \geq 0$ is quasi-convex. Is it also convex? Why?

14. Determine whether the function

$$f(x_1, x_2) = x_1 - x_1 x_2$$

with $x_1 \geq 0, x_2 \geq 0$ is quasi-convex, quasi-concave, convex or neither of the above.

15. Show that the function

$$f(x_1, x_2) = 10 x_1^{1/2} x_2^{1/2}$$

with $x_1 > 0, x_2 > 0$ is quasi-concave.

16. Consider the quadratic function

$$f(\boldsymbol{x}) = \boldsymbol{c}^T \boldsymbol{x} + \frac{1}{2} \boldsymbol{x}^T \boldsymbol{Q} \boldsymbol{x}$$

where $\boldsymbol{Q}$ is an $n \times n$ matrix and $\boldsymbol{c} \in \Re^n$.

(i) If $\boldsymbol{Q} = \begin{pmatrix} 2 & 5 \\ 5 & 6 \end{pmatrix}$, $\boldsymbol{c} = 0$, show that $f(\boldsymbol{x})$ is strictly pseudo-convex on $\Re^2_+$.

(ii) If $\boldsymbol{Q} = \begin{pmatrix} 0 & 1 & 2 \\ 1 & 0 & 1 \\ 2 & 1 & 4 \end{pmatrix}$, $\boldsymbol{c} = 0$, show that $f(\boldsymbol{x})$ is pseudo-concave on $\Re^3_+$.

17. Consider the function

$$f(x_1, x_2) = x_1 \cdot x_2$$

with $x_1 > 0, x_2 > 0$. Show that $f(x_1, x_2)$ is strictly quasi-concave.

18. Let $f(\boldsymbol{x})$ be a differentiable function on an open convex set S of $\Re^n$. Prove that it is concave if and only if

$$(\boldsymbol{x}_2 - \boldsymbol{x}_1)^T [\nabla f(\boldsymbol{x}_2) - \nabla f(\boldsymbol{x}_1)] \leq 0$$

for every two points $\boldsymbol{x}_1 \in S, \boldsymbol{x}_2 \in S$.

19. Let $f_1(x)$, $f_2(x)$ be functions defined on a convex set $S \in \Re^n$ and $f_2(x) \neq 0$ on S. Show:

    (i) If $f_1(x)$ is convex, $f_1(x) \leq 0$, $f_2(x)$ is concave, and $f_2(x) > 0$, then $f_1(x) \cdot f_2(x)$ is quasi-convex on S.

    (ii) If $f_1(x)$ is convex, $f_1(x) \leq 0$, $f_2(x)$ is convex, and $f_2(x) > 0$, then $f_1(x) \cdot f_2(x)$ is quasi-concave on S.

20. What additional conditions are needed in problem 19 so as to have pseudo-convexity in (i) and pseudo-concavity in (ii)?

21. Consider the function

$$f(x, y) = x^a \cdot y^b$$

with $x \geq 0$, and $y \geq 0$. Find the conditions on a and b for which the function $f(x, y)$ is convex (concave).

Chapter 3 Fundamentals of Nonlinear Optimization

This chapter discusses the fundamentals of nonlinear optimization. Section 3.1 focuses on optimality conditions for unconstrained nonlinear optimization. Section 3.2 presents the first-order and second-order optimality conditions for constrained nonlinear optimization problems.

3.1 Unconstrained Nonlinear Optimization

This section presents the formulation and basic definitions of unconstrained nonlinear optimization along with the necessary, sufficient, and necessary and sufficient optimality conditions.

3.1.1 Formulation and Definitions

An unconstrained nonlinear optimization problem deals with the search for a minimum of a nonlinear function $f(x)$ of n real variables $x = (x_1, x_2, \cdots, x_n)$, and is denoted as

$$\left\{ \begin{array}{ll} MIN & f(x) \\ x \in \Re^n & \end{array} \right. \tag{3.1}$$

Each of the n nonlinear variables $x_1, x_2, \cdots, x_n$ are allowed to take any value from $-\infty$ to $+\infty$.

Unconstrained nonlinear optimization problems arise in several science and engineering applications ranging from simultaneous solution of nonlinear equations (e.g., chemical phase equilibrium) to parameter estimation and identification problems (e.g., nonlinear least squares).

Definition 3.1.1 (Local Minimum) $x^\star \in \Re^n$ is called a local optimum of (3.1) if there exists a ball of radius ϵ around $x^\star$, $B_\epsilon(x^\star)$:

$$f(x^\star) \leq f(x) \quad \text{for all } x \in B_\epsilon(x^\star).$$

Definition 3.1.2 (Global Minimum) $x^\star \in \Re^n$ is called a local optimum of (3.1) if

$$f(x^\star) \leq f(x) \quad \text{for all } x \in \Re^n.$$

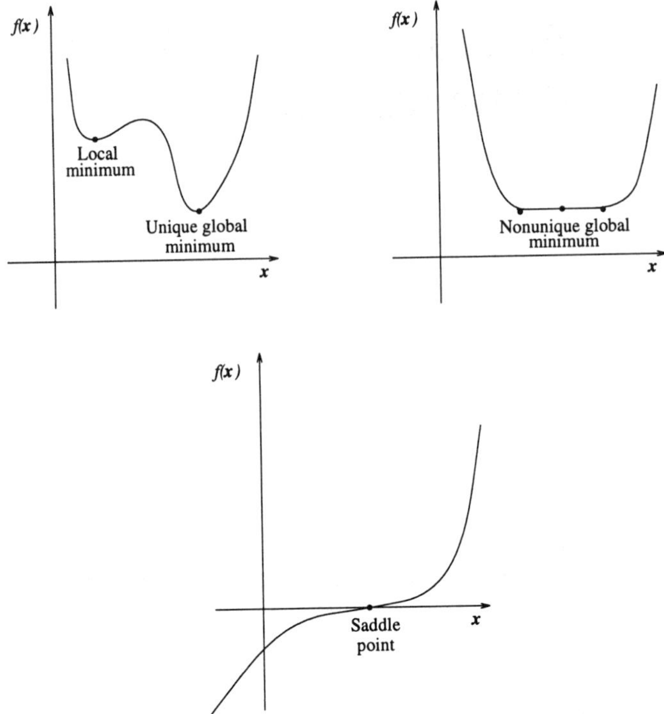

Figure 3.1: Local minimum, global minimum and saddle points

A global minimum is unique if the strict form of the inequality holds.

Definition 3.1.3 (Saddle Point) Let the vector x be partitioned into two subvectors x_a and x_b. $(x_a^\star, x_b^\star)$ is called a saddle point of $f(x_a, x_b)$ if there exists a ball of radius ϵ around $(x_a^\star, x_b^\star)$, $B_\epsilon(x_a^\star, x_b^\star)$:

$$f(x_a, x_b^\star) \leq f(x_a^\star, x_b^\star) \leq f(x_a^\star, x_b) \quad \text{for all } (x_a, x_b) \in B_\epsilon(x_a^\star, x_b^\star).$$

Illustration 3.1.1 Figure 3.1 shows a local minimum, unique global minimum, nonunique global minimum, and a saddle point.

3.1.2 Necessary Optimality Conditions

The necessary optimality conditions are the conditions which must hold at any minimum for a problem.

Theorem 3.1.1 (First-order necessary conditions)
Let $f(x)$ be a differentiable function in $\Re^n$ at x^*. If x^* is a local minimum, then

$$\nabla f(x^*) = 0 \tag{3.2}$$

Note: A point x^* satisfying (3.2) is called a stationary point.

Theorem 3.1.2 (Second-order necessary conditions)
Let $f(x)$ be a twice differentiable function in $\Re^n$ at x^*. If x^* is a local minimum, then

(i) $\nabla f(x^*) = 0$, *and*

(ii) *The Hessian matrix $H(x^*)$, given by*

$$H(x^*) = \frac{\partial^2 f}{\partial x_i \partial x_j}(x^*)$$

is positive semidefinite; that is, $y^T H(x^*) y \geq 0$ for all $y \in \Re^n$.

Illustration 3.1.2 Consider the unconstrained quadratic problem

$$\min_{x \in \Re^n} \; x^T Q x + a^T x.$$

The first order necessary conditions are

$$2Qx^* + a = 0 \quad \Rightarrow x^* = -\frac{1}{2} Q^{-1} a.$$

The second order necessary conditions are

$$\begin{aligned} 2Qx^* + a = 0, &\quad \text{and} \\ Q \quad \text{must be positive semidefinite.} & \end{aligned}$$

3.1.3 Sufficient Optimality Conditions

The sufficient optimality conditions are the conditions which, if satisfied at a point, guarantee that the point is a minimum.

Theorem 3.1.3
Let $f(x)$ be twice differentiable in $\Re^n$ at x^*. If

(i) $\nabla f(x^*) = 0$, *and*

(ii) *$H(x^*)$ is positive semidefinite,*

then x^* is a local minimum.

Remark 1 If condition (ii) becomes $H(x^*)$ is positive definite, then x^* is a strict local minimum.

Illustration 3.1.3

$$f(x) = x_1^3 - 2x_1 - x_1^2 + 6x_2 + x_2^2$$

The stationary points are

$$\begin{aligned} 3x_1^2 - 2 - 2x_1 &= 0 \quad \Rightarrow x_1 = \frac{1+\sqrt{7}}{3}, \text{ or } x_1 = \frac{1-\sqrt{7}}{3}, \\ 2x_2 + 6 &= 0, \quad \Rightarrow x_2 = -3. \end{aligned}$$

The Hessian is

$$H(x) = \begin{pmatrix} 6x_1 - 2 & 0 \\ 0 & 2 \end{pmatrix}.$$

At $(x_1^\star, x_2^\star) = \left(\frac{1+\sqrt{7}}{3}, -3\right)$, the Hessian is

$$H(x_1^\star) = \begin{pmatrix} 2\sqrt{7} & 0 \\ 0 & 2 \end{pmatrix},$$

which is positive definite, and hence $(x_1^\star, x_2^\star)$ is a strict local minimum. However, at $(x_1^\star, x_2^\star) = \left(\frac{1-\sqrt{7}}{3}, -3\right)$, the Hessian is

$$H(x_1^\star) = \begin{pmatrix} -2\sqrt{7} & 0 \\ 0 & 2 \end{pmatrix},$$

which is *not* positive semidefinite.

3.1.4 Necessary and Sufficient Optimality Conditions

Theorem 3.1.4
Let $f(x)$ be pseudoconvex in $\Re^n$ at x^*. Then, x^* is a global minimum if and only if

$$\nabla f(x^*) = 0.$$

Illustration 3.1.4

$$f(x_1, x_2) = 2x_1^2 + 6x_2^2 + 4x_1 - 3x_2.$$

The stationary points are

$$\begin{aligned} 4x_1 + 4 &= 0 \quad \Rightarrow x_1^\star = -1, \\ 12x_2 - 3 &= 0 \quad \Rightarrow x_2^\star = 0.25. \end{aligned}$$

The Hessian is

$$H(x) = \begin{pmatrix} 4 & 0 \\ 0 & 12 \end{pmatrix},$$

which is positive definite (the eigenvalues $\lambda_1 = 4, \lambda_2 = 12$ are positive everywhere and hence at $(x_1^\star, x_2^\star)$). As a result, $f(x_1^\star, x_2^\star)$ is convex and hence pseudoconvex. Thus, the stationary point $(-1, 0.25)$ is a global minimum.

Remark 1 Necessary and sufficient optimality conditions can be also expressed in terms of higher order derivatives assuming that the function $f(x)$ has such higher order derivatives. For instance, a necessary and sufficient condition for $x^* \in \Re^n$ being a local minimum of a univariate function $f(x)$ which has $(k+1)^{th}$ derivative can be stated as

> " $x^* \in \Re^n$ is a local minimum of $f(x)$ if and only if either $f^{(k)}(x^*) = 0$ for all $k = 1, 2, \ldots$ or else there exists an even $k \geq 1$ such that $f^{(k+1)}(x^*) > 0$ while $f^{(k)}(x^*) = 0$ for all $1 \leq k < k+1$ where $f^{(k)}(x^*)$, $f^{(k+1)}(x^*)$ are the kth and $(k+1)^{th}$ order derivatives of $f(x)$."

3.2 Constrained Nonlinear Optimization

This section presents first the formulation and basic definitions of constrained nonlinear optimization problems and introduces the Lagrange function and the Lagrange multipliers along with their interpretation. Subsequently, the Fritz John first-order necessary optimality conditions are discussed as well as the need for first-order constraint qualifications. Finally, the necessary, sufficient Karush-Kuhn-Tucker conditions are introduced along with the saddle point necessary and sufficient optimality conditions.

3.2.1 Formulation and Definitions

A constrained nonlinear programming problem deals with the search for a minimum of a function $f(x)$ of n real variables $x = (x_1, x_2, \ldots, x_n) \in X \subseteq \Re^n$ subject to a set of equality constraints $h(x) = 0$ ($h_i(x) = 0, i = 1, 2, \ldots, m$), and a set of inequality constraints $g(x) \leq 0$ ($g_j(x) \leq 0, j = 1, 2, \ldots, p$), and is denoted as

$$\begin{aligned} & \min \; f(x) \\ s.t. \quad & h(x) = 0 \\ & g(x) \leq 0 \\ & x \in X \subseteq \Re^n \end{aligned} \tag{3.3}$$

If any of the functions $f(x), h(x), g(x)$ is nonlinear, then the formulation (3.3) is called a constrained nonlinear programming problem. The functions $f(x), h(x), g(x)$ can take any form of nonlinearity, and we will assume that they satisfy continuity and differentiability requirements.

Constrained nonlinear programming problems abound in a very large number of science and engineering areas such as chemical process design, synthesis and control; facility location; network design; electronic circuit design; and thermodynamics of atomic/molecular clusters.

Definition 3.2.1 (Feasible Point(s)) A point $x \in X$ satisfying the equality and inequality constraints is called a feasible point. Thus, the set of all feasible points of $f(x)$ is defined as

$$F = \{x \in X \subseteq \Re^n : h(x) = 0, g(x) \leq 0\}.$$

Definition 3.2.2 (Active, inactive constraints) An inequality constraint $g_j(x)$ is called active at a feasible point $\bar{x} \in X$ if $g_j(\bar{x}) = 0$. An inequality constraint $g_j(x)$ is called inactive at a feasible point $\bar{x} \in X$ if $g_j(\bar{x}) < 0$.

Remark 1 The constraints that are active at a feasible point $\bar{x}$ restrict the feasibility domain while the inactive constraints do not impose any restrictions on the feasibility in the neighborhood of $\bar{x}$, defined as a ball of radius ϵ around $\bar{x}$, $B_\epsilon(\bar{x})$.

Definition 3.2.3 (Local minimum) $x^* \in F$ is a local minimum of (3.3) if there exists a ball of radius ϵ around x^*, $B_\epsilon(x^*)$:

$$f(x^*) \leq f(x) \quad \text{for each } x \in B_\epsilon(x^*) \cap F.$$

Definition 3.2.4 (Global minimum) $x^* \in F$ is a global minimum of (3.3) if

$$f(x^*) \leq f(x) \quad \text{for each } x \in F.$$

Definition 3.2.5 (Feasible direction vector) Let a feasible point $\bar{x} \in F$. Then, any point x in a ball of radius ϵ around $\bar{x}$ which can be written as $\bar{x} + d$ is a nonzero vector if and only if $x \neq \bar{x}$. A vector $d \neq 0$ is called a feasible direction vector from $\bar{x}$ if there exists a ball of radius ϵ:

$$(\bar{x} + \lambda d) \in B_\epsilon(\bar{x}) \cap F \quad \text{for all } 0 \leq \lambda \leq \frac{\epsilon}{\| d \|}.$$

The set of feasible direction vectors $d \neq 0$ from $\bar{x}$ is called the cone of feasible directions of F at $\bar{x}$.

Illustration 3.2.1 Figure 3.2 shows the feasible region, a point $\bar{x}$, and feasible direction vectors from $\bar{x}$.

Remark 2 If $\bar{x}$ is a local minimum of (3.3) and d is a feasible direction vector from $\bar{x}$, then for sufficiently small λ, we must have

$$f(\bar{x}) \leq f(\bar{x} + \lambda d).$$

Lemma 3.2.1 Let d be a non-zero feasible direction vector from $\bar{x}$. Then, $\bar{x}$ must satisfy the conditions:

$$\begin{aligned} d^T \nabla h(\bar{x}) &= 0, \\ d^T \nabla g(\bar{x}) &\leq 0 \quad \text{for active } g(\bar{x}). \end{aligned}$$

Definition 3.2.6 (Improving feasible direction vector) A feasible direction vector $d \neq 0$ at $\bar{x}$ is called an improving feasible direction vector at $\bar{x}$ if

$$f(\bar{x} + \lambda d) < f(\bar{x}) \quad \text{for all } 0 \leq \lambda \leq \frac{\epsilon}{\| d \|}.$$

The set of improving feasible direction vectors $d \neq 0$ from $\bar{x}$ is called the cone for improving feasible directions of F at $\bar{x}$.

Remark 3 If $d \neq 0$, and $d^T \nabla f(\bar{x}) < 0$ then d is an improving feasible direction vector at $\bar{x}$.

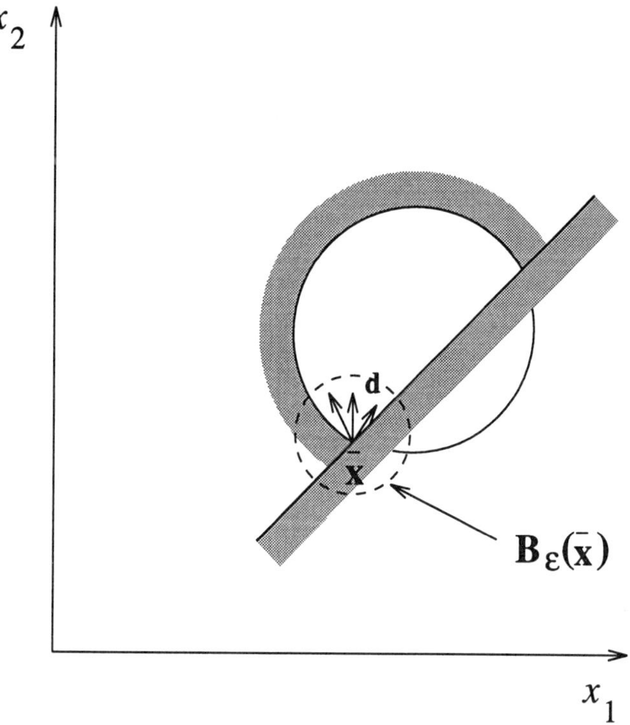

Figure 3.2: Feasible region and feasible direction vectors

3.2.2 Lagrange Functions and Multipliers

A key idea in developing necessary and sufficient optimality conditions for nonlinear constrained optimization problems is to transform them into unconstrained problems and apply the optimality conditions discussed in Section 3.1 for the determination of the stationary points of the unconstrained function . One such transformation involves the introduction of an auxiliary function, called the Lagrange function $L(x, \lambda, \mu)$, defined as

$$L(x, \lambda, \mu) = f(x) + \lambda^T h(x) + \mu^T g(x), \quad \mu \geq 0,$$

where $\lambda^T = (\lambda_1, \lambda_2, \ldots, \lambda_m)$ and $\mu^T = (\mu_1, \mu_2, \ldots, \mu_p)$ are the Lagrange multipliers associated with the equality and inequality constraints, respectively. The multipliers λ associated with the equalities $h(x) = 0$ are unrestricted in sign, while the multipliers μ associated with the inequalities $g(x) \leq 0$ must be nonnegative.

The transformed unconstrained problem then becomes to find the stationary points of the Lagrange function

$$\min_{x, \lambda, \mu \geq 0} L(x, \lambda, \mu) = f(x) + \lambda^T h(x) + \mu^T g(x) \tag{3.4}$$

Remark 1 The implications of transforming the constrained problem (3.3) into finding the stationary points of the Lagrange function are two-fold: (i) the number of variables has increased from n (i.e. the x variables) to $n + m + p$ (i.e. the x, λ and μ variables); and (ii) we need to establish the relation between problem (3.3) and the minimization of the Lagrange function with respect to x for fixed values of the lagrange multipliers. This will be discussed in the duality theory chapter. Note also that we need to identify which of the stationary points of the Lagrange function correspond to the minimum of (3.3).

Illustration 3.2.2 Consider the following two-variable constrained nonlinear programming problem in the form (3.3):

$$\min \; x_1^2 + x_2^2$$

$$\begin{aligned} s.t. \quad x_1^2 - 4x_2 &= 0 \\ (x_1 - 1)^2 + (x_2 - 2)^2 - 9 &\leq 0 \end{aligned}$$

The Lagrange function is

$$L(x_1, x_2, \lambda_1, \mu_1) = x_1^2 + x_2^2 + \lambda_1[x_1^2 - 9x_1] + \mu_1[(x_1 - 1)^2 + (x_2 - 2)^2 - 9], \quad \mu_1 \geq 0,$$

and has four variables x_1, x_2, λ_1, and μ_1.

Illustration 3.2.3 Consider the following quadratic programming problem:

$$\min \; c^T x + x^T Q x$$

$$\begin{aligned} s.t. \quad Ax - b &= 0 \leftarrow \lambda \\ Bx - d &\leq 0 \leftarrow \mu \end{aligned}$$

where A, B, are $(m \times n)$, $(p \times n)$ matrices and Q is the Hessian matrix.

The Lagrange function is

$$L(x, \lambda, \mu) = c^T x + x^T Q x + \lambda^T [Ax - b] + \mu^T [Bx - d], \quad \mu \geq 0.$$

3.2.3 Interpretation of Lagrange Multipliers

The Lagrange multipliers in a constrained nonlinear optimization problem have a similar interpretation to the dual variables or shadow prices in linear programming. To provide such an interpretation, we will consider problem (3.3) with only equality constraints; that is,

$$\min \; f(x) \tag{3.5}$$

$$s.t. \quad h_i(x) = 0 \quad i = 1, 2, \ldots, m < n \tag{3.6}$$

$$x \in X.$$

Let $\bar{x}$ be a global minimum of (3.6) at which the gradients of the equality constraints are linearly independent (i.e., $\bar{x}$ is a regular point). Perturbing the right-hand sides of the equality constraints, we have

$$\min \ f(x) \tag{3.7}$$

$$\begin{aligned} s.t. \quad h_i(x) &= b_i \quad i = 1, 2, \ldots, m < n \\ x &\in X \end{aligned} \tag{3.8}$$

where $b = (b_1, b_2, \ldots, b_m)$ is the perturbation vector.

If the perturbation vector changes, then the optimal solution of (3.8) and its multipliers will change, since in general $x = x(b)$ and $\lambda = \lambda(b)$. Then, the Lagrange function takes the form

$$L(x, \lambda) = L(b) = f(x(b)) + \lambda(b)^T[h(x(b)) - b].$$

Let us assume that the stationary point of L corresponds to the global minimum. Then, $f(\bar{x}) = L(\bar{x}, \bar{\lambda})$.

Taking the gradient of the Lagrange function with respect to the perturbation vector b and rearranging the terms we obtain

$$\nabla_b L = \left[\frac{\partial x}{\partial b}\right]^T \left(\nabla_x f + \left[\frac{\partial h}{\partial x}\right]^T \lambda\right) + \left[\frac{\partial \lambda}{\partial b}\right]^T [h(x) - b] - \lambda,$$

where

$\left[\frac{\partial x}{\partial b}\right]$ is a $n \times m$ matrix,

$\left[\frac{\partial \lambda}{\partial b}\right]$ is a $m \times m$ matrix,

$\left[\frac{\partial h}{\partial x}\right]$ is a $m \times n$ matrix.

Note that the terms within the first and second parentheses correspond to the gradients of the Lagrange function with respect to x and λ, respectively, and hence they are equal to zero due to the necessary conditions $\nabla_x L = \nabla_\lambda L = 0$. Then we have

$$\nabla_b L(\bar{x}, \bar{\lambda}) = -\bar{\lambda}\,.$$

Since $\bar{x}$ is a global minimum, then we have

$$\nabla_b f(\bar{x}) = -\bar{\lambda}\,.$$

Therefore, the Lagrange multipliers $\bar{\lambda}$ provide information on the sensitivity of the objective function with respect to the perturbation vector b at the optimum point $\bar{x}$.

Illustration 3.2.4 Consider the following convex quadratic problem subject to a linear equality constraint:

$$\min \ (x_1 - 5)^2 + (x_2 - 5)^2$$

$$s.t \quad x_1 - x_2 = 0.$$

We consider a perturbation of the right-hand sides of the equality constraint

$$x_1 - x_2 = b.$$

The Lagrange function takes the form

$$L = (x_1 - 5)^2 + (x_2 - 5)^2 + \lambda[x_1 - x_2 - b]$$

The gradients of the Lagrange function with respect to x_1, x_2, and λ are

$$\nabla L = [2(x_1 - 5) + \lambda, 2(x_2 - 5) - \lambda, x_1 - x_2 - b]^T.$$

Then, the stationary point of the Lagrange function is

$$\begin{aligned} 2(x_1 - 5) + \lambda &= 0, \\ 2(x_2 - 5) - \lambda &= 0, \\ x_1 - x_2 - b &= 0; \end{aligned}$$

$$\begin{aligned} \Rightarrow \quad \bar{x}_1 &= \frac{10 + b}{2}, \\ \bar{x}_2 &= \frac{10 - b}{2}, \\ \bar{\lambda} &= -b. \end{aligned}$$

The sensitivity of the objective function with respect to the perturbation vector b is

$$\frac{\partial f(\bar{x})}{\partial b} = -\bar{\lambda} = b,$$

which implies that (i) for $b > 0$, an increase in b would result in an increase of the objective function; (ii) for $b < 0$ an increase in b would represent a decrease of the objective function; and (iii) for $b = 0$, we have $\bar{\lambda} = 0$ and hence the solution of the constrained problem is identical to the unconstrained one $(x_1, x_2) = (5, 5)$.

3.2.4 Existence of Lagrange Multipliers

The existence of Lagrange multipliers depends on the form of the constraints and is not always guaranteed. To illustrate instances in which Lagrange multipliers may not have finite values (i.e., no existence), we will study problem (3.6), and we will assume that we have identified a candidate optimum point, $\boldsymbol{x}'$, which satisfies the equality constraints; that is,

$$h_i(\boldsymbol{x}') = 0, \quad i = 1, 2, \ldots, m.$$

The Lagrange function is

$$L = f(\boldsymbol{x}) + \lambda^T \boldsymbol{h}(\boldsymbol{x}),$$

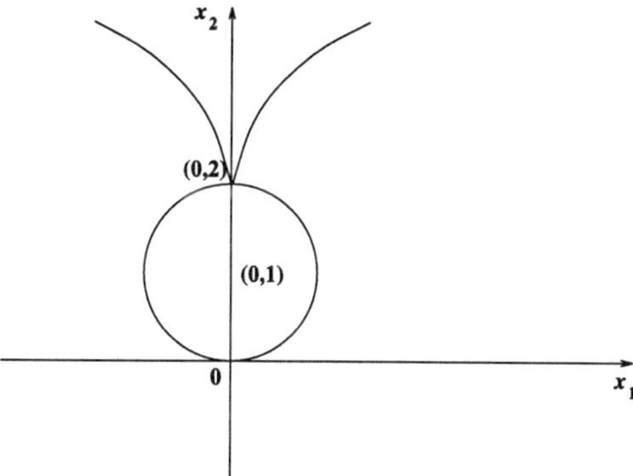

Figure 3.3: Feasible region and objective of illustration 3.2.5

and the stationary point of the Lagrange functions is obtained from

$$\nabla_{\boldsymbol{x}} f(\boldsymbol{x}) + \left[\frac{\partial h(\boldsymbol{x})}{\partial \boldsymbol{x}}\right] \lambda = 0.$$

Then, at the candidate point $\boldsymbol{x}'$, we have

$$\nabla_{\boldsymbol{x}} f(\boldsymbol{x}') + \left[\frac{\partial h(\boldsymbol{x})}{\partial \boldsymbol{x}}\right]^T \Bigg|_{\boldsymbol{x}'} \lambda = 0.$$

To calculate λ, we need to have the matrix $\frac{\partial h(\boldsymbol{x})}{\partial \boldsymbol{x}}$ of full rank (i.e., m) since we have to take its inverse. Hence, if $\frac{\partial h(\boldsymbol{x})}{\partial \boldsymbol{x}}$ is of full rank (i.e., m) then, the Lagrange multipliers have finite values.

Illustration 3.2.5 Consider the minimization of a squared distance subject to one equality constraint:

$$\begin{array}{ll} \min & x_1^2 + (x_2 - 1)^2 \\ s.t. & (x_2 - 2)^3 - x_1^2 = 0 \end{array}$$

and let $\boldsymbol{x}' = (0, 2)$ be the candidate optimum point (see Figure 3.3). The Lagrange function is

$$\begin{aligned} L(x_1, x_2, \lambda) &= x_1^2 + (x_2 - 1)^2 + \lambda\left[(x_2 - 2)^3 - x_1^2\right], \\ \frac{\partial h(\boldsymbol{x})}{\partial x_1} &= -2x_1 \Rightarrow \frac{\partial h(\boldsymbol{x})}{\partial x_1}\Big|_{(0,2)} = 0, \\ \frac{\partial h(\boldsymbol{x})}{\partial x_2} &= 3(x_2 - 2)^2 \Rightarrow \frac{\partial h(\boldsymbol{x})}{\partial x_2}\Big|_{(0,2)} = 0. \end{aligned}$$

Then the rank of

$$\frac{\partial h(\boldsymbol{x})}{\partial \boldsymbol{x}}\Big|_{(0,2)} \quad \text{is} \quad 0 < 1 = m.$$

Thus, the Lagrange multiplier λ cannot take a finite value (i.e., it does not exist).

We can also illustrate it by taking the gradients of the Lagrange function with respect to x_1, x_2, and λ.

$$\begin{aligned}\frac{\partial L}{\partial x_1} &= 2x_1 - 2x_1\lambda = 0,\\ \frac{\partial L}{\partial x_2} &= 2(x_2 - 1) + 3\lambda(x_2 - 2)^2 = 0,\\ \frac{\partial L}{\partial \lambda} &= (x_2 - 2)^3 - x_1^2 = 0.\end{aligned}$$

At $(x_1, x_2) = (0, 2)$ we can see that we cannot find a finite λ that satisfies $\frac{\partial L}{\partial x_2} = 0$.

3.2.5 Weak Lagrange Functions

In the definition of the Lagrange function $L(x, \lambda, \mu)$ (see section 3.2.2) we associated Lagrange multipliers with the equality and inequality constraints only. If, however, a Lagrange multiplier μ_0 is associated with the objective function as well, the definition of the *weak* Lagrange function $L'(x, \lambda, \mu)$ results; that is,

$$L'(x, \lambda, \mu) = \mu_0 f(x) + \sum_{i=1}^{m} \lambda_i h_i(x) + \sum_{j=1}^{p} \mu_j g_j(x).$$

3.2.6 First-Order Necessary Optimality Conditions

In this section we present, under the assumption of differentiability, the first-order necessary optimality conditions for the constrained nonlinear programming problem (3.3) as well as the corresponding geometric necessary optimality conditions.

3.2.6.1 Fritz John Conditions

Let $\bar{x} \in X$ be a feasible solution of (3.3), that is, $\boldsymbol{h}(\bar{x}) = \boldsymbol{0}, \boldsymbol{g}(\bar{x}) \leq \boldsymbol{0}$. Let also $f(x)$ and $\boldsymbol{g}(x)$ be differentiable at $\bar{x}$ and $\boldsymbol{h}(x)$ have continuous first partial derivatives at $\bar{x}$. Then, if $\bar{x}$ is a local solution of (3.3), there exist Lagrange multipliers μ_0, λ and μ:

$$\begin{aligned}\mu_0 \nabla f(\bar{x}) + \lambda^T \nabla \boldsymbol{h}(\bar{x}) + \mu^T \boldsymbol{g}(\bar{x}) &= \boldsymbol{0}\\ \boldsymbol{h}(\bar{x}) &= \boldsymbol{0}\\ \boldsymbol{g}(\bar{x}) &\leq \boldsymbol{0}\\ \mu_j g_j(\bar{x}) &= 0 \quad j = 1, 2, \ldots, p\\ (\mu_0, \mu_j) &\geq (0, 0) \quad j = 1, 2, \ldots, p\\ (\mu_0, \lambda, \mu) &\neq (0, \boldsymbol{0}, \boldsymbol{0})\end{aligned}$$

where $\nabla f(\bar{x})$ is an $(n \times 1)$ vector, $\nabla \boldsymbol{h}(\bar{x})$ is a $(m \times n)$ matrix,
$\nabla \boldsymbol{g}(\bar{x})$ is a $(p \times n)$ matrix,
μ_0 is a scalar, λ is a $(m \times 1)$ vector, and μ is a $(p \times 1)$ vector.

The constraints $\{\mu_j g_j(\bar{x}) = 0,\ j = 1, 2, \ldots, p\}$ are called complementarity constraints.

Remark 1 The corresponding geometric necessary optimality condition is that the set of feasible directions defined by

$$\begin{aligned} Z \;=\; & \{d : d^T\nabla f(\bar{x}) < 0; d^T\nabla h_i(\bar{x}) = 0, \;\; i = 1,2,\ldots,m \\ & d^T\nabla g_j(\bar{x}) < 0 \;\text{ for }\; j : g_j(\bar{x}) = 0, \} \end{aligned}$$

is empty (i.e., $Z = \emptyset$) assuming that $\nabla h_i(\bar{x}), i = 1,2,\ldots,m$ are linearly independent.

Illustration 3.2.6 (Verification of Fritz John necessary conditions) Verify the Fritz John conditions at $\bar{x} = (0,1)$ for

$$\left\{ \begin{array}{lrcl} MIN & f(x) & = & x_1^2 + (x_2 - 2)^2 \\ s.t. & g(x) & = & x_1^2 + x_2^2 - 1 \leq 0 \\ & h(x) & = & x_1 + x_2 - 1 = 0 \end{array} \right.$$

Notice that at the point $\bar{x} = (0,1)$ the inequality constraint is binding.

$$\begin{aligned} \nabla f(x) &= [2x_1, 2(x_2 - 2)]^T, \\ \nabla h(x) &= (1,1)^T, \\ \nabla g(x) &= (2x_1, 2x_2)^T, \\ \nabla f(\bar{x}) &= (0,-2)^T, \\ \nabla h(\bar{x}) &= (1,1)^T, \\ \nabla g(\bar{x}) &= (0,2)^T; \end{aligned}$$

$\bar{x}$ satisfies the $h(\bar{x}) = 0, g(\bar{x}) = 0$;

$$\mu_0 \begin{bmatrix} 0 \\ -2 \end{bmatrix} + \lambda \begin{bmatrix} 1 \\ 1 \end{bmatrix} + \mu \begin{bmatrix} 0 \\ 2 \end{bmatrix} = \begin{bmatrix} 0 \\ 0 \end{bmatrix}$$

$$\Rightarrow \left\{ \begin{array}{c} \lambda = 0 \\ -2\mu_0 + \lambda + 2\mu = 0 \end{array} \right\} \Rightarrow \left\{ \begin{array}{c} \lambda = 0 \\ \mu_0 = \mu. \end{array} \right.$$

Then, the Fritz John conditions are satisfied at $\bar{x} = (0,1)$ for $\mu_0 = \mu = 1$ for instance, since $(\mu_0, \mu) \geq (0,0)$ and $(\mu_0, \lambda, \mu) = (1,0,1) \neq (0,0,0)$.

Remark 2 In the Fritz John first-order necessary optimality conditions, the multiplier μ_0 associated with the objective function can become zero at the considered point $\bar{x}$ without violating the optimality conditions. In such a case, the Lagrange function becomes independent of $f(x)$ and the conditions are satisfied for any differentiable objective function $f(x)$ whether it exhibits a local optimum at $\bar{x}$ or not. This weakness of the Fritz John conditions is illustrated in the following example.

Illustration 3.2.7 Consider the problem

$$\left\{ \begin{array}{lrcl} MIN & f(x) & = & x_1 + x_2 \\ s.t. & g_1(x) & = & (x_1 - 1)^2 + x_2^2 - 1 \leq 0 \\ & g_2(x) & = & (x_1 - 3)^2 + x_2^2 - 1 \leq 0 \end{array} \right.$$

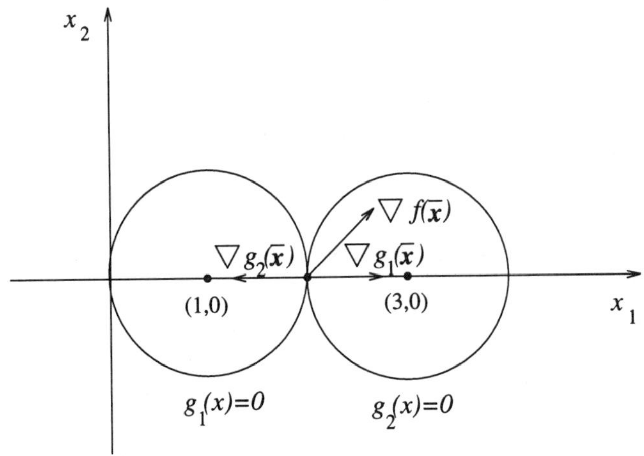

Figure 3.4: Example of a degenerate feasible region

which has only one feasible point $(2,0)$ and its feasible region is shown in Figure 3.4 (degenerate feasible region). Note that at $(2,0)$ both $g_1(x)$ and $g_2(x)$ are active:

$$\begin{aligned} \nabla f(\bar{x}) &= (1,1)^T, \\ \nabla g_1(\bar{x}) &= (2,0)^T, \\ \nabla g_2(\bar{x}) &= (-2,0)^T. \end{aligned}$$

Note that $\nabla g_1(\bar{x}), \nabla g_1(\bar{x})$ are linearly dependent.

$$\mu_0 \begin{bmatrix} 1 \\ 1 \end{bmatrix} + \mu_1 \begin{bmatrix} 2 \\ 0 \end{bmatrix} + \mu_2 \begin{bmatrix} -2 \\ 0 \end{bmatrix} = \mathit{0}$$

$$\Rightarrow \left\{ \begin{aligned} \mu_0 + 2\mu_1 - 2\mu_2 &= 0 \\ \mu_0 \qquad\qquad &= 0 \end{aligned} \right. \Rightarrow \left\{ \begin{aligned} \mu_1 &= \mu_2 \\ \mu_0 &= 0. \end{aligned} \right.$$

For $\mu_1 = \mu_2 = 1$ and $\mu_0 = 0$, for instance, the Fritz John conditions are satisfied at $(2,0)$. In this case, however, the objective function disappears from consideration.

To remove this weakness of the Fritz John necessary conditions, we need to determine the required restrictions under which μ_0 is strictly positive ($\mu_0 > 0$). These restrictions are called *first-order constraint qualifications* and will be discussed in the following section.

3.2.6.2 First-Order Constraint Qualifications

As we have seen from the previous illustration, when μ_0 equals zero, the Fritz John first order necessary optimality conditions do not utilize the gradient of the objective function. As a result, the gradient conditions represent simply a linear combination of the active inequalities and equality constraints that is equal to zero. In such cases, the Fritz-John conditions are not useful in identifying a local optimum of the function $f(x)$. A number of additional conditions are needed to guarantee that $\mu_0 > 0$. These are the first-order constraint qualifications, and a selected number is presented in the following.

Let $\bar{x}$ be a local optimum of (3.3), X be an open set, and J be defined as the set $J = \{j : g_j(\bar{x}) = 0\}$. Let also $g_j(x)$, $j \in J$, $h_i(x)$, $i = 1, 2, \ldots, m$ be continuously differentiable at $\bar{x}$.

Linear Independence Constraint Qualification

The gradients $\nabla g_j(\bar{x})$ for $j \in J$ and $\nabla h_i(\bar{x})$ for $i = 1, 2, \ldots, m$ are linearly independent.

Slater Constraint Qualification

The constraints $g_j(\bar{x})$ for $j \in J$ are pseudo-convex at $\bar{x}$. The constraints $h_i(\bar{x})$ for $i = 1, 2, \ldots, m$ are quasi-convex and quasi-concave. The gradients $\nabla h_i(\bar{x})$ for $i = 1, 2, \ldots, m$ are linearly independent, and there exists an $\tilde{x} \in X$ such that $g_j(\tilde{x}) < 0$ for $j \in J$ and $h_i(\tilde{x}) = 0$ for $i = 1, 2, \ldots, m$.

Kuhn-Tucker Constraint Qualification

There exists a nonzero vector $z \in \Re^n$ for which

$$\{z^T \nabla \boldsymbol{h}(\bar{x}) = 0 \text{ and } z^T \nabla g_j(\bar{x}) \leq 0 \; j \in J\}$$

implies that there exists an n-dimensional vector function $w(\tau)$ on the interval $[0, 1]$:

(i) $w(0) = \bar{x}$,

(ii) $w(\tau) \in X$ for $0 \leq \tau \leq 1$,

(iii) w is once-differentiable at $\tau = 0$, and

$$\frac{dw(0)}{d\tau} = \lambda z \quad \text{for some } \lambda > 0.$$

The Weak Reverse Convex Constraint Qualification

The constraints $\boldsymbol{h}(x)$ and $\boldsymbol{g}(x)$ are continuously differentiable at $\bar{x}$. Each $g_j(\bar{x})$, $j \in J$ is pseudoconcave at $\bar{x}$ or linear. Each $h_i(\bar{x})$, $i = 1, 2, \ldots, m$ is both pseudoconvex and pseudoconcave at $\bar{x}$.

Remark 1 Note that in the Kuhn-Tucker constraint qualification $w(\tau)$ is a once-differentiable arc which starts at $\bar{x}$. Then, the Kuhn-Tucker constraint qualification holds if z is tangent to $w(\tau)$ in the constrained region.

Remark 2 The linear independence constraint qualification as well as the Slater's imply the Kuhn-Tucker constraint qualification.

3.2.6.3 Karush-Kuhn-Tucker Necessary Conditions

Let $\bar{x} \in X$ be a feasible solution of (3.3). Let also $f(x)$ and $g(x)$ be differentiable at $\bar{x}$ and $h(x)$ have continuous first partial derivatives at $\bar{x}$. If $\bar{x}$ is a local optimum of (3.3) and one of the following constraint qualifications:

(i) Linear independence,

(ii) Slater,

(iii) Kuhn-Tucker, or

(iv) Weak reverse convex

is satisfied, then there exist Lagrange multipliers λ, μ:

$$\begin{aligned} \nabla f(\bar{x}) + \lambda^T \nabla h(\bar{x}) + \mu^T \nabla g(\bar{x}) &= 0 \\ h(\bar{x}) &= 0 \\ g(\bar{x}) &\leq 0 \\ \mu_j g_j(\bar{x}) &= 0 \quad j = 1, 2, \ldots, p \\ \mu_j &\geq 0 \quad j = 1, 2, \ldots, p. \end{aligned}$$

Geometric Interpretation of Karush-Kuhn-Tucker Necessary Conditions

From the gradient KKT conditions we have that

$$-\nabla f(\bar{x}) = \lambda^T \nabla h(\bar{x}) + \mu^T \nabla g(\bar{x}),$$

with $\mu_j \geq 0, j = 1, 2, \ldots, p$. The complementarity conditions enforce the Lagrange multipliers of the inactive constraints to take zero values. As a result, $\lambda^T \nabla h(\bar{x}) + \mu^T \nabla g(\bar{x})$ represents a vector that belongs to the cone of the gradients of the active constraints (i.e. equalities and active inequalities). Then the geometrical interpretation of the gradient KKT conditions is that $-\nabla f(\bar{x})$ belongs to the cone of the gradients of the active constraints at the feasible solution $\bar{x}$. If $-\nabla f(\bar{x})$ lies outside the cone of the gradients of the active constraints, then $\bar{x}$ is not a KKT point.

Illustration 3.2.8 Consider the following problem:

$$\left\{ \begin{array}{ll} MIN & x_1^2 + x_2^2 \\ s.t. & x_1^2 - 4x_2 = 0 \\ & (x_1 - 2)^2 + (x_2 - 3)^2 - 4 \leq 0 \end{array} \right.$$

and verify the Karush-Kuhn-Tucker conditions for $\bar{x} = (2, 1)$.

The Lagrange function is

$$\begin{aligned} L &= x_1^2 + x_2^2 + \lambda(x_1^2 - 4x_2) + \mu[(x_1 - 2)^2 + (x_2 - 3)^2 - 4], \\ \nabla f(x) &= (2x_1, 2x_2)^T \Rightarrow \nabla f(\bar{x}) = (4, 2)^T, \\ \nabla h(x) &= (2x_1, -4)^T \Rightarrow \nabla h(\bar{x}) = (4, -4)^T, \\ \nabla g(x) &= (2(x_1 - 2), 2(x_2 - 3))^T \Rightarrow \nabla g(\bar{x}) = (0, -4)^T. \end{aligned}$$

The point $\bar{x} = (2, 1)$ is a feasible point. The gradient conditions of the Lagrange function become

$$\begin{pmatrix} 1 \\ 2 \end{pmatrix} + \lambda \begin{pmatrix} 4 \\ -4 \end{pmatrix} + \mu \begin{pmatrix} 0 \\ -4 \end{pmatrix} = \mathbf{0}$$

$$\Rightarrow \begin{cases} 4 + 4\lambda = 0 \\ 2 - 4\lambda - 4\mu = 0 \end{cases} \Rightarrow \begin{cases} \lambda = -1 \\ \mu = \frac{3}{2} \geq 0 \end{cases}$$

$g(\bar{x}) = 0$ (active) and hence $\mu g(\bar{x}) = 0$ is satisfied. Also, the linear independence constraint qualification is satisfied. Thus, the KKT conditions are satisfied.

Illustration 3.2.9 Consider the example that demonstrated the weakness of the Fritz-John conditions:

$$\begin{cases} MIN & f(x) = x_1 + x_2 \\ s.t. & g_1(x) = (x_1 - 1)^2 + x_2^2 - 1 \leq 0 \\ & g_2(x) = (x_1 - 3)^2 + x_2^2 - 1 \leq 0 \end{cases}$$

at the point $\bar{x} = (2, 0)$ which is feasible and at which both $g_1(\bar{x}), g_2(\bar{x})$ are active. The gradients of the objective function and constraints are :

$$\begin{aligned} \nabla f(\bar{x}) &= (1, 1)^T, \\ \nabla g_1(\bar{x}) &= (2, 0)^T, \\ \nabla g_2(\bar{x}) &= (-2, 0)^T. \end{aligned}$$

The gradient KKT conditions are

$$\begin{bmatrix} 1 \\ 1 \end{bmatrix} + \mu_1 \begin{bmatrix} 2 \\ 0 \end{bmatrix} + \mu_2 \begin{bmatrix} -2 \\ 0 \end{bmatrix} = \mathbf{0}$$

$$\Rightarrow \begin{cases} 1 + 2\mu_1 - 2\mu_2 = 0 \\ 1 = 0 \end{cases}$$

which are not satisfied. Note that the KKT conditions cannot be satisfied because of the linear independence of the active constraints.

3.2.7 First-Order Sufficient Optimality Conditions

In this section, we discuss the first-order sufficient optimality conditions for the constrained nonlinear programming problem (3.3).

3.2.7.1 Karush-Kuhn-Tucker Sufficient Conditions

Let $\bar{x} \in X$ be a feasible solution of (3.3), and let $\bar{x}$ be a KKT point (i.e., it satisfies the gradient conditions, complementarity, nonnegativity of μ's, and constraint qualification). Let also $I^+ = \{i : \lambda_i > 0\}$ and $I^- = \{i : \lambda_i < 0\}$ at $\bar{x}$. If

$f(x)$ is pseudo-convex at $\bar{x}$ with all other feasible points x,

$h_i(x)$ for $i \in I^+$ are quasi-convex at $\bar{x}$ with all other feasible points x,

$h_i(x)$ for $i \in I^-$ are quasi-concave at $\bar{x}$ with all other feasible points x, and

$g_j(x)$ for $j \in J$ are quasi-convex at $\bar{x}$ with all other feasible points x,

then $\bar{x}$ is a global optimum of (3.3). If the above convexity conditions on $f(x), h(x), g(x)$ are restricted within a ball of radius ϵ around $\bar{x}$, then $\bar{x}$ is a local optimum of (3.3).

Illustration 3.2.10 Consider the problem

$$\left\{\begin{array}{lrcl} MIN & f(x) & = & x_1^2 + x_2^2 \\ s.t. & g_1(x) & = & -x_1 x_2 \leq -1 \\ & g_2(x) & = & -x_1 \leq 0 \\ & g_3(x) & = & -x_2 \leq 0 \end{array}\right.$$

and verify the global optimality of the KKT point $\bar{x} = (1,1)$. $f(x)$ is convex, continuous, differentiable and hence pseudo-convex. $g_1(x)$ is quasi-convex, while $g_2(x)$ and $g_3(x)$ are linear and hence quasi-convex. The linear independence constraint qualification is met since we have only one active constraint (i.e., $g_1(\bar{x})$). Thus, $\bar{x} = (1,1)$ is a global optimum.

3.2.8 Saddle Point and Optimality Conditions

This section presents the basic definitions of a saddle point, and discusses the necessary and sufficient saddle point optimality conditions.

Definition 3.2.7 (Saddle point) A real function $\theta(x,y), x \in X \subset \Re^{n_1}, y \in Y \subset \Re^{n_2}$ is said to have a saddle point at $(\bar{x},\bar{y})$ if

$$\theta(\bar{x},y) \leq \theta(\bar{x},\bar{y}) \leq \theta(x,\bar{y})$$

for every $x \in X$ and $y \in Y$.

Remark 1 From this definition, we have that a saddle point is a point that simultaneously minimizes the function θ with respect to x for fixed $y = \bar{y}$ and maximizes the function θ with respect to y for fixed $x = \bar{x}$. Note also that no assumption on differentiability of $\theta(x,y)$ is introduced.

Definition 3.2.8 (Karush-Kuhn-Tucker saddle point) Let the Lagrange function of problem (3.3) be

$$L(x,\lambda,\mu) = f(x) + \lambda^T h(x) + \mu^T g(x), \quad \mu \geq 0$$

A point $(\bar{x},\bar{\lambda},\bar{\mu})$ is called KKT saddle point if

$$L(\bar{x},\lambda,\mu) \leq L(\bar{x},\bar{\lambda},\bar{\mu}) \leq L(x,\bar{\lambda},\bar{\mu})$$

for every $\bar{x} \in X$, λ and μ such that $\mu \geq 0$.

3.2.8.1 Saddle Point Necessary Optimality Conditions

If

(i) $\bar{x} \in X$ is a local optimum solution of (3.3),

(ii) X is a convex set,

(iii) $f(x), g(x)$ are convex functions, and $h(x)$ are affine, and

(v) $g(x) < 0, h(x) = 0$ has a solution,

then there exist $\bar{\lambda}, \bar{\mu}$ with $\bar{\mu} \geq 0$ satisfying $\bar{\mu} g(\bar{x}) = 0$ and

$$L(\bar{x}, \lambda, \mu) \leq L(\bar{x}, \bar{\lambda}, \bar{\mu}) \leq L(x, \bar{\lambda}, \bar{\mu});$$

that is, $(\bar{x}, \bar{\lambda}, \bar{\mu})$ is a saddle point.

Remark 1 If $\bar{x}$ is a KKT point and the additional condition (iii) holds, then $(\bar{x}, \bar{\lambda}, \bar{\mu})$ is a saddle point. Thus, under certain convexity assumptions of f, g, and affinity of h, the Lagrange multipliers in the KKT conditions are also the multipliers in the saddle point criterion.

3.2.8.2 Saddle Point Sufficient Optimality Conditions

If $(\bar{x}, \bar{\lambda}, \bar{\mu})$ is a Karush-Kuhn-Tucker saddle point, that is, there exist $\bar{x} \in X, \bar{\lambda}, \bar{\mu}$ with $\bar{\mu} \geq 0$:

$$L(\bar{x}, \lambda, \mu) \leq L(\bar{x}, \bar{\lambda}, \bar{\mu}) \leq L(x, \bar{\lambda}, \bar{\mu})$$

for every $x \in X$ and all λ, μ with $\mu \geq 0$, then $\bar{x}$ is a solution of problem (3.3).

Remark 1 Note that the saddle point sufficiency conditions do not require either additional convexity assumptions or a constraint qualification like condition. Note also that the saddle point sufficiency conditions do not require any differentiability on the Lagrange function. If in addition, the functions $f(x), h(x), g(x)$ are differentiable, and hence the Lagrange function is differentiable, and $(\bar{x}, \bar{\lambda}, \bar{\mu})$ is a Karush-Kuhn-Tucker Saddle point, then it is a Karush-Kuhn-Tucker point [i.e., it is a solution of (3.3) and it satisfies the constraint qualification].

Remark 2 A KKT saddle point of the Lagrange function implies that the conditions

$$\begin{aligned} \nabla_x L(\bar{x}, \bar{\lambda}, \bar{\mu}) &= 0 \\ h(\bar{x}) &= 0 \\ g(\bar{x}) &\leq 0 \\ \mu_j g_j(\bar{x}) &= 0 \quad j = 1, 2, \ldots, p \\ \mu_j &\geq 0 \quad j = 1, 2, \ldots, p \end{aligned}$$

hold, *without* the need for a constraint qualification. However, an optimal solution of (3.3) $\bar{x}$ does not necessarily imply the existence of $(\bar{\lambda}, \bar{\mu})$ unless a constraint qualification is imposed.

3.2.9 Second-Order Necessary Optimality Conditions

In this section, we discuss the need for second-order optimality conditions, and present the second-order constraint qualification along with the second-order necessary optimality conditions for problem (3.3).

3.2.9.1 Motivation

The first-order optimality conditions utilize information only on the gradients of the objective function and constraints. As a result, the curvature of the functions, measured by the second derivatives, is not taken into account. To illustrate the case in which the first-order necessary optimality conditions do not provide complete information, let us consider the following example suggested by Fiacco and McCormick (1968).

$$\begin{cases} MIN & (x_1 - 1)^2 + x_2^2 \\ s.t. & x_1 - \frac{x_2^2}{k} \leq 0 \end{cases}$$

where the values of the parameter $k > 0$ for which $(0,0)$ is a local minimum are sought.

A constraint qualification required by the KKT conditions is satisfied since we have one constraint and its gradient

$$(+1, -\frac{2x_2}{k})$$

is always nonzero. Considering the gradient KKT conditions at $(0,0)$ of the Lagrange function, we have

$$\begin{pmatrix} -2 \\ 0 \end{pmatrix} + \mu \begin{pmatrix} 1 \\ 0 \end{pmatrix} = \mathbf{0}$$

$$\Rightarrow \quad -2 + \mu = 0 \quad \Rightarrow \quad \mu = 2$$

regardless of the values of the parameter k. However, for $k = 1$ the point (0,0) is not a local minimum while for $k = 4$ it is. Therefore, the first-order necessary KKT conditions indicate that $(0,0)$ is a candidate for a local minimum, but they provide no further information on the appropriate range for the values of k.

3.2.9.2 Second-Order Constraint Qualification

Let $\bar{x} \in X$ be a feasible point of problem (3.3) and let the functions $\boldsymbol{h}(x)$ and $\boldsymbol{g}(x)$ be twice differentiable. Let z be any nonzero vector such that

$$\begin{cases} z^T \nabla h_i(\bar{x}) & = & 0 \quad i = 1, 2, \ldots, m, \quad \text{and} \\ z^T \nabla g_j(\bar{x}) & = & 0 \quad j \in J \equiv \{j : g_j(\bar{x}) = 0\}. \end{cases}$$

Then, the second-order constraint qualification holds at $\bar{x}$ if z is the tangent of a twice differentiable arc $\omega(\tau)$ that starts at $\bar{x}$, along which

$$\begin{aligned} \frac{\partial \omega(0)}{\partial \tau} &= \lambda \cdot z \text{ for some positive } \lambda, \\ \omega(0) &= \bar{x}, \end{aligned}$$

where $0 \le \tau \le \epsilon$ with $\epsilon > 0$,

$$\begin{aligned} h_i(\omega(\tau)) &\equiv 0\,, \quad i = 1, 2, \ldots, m, \text{ and} \\ g_j(\omega(\tau)) &\equiv 0\,, \quad j \in J. \end{aligned}$$

Remark 1 The second-order constraint qualification at $\bar{x}$, as well as the first-order constraint qualification, are satisfied if the gradients

$$\begin{aligned} &\nabla h_i(\bar{x}) \quad , \quad i = 1, 2, \ldots, m \text{ , and} \\ &\nabla g_j(\bar{x}) \quad , \quad j \in J, \end{aligned}$$

are linearly independent. Note, however, that the second-order constraint qualification may not imply the first-order constraint qualification.

Illustration 3.2.11 This example is taken from Fiacco and McCormick (1968), and it demonstrates that the second-order constraint qualification does not imply the first-order Kuhn-Tucker constraint qualification.

$$\begin{aligned} \min \quad & x_1 + x_2 \\ s.t. \quad & g_1(x_1, x_2) = x_1^2 + (x_2 - 1)^2 - 1 \le 0 \\ & g_2(x_1, x_2) = x_1^2 + (x_2 + 1)^2 - 1 \le 0 \\ & g_3(x_1, x_2) = -x_1 \le 0 \end{aligned}$$

It exhibits only one feasible point, $\bar{x} = (0, 0)^T$, which is its solution. The gradients of the active constraints at $\bar{x}$ are

$$\begin{aligned} \nabla g_1(\bar{x}) &= (0, -2)^T, \\ \nabla g_2(\bar{x}) &= (0, 2)^T, \\ \nabla g_3(\bar{x}) &= (-1, 0)^T. \end{aligned}$$

The first-order Kuhn-Tucker constraint qualification is not satisfied since there are no arcs pointing in the constrained region which is a single point, and hence there are no tangent vectors z contained in the constraint region. The second-order constraint qualification holds, however, since there are no nonzero vectors orthogonal to all three gradients.

3.2.9.3 Second-Order Necessary Conditions

Let $\bar{x}$ be a local optimum of problem (3.3), the functions $f(x), h(x), g(x)$ be twice continuously differentiable, and the second-order constraint qualification holds at $\bar{x}$. If there exist Lagrange multipliers $\bar{\lambda}, \bar{\mu}$ satisfying the KKT first-order necessary conditions:

$$\begin{aligned} \nabla f(\bar{x}) + \bar{\lambda}^T \nabla h(\bar{x}) + \bar{\mu}^T \nabla g(\bar{x}) &= 0 \\ h(\bar{x}) &= 0 \\ g(\bar{x}) &\le 0 \\ \bar{\mu}_j g_j(\bar{x}) &= 0 \quad j = 1, 2, \ldots, p \\ \bar{\mu}_j &\ge 0 \quad j = 1, 2, \ldots, p \end{aligned}$$

and if for every nonzero vector z:

$$\begin{cases} z^T \nabla h_i(\bar{x}) &= 0 \quad i = 1, 2, \ldots, m, \quad \text{and} \\ z^T \nabla g_j(\bar{x}) &= 0 \quad j \in J \equiv \{j : g_j(\bar{x}) = 0\} \end{cases}$$

then,

$$z^T \nabla^2 L(\bar{x}, \bar{\lambda}, \bar{\mu}) z \geq 0$$

where

$$L(\bar{x}, \bar{\lambda}, \bar{\mu}) = f(\bar{x}) + \bar{\lambda}^T \boldsymbol{h}(\bar{x}) + \bar{\mu}^T \boldsymbol{g}(\bar{x}).$$

Illustration 3.2.12 This example is the motivating example for the second order necessary optimality conditions:

$$\begin{array}{ll} \min & (x_1 - 1)^2 + x_2^2 \\ s.t. & x_1 - \frac{x_2^2}{k} \leq 0 \end{array}$$

where $k > 0$ and k is a parameter. Let $\bar{x} = (0, 0)$ be a local minimum. The Lagrange function is

$$L(x_1, x_2, \bar{\mu}) = (x_1 - 1)^2 + x_2^2 + \bar{\mu}(x_1 - \frac{x_2^2}{k}).$$

If $\bar{x} = (0, 0)$ is a local minimum, then it satisfies the first-order necessary KKT conditions. From the gradient with respect to x_1, the Lagrange multiplier $\bar{\mu}$ can be obtained:

$$2(\bar{x_1} - 1) + \bar{\mu} = 0 \Rightarrow \bar{\mu} = 2.$$

Then, the Lagrange function becomes

$$L(x_1, x_2, \bar{\mu}) = (x_1 - 1)^2 + x_2^2 + 2(x_1 - \frac{x_2^2}{k}).$$

The Hessian of the Lagrange function at $\bar{x}$, $\nabla^2 L(\bar{x}, \bar{\mu})$ is

$$\nabla^2 L(\bar{x}, \bar{\mu}) = \begin{bmatrix} 2 & 0 \\ 0 & 2 - \frac{4}{k} \end{bmatrix}.$$

The gradient of the constraint at $\bar{x} = (0, 0)$ is $(+1, 0)$. As a result, the nonzero vector z suffices to take the form:

$$z = (0, z_2)^T$$

in checking the second-order necessary conditions:

$$z^T \nabla^2 L(\bar{x}, \bar{\mu}) z = (0, z_2) \begin{bmatrix} 2 & 0 \\ 0 & 2 - \frac{4}{k} \end{bmatrix} \begin{pmatrix} 0 \\ z_2 \end{pmatrix} = z_2^2 (2 - \frac{4}{k}).$$

Hence, if $\bar{x} = (0, 0)$ is a local minimum, then from the second-order necessary optimality conditions the following holds:

$$z_2^2 (2 - \frac{4}{k}) \geq 0,$$

which implies that $(2 - \frac{4}{k}) \geq 0$, or $k \geq 2$. As a result, the second-order necessary optimality conditions hold at $\bar{x} = (0, 0)$ for $k \geq 2$.

3.2.10 Second-Order Sufficient Optimality Conditions

This section presents the second-order sufficient optimality conditions for a feasible point $\bar{x}$ of (3.3) to be a local minimum.

Let $f(x), h(x), g(x)$ be twice differentiable, and assume that there exist Lagrange multipliers $\bar{\lambda}, \bar{\mu}$ satisfying the first-order necessary KKT optimality conditions:

$$\begin{aligned} \nabla f(\bar{x}) + \bar{\lambda}^T \nabla h(\bar{x}) + \bar{\mu}^T \nabla g(\bar{x}) &= 0 \\ h(\bar{x}) &= 0 \\ g(\bar{x}) &\leq 0 \\ \bar{\mu}_j g_j(\bar{x}) &= 0 \quad j = 1, 2, \ldots, p \\ \bar{\mu}_j &\geq 0 \quad j = 1, 2, \ldots, p \end{aligned}$$

If for every nonzero vector z:

$$\begin{cases} z^T \nabla h_i(\bar{x}) = 0 & i = 1, 2, \ldots, m, \quad \text{and} \\ z^T \nabla g_j(\bar{x}) = 0 & j \in J_1 \equiv \{j : g_j(\bar{x}) = 0, \mu_j > 0\}, \\ z^T \nabla g_j(\bar{x}) \leq 0 & j \in J_2 \equiv \{j : g_j(\bar{x}) = 0, \mu_j = 0\}, \end{cases}$$

it follows that

$$z^T \nabla^2 L(\bar{x}, \bar{\lambda}, \bar{\mu}) z > 0,$$

then, $\bar{x}$ is a strict (unique) local minimum of problem (3.3).

Illustration 3.2.13 This example is the one considered in illustration 3.2.12.

$$\begin{aligned} \min \quad & (x_1 - 1)^2 + x_2^2 \\ s.t. \quad & x_1 - \frac{x_2^2}{k} \leq 0 \end{aligned}$$

where the parameter k is strictly positive. In this illustration, the values of the parameter k for which th point $\bar{x} = (0, 0)$ is a local minimum are sought.

$$z^T \nabla^2 L(\bar{x}, \mu) z = (0, z_2) \left[\begin{array}{cc} 2 & 0 \\ 0 & 2 - \frac{4}{k} \end{array} \right) \left(\begin{array}{c} 0 \\ z_2 \end{array} \right] = z_2^2 (2 - \frac{4}{k}).$$

According to the second-order sufficient conditions, if $z^T \nabla^2 L(\bar{x}, \mu) z > 0$, then $\bar{x}$ is a local minimum.

$$z_2^2 (2 - \frac{4}{k}) > 0 \Rightarrow 2 - \frac{4}{k} > 0 \Rightarrow k > 2.$$

In other words, for $k > 2, \bar{x} = (0, 0)$ is a strict local minimum, while for $k \leq 2, \bar{x} = (0, 0)$ is *not* a strict local minimum.

3.2.11 Outline of Nonlinear Algorithmic Methods

The optimality conditions discussed in the previous sections formed the theoretical basis for the development of several algorithms for unconstrained and constrained nonlinear optimization problems. In this section, we will provide a brief outline of the different classes of nonlinear multivariable optimization algorithms.

The algorithms for unconstrained nonlinear optimization problems in multiple dimensions can be broadly classified as : (i) searching without the use of derivatives, (ii) searching using first order derivatives, and (iii) searching using second order derivatives. Typical examples of algorithms that do not use derivatives are the simplex search method, the Hooke and Jeeves method, the Rosenbrock method, and the conjugate direction method. Algorithms which employ first order gradient information include the steepest descent method, the discrete Newton's method, quasi-Newton methods, and conjugate gradient methods. The steepest descent method performs a line search along the direction of the negative gradient of the minimized function. The discrete Newton's method approximates the Hessian matrix by finite-differences of the gradient and deflects the steepest descent direction by premultiplying it by the inverse of the approximated Hessian matrix. Quasi-Newton methods build up an approximation of the curvature of the nonlinear function employing information on the function and its gradient only, and hence avoid the explicit hessian matrix formation. The conjugate gradient methods combine current information about the gradient with the gradients of previous iterations and the previous search direction for the calculation of the new search direction. They generate search directions without the need to store a matrix, and they are very useful when methods based on matrix factorization cannot be applied. Algorithms which make use of second order derivative information include the Newton's method, the Levenberg-Marquardt method, and trust region methods. In the Newton method, the inverse of the Hessian matrix premultiplies the steepest descent direction and a suitable search direction is found for a quadratic approximation of the function. Newton's method converges quadratically if it is initialized appropriately close to a local minimum. Modifications of Newton's method which include trust region methods and the Levenberg-Marquardt method aim at safeguarding the line search so as to consider points during the search at which the quadratic approximation remains valid.

Constrained nonlinear optimization algorithms belong to the classes of (i) exterior penalty function methods, (ii) interior (barrier) function methods, (iii) gradient projection methods, (iv) generalized reduced gradient methods, (v) successive linear programming methods, and (vi) successive quadratic programming methods. In the exterior penalty function methods, a penalty term is added to the objective function for any violation of the equality and inequality constraints. The constrained nonlinear optimization problem is transformed into a single unconstrained one or a sequence of unconstrained problems, a sequence of infeasible points is generated, and as the penalty parameter is made large enough the generated points approach the optimal solution from outside of the feasible region. To avoid ill-conditioning which results from the need to make the penalty parameter infinitely large, exact exterior penalty function methods were developed. These methods aim at recovering an exact optimum for finite values of the penalty parameter. Nondifferentiable exact exterior penalty function methods introduce absolute values of the equality constraints in the transformed objective. Differentiable exact exterior penalty function methods like the augmented lagrange penalty function, add to the ordinary lagrange function quadratic penalty terms for the equality and inequality constraints.

The interior (barrier) function methods transform a constrained problem into an unconstrained one or into a sequence of unconstrained problems and they introduce a barrier (interior penalty) that prevents the generated points from leaving the feasible region. These methods generate a sequence of feasible points whose limit corresponds to an optimal solution of the constrained nonlinear optimization problem. An example of barrier functions are those that introduce the logarithms of the inequalities in the objective function (i.e. logarithmic barrier functions) and require update of a single barrier parameter. The logarithmic barrier methods exhibit severe ill-conditioning problems in a region close to the solution. Modified barrier methods were introduced recently that exhibit finite convergence as opposed to asymptotic convergence of the classical barrier functions. These methods make explicit use of the lagrange multipliers, update the barrier parameter and the multipliers at each iteration with low computational complexity, and attain the solution without the need to drive the barrier parameter to zero.

The gradient projection methods when applied to problems with linear constraints feature the projection of the negative gradient of the objective function onto the null space of the gradients of the equalities and binding inequality constraints and lead to improving feasible directions. In the presence of nonlinear constraints, such a projected gradient may not lead to feasible points and hence a correction needs to be introduced so as to drive it to the feasible region. Generalized reduced gradient methods employ successive linearization of the objective function and constraints, reduce the dimensionality of the problem to an independent subset of variables, determine the search components for the independent variables, define new variables that are normal to the constraints, and express the gradient and search direction in terms of the independent variables.

In each iteration of the successive linear programming methods, a linear problem is formulated based on a first order Taylor series approximation to the objective function and constraints. This linear problem contains bound constraints on the direction vectors for all the variables, and its solution provides the new search direction vector. A criterion, based on an exact penalty function is introduced for accepting or rejecting a new search direction which is accompanied by a modification of the bounds on the variables. This class of methods has the advantage of employing only linear programming problems and is better suited for optimization problems that have predominantly linear constraints. The successive linear programming methods feature a quadratic rate of convergence in the case that the optimum solution is at a vertex of the feasible region. For problems with nonvertex optimum solutions however, they can exhibit slow convergence.

Successive quadratic programming methods aim at improving the convergence behavior of the successive linear programming methods by employing a second order approximation. These methods use Newton's method or Quasi-Newton methods to solve directly the Karush Kuhn Tucker optimality conditions of the original problem. The search direction results from the solution of a quadratic programming problem which has as objective function a quadratic approximation of the Lagrange function with a positive definite approximation of its Hessian and as constraints a linear approximation of the original nonlinear constraints. Employing the BFGS update for the approximated hessian of the Lagrange function was shown to lead to superlinear convergence when the initialization is sufficiently close to a solution point. Introducing an exact penalty function (e.g. absolute value penalty) overcomes the limitation of being sufficiently close to a solution point and leads to successive quadratic programming methods that are shown to be globally convergent under certain mild assumptions.

Summary and Further Reading

This chapter introduces the fundamentals of unconstrained and constrained nonlinear optimization. Section 3.1 presents the formulation and basic definitions of unconstrained nonlinear optimization along with the necessary, sufficient, and necessary and sufficient optimality conditions. For further reading refer to Hestenes (1975), Luenberger (1984), and Minoux (1986).

Section 3.2.1 presents the formulation and basic definitions in constrained nonlinear programming. Sections 3.2.2, 3.2.3, 3.2.4, and 3.2.5 introduce the Lagrange function and its multipliers, provide an interpretation of the Lagrange multipliers along with a condition for their existence, and introduce the weak Lagrange function. Section 3.2.6 discusses first the Fritz John first-order necessary optimality conditions, the need for a first-order constraint qualification, and different types of first-order constraint qualifications. In the sequel, the Karush-Kuhn-Tucker first-order necessary optimality conditions are introduced along with their geometrical interpretation. Section 3.2.7 presents the Karush-Kuhn-Tucker first-order sufficient optimality conditions, while section 3.2.8 introduces the saddle point necessary and sufficient optimality conditions. Further reading in these subjects can be found in Avriel (1976), Bazaraa *et al.* (1993), Hestenes (1975), Mangasarian (1969), and Minoux (1986).

Sections 3.2.9 and 3.2.10 present the second-order necessary and sufficient optimality conditions along with the second-order constraint qualification. The reader is referred to Fiacco and McCormick (1968) and McCormick (1983) for further reading. Section 3.2.11 discusses briefly the different classes of algorithms for unconstrained and constrained nonlinear optimization problems. The reader interested in detailed descriptions of the algorithmic developments in unconstrained and constrained nonlinear programming is referred to the books of Bazaraa *et al.* (1993), Dennis and Schnabel (1983), Edgar and Himmelblau (1988), Fletcher (1987), Gill *et al.* (1981), Luenberger (1984), Minoux (1986), and Reklaitis *et al.* (1983).

Problems on Nonlinear Optimization

1. Determine and classify the stationary points of the following functions:

   (i) $f(x_1, x_2, x_3) = x_1^2 + x_2^2 + x_3^2 - x_1x_2 - x_2x_3 - x_1x_3$

   (ii) $f(x_1, x_2, x_3) = e^{x_1 - x_2} + e^{x_2 - x_1} + e^{x_1} + x_3$

   (iii) $f(x_1, x_2) = e^{x_1 - x_2} + 2 \cdot e^{x_2 - x_1}$

2. Apply the necessary, sufficient, and necessary and sufficient optimality conditions for the functions:

   (i) $f(x) = \frac{1}{x^{12}} - \frac{2}{x^6}$

   (ii) $f(x_1, x_2, x_3, x_4, x_5, x_6) =$
   $\frac{1}{[(x_1-x_4)^2+(x_2-x_5)^2+(x_3-x_6)^2]^6} - \frac{2}{[(x_1-x_4)^2+(x_2-x_5)^2+(x_3-x_6)^2]^3}$

3. Consider problem formulation (3.3) with $f(x), h(x), g(x)$ differentiable functions. Show that $\bar{x}$ is a KKT point if and only if $\bar{x}$ solves the first-order linear programming approximation given by:

$$\begin{array}{rrcl} \min & f(\bar{x}) + \nabla f(\bar{x})^T(x - \bar{x}) & & \\ s.t. & \nabla h_i(\bar{x})^T(x - \bar{x}) = 0 & & i = 1, 2, \ldots, m \\ & g_i(\bar{x}) + \nabla g_i(\bar{x})^T(x - \bar{x}) \leq 0 & & j = 1, 2, \ldots, p \end{array}$$

4. Consider the problem:

$$\begin{array}{ll} \min & x_1^2 + 2(x_2 - 1)^2 \\ s.t. & x_1 + x_2 = 2 \end{array}$$

   Determine a point satisfying the KKT conditions and check whether it is indeed an optimal solution.

5. Consider the problem:

$$\begin{array}{rl} \min & (x_1 - 1)^2 + (x_2 - 3)^2 \\ s.t. & x_1 + x_2 \leq 8 \\ & x_1^2 - 2x_2 \leq 1 \\ & x_1, x_2 \geq 0 \end{array}$$

   Write the KKT optimality conditions, determine a KKT point $\bar{x}$, and provide a geometrical interpretation of the KKT conditions.

6. Apply the KKT optimality conditions to the problem:

$$\begin{array}{rl} & \min\ e^{-(x_1+x_2)} \\ s.t. & e^{x_1} + e^{x_2} \leq 30 \\ & x_1 \geq 0 \\ & x_2 \geq 0 \end{array}$$

   Is its solution a global minimum? Why?

7. Consider the following two problems:

$$(P_1) \quad \begin{cases} \min \ f(x) \\ s.t. \ \ h(x) = 0 \\ \quad\ \ g(x) \leq 0 \end{cases} \qquad\qquad (P_2) \quad \begin{cases} \min \ f(x) \\ s.t. \ \ h(x) = 0 \\ \quad\ g(x) + s = 0 \\ \quad\ s \geq 0 \end{cases}$$

Write the KKT conditions for (P_1) and (P_2), and compare them.

8. Consider the problem:

$$\begin{aligned} \min \ & -x_1 + x_1 x_2 - x_2 \\ s.t. \ & -6x_1 + 8x_2 \leq 3 \\ & 3x_1 - x_2 \leq 3 \\ & 0 \leq x_1, x_2 \leq 3 \end{aligned}$$

Write the KKT conditions, and check the optimality of the points $\bar{x} = (0.916, 1.062)$ and $\bar{x} = (1.167, 0.5)$.

9. Consider the following problem:

$$\begin{aligned} \min \ & x_2^2 - x_1^2 + 6x_1 - 4x_2 \\ s.t. \ & x_2 - 2x_1 \leq 2 \\ & 2x_1 - x_2 \leq 8 \\ & -x_1 - 2x_2 \leq -4 \\ & x_2 \leq 4 \\ & x_1, x_2 \geq 0 \end{aligned}$$

(i) Is this problem convex? Why?

(ii) Write the KKT conditions, and check the optimality of the point $\bar{x} = (0, 2)$.

10. Consider the following problem:

$$\begin{aligned} \min \ & x_1 + x_2 \\ s.t. \ & x_1^2 + x_2^2 \leq 4 \\ & x_1^2 + x_2^2 \geq 1 \\ & x_1 - x_2 - 1 \leq 0 \\ & x_2 - x_1 - 1 \leq 0 \\ & -2 \leq x_1 \leq 2 \\ & -2 \leq x_2 \leq 2 \end{aligned}$$

(i) Is this problem convex? Why?

(ii) Write the KKT conditions, and check the point $\bar{x} = (-1.414, -1.414)$.

11. Consider the problem:

$$\begin{aligned} \min\ & 3x_1x_2 - x_1 - x_2 \\ s.t.\ & -x_1 \leq -1 \\ & -x_2 \leq -1 \end{aligned}$$

(i) Write the first-order necessary KKT conditions.

(ii) Does $\bar{x} = (2, 1)^T$ satisfy the necessary conditions?

(iii) Is $\bar{x} = (1, 1)^T$ a global minimum?

12. Consider the problem:

$$\begin{aligned} \min\ & (x_1 - 3)^2 + (x_2 - 3)^2 \\ s.t.\ & x_1^2 - x_2 \leq 0 \\ & x_1 + x_2 + x_3 \leq 0 \end{aligned}$$

(i) Write the first-order necessary KKT conditions.

(ii) Check whether the KKT point is a global minimum.

13. Consider the problem:

$$\begin{aligned} \min\ & (x_1 - x_2 + x_3)^2 \\ s.t.\ & x_1 + 2x_2 - x_3 = 5 \\ & x_1 - x_2 - x_3 = -1 \end{aligned}$$

Check whether the point $\bar{x} = (1.5, 2, 0.5)$ satisfies the KKT conditions.

14. Consider the problem:

$$\begin{aligned} \min\ & -x_1 \\ s.t.\ & x_1^2 + x_2^2 \leq 1 \\ & (x_1 - 1)^3 - x_2 \leq 0 \end{aligned}$$

(i) Does the first-order KKT constraint qualification hold at $\bar{x} = (1, 0)^T$?

(ii) Is $\bar{x} = (1, 0)^T$ a KKT point?

Chapter 4 Duality Theory

Nonlinear optimization problems have two different representations, the primal problem and the dual problem. The relation between the primal and the dual problem is provided by an elegant duality theory. This chapter presents the basics of duality theory. Section 4.1 discusses the primal problem and the perturbation function. Section 4.2 presents the dual problem. Section 4.3 discusses the weak and strong duality theorems, while section 4.4 discusses the duality gap.

4.1 Primal Problem

This section presents the formulation of the primal problem, the definition and properties of the perturbation function, the definition of stable primal problem, and the existence conditions of optimal multiplier vectors.

4.1.1 Formulation

The primal problem (P) takes the form:

$$(P)\left\{\begin{array}{lll} \min & f(\boldsymbol{x}) & \\ \text{s.t.} & \boldsymbol{h}(\boldsymbol{x}) & = \boldsymbol{0} \\ & \boldsymbol{g}(\boldsymbol{x}) & \leq \boldsymbol{0} \\ & \boldsymbol{x} & \in \boldsymbol{X} \end{array}\right.$$

where $\boldsymbol{x}$ is a vector of n variables;

$\boldsymbol{h}(\boldsymbol{x})$ is a vector of m real valued functions;
$\boldsymbol{g}(\boldsymbol{x})$ is a vector of p real valued functions;
$f(\boldsymbol{x})$ is a real valued function; and
$\boldsymbol{X}$ is a nonempty convex set.

4.1.2 Perturbation Function and Its Properties

The perturbation function, $v(y)$, that is associated with the primal problem (P) is defined as:

$$v(y) = \begin{cases} \inf & f(x) \\ \text{s.t.} & h(x) \quad = 0 \\ & g(x) \quad \leq y \\ & \quad\;\; x \quad \in X \end{cases}$$

and y is an p-dimensional perturbation vector.

Remark 1 For $y = 0$, $v(0)$ corresponds to the optimal value of the primal problem (P). Values of the perturbation function $v(y)$ at other points different than the origin $y = 0$ are useful on the grounds of providing information on sensitivity analysis or parametric effects of the perturbation vector y.

Property 4.1.1 (Convexity of $v(y)$)
Let Y be the set of all vectors y for which the perturbation function has a feasible solution, that is,

$$Y = \{y \in R^p |\; h(x) = 0,\; g(x) \leq y \text{ for some } x \in X\}.$$

If Y is a convex set, then $v(y)$ is a convex function on Y.

Remark 2 $v(y) = \infty$ if and only if $y \notin Y$.

Remark 3 Y is a convex set if $h(x)$ are linear and $g(x)$ are convex on the convex set X.

Remark 4 The convexity property of $v(y)$ is the fundamental element for the relationship between the primal and dual problems. A number of additional properties of the perturbation function $v(y)$ that follow easily from its convexity are

(i) If Y is a finite set then $v(y)$ is continuous on Y.

(ii) the directional derivative of $v(y)$ exists in every direction at every point at which $v(y)$ is finite.

(iii) $v(y)$ has a subgradient at every interior point of Y at which $v(y)$ is finite.

(iv) $v(y)$ is differentiable at a point $\bar{y}$ in Y if and only if it has a unique subgradient at $\bar{y}$.

Note also that these additional properties hold for any convex function defined on a convex set.

4.1.3 Stability of Primal Problem

Definition 4.1.1 The primal problem (P) is stable if $v(0)$ is finite and there exists a scalar $L > 0$ such that

$$\frac{v(0) - v(y)}{||y||} \leq L \text{ for all } y \neq 0.$$

Remark 1 The above definition of stability does not depend on the particular norm $||y||$ that is used. In fact, it is necessary and sufficient to consider a *one- dimensional* choice of y.

Remark 2 The property of stability can be interpreted as a Lipschitz continuity condition on the perturbation function $v(y)$.

Remark 3 If the stability condition does not hold, then $[v(0) - v(y)]$ can be made as large as desired even with small perturbation of the vector y.

4.1.4 Existence of Optimal Multipliers

Theorem 4.1.1
Let the primal problem (P) have an optimum solution x^*. Then, an optimal multiplier vector $(\bar{\lambda}, \bar{\mu})$ exists if and only if (P) is stable. Furthermore, $(\bar{\lambda}, \bar{\mu})$ is an optimal multiplier vector for (P) if and only if $(-\bar{\lambda}, -\bar{\mu})$ is a subgradient of $v(y)$ at $y = 0$.

Remark 1 This theorem points out that stability is not only a necessary but also sufficient constraint qualification, and hence it is implied by any constraint qualification used to prove the necessary optimality conditions.

Remark 2 If the objective function $f(x)$ is sufficiently well behaved, then the primal problem (P) will be stable regardless of how poorly behaved the constraints are. For instance, if $f(x) = c$, then (P) is stable as long as the constraints are feasible.

Remark 3 If the constraints are linear, it is still possible for (P) to be unstable. Consider the problem:

Illustration 4.1.1 Consider the problem:

$$\left\{ \begin{array}{rcl} \min f(x) & = & -x^{\frac{1}{3}} \\ \text{s.t. } g(x) & = & x \leq 0 \\ x \in X & = & \{x | x \geq 0\} \end{array} \right.$$

Note that as x approaches zero, the objective function has infinite steepness.

By perturbing the right-hand side in the positive direction, we obtain

$$\frac{0 - (-y^{\frac{1}{3}})}{|y|} = \frac{1}{y^{\frac{2}{3}}},$$

which can be made as large as desired by making the perturbation y sufficiently small.

4.2 Dual Problem

This section presents the formulation of the dual problem, the definition and key properties of the dual function, and a geometrical interpretation of the dual problem.

4.2.1 Formulation

The dual problem of (P), denoted as (D), takes the following form:

$$(D)\begin{cases} \max\limits_{\substack{\lambda \\ \mu \geq 0}} \inf\limits_{x \in X} L(x,\ \lambda,\ \mu) \\ \text{s.t. } L(x,\ \lambda,\ \mu) = f(x) + \lambda^T \mathbf{h}(x) + \mu^T \mathbf{g}(x) \end{cases}$$

where $L(x,\ \lambda,\ \mu)$ is the Lagrange function, λ is the m-vector of Lagrange multipliers associated with the equality constraints, and μ is the p-vector of Lagrange multipliers associated with the inequality constraints.

Remark 1 Note that the inner problem of the dual

$$\inf_{x \in X} L(x,\ \lambda,\ \mu)$$

is a function of λ and μ (i.e., it is parametric in λ and μ). Hence it may take the value of $(-\infty)$ for some λ and μ. If the infimum is attained and is finite for all (λ, μ) then the inner problem of the dual can be written as

$$\min_{x \in X} L(x,\ \lambda,\ \mu)$$

Remark 2 The dual problem consists of (i) an inner minimization problem of the Lagrange function with respect to $x \in X$ and (ii) on outer maximization problem with respect to the vectors of the Lagrange multipliers (unrestricted λ, and $\mu \geq 0$). The inner problem is parametric in λ and μ. For fixed x at the infimum value, the outer problem becomes linear in λ and μ.

4.2.2 Dual Function and Its Properties

Definition 4.2.1 (Dual function) The dual function, denoted as $\phi(\lambda,\ \mu)$, is defined as the inner problem of the dual:

$$\phi(\lambda,\ \mu) = \inf_{x \in X} L(x,\ \lambda,\ \mu)$$

Property 4.2.1 (Concavity of $\phi(\lambda,\ \mu)$)
Let $f(x)$, $h(x)$, $g(x)$ be continuous and X be a nonempty compact set in R^n. Then the dual function $\phi(\lambda,\ \mu)$ is concave.

Remark 1 Note that the dual function $\phi(\lambda, \mu)$ is concave without assuming any type of convexity for the objective function $f(x)$ and constraints $h(x)$, $g(x)$.

Remark 2 Since $\phi(\lambda,\ \mu)$ is concave, and the outer problem is a maximization problem over λ and $\mu \geq 0$, then a local optimum of $\phi(\lambda,\ \mu)$ is also a global one. The difficulty that arises though is that we only have $\phi(\lambda,\ \mu)$ as a parametric function of λ and μ and not as an explicit functionality of λ and μ.

Remark 3 The dual function $\phi(\lambda,\ \mu)$ is concave since it is the pointwise infimum of a collection of functions that are linear in λ and μ.

Remark 4 Since the dual function of $\phi(\lambda,\ \mu)$ is concave, it has a subgradient at $(\bar{\lambda},\ \bar{\mu})$ that is defined as the vector d_1, d_2:

$$\phi(\lambda,\ \mu) \leq \phi(\bar{\lambda},\ \bar{\mu}) + d_1^T(\lambda - \bar{\lambda}) + d_2^T(\mu - \bar{\mu})$$

for all λ, and $\mu \geq \mathbf{0}$.

Property 4.2.2 (Subgradient of dual function)
Let $f(x)$, $h(x)$, $g(x)$ be continuous, and X be a nonempty, compact set in R^n. Let

$$Y(\lambda,\ \mu) = \{x^* : x^* \text{ minimizes } L(x,\ \lambda,\ \mu) \text{ over } x \in X\}$$

If for any $(\bar{\lambda},\ \bar{\mu})$, $Y(\bar{\lambda},\ \bar{\mu})$ is nonempty, and $x^* \in Y(\bar{\lambda},\ \bar{\mu})$, then

$$(h(x^*), g(x^*)) \text{ is a subgradient of } \phi(\lambda, \mu) \text{ at } (\bar{\lambda}, \bar{\mu}).$$

Remark 5 This property provides a sufficient condition for a subgradient.

Property 4.2.3 (Differentiability of dual function)
Let $f(x)$, $h(x)$, $g(x)$ be continuous, and X be a nonempty compact set. If $Y(\bar{\lambda},\ \bar{\mu})$ reduces to a single element at the point $(\bar{\lambda},\ \bar{\mu})$, then the dual function $\phi(\lambda, \mu)$ is differentiable at $(\bar{\lambda},\ \bar{\mu})$ and its gradient is

$$\nabla\phi(\bar{\lambda}, \bar{\mu}) = (h(x^*), g(x^*)).$$

4.2.3 Illustration of Primal-Dual Problems

Consider the following constrained problem:

$$\left\{\begin{array}{lrl} \min & (x_1 - 1)^2 + (x_2 - 1)^2 + (x_3 - 1)^2 & \\ \text{s.t.} & 2x_1 + 4x_2 & = 10 \\ & 2x_1 + 2x_2 - 4x_3 & \leq 0 \\ & x_1, x_2, x_3 & \geq 0 \end{array}\right.$$

where

$$\begin{aligned} f(x) &= (x_1 - 1)^2 + (x_2 - 1)^2 + (x_3 - 1)^2, \\ h(x) &= 2x_1 + 4x_2 - 10, \\ g(x) &= 2x_1 + 2x_2 - 4x_3, \text{ and} \\ X &= \{x_1, x_2, x_3 | x_1 \geq 0, x_2 \geq 0, x_3 \geq 0\}. \end{aligned}$$

The Lagrange function takes the form:

$$\begin{aligned} L(x_1, x_2, x_3, \lambda, \mu) &= (x_1 - 1)^2 + (x_2 - 1)^2 + (x_3 - 1)^2 \\ &\quad + \lambda\,(2x_1 + 4x_2 - 10) \\ &\quad + \mu\,(2x_1 + 2x_2 - 4x_3). \end{aligned}$$

The minimum of the $L(x_1, x_2, x_3, \lambda, \mu)$ is given by

$$\begin{aligned}
\frac{\partial L}{\partial x_1} &= 2(x_1 - 1) + 2\lambda + 2\mu &= 0 \Rightarrow x_1 = 1 - \lambda - \mu, \\
\frac{\partial L}{\partial x_2} &= 2(x_2 - 1) + 4\lambda + 2\mu &= 0 \Rightarrow x_2 = 1 - 2\lambda - \mu, \\
\frac{\partial L}{\partial x_3} &= 2(x_3 - 1) - 4\mu &= 0 \Rightarrow x_3 = 1 + 2\mu.
\end{aligned}$$

The dual function $\phi(\lambda, \mu)$ then becomes

$$\begin{aligned}
\phi(\lambda, \mu) = & \; (-\lambda - \mu)^2 + (-2\lambda - \mu)^2 + (2\mu)^2 \\
& + \lambda[2(1 - \lambda - \mu) + 4(1 - 2\lambda - \mu) - 10] \\
& + \mu[2(1 - \lambda - \mu) + 2(1 - 2\lambda - \mu) - 4(1 + 2\mu)] \\
= & \; -5\lambda^2 - 6\mu^2 - 6\lambda\mu - 4\lambda.
\end{aligned}$$

This dual function is concave in λ and μ. (The reader can verify it by calculating the eigenvalues of the Hessian of this quadratic function in λ and μ, which are both negative.)

The maximum of the dual function $\phi(\lambda, \mu)$ can be obtained when

$$\begin{aligned}
\frac{\partial \phi(\lambda,\mu)}{\partial \lambda} &= -10\lambda - 6\mu - 4 = 0, \\
\frac{\partial \phi(\lambda,\mu)}{\partial \mu} &= -12\mu - 6\lambda = 0,
\end{aligned}$$

which results in

$$\begin{aligned}
\lambda^* &= -\tfrac{4}{7}, \\
\mu^* &= \tfrac{2}{7}, \\
\text{and } \phi(\lambda^*, \mu^*) &= (8/7).
\end{aligned}$$

$$\text{Also, } \left\{ \begin{array}{l} x_1^* = (9/7) \\ x_2^* = (13/7) \\ x_3^* = (11/7) \end{array} \right\} \text{ and } f(x_1^*, x_2^*, x_3^*) = (8/7).$$

4.2.4 Geometrical Interpretation of Dual Problem

The geometrical interpretation of the dual problem provides important insight with respect to the dual function, perturbation function, and their properties. For illustration purposes, we will consider the primal problem (P) consisting of an objective function $f(x)$ subject to constraints $g_1(x) \leq 0$ and $g_2(x) \leq 0$ in a single variable x.

The geometrical portrayal is based upon the image of $\boldsymbol{X}$ under $f(x), g_1(x)$, and $g_2(x)$ that is represented by the image set I:

$$\begin{aligned}
I = \{(z_1, z_2, z_3) \in R^3 : z_1 \geq g_1(x), z_2 \geq g_2(x) \text{ and} \\
z_3 = f(x) \text{ for some } x \in \boldsymbol{X}\},
\end{aligned}$$

which is shown in Figure 4.1.

Geometrical interpretation of primal problem (P):

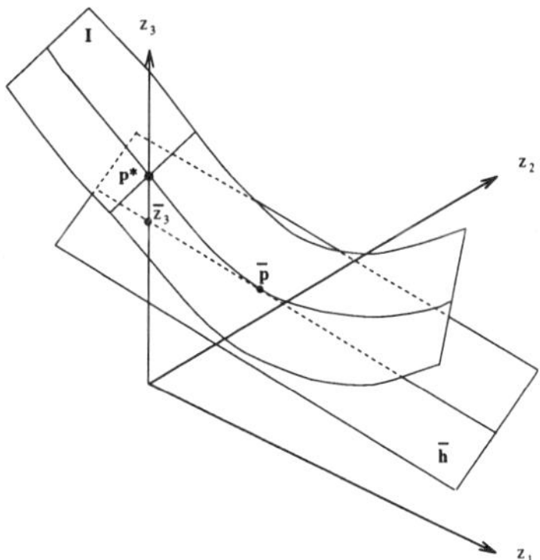

Figure 4.1: Image set I

Note that the intersection point of I and z_3, denoted as (P^*) in the figure, is the image of the optimal solution of the primal problem (P),

$$P^* = [g_1(x^*), g_2(x^*), f(x^*)],$$

where x^* is the minimizer of the primal problem (P). Hence the primal problem (P) can be explained as follows: Determine the point in the image set I which minimizes z_3 subject to $z_1 \leq 0$ and $z_2 \leq 0$.

For the geometrical interpretation of the dual problem, we will consider particular values for the Lagrange multipliers μ_1, μ_2 associated with the two inequality constraints $(\mu_1 \geq 0, \mu_2 \geq 0)$, denoted as $\bar{\mu_1}, \bar{\mu_2}$.

To evaluate the dual function at $\bar{\mu_1}, \bar{\mu_2}$ (i.e., the maximum of (D) at $\bar{\mu_1}, \bar{\mu_2}$), we have

$$\begin{aligned} &\min \{f(x) + \bar{\mu_1} g_1(x) + \bar{\mu_2} g_2(x)\} \\ &s.t.\ x \in X. \end{aligned}$$

This is equivalent to

$$\begin{aligned} \min \quad & z_3 + \bar{\mu_1} z_1 + \bar{\mu_2} z_2 \\ s.t. \quad & (z_1, z_2, z_3) \in I \end{aligned}$$

Note that the objective function

$$z_3 + \bar{\mu_1} z_1 + \bar{\mu_2} z_2$$

is a plane in (z_1, z_2, z_3) with slope $(-\bar{\mu_1}, -\bar{\mu_2})$ as illustrated in Figure 4.1.

Geometrical Interpretation of Dual Function at $(\bar{\mu_1}, \bar{\mu_2})$:

The dual function at $(\bar{\mu}_1, \bar{\mu}_2)$ corresponds to determining the lowest plane with slope $(-\bar{\mu}_1, -\bar{\mu}_2)$ which intersects the image set I. This corresponds to the supporting hyperplane $\bar{h}$ which is tangent to the image set I at the point $\bar{P}$, as shown in Figure 4.1.

Note that the point $\bar{P}$ is the image of

$$\begin{array}{ll} \min & f(x) + \mu_1 g_1(x) + \mu_2 g_2(x) \\ \text{s.t.} & x \in X \end{array}$$

The minimum value of this problem is the value of z_3 where the supporting hyperplane $\bar{h}$ intersects the ordinate, denoted as $\bar{z}_3$ in Figure 4.1.

Geometrical Interpretation of Dual Problem:

Determine the value of $(\bar{\mu_1}, \bar{\mu_2})$, which defines the slope of a supporting hyperplane to the image set I, such that it intersects the ordinate at the highest possible value. In other words, identify $(\bar{\mu_1}, \bar{\mu_2})$ so as to maximize $\bar{z_3}$.

Remark 1 The value of $(\bar{\mu_1}, \bar{\mu_2})$ that intersects the ordinate at the maximum possible value in Figure 4.1 is the supporting hyperplane of I that goes through the point P^*, which is the optimal solution to the primal problem (P).

Remark 2 It is not always possible to obtain the optimal value of the dual problem being equal to the optimal value of the primal problem. This is due to the form that the image set I can take for different classes of mathematical problems (i.e., form of objective function and constraints). This serves as a motivation for the weak and strong duality theorems to be presented in the following section.

4.3 Weak and Strong Duality

In the previous section we have discussed geometrically the nature of the primal and dual problems. In this section, we will present the weak and strong duality theorems that provide the relationship between the primal and dual problem.

Theorem 4.3.1 (Weak duality)
Let $\bar{x}$ be a feasible solution to the primal problem (P), and $(\bar{\lambda}, \bar{\mu})$ be a feasible solution to the dual problem (D). Then, the objective function of (P) evaluated at $\bar{x}$ is greater or equal to the objective function of (D) evaluated at $(\bar{\lambda}, \bar{\mu})$; that is,

$$f(\bar{x}) \geq \phi(\bar{\lambda}, \bar{\mu}).$$

Remark 1 Any feasible solution of the dual problem (D) represents a lower bound on the optimal value of the primal problem (P).

Remark 2 Any feasible solution of the primal problem (P) represents an upper bound on the optimal value of the dual problem (D).

Remark 3 This lower-upper bound feature between the dual and primal problems is very important in establishing termination criteria in computational algorithms. In particular applications , if at some iteration feasible solutions exist for both the primal and the dual problems and are close to each other in value, then they can be considered as being *practically* optimal for the problem under consideration.

Remark 4 This important lower-upper bound result for the dual-primal problems that is provided by the weak duality theorem, is *not* based on any convexity assumption. Hence, it is of great use for nonconvex optimization problems as long as the dual problem can be solved efficiently.

Remark 5 If the optimal value of the primal problem (P) is $-\infty$, then the dual problem must be infeasible (i.e., essentially infeasible).

Remark 6 If the optimal value of the dual problem (D) is $+\infty$, then the primal problem (P) must be infeasible.

The weak duality theorem provides the lower-upper bound relationship between the dual and the primal problem. The conditions needed so as to attain equality between the dual and primal solutions are provided by the following strong duality theorem.

Theorem 4.3.2 (Strong duality)
Let $f(\boldsymbol{x}), \boldsymbol{g}(\boldsymbol{x})$ be convex, $\boldsymbol{h}(\boldsymbol{x})$ be affine, and X be a nonempty convex set in R^n. If the primal problem (P) is stable, then

(i) *The dual problem (D) has an optimal solution.*

(ii) *The optimal values of the primal problem (P) and dual problem (D) are equal.*

(iii) *$(\bar{\lambda}, \bar{\mu})$ are an optimal solution of the dual problem if and only if $(-\bar{\lambda}, -\bar{\mu})$ is a subgradient of the perturbation function $v(\boldsymbol{y})$ at $\boldsymbol{y} = \boldsymbol{0}$.*

(iv) *Every optimal solution $(\bar{\lambda}, \bar{\mu})$ of the dual problem (D) characterizes the set of all optimal solutions (if any) of the primal problem (P) as the minimizers of the Lagrange function:*

$$L(\boldsymbol{x}, \bar{\lambda}, \bar{\mu}) = f(\boldsymbol{x}) + \bar{\lambda}^T \boldsymbol{h}(\boldsymbol{x}) + \bar{\mu}^T \boldsymbol{g}(\boldsymbol{x})$$

over $\boldsymbol{x} \in X$ which also satisfy the feasibility conditions

$$\boldsymbol{h}(\boldsymbol{x}) = \boldsymbol{0}$$
$$\boldsymbol{g}(\boldsymbol{x}) \leq \boldsymbol{0}$$

and the complementary slackness condition

$$\bar{\mu}^T \boldsymbol{g}(\boldsymbol{x}) = 0.$$

Remark 1 Result (ii) precludes the existence of a gap between the primal problem and dual problem values which is denoted as duality gap. It is important to note that nonexistence of duality gap is guaranteed under the assumptions of convexity of $f(\boldsymbol{x}), \boldsymbol{g}(\boldsymbol{x})$, affinity of $\boldsymbol{h}(\boldsymbol{x})$, and stability of the primal problem (P).

Remark 2 Result (iii) provides the relationship between the perturbation function $v(\boldsymbol{y})$ and the set of optimal solutions $(\bar{\lambda}, \bar{\mu})$ of the dual problem (D).

Remark 3 If the primal problem (P) has an optimal solution and it is stable, then using the theorem of existence of optimal multipliers (see section 4.1.4), we have an alternative interpretation of the optimal solution $(\bar{\lambda}, \bar{\mu})$ of the dual problem (D): that $(\bar{\lambda}, \bar{\mu})$ are the optimal Lagrange multipliers of the primal problem (P).

Remark 4 Result (iii) holds also under a weaker assumption than stability; that is, if $v(\mathbf{0})$ is finite and the optimal values of (P) and (D) are equal.

Remark 5 Result (iv) can also be stated as
If $(\bar{\lambda}, \bar{\mu})$ are optimal in the dual problem (D), then x is optimal in the primal problem (P) if and only if $(x, \bar{\lambda}, \bar{\mu})$ satisfies the optimality conditions of (P).

Remark 6 The geometrical interpretation of the primal and dual problems clarifies the weak and strong duality theorems. More specifically, in the vicinity of $y = 0$, the perturbation function $v(y)$ becomes the z_3-ordinate of the image set I when z_1 and z_2 equal y. In Figure 4.1, this ordinate does not decrease infinitely steeply as y deviates from zero. The slope of the supporting hyperplane to the image set I at the point P^*, $(-\bar{\mu}_1, -\bar{\mu}_2)$, corresponds to the subgradient of the perturbation function $v(y)$ at $y = 0$.

Remark 7 An instance of unstable problem (P) is shown in Figure 4.2 The image set I is tangent to the ordinate z_3 at the point P^*. In this case, the supporting hyperplane is perpendicular, and the value of the perturbation function $v(y)$ decreases infinitely steeply as y begins to increase above zero. Hence, there does not exist a subgradient at $y = 0$. In this case, the strong duality theorem does not hold, while the weak duality theorem holds.

4.3.1 Illustration of Strong Duality

Consider the problem

$$\left\{\begin{array}{rl} \min f(x) = & (x_1 - 1)^2 + (x_2 - 1)^2 + (x_3 - 1)^2 \\ \text{s.t. } h(x) = & 2x_1 + 4x_2 - 10 = 0 \\ g(x) = & 2x_1 + 2x_2 - 4x_3 \leq 0 \\ & x_1, x_2, x_3 \geq 0 \end{array}\right.$$

that was used to illustrate the primal-dual problems in section 4.2.3.

The objective function $f(x)$ and the inequality constraint $g(x)$ are convex since $f(x)$ is separable quadratic (sum of quadratic terms, each of which is a linear function of x_1, x_2, x_3, respectively) and $g(x)$ is linear. The equality constraint $h(x)$ is linear. The primal problem is also stable since $v(\mathbf{0})$ is finite and the additional stability condition (Lipschitz continuity-like) is satisfied since $f(x)$ is well behaved and the constraints are linear. Hence, the conditions of the strong duality theorem are satisfied. This is why

$$\phi(\lambda^*, \mu^*) = f(x_1^*, x_2^*, x_3^*) = (8/7),$$

as was calculated in section 4.2.3.

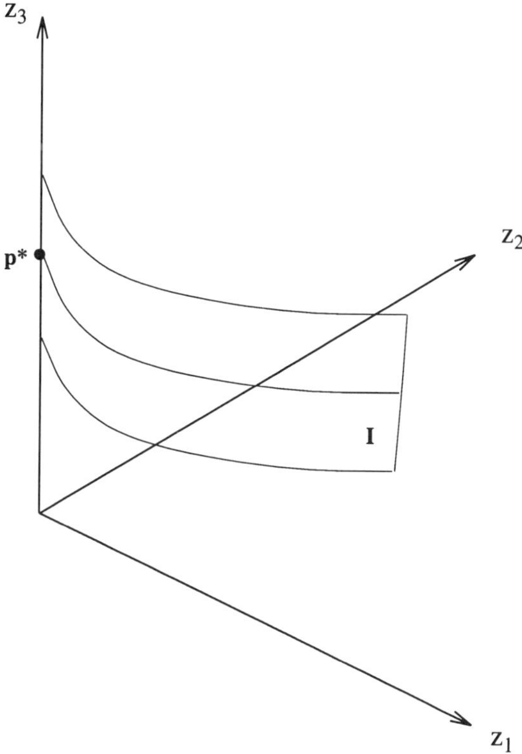

Figure 4.2: Unstable problem (P)

Also notice that since $\phi(\lambda, \mu)$ is differentiable at $(\lambda^*, \mu^*) = \left(-\frac{4}{7}, \frac{2}{7}\right)$ and its gradient is

$$\nabla\phi(\lambda^*, \mu^*) = [h(x^*), g(x^*)].$$

4.3.2 Illustration of Weak and Strong Duality

Consider the following bilinearly (quadratically) constrained problem

$$\left\{\begin{array}{ll} \min & -x_1 - x_2 \\ \text{s.t.} & x_1 x_2 \leq 4 \\ & 0 \leq x_1 \leq 4 \\ & 0 \leq x_2 \leq 8 \end{array}\right.$$

where

$$\begin{aligned} f(x) = &\ -x_1 - x_2, \\ g(x) = &\ x_1 x_2 - 4, \text{ and} \\ X = &\ \{(x_1, x_2) : 0 \leq x_1 \leq 4, 0 \leq x_2 \leq 8\}. \end{aligned}$$

This problem is nonconvex due to the bilinear constraint and its global solution is $(x_1^*, x_2^*) = (0.5, 8)$, $f(x_1^*, x_2^*) = -8.5$.

The Lagrange function takes the form (by dualizing only the bilinear constraint):

$$L(x_1, x_2, \mu) = -x_1 - x_2 + \mu(x_1 x_2 - 4),$$

The minimum of the Lagrange function with respect to x_1, x_2 is given by

$$\begin{aligned} \frac{\partial L}{\partial x_1} &= -1 + \mu x_2 = 0 \Rightarrow x_2 = \frac{1}{\mu}, \\ \frac{\partial L}{\partial x_2} &= -1 + \mu x_1 = 0 \Rightarrow x_1 = \frac{1}{\mu}. \end{aligned}$$

The dual function then becomes

$$\begin{aligned} \phi(\mu) &= -\tfrac{1}{\mu} - \tfrac{1}{\mu} + \mu\left(\tfrac{1}{\mu} \cdot \tfrac{1}{\mu} - 4\right) \\ &= -\tfrac{1}{\mu} - 4\mu. \end{aligned}$$

The maximum of the dual function can be obtained when

$$\frac{\partial \phi}{\partial \mu} = \frac{1}{\mu^2} - 4 = 0 \Rightarrow \bar{\mu} = 0.5.$$

Then, $\phi(\bar{\mu}) = -4$

$$\begin{aligned} \bar{x}_1 &= \tfrac{1}{\mu} = 2, \\ \bar{x_2} &= \tfrac{1}{\mu} = 2, \\ &\Downarrow \\ f(\bar{x}_1, \bar{x}_2) &= -4 \geq \phi(\bar{\mu}) = -4. \end{aligned}$$

4.3.3 Illustration of Weak Duality

Consider the following constrained problem:

$$\left\{ \begin{array}{ll} \min & x_1^{0.5} + x_2^{0.5} - 6x_1 \\ s.t. & x_2 - 3x_1 = 0 \\ & x_2 \leq 4 \\ & x_1, x_2 \geq 0 \end{array} \right.$$

where

$$\begin{aligned} f(x) &= x_1^{0.5} + x_2^{0.5} - 6x_1, \\ h(x) &= x_2 - 3x_1, \\ g(x) &= x_2 - 4, \\ X &= \{(x_1, x_2) : x_1 \geq 0, x_2 \geq 0\}. \end{aligned}$$

The global optimum solution is obtained at $(x_1, x_2) = (4/3, 4)$ and the objective function value is $(\frac{2}{\sqrt{3}} - 6)$.

The Lagrange function takes the form

$$L(x_1, x_2, \lambda, \mu) = x_1^{0.5} + x_2^{0.5} - 6x_1 + \lambda(x_2 - 3x_1) + \mu(x_2 - 4).$$

The minimum of the $L(x_1, x_2, \lambda, \mu)$ over $\boldsymbol{x} \in \boldsymbol{X}$ is given by

$$\begin{aligned} \frac{\partial L}{\partial x_1} &= 0.5x_1^{-0.5} - 6 - 3\lambda = 0 \Rightarrow x_1^{0.5} = \frac{1}{6(\lambda+2)}, \\ \frac{\partial L}{\partial x_2} &= 0.5x_2^{-0.5} + \lambda + \mu = 0 \Rightarrow x_2^{0.5} = -\frac{1}{2(\lambda+\mu)}. \end{aligned}$$

Since $x_1, x_2 \geq 0$, we must have $\lambda + 2 \geq 0$ and $\lambda + \mu \leq 0$.

Since $L(x_1, x_2, \lambda, \mu)$ is strictly concave in x_1, x_2, it achieves its minimum at boundary. Now,

$$L(x_1, x_2, \lambda, \mu) = \sqrt{x_1} - (3\lambda + 6)x_1 + \sqrt{x_2} + (\lambda + \mu)\, x_2 - 4\mu.$$

Clearly if $3\lambda + 6 > 0$ or $\lambda + \mu < 0$, $\phi(\lambda, \mu) = -\infty$ and when $3\lambda + 6 \leq 0$, and $\lambda + \mu \geq 0$, the minimum of L is achieved at $x_1 = x_2 = 0$, with $\phi(\lambda, \mu) = -4\mu$. Notice that $\lambda = -2 \Rightarrow \mu \geq -\lambda \geq 2$:

$$\begin{aligned} \phi(\lambda, \mu) &\leq -8, \\ \max_{\lambda, \mu \geq 0} \phi(\lambda, \mu) &= -8 < \frac{2}{\sqrt{3}} - 6. \end{aligned}$$

4.4 Duality Gap and Continuity of Perturbation Function

Definition 4.4.1 (Duality gap) If the weak duality inequality is strict:

$$f(\bar{x}) > \phi(\bar{\lambda}, \bar{\mu}),$$

then, the difference $f(\bar{x}) - \phi(\bar{\lambda}, \bar{\mu})$ is called a duality gap.

Remark 1 The difference in the optimal values of the primal and dual problems can be due to a lack of continuity of the perturbation function $v(y)$ at $y = 0$. This lack of continuity does not allow the existence of supporting hyperplanes described in the geometrical interpretation section.

Remark 2 The perturbation function $v(y)$ is a convex function if Y is a convex set (see section 4.1.2). A convex function can be discontinuous at points on the boundary of its domain. For $v(y)$, the boundary corresponds to $y = 0$. The conditions that provide the relationship between gap and continuity of $v(y)$ are presented in the following theorem.

Theorem 4.4.1 (Continuity of perturbation function)
Let the perturbation function $v(y)$ be finite at $y = 0$, that is, $v(0)$ is finite. The optimal values of the primal and dual problems are equal (i.e., there is no duality gap) if and only if $v(y)$ is lower semicontinuous at $y = 0$.

Remark 3 Conditions for $v(y)$ to be lower semicontinuous at $y = 0$ are

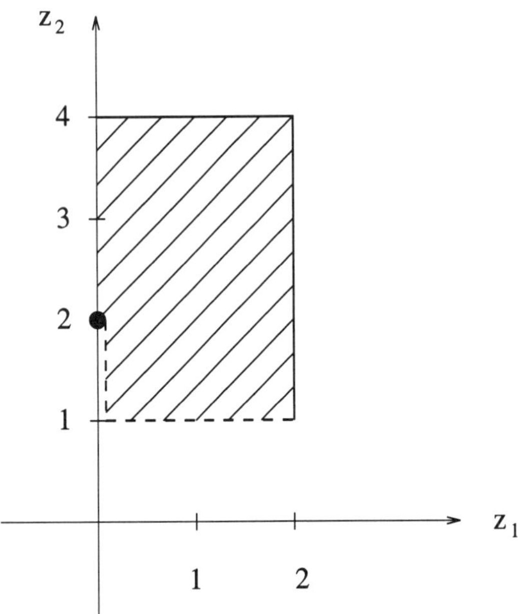

Figure 4.3: Illustration of duality gap

1. X is closed, and
2. $f(x), g(x)$ are continuous on X, and
3. $f(x^*)$ is finite, and
4. $\{x \in X : h(x) = 0, g(x) \leq 0, \text{ and } f(x) \leq \alpha\}$ is a bounded and nonempty convex set for some scalar $\alpha \geq v(0)$.

4.4.1 Illustration of Duality Gap

Consider the following problem

$$\left\{ \begin{array}{c} \min\limits_{(x_1, x_2) \in X} \quad x_2 \\ s.t. \quad x_1 \leq 0 \\ X = \{(x_1, x_2) : 0 \leq x_1 \leq 2, 1 < x_2 \leq 4, \text{ and} \\ x_2 \geq 2 \text{ if } x_1 = 0\} \end{array} \right.$$

where $z_1 = x_1$ and $z_2 = x_2$, and the plot of z_1 vs z_2 is shown in Figure 4.3.

The optimal solution of the primal problem is attained at $x_2 = 2, x_1 = 0$.

Notice though that the optimal value of the dual problem cannot equal that of the primal due to the loss of lower semicontinuity of the perturbation function $v(y)$ at $y = 0$.

Summary and Further Reading

This chapter introduces the fundamentals of duality theory. Section 4.1 presents the formulation of the primal problem, defines the perturbation function associated with the primal problem and discuss its properties, and establishes the relationship between the existence of optimal multipliers and the stability of the primal problem. Section 4.2 presents the formulation of the dual problem and introduces the dual function and its associated properties along with its geometrical interpretation. Section 4.3 presents the weak and strong duality theorems, while section 4.4 defines the duality gap and establishes the connection between the continuity of the perturbation function and the existence of the duality gap. Further reading in these subjects can be found in Avriel (1976), Bazaraa *et al.* (1993), Geoffrion (1971), Geoffrion (1972b), Minoux (1986), and Walk (1989).

Problems on Duality Theory

1. Consider the problem:

$$\begin{cases} \min & -x^{1/2} \\ \text{s.t.} & x \leq 0 \\ & x \in X = \{x \mid x \geq 0\} \end{cases}$$

Is this problem stable?

2. Consider the problem:

$$\begin{cases} \min & (x_1 - 5)^2 + x_2^2 \\ \text{s.t.} & x_1 + x_2 \leq 5 \\ & x_1 \geq 0, \ x_2 \geq 0 \end{cases}$$

   (a) Determine the dual function,

   (b) Verify the nonexistence of duality gap, and

   (c) Formulate and solve the dual problem.

3. Consider the linear problem:

$$\begin{cases} \min & c^T x \\ \text{s.t.} & Ax = b \\ & x \geq 0 \end{cases}$$

Determine the dual function, and the dual problem.

4. Consider the quadratic problem:

$$\begin{cases} \min & c^T x + \frac{1}{2} x^T Q x \\ \text{s.t.} & Ax \leq b \end{cases}$$

   (a) Formulate the dual function,

   (b) If Q is symmetric and positive semi-definite, show that the dual problem becomes

$$\begin{cases} \max & -\frac{1}{2} x^T Q x - b^T \mu \\ \text{s.t.} & Qx + A^T \mu = -c \\ & \mu \leq 0 \end{cases}$$

   (c) If Q is positive definite, show that the dual function becomes

$$\begin{aligned} \phi(\mu) &= \frac{1}{2} \mu^T (-AQ^{-1} A^T) \mu + \mu^T (-b - AQ^{-1} c) \\ &\quad - \frac{1}{2} c^T Q^{-1} c \end{aligned}$$

   and formulate the dual problem.

5. Verify the strong duality theorem for the problem

$$\begin{cases} \min & x_1^2 + e^{x_2} \\ \text{s.t.} & x_1 + x_2 \leq 10 \\ & x_1, x_2 \geq 0 \end{cases}$$

6. Consider the following posynomial geometric programming problem:

$$\begin{cases} \min & 40x_1^{-1}x_2^{-1}x_3^{-1} + 40x_2x_3 \\ \text{s.t} & 0.5x_1x_3 + 0.25x_1x_2 \leq 1 \\ & x_1, x_2, x_3 \geq 0 \end{cases}$$

   (a) Derive its dual problem formulation.

   (b) Is there a duality gap? Why?

Part 2

Fundamentals of Mixed-Integer Optimization

Chapter 5 Mixed-Integer Linear Optimization

This chapter provides an introduction to the basic notions in Mixed-Integer Linear Optimization. Sections 5.1 and 5.2 present the motivation, formulation, and outline of methods. Section 5.3 discusses the key ideas in a branch and bound framework for mixed-integer linear programming problems.

5.1 Motivation

A large number of optimization models have continuous and integer variables which appear linearly, and hence separably, in the objective function and constraints. These mathematical models are denoted as Mixed–Integer Linear Programming MILP problems. In many applications of MILP models the integer variables are $0-1$ variables (i.e., binary variables), and in this chapter we will focus on this sub–class of MILP problems.

A wide range of applications can be modeled as mixed-integer linear programming MILP problems. These applications have attracted a lot of attention in the field of *Operations Research* and include facility location and allocation problems, scheduling problems, and fixed-charge network problems. The excellent books of Nemhauser and Wolsey (1988), and Parker and Rardin (1988) provide not only an exposition to such applications but also very thorough presentation of the theory of discrete optimization. Applications of MILP models in Chemical Engineering have also received significant attention particularly in the areas of *Process Synthesis, Design, and Control.* These applications include (i) the minimum number of matches in heat exchanger synthesis (Papoulias and Grossmann, 1983; see also chapter 8) (ii) heat integration of sharp distillation sequences (Andrecovich and Westerberg, 1985); (iii) multicomponent multiproduct distillation column synthesis (Floudas and Anastasiadis, 1988); (iv) multiperiod heat exchanger network, and distillation system synthesis (Floudas and Grossmann, 1986; Paules and Floudas, 1988); flexibility analysis of chemical processes (Grossmann and Floudas, 1987); (v) structural properties of control systems (Georgiou and Floudas, 1989, 1990); (vi) scheduling of batch processes (e.g., Rich and Prokapakis, 1986, 1986; Kondili *et al.*, 1993; Shah *et al.*, 1993; Voudouris and Grossmann, 1992, 1993); and (vii) planning and scheduling of batch processes (Shah and Pantelides, 1991, Sahinidis *et al.*, 1989, Sahinidis and Grossmann, 1991). In addition to the above applications, MILP models are employed as subproblems in the mixed-integer nonlinear optimization approaches which we will discuss in the next chapter.

5.2 Formulation

In this section, we will present the formulation of Mixed-Integer Linear Programming MILP problems, discuss the complexity issues, and provide a brief overview of the solution methodologies proposed for MILP models.

5.2.1 Mathematical Description

The MILP formulation with $0-1$ variables is stated as

$$\begin{aligned} \min \quad & c^T x + d^T y \\ s.t. \quad & Ax + By \leq b \\ & x \geq 0 \ , \ x \in X \subseteq \Re^n \\ & y \in \{0,1\}^q \end{aligned} \qquad (1)$$

where x is a vector of n continuous variables ,
y is a vector of q $0-1$ variables ,
c, d are $(n \times 1)$ and $(q \times 1)$ vectors of parameters ,
A, B are matrices of appropriate dimension ,
b is a vector of p inequalities.

Remark 1 Formulation (1) has a linear objective function and linear constraints in x and y. It also contains a mixed set of variables (i.e., continuous x and $0-1$ y variables). Note that if the vector c is zero and the matrix A consists of zero elements then (1) becomes an integer linear programming ILP problem. Similarly, if the vector d is zero and the matrix B has all elements zero, then (1) becomes a linear LP problem.

In this chapter we will discuss briefly the basics of the mixed-integer linear programming MILP model with $0-1$ variables. For exposition to integer linear programming ILP with respect to all approaches the reader is referred to the excellent books of Nemhauser and Wolsey (1988), Parker and Rardin (1988), and Schrijver (1986).

5.2.2 Complexity Issues in MILP

The major difficulty that arises in mixed-integer linear programming MILP problems for the form (1) is due to the combinatorial nature of the domain of y variables. Any choice of 0 or 1 for the elements of the vector y results in a LP problem on the x variables which can be solved for its best solution.

One may follow the brute-force approach of enumerating fully all possible combinations of $0-1$ variables for the elements of the y vector. Unfortunately, such an approach grows exponentially in time with respect to its computational effort. For instance, if we consider one hundred $0-1$ y variables then we would have 2^{100} possible combinations.

As a result, we would have to solve 2^{100} LP problems. Hence, such an approach that involves complete enumeration becomes prohibitive.

Nemhauser and Wolsey (1988) provide a summary of the complexity analysis results for several classes of mixed-integer programming problems. The MILP problem (1) (i.e., with $0-1$ y variables) belongs to the class of $\mathcal{NP}$–complete problems (Vavasis (1991)).

Despite the complexity analysis results for the combinatorial nature of MILP models of the form (1), several major algorithmic approaches have been proposed and applied successfully to medium and large size application problems. In the sequel, we will briefly outline the proposed approaches and subsequently concentrate on one of them, namely, the branch and bound approach.

5.2.3 Outline of MILP Algorithms

The proposed algorithmic approaches for MILP problems can be classified as

(i) Branch and bound methods (Land and Doig, 1960; Dakin, 1965; Driebeek, 1966; Beale and Tomlin, 1970; Tomlin, 1971; Beale and Forrest, 1976; Beale, 1968, 1979; Martin and Schrage, 1985; Pardalos and Rosen, 1987);

(ii) Cutting plane methods (Gomory, 1958, 1960; Crowder *et al.*, 1983; Van Roy and Wolsey, 1986, 1987; Padberg and Rinaldi, 1991; Boyd, 1994);

(iii) Decomposition methods (Benders, 1962; Fisher, 1981; Geoffrion and McBride, 1978; Magnanti and Wong, 1981; Van Roy, 1983, 1986);

(iv) Logic-based methods (Balas, 1975, 1979; Jeroslow, 1977; Hooker, 1988; Jeroslow and Lowe, 1984, 1985; Jeroslow and Wang, 1990; Raman and Grossmann, 1992).

In the branch and bound algorithms, a binary tree is employed for the representation of the $0-1$ combinations, the feasible region is partitioned into subdomains systematically, and valid upper and lower bounds are generated at different levels of the binary tree.

In the cutting plane methods, the feasible region is not divided into subdomains but instead new constraints, denoted as cuts, are generated and added which reduce the feasible region until a $0-1$ optimal solution is obtained.

In the decomposition methods, the mathematical structure of the models is exploited via variable partitioning, duality, and relaxation methods.

In the logic-based methods, disjunctive constraints or symbolic inference techniques are utilized which can be expressed in terms of binary variables.

An exposition of theoretical, algorithmic, and computational issues of the above classes of algorithms is provided in Nemhauser and Wolsey (1988) and Parker and Rardin (1988).

In the next section we will focus only on describing the basic principles of the branch and bound methods which are the most commonly used algorithms in large-scale mixed-integer linear programming solvers.

5.3 Branch and Bound Method

In this section, (i) we will discuss the basic notions of separation, relaxation, and fathoming employed in a branch and bound method, (ii) we will present a general branch and bound algorithm, and (iii) we will discuss a branch and bound method which is based on linear programming relaxation.

5.3.1 Basic Notions

A general branch and bound method for MILP problems of the form (1) is based on the key ideas of separation, relaxation, and fathoming which are outlined in the following:

5.3.1.1 Separation

Definition 5.3.1 Let an MILP model of form (1) be denoted as *(P)* and let its set of feasible solutions be denoted as $FS(P)$. A set of subproblems $(P_1), (P_2), \ldots, (P_n)$ of (P) is defined as a separation of (P) if the following conditions hold:

(i) A feasible solution of any of the subproblems $(P_1), (P_2), \ldots, (P_n)$ is a feasible solution of (P); and

(ii) Every feasible solution of (P) is a feasible solution of *exactly one* of the subproblems.

Remark 1 The above conditions (i) and (ii) imply that the feasible solutions of the subproblems $FS(P_1), FS(P_2), \ldots, FS(P_n)$ are a partition of the feasible solutions of the problems (P), that is $FS(P)$. As a result, the original problem (P) is called a parent node problem while the subproblems $(P_1), (P_2), \ldots, (P_n)$ are called the children node problems.

Remark 2 An important question that arises in a branch and bound framework is how one decides about generating a separation of problem (P) into subproblems $(P_1), (P_2), \ldots, (P_n)$. One frequently used way of generating a separation in MILP problems of form (1) is by considering contradictory constraints on a single binary variable. For instance the following MILP problem:

$$\begin{aligned} \min \quad & -3y_1 - 2y_2 - 3y_3 + 2x_2 \\ s.t. \quad & y_1 + y_2 + y_3 + x_1 \geq 2 \\ & 5y_1 + 3y_2 + 4y_3 + 10x_1 \leq 10 \\ & x_1 \geq 0 \\ & y_1, y_2, y_3 = 0, 1 \end{aligned}$$

can be separated into two subproblems by branching (i.e., separating) on the variable y_1; that is, subproblem 1 will have as additional constraint $y_1 = 0$, while subproblem 2 will have $y_1 = 1$.

Another way of generating a separation which is applicable to MILP problems that have constraints of the form (called generalized upper bound constraints):

$$\begin{aligned} \sum_{i \in S} y_i &= 1, \\ y_i &= 0-1, \;\; i \in S, \end{aligned}$$

is by selecting a part of the summation, set it equal to zero and use mutually exclusive solutions. For instance if an MILP model includes the following multiple choice constraint

$$y_1 + y_2 + y_3 + y_4 = 1,$$

we can construct a separation by including $y_1 + y_2 = 0$ in subproblem 1 and $y_3 + y_4 = 0$ in subproblem 2. This type of separation has been reported to be very effective in such special structured problems.

5.3.1.2 Relaxation

Definition 5.3.2 An optimization problem, denoted as *(RP)*, is defined as a relaxation of problem *(P)* if the set of feasible solutions of *(P)* is a subset of the set of feasible solutions of *(RP)*; that is,

$$FS(P) \subseteq FS(RP)$$

Remark 1 The above definition of relaxation implies the following relationships between problem *(P)* and the relaxed problem *(RP)*:

(i) If *(RP)* has no feasible solution, then *(P)* has no feasible solution;

(ii) Let the optimal solution of *(P)* be z_P and the optimal solution of *(RP)* be z_{RP}. Then,

$$z_{RP} \leq z_P$$

; that is, the solution of the relaxed problem *(RP)* provides a lower bound on the solution of problem *(P)*;

(iii) If an optimal solution of *(RP)* is feasible for problem *(P)*, then it is an optimal solution of *(P)*.

Remark 2 A key issue in a branch and bound algorithm is how one generates a relaxation of problem *(P)* which is an MILP of form (1). One way of relaxation is by simply omitting one or several constraints of problem *(P)*. Another way is by setting one or more positive coefficients of binary variables of the objective function, which are still free, equal to zero. Another alternative of generating a valid relaxation is by replacing the integrality conditions on the *y* variables by $0 \leq y \leq 1$. This type of relaxation results in a linear programming problem, and it is denoted as linear programming relaxation. It is the most frequently used relaxation.

Remark 3 The selection of a relaxation among a number of alternatives is based on the trade-off between two competing criteria. The first criterion corresponds to the capability of solving the relaxed problem easily. The second criterion is associated with the type and quality of lower bound that the relaxed problem produces for problem *(P)*. In general, the easier the relaxed problem *(RP)* is to solve, the greater the gap is between the optimal solution of *(P)* and the lower bound provided by *(RP)*.

5.3.1.3 Fathoming

Let (CS) be a candidate subproblem in solving (P). We would like to determine whether the feasible region of (CS), $F(CS)$, contains an optimal solution of (P) and find it if it does.

Definition 5.3.3 A candidate subproblem (CS) will be considered that has been fathomed if one of the following two conditions takes place:

(i) It can be ascertained that the feasible solution $F(CS)$ cannot contain a better solution than the best solution found so far (i.e., the incumbent); or

(ii) An optimal solution of (CS) is found.

In either case, the candidate problem has been considered and needs no further separation.

Remark 1 (Fathoming criteria):

There are three general fathoming criteria in a branch and bound algorithm that are based on relaxation. Before stating these fathoming criteria let us denote as

(RCS), a relaxation of a candidate subproblem (CS),

z_{RCS}, the optimal value of (RCS),

z_{CS}, the optimal value of (CS),

z^*, the value of the incumbent (i.e., best solution so far) ($z^* = +\infty$ if no feasible solution of (P) has been found so far.

Fathoming Criterion 1 – $FC1$: If (RCS) has no feasible solution, then (CS) has no feasible solution and hence can be fathomed.

Note that $FC1$ is based on condition (i) of remark 1.

Fathoming Criterion 2 – $FC2$: If the optimal solution of (RCS) is greater or equal to the incumbent, i.e.,

$$z_{RCS} \geq z^*,$$

then (CS) can be fathomed.

Note that it is possible to have

$$z_{CS} \geq z^* > z_{RCS}.$$

In this case the candidate subproblem (CS) is not fathomed. Also note that the tighter the lower bound provided by the relaxation is the more frequently this fathoming criterion will be employed effectively.

Fathoming Criterion 3 – $FC3$: If an optimal solution of (RCS) is found to be feasible in (CS), then it must be an optimal solution of (CS). Hence (CS) is fathomed.

Note that this solution is also feasible in (P) and therefore we can update the incumbent if its value is less than z^*.

A number of proposed branch and bound algorithms solve the relaxation subproblems (RCS) to optimality first and subsequently apply the aforementioned fathoming tests. There exist, however, algorithms which either do not solve the (RCS) to optimality but instead apply sufficient conditions for the fathoming criterion (e.g., use of good suboptimal solutions of dual) or not only solve the (RCS) to optimality but also apply a post optimality test aiming at improving the lower bounds obtained by the relaxation.

5.3.2 General Branch and Bound Framework

The main objective in a general branch and bound algorithm is to perform an enumeration of the alternatives without examining all $0-1$ combinations of the y–variables. A key element in such an enumeration is the representation of alternatives via a binary tree. A binary tree representation is shown in Figure 5.1 for an MILP model with three binary y–variables. As shown in Figure 5.1, the binary tree consists of three levels, each level has a number of nodes, and there exists a specific connection between nodes of succeeding levels via arcs. At level $l = 0$, there is one node, called the root node. At level $l = 1$, there are two nodes and the branching is based on setting $y_1 = 0$ for the one node and $y_1 = 1$ for the second node. The root node is the parent node for the two nodes of level 1, and hence at level 1, we have two candidate subproblems. At level $l = 2$, there are four nodes and the branching is based on the y_2 variable. Note that the parent nodes are the two nodes of level 1. At level $l = 3$, there are eight nodes and the branching is based on the y_2 variable. Each node of level 2 is the parent node of two children nodes of level 3. As a result of this binary tree representation we have generated a number of candidate subproblems at each level, namely, two at level 1, four at level 2, and eight at level 3. Let us denote such candidate subproblems, as $(CS)^l_{n(l)}$, where l is the level and $n(l)$ is the number of candidate subproblem in level l.

The basic ideas in a branch and bound algorithm are outlined in the following. First we make a reasonable effort to solve the original problem (e.g., considering a relaxation of it). If the relaxation does not result in a $0-1$ solution for the y–variables, then we separate the root node into two or more candidate subproblems at level 1 and create a list of candidate subproblems. We select one of the candidate subproblems of level 1, we attempt to solve it, and if its solution is integral, then we return to the candidate list of subproblems and select a new candidate subproblem. Otherwise, we separate the candidate subproblem into two or more subproblems at level 2 and add its children nodes to the list of candidate subproblems. We continue this procedure until the candidate list is exhausted and report as optimal solution the current incumbent. Note that the finite termination of such a procedure is attained if the set of feasible solutions of the original problem (P), denoted as $FS(P)$ is finite.

To avoid the enumeration of all candidate subproblems we employ the fathoming tests discussed in section 5.3.1.3. These tests allow us to eliminate from further consideration not only nodes of the binary tree but also branches of the tree which correspond to their children nodes. The success of the branch and bound algorithm is based on the percentage of eliminated nodes and the effort required to solve the candidate subproblems.

A general branch and bound algorithm can be stated as follows:

Step 1 – Initialization: Initialize the list of candidate subproblems to consist of the MILP alone, and set $z^* = +\infty$.

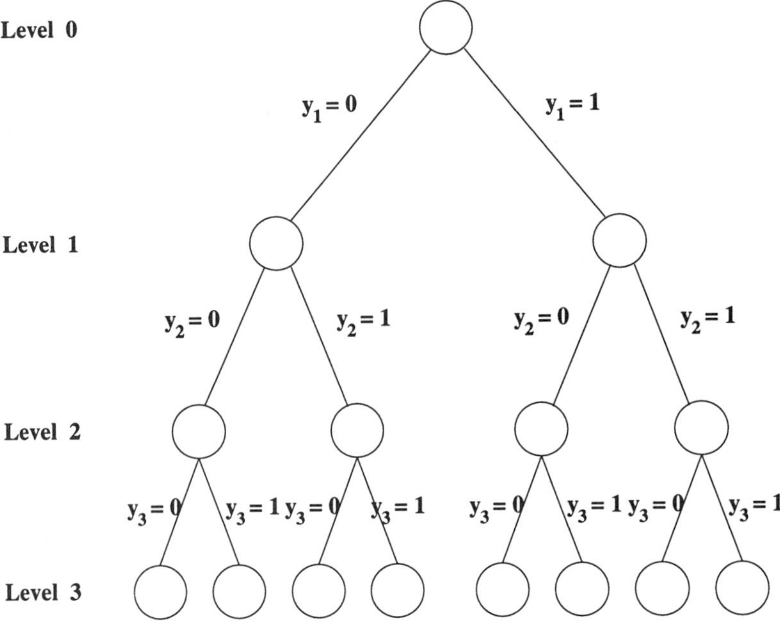

Figure 5.1: Binary tree representation

Step 2 – Termination: If the list of candidate subproblems is empty, then terminate with optimal solution the current incumbent. If an incumbent does not exist, then the MILP problem is infeasible.

Step 3 – Selection of candidate subproblem: Select one of the subproblems in the candidate list to become the current candidate subproblem (CS).

Step 4 – Relaxation: Select a relaxation (RCS) of the current candidate subproblem (CS), solve if and denote its solution by z_{RCS}.

Step 5 – Fathoming: Apply the three fathoming criteria:

$FC1$: If (RCS) is infeasible, then the current (CS) has no feasible solution. Go to step 2.

$FC2$: If $z_{RCS} \geq z^*$, then the current (CS) has no feasible solution better than the incumbent. Go to step 2.

$FC3$: If the optimal solution of (RCS) is feasible for (CS) (e.g., integral), then it is an optimal solution of (CS). If $z_{RCS} \geq z^*$, then record this solution as the new incumbent; that is, $z^* = z_{RCS}$. Go to step 2.

Step 6 – Separation: Separate the current candidate subproblem (CS) and add its children nodes to the list of candidate subproblems. Go to step 2.

Remark 1 In Step 1 of the general branch and bound algorithm, we can use prior knowledge on an upper bound and initialize the incumbent to such an upper bound. We can also start with a list of candidate subproblems based on prior separations of the MILP problem.

Remark 2 –Node selection: Different criteria for the selection of the next candidate subproblems may lead to different ways of enumerating the binary tree. There are three main alternative approaches for such a selection: (i) the Last-In-First-Out (LIFO) which is known as a depth-first search with backtracking, (ii) the breadth-first search, and (iii) the best bound search. In (i), the next candidate subproblem selected is one of the children nodes of the current one and as a result it can be the last one added to the list of candidate subproblems. Backtracking takes place when a current node is fathomed and at that point we go back from this node toward the root node until we identify a node that has a child node not considered. Such an approach has as primary advantages the easy reoptimization, compact bookkeeping, and that on the average feasible solutions are likely to be obtained deep in the tree. It may require, however, a large number of candidate subproblems so as to prove optimality. The depth-first with backtracking is the standard option in most commercial codes. In (ii), all the nodes at a given level have to be considered before considering a node in the a subsequent level. It is used extensively as a basis for heuristic selection of nodes or for providing estimates of fast improvements of the bounds. In (iii), the next candidate subproblem is the one stored in the list that has the least lower bound. This node selection alternative results in fewer candidate subproblems. Note that there is a trade-off between the different node-selection strategies, and there is no rule that dominates in producing good incumbent solutions. As a result, we have to make a compromise empirically among the different rules.

Remark 3 –Branching variable selection: The choice of variable upon which we branch at a particular level have been shown to be very important from the computational viewpoint. Since robust methods for selecting the branching variables are not available, a common practice is to establish a set of priorities provided by the user. Other alternatives include the use of pseudo-costs which are based on pricing infeasibility, and the use of penalties.

5.3.3 Branch and Bound Based on Linear Programming Relaxation

The linear programming LP relaxation of the MILP model is the most frequently used type of relaxation in branch and bound algorithms. In the root node of a binary tree, the LP relaxation of the MILP model of (1) takes the form:

$$\begin{aligned} \min \quad & c^T x + d^T y \\ s.t. \quad & Ax + By \leq b \\ & x \geq 0 \ , \ x \in X \subseteq \Re^n \\ & 0 \leq y \leq 1 \end{aligned}$$

in which all binary y–variables have been relaxed to continuous variables with lower and upper bounds of zero and one, respectively. Based on the notation of candidate subproblems, which are denoted $(CS)^l_{n(l)}$ where l is the level and $n(l)$ is the number at level l, the LP relaxation at the root node will be denoted as $(RCS)^0_1$, and its optimal solution as $z_{(RCS)^0_1}$.

Note that $z_{(RCS)_1^0}$ is a lower bound on the optimal solution of the MILP model of (1). Also note that if the solution of $(RCS)_1^0$ turns out to have all the y–variables at $0 - 1$ values, then we can terminate since the LP relaxation satisfies the integrality conditions.

At subsequent levels of the binary tree, the LP relaxation of the candidate subproblem will have only a subset of the y–variables set to zero or one while the rest of the y-variables will be treated as continuous variables with bounds between zero and one. For instance, at level 1 of the binary tree shown in Figure 5.1, at which we have two candidate subproblems $(CS)_1^1$, $(CS)_2^1$, the LP relaxation of $(CS)_1^1$ takes the form:

$$\begin{aligned} \min \quad & c^T x + d^T y \\ s.t. \quad & Ax + By \leq b \\ & x \geq 0 \ , \ x \in X \subseteq \Re^n \\ & y_1 = 0 \ , \ 0 \leq y_2 \leq 1 \ , \ 0 \leq y_3 \leq 1 \end{aligned}$$

Note that the solution of $(RCS)_1^0$ shown above is an upper bound on the solution of $(RCS)_1^0$ since it has a fixed $y_1 = 0$.

In a similar fashion the LP relaxations at level 2 of the binary tree shown in Figure 5.1 which has four candidate subproblems $(CS)_1^2$, $(CS)_2^2$, $(CS)_3^2$, $(CS)_4^2$ will feature y_1 and y_2 fixed to either zero or one values while the y_3 variable will be treated as continuous with bounds of zero and one.

The algorithmic statement of a branch and bound algorithm which is based on LP relaxation of the candidate subproblems is the same as the general branch and bound algorithm described in Section 5.3.2 with the only difference that the relaxation of the LP relaxation outlined above. We will illustrate the LP relaxation branch and bound approach via the following example.

5.3.3.1 Illustration

This MILP example features three binary variables, one continuous variable and is of the form:

$$\begin{aligned} \min \quad & 2x_1 - 3y_1 - 2y_2 - 3y_3 \\ s.t. \quad & x_1 + y_1 + y_2 + y_3 \geq 2 \\ & 10x_1 + 5y_1 + 3y_2 + 4y_3 \leq 10 \\ & x_1 \geq 0 \\ & y_1 \ , \ y_2 \ , \ y_3 \ = 0, 1 \end{aligned}$$

The optimal solution of this MILP problem has an objective value of $z^* = -6$ and $x = 0$, $(y_1, y_2, y_3) = (1, 0, 1)$.

Solving the linear programming relaxation at the root node, we obtain as solution

$$\begin{aligned} z &= -6.8, \\ (y_1, y_2, y_3) &= (0.6, 1, 1), \\ x &= 0. \end{aligned}$$

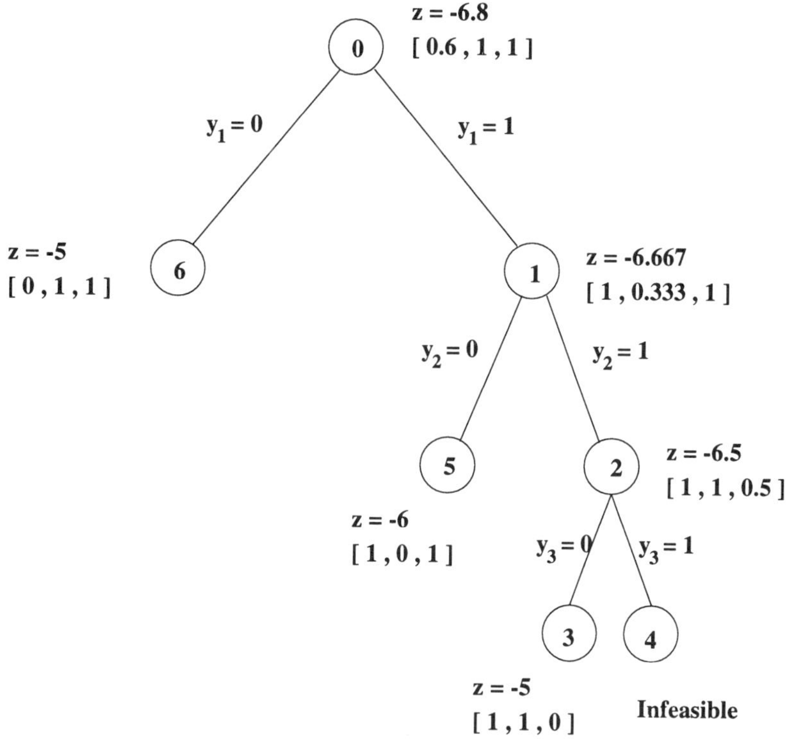

Figure 5.2: Binary tree for depth first search with backtracking

Note that -6.8 is a lower bound on the optimal solution of the MILP of -6.

The binary trees for (i) depth first search with backtracking and (ii) breadth first search are shown in Figures 5.2 and 5.3 respectively. The number within the nodes indicate the sequence of candidate subproblems for each type of search.

Using the depth first search with backtracking, we obtain the optimal solution in node 5 as shown in Figure 5.2, and we need to consider 6 nodes plus the root node of the tree.

Using the breadth first search the optimal solution is obtained in node 3 as shown in Figure 5.3, and we need to consider the same number of nodes as in Figure 5.2.

At level 1, the lower and upper bounds for the depth first search are $(-6.667, +\infty)$ while for the breadth first search are $(-6.667, -5)$. At level 2, the lower and upper bounds for the depth first search are $(-6.667, +\infty)$ while for the breadth first search are $(-6.667, -6)$. At level 3, the lower and upper bounds for the depth first search are $(-6.5, -5)$ while for the breadth first search we have reached termination at -6 since there are no other candidate subproblems in the list. When the backtracking begins for the depth first search we find in node 5 the upper bound of -6, subsequently we check node 6 and terminate with the least upper bound of -6 as the optimal solution.

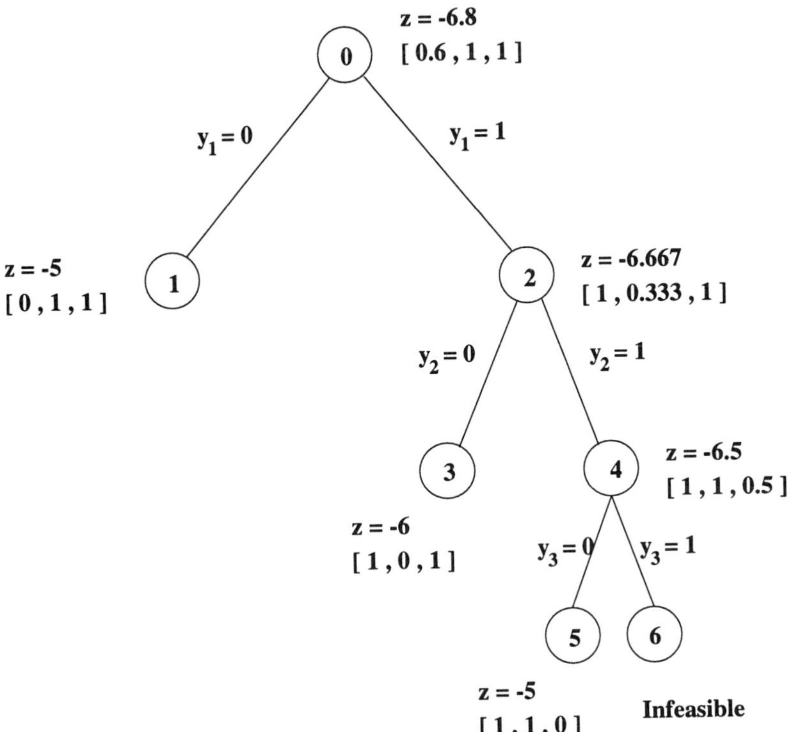

Figure 5.3: Binary tree for breadth first search

Summary and Further Reading

In this chapter we have briefly introduced the basic notions of a branch and bound algorithmic framework, described a general branch and bound algorithm and a linear relaxation based branch and bound approach, and illustrated these ideas with a simple example. This material is intended only as a basic introduction to mixed–integer linear programming MILP problems. These MILP problems are employed as subproblems in the MINLP approaches that are discussed extensively in Chapter 6. The reader who is interested in detailed theoretical, algorithmic and computational exposition of MILP problems is directed to the excellent books of Nemhauser and Wolsey (1988), Parker and Rardin (1988), and Schrijver (1986).

Chapter 6 Mixed-Integer Nonlinear Optimization

This chapter presents the fundamentals and algorithms for mixed-integer nonlinear optimization problems. Sections 6.1 and 6.2 outline the motivation, formulation, and algorithmic approaches. Section 6.3 discusses the Generalized Benders Decomposition and its variants. Sections 6.4, 6.5 and 6.6 presents the Outer Approximation and its variants with Equality Relaxation and Augmented Penalty. Section 6.7 discusses the Generalized Outer Approximation while section 6.8 compares the Generalized Benders Decomposition with the Outer Approximation. Finally, section 6.9 discusses the Generalized Cross Decomposition.

6.1 Motivation

A wide range of nonlinear optimization problems involve integer or discrete variables in addition to the continuous variables. These classes of optimization problems arise from a variety of applications and are denoted as Mixed-Integer Nonlinear Programming MINLP problems.

The integer variables can be used to model, for instance, sequences of events, alternative candidates, existence or nonexistence of units (in their zero-one representation), while discrete variables can model, for instance, different equipment sizes. The continuous variables are used to model the input-output and interaction relationships among individual units/operations and different interconnected systems.

The nonlinear nature of these mixed–integer optimization problems may arise from (i) nonlinear relations in the integer domain exclusively (e.g., products of binary variables in the quadratic assignment model), (ii) nonlinear relations in the continuous domain only (e.g., complex nonlinear input-output model in a distillation column or reactor unit), (iii) nonlinear relations in the joint integer-continuous domain (e.g., products of continuous and binary variables in the scheduling/planning of batch processes, and retrofit of heat recovery systems). In this chapter, we will focus on nonlinearities due to relations (ii) and (iii). An excellent book that studies mixed-integer linear optimization, and nonlinear integer relationships in combinatorial optimization is the one by Nemhauser and Wolsey (1988).

The coupling of the integer domain with the continuous domain along with their associated nonlinearities make the class of MINLP problems very challenging from the theoretical, algorithmic, and computational point of view. Apart from this challenge, however, there exists a broad spectrum of applications that can be modeled as mixed-integer nonlinear programming problems. These applications have a prominent role in the area of *Process Synthesis* in chemical engineering and include: (i) the synthesis of grassroot heat recovery networks (Floudas and Ciric, 1989; Ciric and Floudas, 1990; Ciric and Floudas, 1989; Yee *et al.*, 1990a; Yee and Grossmann, 1990; Yee *et al.*, 1990b); (ii) the retrofit of heat exchanger systems (Ciric and Floudas, 1990; Papalexandri and Pistikopoulos, 1993); (iii) the synthesis of distillation-based separation systems (Paules and Floudas, 1988; Viswanathan and Grossmann, 1990; Aggarwal and Floudas, 1990; Aggarwal and Floudas, 1992; Paules and Floudas, 1992); (iv) the synthesis of complex reactor networks (Kokossis and Floudas, 1990; Kokossis and Floudas, 1994); (v) the synthesis of reactor-separator-recycle systems (Kokossis and Floudas, 1991); (vi) the synthesis of utility systems (Kalitventzeff and Marechal, 1988); and the synthesis of total process systems (Kocis and Grossmann, 1988; Kocis and Grossmann, 1989a; Kravanja and Grossmann, 1990). An excellent review of the mixed–integer nonlinear optimization frameworks and applications in *Process Synthesis* are provided in Grossmann (1990). Algorithmic advances for logic and global optimization in Process Synthesis are reviewed in Floudas and Grossmann (1994).

Key applications of MINLP approaches have also emerged in the area of *Design, Scheduling, and Planning of Batch Processes* in chemical engineering and include: (i) the design of multiproduct plants (Grossmann and Sargent, 1979; Birewar and Grossmann, 1989; Birewar and Grossmann, 1990); and (ii) the design and scheduling of multipurpose plants (Vaselenak *et al.*, 1987; Vaselenak *et al.*, 1987; Faqir and Karimi, 1990; Papageorgaki and Reklaitis, 1990; Papageorgaki and Reklaitis, 1990; Wellons and Reklaitis, 1991; Wellons and Reklaitis, 1991; Sahinidis and Grossmann, 1991; Fletcher *et al.*, 1991). Excellent recent reviews of the advances in the design, scheduling, and planning of batch plants can be found in Reklaitis (1991), and Grossmann *et al.* (1992).

Another important applications of MINLP models have recently been reported for (i) the computer-aided molecular design aspects of selecting the best solvents (Odele and Macchietto, 1993); and (ii) the interaction of design and control of chemical processes (Luyben and Floudas (1994a), Luyben and Floudas (1994b)).

MINLP applications received significant attention in other engineering disciplines. These include (i) the facility location in a multiattribute space (Ganish *et al.*, 1983); (ii) the optimal unit allocation in an electric power system (Bertsekas *et al.*, 1983); (iii) the facility planning of an electric power generation (Bloom, 1983; Rouhani *et al.*, 1985); (iv) the topology optimization of transportation networks (Hoang, 1982); and (v) the optimal scheduling of thermal generating units (Geromel and Belloni, 1986).

6.2 Formulation

The primary objective in this section is to present the general formulation of MINLP problems, discuss the difficulties and present an overview of the algorithmic approaches developed for the solution of such models.

6.2.1 Mathematical Description

The general MINLP formulation can be stated as

$$\begin{aligned} \min_{\boldsymbol{x},\boldsymbol{y}} \quad & f(\boldsymbol{x},\boldsymbol{y}) \\ s.t. \quad & \boldsymbol{h}(\boldsymbol{x},\boldsymbol{y}) = \boldsymbol{0} \\ & \boldsymbol{g}(\boldsymbol{x},\boldsymbol{y}) \leq \boldsymbol{0} \\ & \boldsymbol{x} \in \boldsymbol{X} \subseteq \Re^n \\ & \boldsymbol{y} \in \boldsymbol{Y} \text{ integer} \end{aligned} \tag{6.1}$$

Here $\boldsymbol{x}$ represents a vector of n continuous variables (e.g., flows, pressures, compositions, temperatures, sizes of units), and $\boldsymbol{y}$ is a vector of integer variables (e.g., alternative solvents or materials); $\boldsymbol{h}(\boldsymbol{x},\boldsymbol{y}) = \mathbf{0}$ denote the m equality constraints (e.g., mass, energy balances, equilibrium relationships); $\boldsymbol{g}(\boldsymbol{x},\boldsymbol{y}) \leq \mathbf{0}$ are the p inequality constraints (e.g., specifications on purity of distillation products, environmental regulations, feasibility constraints in heat recovery systems, logical constraints); $f(\boldsymbol{x},\boldsymbol{y})$ is the objective function (e.g., annualized total cost, profit, thermodynamic criteria).

Remark 1 The integer variables $\boldsymbol{y}$ with given lower and upper bounds,

$$\boldsymbol{y}^L \leq \boldsymbol{y} \leq \boldsymbol{y}^U,$$

can be expressed through 0–1 variables (i.e., binary) denoted as $\mathbf{z}$, by the following formula:

$$y = y^L + z_1 + 2z_2 + 4z_3 + \ldots + 2^{N-1} z_N,$$

where N is the minimum number of 0–1 variables needed. This minimum number is given by

$$N = 1 + INT\left\{\frac{\log\left(y^U - y^L\right)}{\log 2}\right\},$$

where the INT function truncates its real argument to an integer value. This approach however may not be practical when the bounds are large.

Then, formulation (6.1) can be written in terms of 0–1 variables:

$$\begin{aligned} \min_{\boldsymbol{x},\boldsymbol{y}} \quad & f(\boldsymbol{x},\boldsymbol{y}) \\ \text{s.t.} \quad & \boldsymbol{h}(\boldsymbol{x},\boldsymbol{y}) = 0 \\ & \boldsymbol{g}(\boldsymbol{x},\boldsymbol{y}) \leq 0 \\ & \boldsymbol{x} \in \boldsymbol{X} \subseteq \Re^n \\ & \boldsymbol{y} \in \boldsymbol{Y} = \{0,1\}^q \end{aligned} \tag{6.2}$$

where $\boldsymbol{y}$ now is a vector of q 0–1 variables (e.g., existence of a process unit ($y_i = 1$) or non-existence ($y_i = 0$). We will focus on (6.2) in the majority of the subsequent developments.

6.2.2 Challenges/Difficulties in MINLP

Dealing with mixed–integer nonlinear optimization models of the form (6.1) or (6.2) present two major challenges/difficulties. These difficulties are associated with the nature of the problem, namely, the combinatorial domain (y–domain) and the continuous domain (x–domain).

As the number of binary variables y in (6.2) increase, one faces with a large combinatorial problem, and the complexity analysis results characterize the MINLP problems as NP-complete (Nemhauser and Wolsey, 1988). At the same time, due to the nonlinearities the MINLP problems are in general nonconvex which implies the potential existence of multiple local solutions. The determination of a global solution of the nonconvex MINLP problems is also NP-hard (Murty and Kabadi, 1987), since even the global optimization of constrained nonlinear programming problems can be NP-hard (Pardalos and Schnitger, 1988), and even quadratic problems with one negative eigenvalue are NP-hard (Pardalos and Vavasis, 1991). An excellent book on complexity issues for nonlinear optimization is the one by Vavasis (1991).

Despite the aforementioned discouraging results from complexity analysis which are worst-case results, significant progress has been achieved in the MINLP area from the theoretical, algorithmic, and computational perspective. As a result, several algorithms have been proposed, their convergence properties have been investigated, and a large number of applications now exist that cross the boundaries of several disciplines. In the sequel, we will discuss these developments.

6.2.3 Overview of MINLP Algorithms

A representative collection of algorithms developed for solving MINLP models of the form (6.2) or restricted classes of (6.2) includes, in chronological order of development, the following:

1. Generalized Benders Decomposition, **GBD** (Geoffrion, 1972; Paules and Floudas, 1989; Floudas *et al.*, 1989);
2. Branch and Bound, **BB** (Beale, 1977; Gupta, 1980; Ostrovsky *et al.*, 1990; Borchers and Mitchell, 1991);
3. Outer Approximation, **OA** (Duran and Grossmann, 1986a);
4. Feasibility Approach, **FA** (Mawengkang and Murtagh, 1986);
5. Outer Approximation with Equality Relaxation, **OA/ER** (Kocis and Grossmann, 1987);
6. Outer Approximation with Equality Relaxation and Augmented Penalty, **OA/ER/AP** (Viswanathan and Grossmann, 1990)
7. Generalized Outer Approximation, **GOA** (Fletcher and Leyffer, 1994)
8. Generalized Cross Decomposition, **GCD** (Holmberg, 1990);

In the pioneering work of Geoffrion (1972) on the Generalized Benders Decomposition **GBD** two sequences of updated upper (nonincreasing) and lower (nondecreasing) bounds are created that converge within ϵ in a finite number of iterations. The upper bounds correspond to solving subproblems in the x variables by fixing the y variables, while the lower bounds are based on duality theory.

The branch and bound **BB** approaches start by solving the continuous relaxation of the MINLP and subsequently perform an implicit enumeration where a subset of the 0–1 variables is fixed at each node. The lower bound corresponds to the NLP solution at each node and it is used to expand on the node with the lowest lower bound (i.e., breadth first enumeration), or it is used to eliminate nodes if the lower bound exceeds the current upper bound (i.e., depth first enumeration). If the continuous relaxation NLP of the MINLP has 0–1 solution for the y variables, then the **BB** algorithm will terminate at that node. With a similar argument, if a tight NLP relaxation results in the first node of the tree, then the number of nodes that would need to be eliminated can be low. However, loose NLP relaxations may result in having a large number of NLP subproblems to be solved which do not have the attractive update features that LP problems exhibit.

The Outer Approximation **OA** addresses problems with nonlinear inequalities, and creates sequences of upper and lower bounds as the **GBD**, but it has the distinct feature of using primal information, that is the solution of the upper bound problems, so as to linearize the objective and constraints around that point. The lower bounds in **OA** are based upon the accumulation of the linearized objective function and constraints, around the generated primal solution points.

The feasibility approach **FA** rounds the relaxed NLP solution to an integer solution with the least local degradation by successively forcing the superbasic variables to become nonbasic based on the reduced cost information.

The **OA/ER** algorithm, extends the **OA** to handle nonlinear equality constraints by relaxing them into inequalities according to the sign of their associated multipliers.

The **OA/ER/AP** algorithm introduces an augmented penalty function in the lower bound subproblems of the **OA/ER** approach.

The Generalized Outer Approximation **GOA** extends the **OA** to the MINLP problems of types (6.1),(6.2) and introduces exact penalty functions.

The Generalized Cross Decomposition **GCD** simultaneously utilizes primal and dual information by exploiting the advantages of Dantzig-Wolfe and Generalized Benders decomposition.

In the subsequent sections, we will concentrate on the algorithms that are based on decomposition and outer approximation, that is on 1., 3., 5., 6., 7., and 8.. This focus of our study results from the existing evidence of excellent performance of the aforementioned decomposition-based and outer approximation algorithms compared to the branch and bound methods and the feasibility approach.

6.3 Generalized Benders Decomposition, GBD

6.3.1 Formulation

Geoffrion (1972) generalized the approach proposed by Benders (1962), for exploiting the structure of mathematical programming problems (6.2), to the class of optimization problems stated as

$$\begin{aligned} \min_{x,y} \quad & f(x,y) \\ s.t. \quad & h(x,y) = 0 \\ & g(x,y) \leq 0 \\ & x \in X \subseteq \Re^n \\ & y \in Y = \{0,1\}^q \end{aligned}$$

under the following conditions:

C1: X is a nonempty, convex set and the functions

$$\begin{aligned} f : \Re^n \times \Re^q & \rightarrow \Re, \\ g : \Re^n \times \Re^q & \rightarrow \Re^p, \end{aligned}$$

are convex for each fixed $y \in Y = \{0,1\}^q$, while the functions $h : \Re^n \times \Re^l \rightarrow \Re^m$ are linear for each fixed $y \in Y = \{0,1\}^q$.

C2: The set

$$Z_y = \{z \in \Re^p : h(x,y) = 0,\ g(x,y) \leq z \ \text{ for some } \ x \in X\},$$

is closed for each fixed $y \in Y$.

C3: For each fixed $y \in Y \cap V$, where

$$V = \{y : h(x,y) = 0,\ g(x,y) \leq 0,\ \text{for some } \ x \in X\},$$

one of the following two conditions holds:

(i) the resulting problem (6.2) has a finite solution and has an optimal multiplier vector for the equalities and inequalities.

(ii) the resulting problem (6.2) is unbounded, that is, its objective function value goes to $-\infty$.

Remark 1 It should be noted that the above stated formulation (6.2) is, in fact, a subclass of the problems for which the **GBD** of Geoffrion (1972) can be applied. This is due to the specification of $y = \{0,1\}$, while Geoffrion (1972) investigated the more general case of $Y \subseteq \Re^q$, and defined the vector of y variables as "complicating" variables in the sense that if we fix y, then:

(a) Problem (6.2) may be decomposed into a number of independent problems, each involving a different subvector of $\boldsymbol{x}$; or

(b) Problem (6.2) takes a well known special structure for which efficient algorithms are available; or

(c) Problem (6.2) becomes convex in $\boldsymbol{x}$ even though it is nonconvex in the joint $\boldsymbol{x}$–$\boldsymbol{y}$ domain; that is, it creates special structure.

Case (a) may lead to parallel computations of the independent subproblems. Case (b) allows the use of special-purpose algorithms (e.g., generalized network algorithms), while case (c) invokes special structure from the convexity point of view that can be useful for the decomposition of non-convex optimization problems Floudas *et al.* (1989).

In the sequel, we concentrate on $Y = \{0,1\}^q$ due to our interest in MINLP models. Note also that the analysis includes the equality constraints $\boldsymbol{h}(\boldsymbol{x},\boldsymbol{y}) = \boldsymbol{0}$ which are not treated explicitly in Geoffrion (1972).

Remark 2 Condition C2 is not stringent, and it is satisfied if one of the following holds (in addition to C1,C3):

(i) $\boldsymbol{x}$ is bounded and closed and $\boldsymbol{h}(\boldsymbol{x},\boldsymbol{y}), \boldsymbol{g}(\boldsymbol{x},\boldsymbol{y})$ are continuous on $\boldsymbol{x}$ for each fixed $\boldsymbol{y} \in Y$.

(ii) There exists a point $z_{\boldsymbol{y}}$ such that the set

$$\{\boldsymbol{x} \in X : \ \boldsymbol{h}(\boldsymbol{x},\boldsymbol{y}) = \boldsymbol{0}, \ \boldsymbol{g}(\boldsymbol{x},\boldsymbol{y}) \leq z_{\boldsymbol{y}}\}$$

is bounded and nonempty.

Note though that mere continuity of $\boldsymbol{h}(\boldsymbol{x},\boldsymbol{y}),\ \boldsymbol{g}(\boldsymbol{x},\boldsymbol{y})$ on X for each fixed $\boldsymbol{y} \in Y$ does not imply that condition C2 is satisfied. For instance, if $X = [1,\infty]$ and $h(x,y) = x + y$, $g(x,y) = -\frac{1}{x}$, then $z_y = (-\infty, 0)$ which is not closed since for $x \to \infty, g(x,y) \to -\infty$.

Remark 3 Note that the set V represents the values of $\boldsymbol{y}$ for which the resulting problem (6.2) is feasible with respect to $\boldsymbol{x}$. In others words, V denotes the values of $\boldsymbol{y}$ for which there exists a feasible $\boldsymbol{x} \in X$ for $\boldsymbol{h}(\boldsymbol{x},\boldsymbol{y}) = 0, \boldsymbol{g}(\boldsymbol{x},\boldsymbol{y}) \leq 0$. Then, the intersection of $\boldsymbol{y}$ and V, $Y \cap V$, represents the projection of the feasible region of (2) onto the $\boldsymbol{y}$–space.

Remark 4 Condition C3 is satisfied if a first-order constraint qualification holds for the resulting problem (6.2) after fixing $\boldsymbol{y} \in Y \cap V$.

6.3.2 Basic Idea

The basic idea in Generalized Benders Decomposition **GBD** is the generation, at each iteration, of an upper bound and a lower bound on the sought solution of the MINLP model. The upper bound results from the *primal* problem, while the lower bound results from the *master* problem. The primal problem corresponds to problem (6.2) with fixed $\boldsymbol{y}$–variables (i.e., it is in the $\boldsymbol{x}$–space only), and its solution provides information about the upper bound and the Lagrange multipliers associated with the equality and inequality constraints. The master problem is derived via nonlinear duality theory, makes use of the Lagrange multipliers obtained in the primal problem, and its solution provides information about the lower bound, as well as the next set of fixed y–variables to be used subsequently in the primal problem. As the iterations proceed, it is shown that the sequence of updated upper bounds is nonincreasing, the sequence of lower bounds is nondecreasing, and that the sequences converge in a finite number of iterations.

6.3.3 Theoretical Development

This section presents the theoretical development of the Generalized Benders Decomposition **GBD**. The primal problem is analyzed first for the feasible and infeasible cases. Subsequently, the theoretical analysis for the derivation of the master problem is presented.

6.3.3.1 The Primal Problem

The primal problem results from fixing the y variables to a particular 0–1 combination, which we denote as y^k where k stands for the iteration counter. The formulation of the primal problem $P(y^k)$, at iteration k is

$$\left[\begin{array}{lrcl} \min\limits_{x} & f(x,y^k) & & \\ s.t. & h(x,y^k) & = & 0 \\ & g(x,y^k) & \leq & 0 \\ & x \in X & \subseteq & \Re^n \end{array}\right. \qquad (P(y^k))$$

Remark 1 Note that due to conditions C1 and C3(i), the solution of the primal problem $P(y^k)$ is its global solution.

We will distinguish the two cases of (i) feasible primal, and (ii) infeasible primal, and describe the analysis for each case separately.

Case (i): Feasible Primal

If the primal problem at iteration k is feasible, then its solution provides information on x^k, $f(x^k, y^k)$, which is the upper bound, and the optimal multiplier vectors λ^k, μ^k for the equality and inequality constraints. Subsequently, using this information we can formulate the Lagrange function as

$$L(x, y, \lambda^k, \mu^k) = f(x, y) + {\lambda^k}^T h(x, y) + {\mu^k}^T g(x, y).$$

Case (ii): Infeasible Primal

If the primal is detected by the NLP solver to be infeasible, then we consider its constraints

$$\begin{array}{rcl} h(x,y^k) & = & 0 \\ g(x,y^k) & \leq & 0 \\ x \in X & \subseteq & \Re^n \end{array}$$

where the set X, for instance, consists of lower and upper bounds on the x variables. To identify a feasible point we can minimize an l_1 or l_∞ sum of constraint violations. An l_1-minimization problem can be formulated as

$$\begin{aligned} \min_{x \in X} \quad & \sum_{i=1}^{p} \alpha_i \\ s.t. \quad h(x, y^k) &= 0 \\ g_i(x, y^k) &\leq \alpha_i \quad i = 1, 2, \ldots, p \\ \alpha_i &\geq 0 \quad i = 1, 2, \ldots, p \end{aligned}$$

Note that if $\sum_{i=1}^{p} \alpha_i = 0$, then a feasible point has been determined.

Also note that by defining as

$$\begin{aligned} \alpha^+ &= \max(0, \alpha) \quad \text{and} \\ g_i^+(x, y^k) &= \max[0, g_i(x, y^k)] \end{aligned}$$

the l_1–minimization problem is stated as

$$\begin{aligned} \min_{x \in X} \quad & \sum_{i=1}^{P} g_i^+ \\ s.t. \quad h(x, y^k) &= 0 \end{aligned}$$

An l_∞–minimization problem can be stated similarly as:

$$\begin{aligned} \min_{x \in X} \max_{1,2,\ldots,p} \quad & g_i^+(x, y^k) \\ s.t. \quad h(x, y^k) &= 0 \end{aligned}$$

Alternative feasibility minimization approaches aim at keeping feasibility in any constraint residual once it has been established. An l_1-minimization in these approaches takes the form:

$$\begin{aligned} \min_{x \in X} \quad & \sum_{i \in I'} g_i^+(x, y^k) \\ s.t. \quad h(x, y^k) &= 0 \\ g_i(x, y^k) &\leq 0, \quad i \in I \end{aligned}$$

where I is the set of feasible constraints; and I' is the set of infeasible constraints. Other methods seek feasibility of the constraints one at a time whilst maintaining feasibility for inequalities indexed by $i \in I$. This feasibility problem is formulated as

$$\min_{x \in X} \sum_{i \in I'} w_i \, g_i^+(x, y^k)$$
$$\text{s.t.} \quad h(x, y^k) = 0$$
$$g_i(x, y^k) \leq 0, \quad i \in I$$

and it is solved at any one time.

To include all mentioned possibilities Fletcher and Leyffer (1994) formulated a general feasibility problem **(FP)** defined as

$$\left[\begin{array}{ll} \min_{x \in X} & \sum_{i \in I'} w_i \, g_i^+(x, y^k) \\ \text{s.t.} & h(x, y^k) = 0 \\ & g_i(x, y^k) \leq 0, \quad i \in I \end{array} \right. \qquad \textbf{(FP)}$$

The weights w_i are non-negative and not all are zero. Note that with $w_i = 1 \;\; i \in I'$ we obtain the l_1–minimization. Also in the l_∞–minimization, there exist nonnegative weights at the solution such that

$$\sum_{i \in I'} w_i = 1,$$

and $w_i = 0$ if $g_i(x, y^k)$ does not attain the maximum value.

Note that infeasibility in the primal problem is detected when a solution of **(FP)** is obtained for which its objective value is greater than zero.

The solution of the feasibility problem **(FP)** provides information on the Lagrange multipliers for the equality and inequality constraints which are denoted as $\bar{\lambda}^k, \bar{\mu}^k$ respectively. Then, the Lagrange function resulting from on infeasible primal problem at iteration k can be defined as

$$\bar{L}^k(x, y, \bar{\lambda}^k, \bar{\mu}^k) = \bar{\lambda}^{k^T} \, h(x, y) + \bar{\mu}^{k^T} \, g(x, y).$$

Remark 2 It should be noted that two different types of Lagrange functions are defined depending on whether the primal problem is feasible or infeasible. Also, the upper bound is obtained only from the feasible primal problem.

6.3.3.2 The Master Problem

The derivation of the master problem in the **GBD** makes use of nonlinear duality theory and is characterized by the following three key ideas:

(i) Projection of problem (6.2) onto the y–space;

(ii) Dual representation of V; and

(iii) Dual representation of the projection of problem (6.2) on the y–space.

In the sequel, the theoretical analysis involved in these three key ideas is presented.

(i) Projection of (6.2) onto the y–space

Problem (6.2) can be written as

$$\begin{aligned} \min_{y} \inf_{x} \quad & f(x,y) \\ s.t. \quad & h(x,y) = 0 \\ & g(x,y) \leq 0 \\ & x \in X \\ & y \in Y = \{0,1\}^q \end{aligned} \tag{6.3}$$

where the min operator has been written separately for y and x. Note that it is infimum with respect to x since for given y the inner problem may be unbounded. Let us define $v(y)$ as

$$\begin{aligned} v(y) = \inf_{x} \; & f(x,y) \\ s.t. \quad & h(x,y) = 0 \\ & g(x,y) \leq 0 \\ & x \in X \end{aligned} \tag{6.4}$$

Remark 1 Note that $v(y)$ is parametric in the y variables and therefore, from its definition corresponds to the optimal value of problem (6.2) for fixed y (i.e., the primal problem $P(y^k)$ for $y = y^k$).

Let us also define the set V as

$$V = \{y : h(x,y) = 0,\; g(x,y) \leq 0 \;\text{ for some }\; x \in X\}\,. \tag{6.5}$$

Then, problem (6.3) can be written as

$$\begin{aligned} \min_{y} \quad & v(y) \\ s.t. \quad & y \in Y \cap V \end{aligned} \tag{6.6}$$

where $v(y)$ and V are defined by (6.4) and (6.5), respectively.

Remark 2 Problem (6.6) is the projection of problem (6.2) onto the y–space. Note also that in (6.5) $y \in Y \cap V$ since the projection needs to satisfy the feasibility considerations.

Having defined the projection problem of (6.2) onto the y–space, we can now state the theoretical result of Geoffrion (1972).

Theorem 6.3.1 (Projection)

(i) *If* (x^*, y^*) *is optimal in (6.2), then* y^* *is optimal in (6.6).*

(ii) *If (6.2) is infeasible or has unbounded solution, then the same is true for (6.6) and vice versa.*

Remark 3 Note that the difficulty in (6.6) is due to the fact that $v(y)$ and V are known only implicitly via (6.4) and (6.5).

To overcome the aforementioned difficulty we have to introduce the dual representation of V and $v(y)$.

(ii) Dual Representation of V

The dual representation of V will be invoked in terms of the intersection of a collection of regions that contain it, and it is described in the following theorem of Geoffrion (1972).

Theorem 6.3.2 (Dual Representation of V **)**
Assuming conditions C1 and C2 a point $y \in Y$ *belongs also to the set* V *if and only if it satisfies the system:*

$$0 \geq \inf \bar{L}(x, y, \bar{\lambda}, \bar{\mu}), \quad \forall \bar{\lambda}, \bar{\mu} \in \Lambda, \tag{6.7}$$

$$\textit{where} \quad \Lambda = \left\{ \bar{\lambda} \in \Re^m, \bar{\mu} \in \Re^p : \bar{\mu} \geq 0, \sum_{i=1}^{p} \bar{\mu}_i = 1 \right\}.$$

Remark 4 Note that (6.7) is an infinite system because it has to be satisfied for all $\bar{\lambda}, \bar{\mu} \in \Lambda$.

Remark 5 The dual representation of the set V needs to be invoked so as to generate a collection of regions that contain it (i.e., system (6.7) corresponds to the set of constraints that have to be incorporated for the case of infeasible primal problems.

Remark 6 Note that if the primal is infeasible and we make use of the l_1–minimization of the type:

$$\begin{aligned} \min_{x} \quad & \sum_{i \in I} \alpha_i \\ \text{s.t.} \quad h(x, y^k) &= 0 \\ g_i(x, y^k) &\leq \alpha_i, \quad i \in I \\ x &\in X \end{aligned} \tag{6.8}$$

then the set Λ results from a straightforward application of the **KKT** gradient conditions to problem (6.8) with respect to α_i.

Having introduced the dual representation of the set V, which corresponds to infeasible primal problems, we can now invoke the dual representation of $v(y)$.

(iii) Dual Representation of $v(y)$

The dual representation of $v(y)$ will be in terms of the pointwise infimum of a collection of functions that support it, and it is described in the following theorem due to Geoffrion (1972).

Theorem 6.3.3 (Dual of $v(y)$)

$$
v(y) = \left[\begin{array}{lrcl} \inf\limits_{x} & f(x,y) & & \\ \text{s.t.} & h(x,y) & = & 0 \\ & g(x,y) & \leq & 0 \\ & x & \in & X \end{array} \right] = \left[\sup_{\lambda,\mu \geq 0} \quad \inf_{x \in X} L(x,y,\lambda,\mu) \right], \forall y \in Y \cap V
$$

$$
\text{where} \quad L(x,y,\lambda,\mu) = f(x,y) + \lambda^T h(x,y) + \mu^T g(x,y). \tag{6.9}
$$

Remark 7 The equality of $v(y)$ and its dual is due to having the strong duality theorem satisfied because of conditions C1, C2, and C3.

Substituting (6.9) for $v(y)$ and (6.7) for $y \in Y \cap V$ into problem (6.6), which is equivalent to (6.3), we obtain

$$
\begin{array}{ll} \min\limits_{y \in Y} & \sup\limits_{\lambda,\mu \geq 0} \inf\limits_{x \in X} L(x,y,\lambda,\mu) \\ s.t. & 0 \geq \inf\limits_{x \in X} \bar{L}(x,y,\bar{\lambda},\bar{\mu}) \end{array}
$$

Using the definition of supremum as the lowest upper bound and introducing a scalar μ_B we obtain:

$$
\left[\begin{array}{llll} \min\limits_{y \in Y, \mu_B} & \mu_B & & \\ s.t. & \mu_B & \geq \inf\limits_{x \in X} L(x,y,\lambda,\mu), & \forall \lambda, \forall \mu \geq 0 \\ & 0 & \geq \inf\limits_{x \in X} \bar{L}(x,y,\bar{\lambda},\bar{\mu}), & \forall\, (\bar{\lambda},\bar{\mu}) \in \Lambda \end{array} \right. \tag{M}
$$

$$
\begin{array}{ll} \text{where} & L(x,y,\lambda,\mu) = f(x,y) + \lambda^T h(x,y) + \mu^T g(x,y) \\ & L(x,y,\bar{\lambda},\bar{\mu}) = \bar{\lambda}^T h(x,y) + \bar{\mu}^T g(x,y) \end{array}
$$

which is called the *master* problem and denoted as **(M)**.

Remark 8 If we assume that the optimum solution of $v(y)$ in (6.4) is bounded for all $y \in Y \cap V$, then we can replace the infimum with a minimum. Subsequently, the master problem will be as follows:

$$\begin{aligned}
\min_{y \in Y, \mu_B} \quad & \mu_B \\
\text{s.t.} \quad & \mu_B \geq \min_{x \in X} L(x, y, \lambda, \mu), \quad \forall \lambda, \mu \geq 0 \\
& 0 \geq \min_{x \in X} \bar{L}(x, y, \bar{\lambda}, \bar{\mu}), \quad \forall (\bar{\lambda}, \bar{\mu}) \in \Lambda
\end{aligned}$$

where $L(x, y, \lambda, \mu)$ and $\bar{L}(x, y, \bar{\lambda}, \bar{\mu})$ are defined as before.

Remark 9 Note that the master problem **(M)** is equivalent to (6.2). It involves, however, an infinite number of constraints, and hence we would need to consider a relaxation of the master (e.g., by dropping a number of constraints) which will represent a lower bound on the original problem. Note also that the master problem features an outer optimization problem with respect to $y \in Y$ and inner optimization problems with respect to x which are in fact parametric in y. It is this outer-inner nature that makes the solution of even a relaxed master problem difficult.

Remark 10 (Geometric Interpretation of the Master Problem) The inner minimization problems

$$\begin{aligned}
& \min_{x \in X} L(x, y, \lambda, \mu), \quad && \forall \lambda, \forall \mu \geq 0, \\
& \min_{x \in X} \bar{L}(x, y, \bar{\lambda}, \bar{\mu}), \quad && \forall (\bar{\lambda}, \bar{\mu}) \in \Lambda,
\end{aligned}$$

are functions of y and can be interpreted as support functions of $v(y)$ [$\xi(y)$ is a support function of $v(y)$ at point y_o if and only if $\xi(y_o) = v(y_o)$ and $\xi(y) \leq v(y) \; \forall \, y \neq y_o$]. If the support functions are linear in y, then the master problem approximates $v(y)$ by tangent hyperplanes and we can conclude that $v(y)$ is convex in y. Note that $v(y)$ can be convex in y even though problem (6.2) is nonconvex in the joint x–y space Floudas and Visweswaran (1990).

In the sequel, we will define the aforementioned minimization problems in terms of the notion of support functions; that is

$$\begin{aligned}
\xi(y; \lambda, \mu) &= \min_{x \in X} L(x, y, \lambda, \mu), \quad \forall \lambda, \; \forall \mu \geq 0 \\
\bar{\xi}(y; \bar{\lambda}, \bar{\mu}) &= \min_{x \in X} \bar{L}(x, y, \bar{\lambda}, \bar{\mu}), \quad \forall (\bar{\lambda}, \bar{\mu}) \in \Lambda
\end{aligned}$$

6.3.4 Algorithmic Development

In the previous section we discussed the primal and master problem for the **GBD**. We have the primal problem being a (linear or) nonlinear programming NLP problem that can be solved via available local NLP solvers (e.g., MINOS 5.3). The master problem, however, consists of outer and inner optimization problems, and approaches towards attaining its solution are discussed in the following.

6.3.4.1 How to Solve the Master Problem

The master problem has as constraints the two inner optimization problems (i.e., for the case of feasible primal and infeasible primal problems) which, however, need to be considered for all λ and all $\mu \geq 0$ (i.e., feasible primal) and all $(\bar{\lambda}, \bar{\mu}) \in \Lambda$ (i.e., infeasible). This implies that the master problem has a very large number of constraints.

The most natural approach for solving the master problem is *relaxation* (Geoffrion, 1972). The basic idea in the relaxation approach consists of the following: (i) ignore all but a few of the constraints that correspond to the inner optimization problems (e.g., consider the inner optimization problems for specific or fixed multipliers (λ^1, μ^1) or $(\bar{\lambda}^1, \bar{\mu}^1)$); (ii) solve the relaxed master problem and check whether the resulting solution satisfies all of the ignored constraints. If not, then generate and add to the relaxed master problem one or more of the violated constraints and solve the new relaxed master problem again; (iii) continue until a relaxed master problem satisfies all of the ignored constraints, which implies that an optimal solution at the master problem has been obtained or until a termination criterion indicates that a solution of acceptable accuracy has been found.

6.3.4.2 General Algorithmic Statement of GBD

Assuming that the problem (6.2) has a finite optimal value, Geoffrion (1972) stated the following general algorithm for **GBD**:

Step 1: Let an initial point $y^1 \in Y \cap V$ (i.e., by fixing $y = y^1$, we have a feasible primal). Solve the resulting primal problem $P(y^1)$ and obtain an optimal primal solution x^1 and optimal multipliers; vectors λ^1, μ^1. Assume that you can find, somehow, the support function $\xi(y; \lambda^1, \mu^1)$ for the obtained multipliers λ^1, μ^1. Set the counters $k = 1$ for feasible and $l = 1$ for infeasible and the current upper bound $UBD = v(y^1)$. Select the convergence tolerance $\epsilon \geq 0$.

Step 2: Solve the relaxed master problem, **(RM)**:

$$\begin{aligned} \min_{y \in Y, \mu_B} \quad & \mu_B \\ s.t. \quad & \mu_B \geq \xi(y; \lambda^k, \mu^k), \quad k = 1, 2, \ldots, K \\ & 0 \geq \bar{\xi}(y; \bar{\lambda}^l, \bar{\mu}^l), \quad l = 1, 2, \ldots, \Lambda \end{aligned}$$

Let $(\hat{y}, \hat{\mu}_B)$ be an optimal solution of the above relaxed master problem. $\hat{\mu}_B$ is a lower bound on problem (6.2); that is, the current lower bound is $LBD = \hat{\mu}_B$. If $UBD - LBD \leq \epsilon$, then terminate.

Step 3: Solve the primal problem for $y = \hat{y}$, that is the problem $P(\hat{y})$. Then we distinguish two cases; feasible and infeasible primal:

Step 3a - Feasible Primal $P(\hat{y})$

The primal has $v(\hat{y})$ finite with an optimal solution $\hat{x}$ and optimal multiplier vectors $\hat{\lambda}, \hat{\mu}$. Update the upper bound $UBD = \min\{UBD, v(\hat{y})\}$. If $UBD - LBD \leq \epsilon$, then terminate. Otherwise, set $k = k + 1$, $\lambda^k = \hat{\lambda}$, and $\mu^k = \hat{\mu}$. Return to step 2, assuming we can somehow determine the support function $\xi(y; \lambda^{k+1}, \mu^{k+1})$.

Step3b - Infeasible Primal $P(\hat{y})$

The primal does not have a feasible solution for $\boldsymbol{y} = \hat{\boldsymbol{y}}$. Solve a feasibility problem (e.g., the l_1–minimization) to determine the multiplier vectors $\bar{\hat{\lambda}}, \bar{\hat{\mu}}$ of the feasibility problem.

Set $l = l + 1$, $\bar{\lambda}^l = \bar{\hat{\lambda}}$, and $\bar{\mu}^l = \bar{\hat{\mu}}$. Return to step 2, assuming we can somehow determine the support function $\bar{\xi}(\boldsymbol{y}; \bar{\lambda}^{l+1}, \bar{\mu}^{l+1})$.

Remark 1 Note that a feasible initial primal is needed in step 1. However, this does not restrict the **GBD** since it is possible to start with an infeasible primal problem. In this case, after detecting that the primal is infeasible, step 3b is applied, in which a support function $\bar{\xi}$ is employed.

Remark 2 Note that step 1 could be altered, that is instead of solving the primal problem we could solve a continuous relaxation of problem (6.2) in which the $\boldsymbol{y}$ variables are treated as continuous bounded by zero and one:

$$\begin{aligned} \min_{\boldsymbol{x}, \boldsymbol{y}} \quad & f(\boldsymbol{x}, \boldsymbol{y}) \\ s.t. \quad & \boldsymbol{h}(\boldsymbol{x}, \boldsymbol{y}) = \boldsymbol{0} \\ & \boldsymbol{g}(\boldsymbol{x}, \boldsymbol{y}) \leq \boldsymbol{0} \\ & \boldsymbol{x} \in \boldsymbol{X} \\ & \boldsymbol{0} \leq \boldsymbol{y} \leq \mathbf{1} \end{aligned} \tag{6.10}$$

If the solution of (6.10) is integral, then we terminate. If there exist fractional values of the $\boldsymbol{y}$ variables, then these can be rounded to the closest integer values, and subsequently these can be used as the starting $\boldsymbol{y}^1$ vector with the possibility of the resulting primal problem being feasible or infeasible.

Remark 3 Note also that in step 1, step 3a, and step 3b a rather important assumption is made; that is, we can find the support functions ξ and $\bar{\xi}$ for the given values of the multiplier vectors (λ, μ) and $(\bar{\lambda}, \bar{\mu})$. The determination of these support functions cannot be achieved in general, since these are parametric functions of $\boldsymbol{y}$ and result from the solution of the inner optimization problems. Their determination in the general case requires a global optimization approach as the one proposed by (Floudas and Visweswaran, 1990; Floudas and Visweswaran, 1993). There exist however, a number of special cases for which the support functions can be obtained explicitly as functions of the $\boldsymbol{y}$ variables. We will discuss these special cases in the next section. If however, it is not possible to obtain explicitly expressions of the support functions in terms of the $\boldsymbol{y}$ variables, then assumptions need to be introduced for their calculation. These assumptions, as well as the resulting variants of **GBD** will be discussed in the next section. The point to note here is that the validity of lower bounds with these variants of **GBD** will be limited by the imposed assumptions.

Remark 4 Note that the relaxed master problem (see step 2) in the first iteration will have as a constraint one support function that corresponds to feasible primal and will be of the form:

$$\begin{aligned} \min_{\boldsymbol{y} \in \boldsymbol{Y}, \mu_B} \quad & \mu_B \\ s.t. \quad & \mu_B \geq \xi(\boldsymbol{y}; \lambda^1, \mu^1) \end{aligned} \tag{6.11}$$

In the second iteration, if the primal is feasible and (λ^2, μ^2) are its optimal multiplier vectors, then the relaxed master problem will feature two constraints and will be of the form:

$$\begin{aligned} \min_{y \in Y, \mu_B} \quad & \mu_B \\ s.t. \quad & \mu_B \geq \xi(y; \lambda^1, \mu^1) \\ & \mu_B \geq \xi(y; \lambda^2, \mu^2) \end{aligned} \tag{6.12}$$

Note that in this case, the relaxed master problem (6.12) will have a solution that is greater or equal to the solution of (6.11). This is due to having the additional constraint. Therefore, we can see that the sequence of lower bounds that is created from the solution of the relaxed master problems is nondecreasing. A similar argument holds true in the case of having infeasible primal in the second iteration.

Remark 5 Note that since the upper bounds are produced by fixing the y variables to different 0-1 combinations, there is no reason for the upper bounds to satisfy any monotonicity property. If we consider however the updated upper bounds (i.e., $UBD = \min_k v(y^k)$), then the sequence for the updated upper bounds is monotonically nonincreasing since by their definition we always keep the best(least) upper bound.

Remark 6 The termination criterion for **GBD** is based on the difference between the updated upper bound and the current lower bound. If this difference is less than or equal to a prespecified tolerance $\varepsilon \geq 0$ then we terminate. Note though that if we introduce in the relaxed master integer cuts that exclude the previously found 0–1 combinations, then the termination criterion can be met by having found an infeasible master problem (i.e., there is no 0–1 combination that makes it feasible).

6.3.4.3 Finite Convergence of GBD

For formulation (6.2), Geoffrion (1972) proved finite convergence of the **GBD** algorithm stated in section 6.3.4.2, which is as follows:

Theorem 6.3.4 (Finite Convergence)
If C1, C2, C3 hold and Y *is a discrete set, then the* **GBD** *algorithm terminates in a finite number of iterations for any given* $\epsilon > 0$ *and even for* $\epsilon = 0$.

Note that in this case exact convergence can be obtained in a finite number of iterations.

6.3.5 Variants of GBD

In the previous section we discussed the general algorithmic statement of **GBD** and pointed out (see remark 3) a key assumption made with respect to the calculation of the support functions $\xi(y; \lambda, \mu)$ and $\bar{\xi}(y; \bar{\lambda}, \bar{\mu})$ from the feasible and infeasible primal problems, respectively. In this section, we will discuss a number of variants of **GBD** that result from addressing the calculation of the aforementioned support functions either rigorously for special cases or making assumptions that may not provide valid lower bounds in the general case.

6.3.5.1 Variant 1 of GBD, v1–GBD

This variant of **GBD** is based on the following assumption that was denoted by Geoffrion (1972) as Property (**P**):

Theorem 6.3.5 (Property (P))
For every λ and $\mu \geq 0$, the infimum of $L(\boldsymbol{x}, \boldsymbol{y}, \lambda, \mu)$ with respect to $\boldsymbol{x} \in \boldsymbol{X}$ (i.e. the support $\xi(y; \lambda, \mu)$) can be taken independently of $\boldsymbol{y}$ so that the support function $\xi(y; \lambda, \mu)$ can be obtained explicitly with little or no more effort than is required to evaluate it at a single value of $\boldsymbol{y}$. Similarly, the support function $\bar{\xi}(y; \bar{\lambda}, \bar{\mu})$, $(\bar{\lambda}, \bar{\mu}) \in \Lambda$ can be obtained explicitly.

Geoffrion (1972) identified the following two important classes of problems where Property (P) holds:

Class 1: $f, \boldsymbol{h}, \boldsymbol{g}$ are linearly separable in $\boldsymbol{x}$ and $\boldsymbol{y}$.

Class 2: Variable factor programming

Geromel and Belloni (1986) identified a similar class to variable factor programming that is applicable to the unit commitment of thermal systems problems.

In class 1 problems, we have

$$\begin{aligned} f(\boldsymbol{x},\boldsymbol{y}) &= f_1(\boldsymbol{x}) + f_2(\boldsymbol{y}), \\ \boldsymbol{h}(\boldsymbol{x},\boldsymbol{y}) &= \boldsymbol{h}_1(\boldsymbol{x}) + \boldsymbol{h}_2(\boldsymbol{y}), \\ \boldsymbol{g}(\boldsymbol{x},\boldsymbol{y}) &= \boldsymbol{g}_1(\boldsymbol{x}) + \boldsymbol{g}_2(\boldsymbol{y}). \end{aligned}$$

In class 2 problems, we have

$$\begin{aligned} f(\boldsymbol{x},\boldsymbol{y}) &= -\sum_i f_i(\boldsymbol{x}^i) y_i, \\ g(\boldsymbol{x},\boldsymbol{y})_j &= \sum_i \boldsymbol{x}^i y_i - c. \end{aligned}$$

In Geromel and Belloni (1986) problems, we have

$$\begin{aligned} f(\boldsymbol{x},\boldsymbol{y}) &= \sum_k \sum_i f_i(x_i(k)) y_i + \sum_i g_i(y_i), \\ g(\boldsymbol{x},\boldsymbol{y})_j &= -\sum_i x_i(k) y_i - L(k). \end{aligned}$$

In the sequel, we will discuss the **v1–GBD** for class 1 problems since this by itself defines an interesting mathematical structure for which other algorithms (e.g., Outer Approximation) has been developed.

v1–GBD under separability

Under the separability assumption, the support functions $\xi(y; \lambda^k, \mu^k)$ and $\bar{\xi}(y; \bar{\lambda}^l, \bar{\mu}^l)$ can be obtained as explicit functions of $\boldsymbol{y}$ since

$$\begin{aligned}
\xi(y;\lambda^k,\mu^k) &= \min_{x\in X} L(x,y,\lambda^k\mu^k) \\
&= \min_{x\in X} \{f(x,y)) + \lambda^{k^T}h(x,y) + \mu^{kT}g(x,y)\} \\
&= \min_{x\in X} \{f_1(x) + f_2(y) + \lambda^{kT}(h_1(x) + h_2(y)) + \mu^{kT}(g_1(x) + g_2(y))\} \\
&= f_2(y) + \lambda^{kT}h_2(y) + \mu^{kT}g_2(x) + \min_{x\in X}[f_1(x) + \lambda^{kT}h_1(x) + \mu^{kT}g_1(x)].
\end{aligned}$$

Remark 1 Note that due to separability we end up with an explicit function of y and a problem only in x that can be solved independently.

Similarly, the support function $\bar{\xi}(y;\bar{\lambda}^l,\bar{\mu}^l)$ is

$$\begin{aligned}
\bar{\xi}(y;\bar{\lambda}^l,\bar{\mu}^l) &= \min_{x\in X} \bar{L}(x,y,\bar{\lambda}^l,\bar{\mu}^l) \\
&= \min_{x\in X} \left\{\bar{\lambda}^{l^T}h(x,y) + \bar{\mu}^{l^T}g(x,y)\right\} \\
&= \min_{x\in X} \left\{\bar{\lambda}^{l^T}(h_1(x,y) + h_2(x,y)) + \bar{\mu}^{l^T}(g_1(x,y) + g_2(x,y))\right\} \\
&= \bar{\lambda}^{l^T}h_2(y) + \bar{\mu}^{l^T}g_2(y) + \min_{x\in X}\left[\bar{\lambda}^{l^T}h_1(x) + \bar{\mu}^{l^T}g_1(x)\right].
\end{aligned}$$

Remark 2 Note that to solve the independent problems in x, we need to know the multiplier vectors (λ^k,μ^k) and $(\bar{\lambda}^l,\bar{\mu}^l)$ from feasible and infeasible primal problems, respectively.

Under the separability assumption, the primal problem for fixed $y = y^k$ takes the form

$$\begin{aligned}
\min_{x\in X} \quad & f_1(x) + f_2(y^k) \\
s.t. \quad & h_1(x) = -h_2(y^k) \\
& g_1(x) \leq -g_2(y^k)
\end{aligned}$$

Now, we can state the algorithmic procedure for the **v1–GBD** under the separability assumption.

v1–GBD Algorithm

Step 1: Let an initial point $y^1 \in Y \cap V$. Solve the primal $P(y^1)$ and obtain an optimal solution x^1, and multiplier vectors λ^1,μ^1. Set the counters $k = 1, l = 1$, and $UBD = v(y^1)$. Select the convergence tolerance $\epsilon \geq 0$.

Step 2: Solve the relaxed master problem:

$$\begin{aligned}
\min_{y\in Y,\mu_B} \quad & \mu_B \\
s.t. \quad & \mu_B \geq f_2(y) + \lambda^{k^T}h_2(y) + \mu^{k^T}g_2(y) + L_1^k, \quad k = 1,2,\ldots,K \\
& 0 \geq \mu_B\bar{\lambda}^{l^T}h_2(y) + \bar{\mu}^{l^T}g_2(y) + L_1^l, \quad l = 1,2,\ldots,\Lambda
\end{aligned}$$

$$\text{where} \quad L_1^k = \min_{x \in X} \left[f_1(x) + \lambda^{k^T} h_1(x) + \mu^{k^T} g_1(x) \right]$$

$$\bar{L}_1^k = \min_{x \in X} \left[f_1(x) + \bar{\lambda}^{l^T} h_1(x) + \bar{\mu}^{l^T} g_1(x) \right]$$

are solutions of the above stated independent problems.

Let $(\hat{y}, \hat{\mu}_B)$ be an optimal solution. $\hat{\mu}_B$ is a lower bound, that is $LBD = \hat{\mu}_B$. If $UBD - LBD \leq \epsilon$, then terminate.

Step 3: As in section 6.3.4.2.

Remark 3 Note that if in addition to the separability of x and y, we assume that y participates linearly (i.e., conditions for Outer Approximation algorithm), then we have

$$\begin{aligned} f_2(y) &= c^T y, \\ h_2(y) &= Ay, \\ g_2(y) &= By, \end{aligned}$$

in which case, the relaxed master problem of step 2 of **v1–GBD** will be a linear 0–1 programming problem with an additional scalar μ_B, which can be solved with available solvers (e.g., CPLEX, ZOOM, SCICONIC).

If the y variables participate separably but in a nonlinear way, then the relaxed master problem is of 0–1 nonlinear programming type.

Remark 4 Note that due to the strong duality theorem we do not need to solve the problems for $L_1^k, \bar{L}_1^l$ since their optimum solutions are identical to the ones of the corresponding feasible and infeasible primal problems with respect to x, respectively.

Illustration 6.3.1 This example is a modified version of example 1 of Kocis and Grossmann (1987) and can be stated as follows:

$$\begin{aligned} \min_{x_1, y} \quad & -y + 2x_1 - \ln(0.5x_1) \\ \text{subject to} \quad & -x_1 - \ln(0.5x_1) + y \leq 0 \\ & 0.5 \leq x_1 \leq 1.4 \\ & y = \{0, 1\} \end{aligned}$$

Note that it features separability in x and y and linearity in y.

$$\begin{aligned} f_1(x) &= 2x_1 - \ln(0.5x_1), \\ f_2(y) &= -y, \\ g_1(x) &= -x_1 - \ln(0.5x_1), \\ g_2(y) &= y. \end{aligned}$$

Also note that $f_1(x), g_1(x)$ are convex functions in x_1, and hence the required convexity conditions are satisfied.

Based on the previously presented analysis for the **v1–GBD** under the separability assumption, we can now formulate the relaxed master problem in an explicit form.

Relaxed Master Problem

$$\begin{aligned} \min_{y,\mu_B} \quad & \mu_B \\ \text{subject to} \quad & \mu_B \geq -y + \mu^k y + L_1^k, \quad k = 1, 2, \ldots, K \\ & 0 \geq \bar{\mu}^l y + \bar{L}_1^l, \quad l = 1, 2, \ldots, L \end{aligned}$$

$$\text{where} \quad \begin{aligned} L_1^k &= \min_{0.5 \leq x_1 \leq 1.4} 2x_1 - \ln(0.5x_1) + \mu^k(-x1 - \ln(0.5x_1)) \\ \bar{L}_1^l &= \min_{0.5 \leq x_1 \leq 1.4} \bar{\mu}^l(-x1 - \ln(0.5x_1)) \end{aligned}$$

Now we can apply the **v1–GBD** algorithm.

Step 1: Select $y^1 = 0$.
Solve the following primal problem $P(y^1)$.

$$\begin{aligned} \min_{x_1} \quad & 2x_1 - 2\ln(0.5x_1) \\ \text{subject to} \quad & -x_1 - \ln(0.5x_1) \leq 0 \\ & 0.5 \leq x_1 \leq 1.4 \end{aligned}$$

which has as solution

$$\begin{aligned} x_1 &= 0.353, \\ \mu^1 &= 0.381, \end{aligned}$$

and the upper bound is $UBD = 2.558$.

Step 2:

$$\begin{aligned} L_1^1 &= \min_{0.5 \leq x_1 \leq 1.4} 2x_1 - \ln(0.5x_1) + 0.381(-x_1 - \ln(0.5x_1)) \\ &= 2.558. \end{aligned}$$

Note that we do not not need to solve for L_1^k since due to strong duality its solution is identical to the one of the corresponding primal problem $P(y^1)$.

Then, the relaxed master problem is of the form:

$$\begin{aligned} \min_{y,\mu_B} \quad & \mu_B \\ \text{subject to} \quad & \mu_B \geq -y + 0.381y + 2.558 \\ & y = 0, 1 \end{aligned}$$

Its solution is $y^2 = 1$ and the lower bound is

$$LBD = 1.939.$$

Step 3: Solve the Primal for $y^2 = 1,\ P(y^2)$ which has as solution:

$$\begin{aligned} x_1 &= 1.375, \\ \text{objective} &= 2.124, \\ \mu &= 0.73684. \end{aligned}$$

The new upper bound is

$$UBD = \min(2.558, 2.124) = 2.124,$$

and this is the optimal solution since we have examined all 0–1 combinations.

6.3.5.2 Variant 2 of GBD, v2–GBD

This variant of **GBD** is based on the assumption that we can use the optimal solution x^k of the primal problem $P(y^k)$ along with the multiplier vectors for the determination of the support function $\xi(y; \lambda^k, \mu^k)$.

Similarly, we assume that we can use the optimal solution of the feasibility problem (if the primal is infeasible) for the determination of the support function $\bar{\xi}(y; \lambda^k, \mu^k)$.

The aforementioned assumption fixes the x vector to the optimal value obtained from its corresponding primal problem and therefore eliminates the inner optimization problems that define the support functions. It should be noted that fixing x to the solution of the corresponding primal problem may not necessarily produce valid support functions in the sense that there would be no theoretical guarantee for obtaining lower bounds to solution of (6.2) can be claimed in general.

v2–GBD Algorithm

The **v2–GBD** algorithm can be stated as follows:

Step 1: Let an initial point $y^1 \in Y \cap V$.
Solve the primal problem $P(y^1)$ and obtain an optimal solution x^1 and multiplier vectors λ^1, μ^1. Set the counters $k = 1, l = 1$, and $UBD = v(y^1)$ Select the convergence tolerance $\epsilon \geq 0$.

Step 2: Solve the relaxed master problem:

$$\begin{aligned} \min_{y \in Y, \mu_B} \quad & \mu_B \\ \text{s.t.} \quad \mu_B &\geq L(x^k, y, \lambda^k, \mu^k), \quad k = 1, 2, \ldots, K \\ 0 &\geq \bar{L}(x^l, y, \bar{\lambda}^l, \bar{\mu}^l), \quad l = 1, 2, \ldots, \Lambda \end{aligned}$$

$$\text{where} \quad L(x^k, y, \lambda^k, \mu^k) = f(x^k, y) + \lambda^{k^T} h(x^k, y) + \mu^{k^T} g(x^k, y)$$
$$\bar{L}(\bar{x}^l, y, \bar{\lambda}^l, \bar{\mu}^k) = \bar{\lambda}^{k^T} h(x^l, y) + \bar{\mu}^{k^T} g(x^l, y)$$

are the Lagrange functions evaluated at the optimal solution x^k of the primal problem.

Let $(\hat{y}, \hat{\mu}_B)$ be an optimal solution. $\hat{\mu}_B$ is a lower bound, that is, $LBD = \hat{\mu_B}$. If $UBD - LBD \leq \epsilon$, then terminate.

Step 3: As in section 6.3.4.2.

Remark 1 Note that since $y \in Y = \{0-1\}$, the master problem is a 0–1 programming problem with one scalar variable μ_B. If the y variables participate linearly, then it is a 0–1 linear problem which can be solved with standard branch and bound algorithms. In such a case, we can introduce integer cuts of the form:

$$\sum_{i \in B} y_i - \sum_{i \in NB} y_i \leq |B| - 1,$$
$$\text{where} \quad B = \{i : y_i = 1\},$$
$$NB = \{i : y_i = 0\},$$
$$|B| \text{ is the cardinality of B},$$

which eliminate the already found 0–1 combinations. If we employ such a scheme, then an alternative termination criterion is that of having infeasible relaxed master problems. This of course implies that all 0–1 combinations have been considered.

Remark 2 It is of considerable interest to identify the conditions which if satisfied make the assumption in **v2–GBD** a valid one. The assumption in a somewhat different restated form is that:

$$\xi(y; \lambda^k, \mu^k) = \min_{x \in X} L(x, y, \lambda^k, \mu^k) \geq L(x^k, y, \lambda^k, \mu^k), \quad k = 1, 2, \ldots, K,$$
$$\bar{\xi}(y; \bar{\lambda}^l, \mu^l) = \min_{x \in X} \bar{L}(x, y, \bar{\lambda}^l, \bar{\mu}^l) \geq \bar{L}(x^l, y, \bar{\lambda}^l, \bar{\mu}^l), \quad l = 1, 2, \ldots, \Lambda,$$

that is, we assume that the Lagrange function evaluated at the solution of the corresponding primal are valid underestimators of the inner optimization problems with respect to $x \in X$.

Due to condition C1 the Lagrange functions $L(x, y, \lambda^k, \mu^k)$, $\bar{L}(x, y, \bar{\lambda}^l, \bar{\mu}^l)$ are convex in x for each fixed y since they are linear combinations of convex functions in x.

$L(x, y, \lambda^k, \mu^k)$, $\bar{L}(\bar{x}^l, y, \bar{\lambda}^l, \bar{\mu}^l)$ represent local linearizations around the points x^k and $\bar{x}^k$ of the support functions $\xi(y; \lambda^k, \mu^k)$, $\bar{\xi}(y; \bar{\lambda}^l, \mu^l)$, respectively. Therefore, the aforementioned assumption is valid if the projected problem $v(y)$ is convex in y. If, however, the projected problem $v(y)$ is nonconvex, then the assumption does not hold, and the algorithm may terminate at a local (nonglobal) solution or even at a nonstationary point. This analysis was first presented by Floudas and Visweswaran (1990) and later by Sahinidis and Grossmann (1991), and Bagajewicz and Manousiouthakis (1991). Figure 6.1 shows the case in which the assumption is valid, while Figure 6.2 shows a case for which a local solution or a nonstationary point may result.

Note that in the above analysis we did not assume that $Y = \{0, 1\}^q$, and hence the argument is applicable even when the y–variables are continuous. In fact, Figures 6.1 and 6.2 represent continuous y variables.

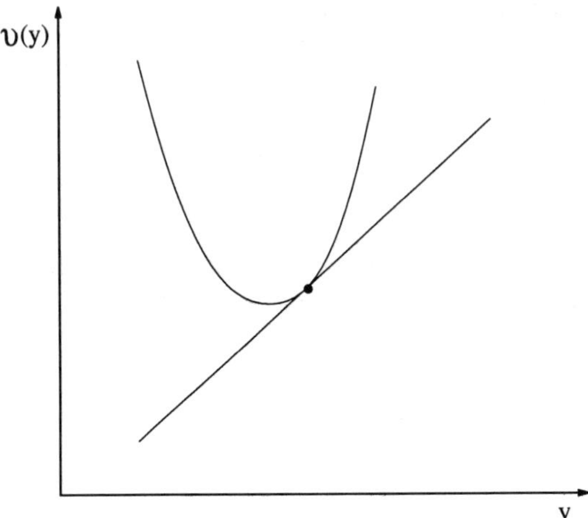

Figure 6.1: Valid support

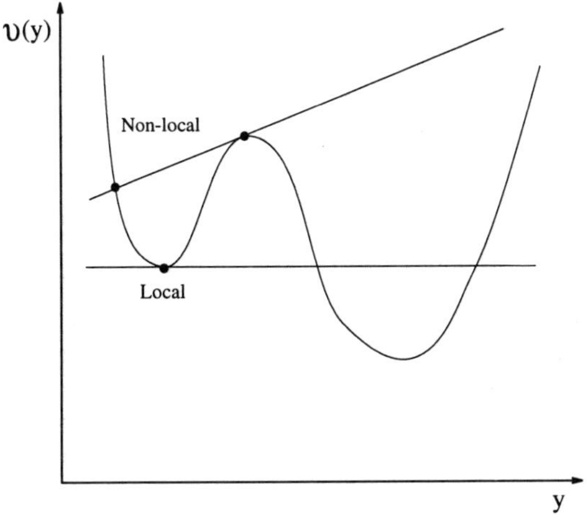

Figure 6.2: Invalid support

Remark 3 It is also very interesting to examine the validity of the assumption made in **v2–GBD** under the conditions of separability of $\boldsymbol{x}$ and $\boldsymbol{y}$ and linearity in $\boldsymbol{y}$ (i.e., **OA** conditions). In this case we have:

$$\begin{aligned} f(\boldsymbol{x},\boldsymbol{y}) &= \mathbf{c}^T\boldsymbol{y} + f_1(\boldsymbol{x}), \\ \boldsymbol{h}(\boldsymbol{x},\boldsymbol{y}) &= A\boldsymbol{y} + \boldsymbol{h}_1(\boldsymbol{x}), \\ \boldsymbol{g}(\boldsymbol{x},\boldsymbol{y}) &= B\boldsymbol{y} + \boldsymbol{g}_1(\boldsymbol{x}). \end{aligned}$$

Then the support function for feasible primal becomes

$$\begin{aligned} \xi(\boldsymbol{y};\lambda^k,\mu^k) &= \mathbf{c}^T\boldsymbol{y} + \lambda^{k^T}(A\boldsymbol{y}) + \mu^{k^T}(B\boldsymbol{y}) \\ &\quad + \min_{\boldsymbol{x}\in X} f_1(\boldsymbol{x}) + \lambda^{k^T}\boldsymbol{h}_1(\boldsymbol{x}) + \mu^{k^T}\boldsymbol{g}_1(\boldsymbol{x}), \end{aligned}$$

which is linear in $\boldsymbol{y}$ and hence convex in $\boldsymbol{y}$. Note also that since we fix $\boldsymbol{x} = \boldsymbol{x}^k$, the $\min_{\boldsymbol{x}\in X}$ is in fact an evaluation at $\boldsymbol{x}^k$. Similarly the case for $\bar{\xi}(\boldsymbol{y};\bar{\lambda}^k,\mu^k)$ can be analyzed.

Therefore, the assumption in **v2–GBD** holds true if separability and linearity hold which covers also the case of linear 0–1 $\boldsymbol{y}$ variables. This way under conditions C1, C2, C3 the **v2–GBD** determined the global solution for separability in $\boldsymbol{x}$ and $\boldsymbol{y}$ and linearity in $\boldsymbol{y}$ problems.

Illustration 6.3.2 This example is taken from Sahinidis and Grossmann (1991) and has three 0–1 variables.

$$\begin{aligned} \min \quad y_1 + y_2 &+ y_3 + 5x^2 \\ s.t. \quad 3x - y_1 - y_2 &\leq 0 \\ -x + 0.1y_2 + 0.25y_3 &\leq 0 \\ y_1 + y_2 + y_3 &\geq 2 \\ y_1 + y_2 + 2(y_3 - 1) &\geq 0 \\ 0.2 \leq x &\leq 1 \\ y_1, y_2, y_3 &= 0,1 \end{aligned}$$

Note that the third and fourth constraint have only 0–1 variables and hence can be moved directly to the relaxed master problem.

Also, note that this example has separability in $\boldsymbol{x}$ and $\boldsymbol{y}$, linearity in $\boldsymbol{y}$, and convexity in $\boldsymbol{x}$ for fixed $\boldsymbol{y}$. Thus, we have

$$\begin{aligned} \xi(\boldsymbol{y};\lambda^k,\mu^k) &= y_1 + y_2 + y_3 + \mu_1^k(-y_1 - y_2) + \mu_2^k(0.1y_2 + 0.25y_3), \\ &\quad + \left[5{x^k}^2 + \mu_1^k(3x^k) + \mu_2^k(-x^k)\right] \\ \bar{\xi}(\boldsymbol{y};\bar{\lambda}^k,\bar{\mu}^k) &= +\mu_1^l(-y_1 - y_2) + \mu_2^l(0.1y_2 + 0.25y_3) + [\bar{\mu}_1^l(3x^l) + \bar{\mu}_2^l(-x^l)]. \end{aligned}$$

Iteration 1

Step 1: Set $(y_1, y_2, y_3) = (1, 1, 1)$
The primal problem becomes

$$\begin{aligned} \min \quad 3 &+ 5x^2 \\ s.t. \quad 3x - 2 &\leq 0 \\ -x + 0.35 &\leq 0 \\ 0.2 \leq x &\leq 1 \end{aligned}$$

and its solution is

$$\begin{aligned} x^1 &= 0.35, \\ \mu_1^1 &= 0, \\ \mu_2^1 &= 3.5, \end{aligned}$$

with objective equal to the $UBD = 3.6125$.

Step 2: The relaxed master problem is

$$\begin{aligned} \min_{y_1, y_2, y_3, \mu_B} \quad & \mu_B \\ s.t. \quad \mu_B &\geq y_1 + y_2 + y_3 + 0\,(-y_1 - y_2)\,3.5\,(0.1y_2 + 0.25y_3) + 0.6125 \\ y_1 + y_2 + y_3 &\geq 2 \\ y_1 + y_2 + 2(y_3 - 1) &\geq 0 \\ y_1, y_2, y_3 &= 0, 1 \end{aligned}$$

which has as solution $y^2 = (1, 1, 0)$ and $\mu_B = LBD = 1.7375$. Since $UBD - LBD = 3.6125 - 1.735 = 1.8775$, we continue with $y = y^2$. Note that $5(0.35)^2 + (0)(30.35) + (3.5)(-0.35) = -0.6125$.

Iteration 2

Step 1: Set $y^2 = (1, 1, 0)$
Solve the primal $P(y^2)$

$$\begin{aligned} \min \quad 2 &+ 5x^2 \\ s.t. \quad 3x - 2 &\leq 0 \\ -x + 0.1 &\leq 0 \\ 0.2 \leq x &\leq 1 \end{aligned}$$

and its solution is

$$\begin{aligned} x^2 &= 0.2, \\ \mu_1^2 &= 0, \\ \mu_2^2 &= 0, \end{aligned}$$

and the objective function is 2.2. The updated upper bound is $UBD = \min\,(3.6125, 2.2) = 2.2$.

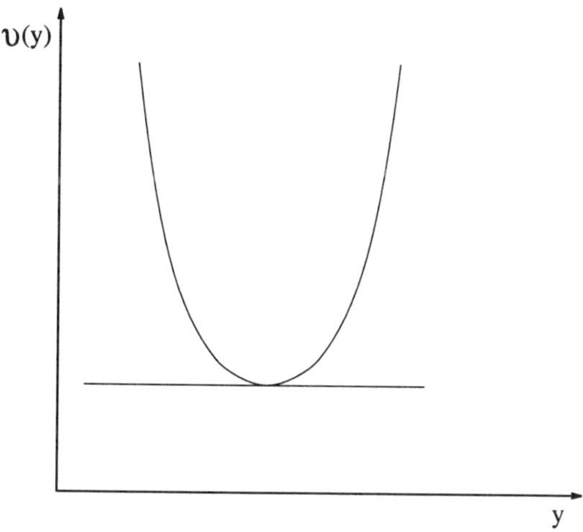

Figure 6.3: Termination of GBD in one iteration

Step 2: The relaxed master problem is:

$$\begin{aligned} & \min \quad \mu_B \\ s.t. \quad \mu_B &\geq y_1 + y_2 + y_3 + 3.5\,(0.1y_2 + 0.25y_3) - 0.6125 \\ \mu_B &\geq y_1 + y_2 + y_3 + 0.2 \\ y_1 + y_2 + y_3 &\geq 2 \\ y_1 + y_2 + 2(y_3 - 1) &\geq 0 \\ y_1, y_2, y_3 &= 0, 1 \end{aligned}$$

which has as solution $(y_1, y_2, y_3) = (1, 1, 0)$ and $\mu_B = LBD = 2.2$. Since $UBD - LBD = 0$, we terminate with $(y_1, y_2, y_3) = (1, 1, 0)$ as the optimal solution.

Remark 4 Note that if we had selected as the starting point the optimal solution; that is $(1, 1, 0)$, then the **v2–GBD** would have terminated in one iteration. This can be explained in terms of Remark 3. Since $v(y)$ is convex, then the optimal point corresponds to the global minimum and the tangent plane to this minimum provides the tightest lower bound which by strong duality equals the upper bound. This is illustrated in Figure 6.3.

6.3.5.3 Variant 3 of GBD, v3–GBD

This variant was proposed by Floudas *et al.* (1989) and denoted as Global Optimum Search **GOS** and was applied to continuous as well as 0–1 set Y. It uses the same assumption as the one in **v2–GBD** but *in addition* assumes that

(i) $f(x,y), g(x,y)$ are convex functions in y for every fixed x, and

(ii) $h(x,y)$ are linear functions in y for every x.

This additional assumption was made so as to create special structure not only in the primal but also in the relaxed master problem. The type of special structure in the relaxed master problem has to do with its convexity characteristics.

The basic idea in **GOS** is to select the $\boldsymbol{x}$ and $\boldsymbol{y}$ variables in a such a way that the primal and the relaxed master problem of the **v2–GBD** satisfy the appropriate convexity requirements and hence attain their respective global solutions.

We will discuss **v3–GBD** first under the separability of $\boldsymbol{x}$ and $\boldsymbol{y}$ and then for the general case.

v3–GBD with separability

Under the separability assumption we have

$$\begin{aligned} f(\boldsymbol{x},\boldsymbol{y}) &= f_1(\boldsymbol{x}) + f_2(\boldsymbol{y}), \\ \boldsymbol{h}(\boldsymbol{x},\boldsymbol{y}) &= \boldsymbol{h}_1(\boldsymbol{x}) + \boldsymbol{h}_2(\boldsymbol{y}), \\ \boldsymbol{g}(\boldsymbol{x},\boldsymbol{y}) &= \boldsymbol{g}_1(\boldsymbol{x}) + \boldsymbol{g}_2(\boldsymbol{y}). \end{aligned}$$

The additional assumption that makes **v3–GBD** different than **v2–GBD** implies that

(i) $f_2(\boldsymbol{y}), \boldsymbol{g}_2(\boldsymbol{y})$ are convex in $\boldsymbol{y}$, and

(ii) $\boldsymbol{h}_2(\boldsymbol{y})$ are linear in $\boldsymbol{y}$.

Then, the relaxed master problem will be

$$\begin{aligned} \min_{\boldsymbol{y},\mu_B} \quad & \mu_B \\ s.t. \quad \mu_B &\geq f_2(\boldsymbol{y}) + \lambda^{k^T} \boldsymbol{h}_2(\boldsymbol{y}) + \mu^{k^T} g_2(\boldsymbol{y}) \\ &\quad + \left[f_1(\boldsymbol{x}^k) + \lambda^{k^T} \boldsymbol{h}_1(\boldsymbol{x}^k) + \mu^{k^T} g_1(\boldsymbol{x}^k) \right], \quad k = 1, 2, \ldots, K \\ 0 &\geq \bar{\lambda}^{l^T} \boldsymbol{h}_2(\boldsymbol{y}) + \bar{\mu}^{l^T} g_2(\boldsymbol{y}) + \left[\bar{\lambda}^{l^T} \boldsymbol{h}_1(\bar{\boldsymbol{x}}^l) + \bar{\mu}^{l^T} g_1(\bar{\boldsymbol{x}}^l) \right], \quad l = 1, 2, \ldots, L \end{aligned}$$

Remark 1 Note that the additional assumption makes the problem convex in $\boldsymbol{y}$ if $\boldsymbol{y}$ represent continuous variables. If $\boldsymbol{y} \in \boldsymbol{Y} = \{0,1\}^q$, and the $\boldsymbol{y}$–variables participate linearly (i.e. f_2, g_2 are linear in $\boldsymbol{y}$), then the relaxed master is convex. Therefore, this case represents an improvement over **v3–GBD**, and application of **v3–GBD** will result in valid support functions, which implies that the global optimum of (6.2) will be obtained.

v3–GBD without separability

The Global Optimum Search **GOS** aimed at exploiting and invoking special structure for nonconvex nonseparable problems of the type (6.2).

$$\min \quad f(\boldsymbol{x},\boldsymbol{y})$$

$$
\begin{aligned}
s.t. \quad h(x,y) &= 0 \\
g(x,y) &\leq 0 \\
x &\in X \subseteq \Re^n \\
y &\in Y \subseteq \Re^q
\end{aligned}
$$

under the conditions C1, C2, C3 and the additional condition:

(i) $f(x,y)$, $g(x,y)$ are convex functions in y for every fixed x,

(ii) $h(x,y)$ are linear functions in y for every x;

so that both the primal and the relaxed problems attain their respective global solutions.

Remark 2 Note that since x and y are not separable, then the **GOS** cannot provide theoretically valid functions in the general case, but only if the $v(y)$ is convex (see **v2–GBD**).

Despite this theoretical limitation, it is instructive to see how, for $Y \subseteq \Re^n$, the convex primal and relaxed master problems are derived. This will be illustrated in the following.

Illustration 6.3.3 This is an example taken from Floudas *et al.* (1989).

$$
\begin{aligned}
\min \quad & -12x_1 - 7x_2 + x_2^2 \\
s.t. \quad & -2x_1^4 - x_2 + 2 = 0 \\
& 0 \leq x_1 \leq 2 \\
& 0 \leq x_2 \leq 3
\end{aligned}
$$

Note that the objective function is convex since it has linear and positive quadratic terms. The only nonlinearities come from the equality constraint. By introducing three new variables w_1, w_2, w_3, and three equalities:

$$
\begin{aligned}
w_1 - x_1 &= 0, \\
w_2 - x_1 w_1 &= 0, \\
w_3 - x_1 w_2 &= 0,
\end{aligned}
$$

we can write an equivalent formulation of

$$
\begin{aligned}
\min \quad & -12x_1 - 7x_2 + x_2^2 \\
s.t. \quad & -2w_3 x_1 - x_2 + 2 = 0 \\
& w_1 - x_1 = 0 \\
& w_2 - x_1 w_1 = 0 \\
& w_3 - x_1 w_2 = 0 \\
& 0 \leq x_1 \leq 2 \\
& 0 \leq x_2 \leq 3
\end{aligned}
$$

Note that if we select as

$$\begin{aligned} y &= x_1, \\ x &= (x_2, w_1, w_2, w_3), \end{aligned}$$

all the imposed convexity conditions are satisfied and hence the primal and the relaxed master problems are convex and attain their respective global solutions.

The primal problem for $y = y^k$ is

$$\begin{aligned} \min \quad & -12y^k - 7x_2 + x_2^2 \\ \text{s.t.} \quad & -2w_3 y^k - x_2 + 2 = 0 \\ & w_1 - y^k = 0 \\ & w_2 - y^k w_1 = 0 \\ & w_3 - y^k w_2 = 0 \\ & 0 \leq x_2 \leq 3 \end{aligned}$$

The relaxed master problem is

$$\begin{aligned} \min_{y,\mu_B} \quad & \mu_B \\ \text{s.t.} \quad & \mu_B \geq L(x^k, y, \lambda^k, \mu^k), \quad k = 1, 2, \ldots, K \\ & 0 \geq \bar{L}(\bar{x}^l, y, \bar{\lambda}^l, \bar{\mu}^l), \quad l = 1, 2, \ldots, L \\ & 0 \leq y \leq 2 \end{aligned}$$

$$\begin{aligned} \text{where} \quad L(x^k, y, \lambda^k, \mu^k) = \quad & -12y - 7x_2^k + (x_2^k)^2 \\ & + \lambda_1^k (-2w_3^k y - x_2^k + 2) \\ & + \lambda_2^k (w_1^k - y) \\ & + \lambda_3^k (w_2^k - y w_1^k) \\ & + \lambda_4^k (w_3^k - y w_2^k); \end{aligned}$$

$$\begin{aligned} L(\bar{x}^l, y, \bar{\lambda}^l, \bar{\mu}^l) = \quad & \bar{\lambda}_1^l (-2\bar{w}_3^l y - \bar{x}_2^l + 2) \\ + \quad & \bar{\lambda}_2^l (\bar{w}_1^l - y) \\ + \quad & \bar{\lambda}_3^l (\bar{w}_2^l - y\bar{w}_1^l) \\ + \quad & \bar{\lambda}_4^l (\bar{w}_3^l - y\bar{w}_2^l). \end{aligned}$$

Remark 3 The primal is convex in x, while the relaxed master is linear in y.

Application of **v3–GBD** from several starting points determines the global solution which is:

$$\begin{aligned} y &= 0.718, \\ x_2 &= 1.47, \\ \text{Objective} &= -16.7389. \end{aligned}$$

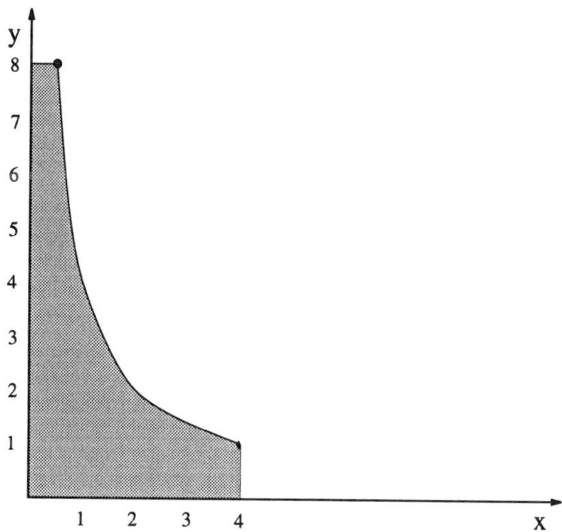

Figure 6.4: Feasible region of illustration 6.3.4

Remark 4 If we substitute x_2 from the equality constraint, then we obtain the equivalent formulation:

$$\begin{array}{ll} \min & -12x_1 - 14(1 - x_1^4) + 4(1 - x_1^4)^2 \\ \text{s.t.} & 0 \leq x_1 \leq 2 \end{array}$$

Eigenvalue analysis on this formulation shows that it is a convex problem. As a result, the projected $v(y)$ problem in our example is convex, and hence the **v3–GBD** converges to the global solution from any point.

Illustration 6.3.4 This example is taken from Floudas and Visweswaran (1990) and served as a motivating example for global optimization.

$$\begin{array}{rl} \min & -x - y \\ \text{s.t.} & xy \leq 4 \\ & 0 \leq x \leq 4 \\ & 0 \leq y \leq 8 \end{array}$$

The feasible region is depicted in Figure 6.4 and is nonconvex due to the bilinear inequality constraint. This problem exhibits a strong local minimum at $(x, y) = (4, 1)$ with objective equal to -5, and a global minimum at $(x, y) = (0.5, 8)$ with objective equal to -8.5.

The projection onto the y space, $v(y)$, is depicted in Figure 6.5.

Note that $v(y)$ is nonconvex and if we select as a starting point in **v3–GBD** $y^1 = 2$, then the algorithm will terminate with this as the solution, which is in fact not even a local solution. This is due to the common assumption of **v2–GBD** and **v3–GBD**.

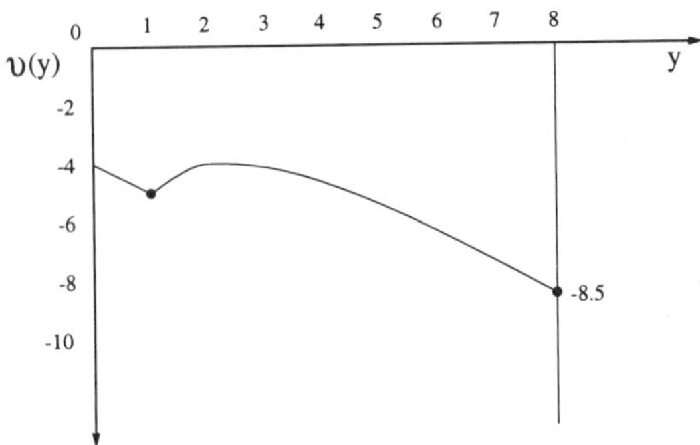

Figure 6.5: Projection onto the y–space for illustration 6.3.4

Remark 5 The global optimization approach (**GOP**) (Floudas and Visweswaran, 1990; Floudas and Visweswaran, 1993) overcomes this fundamental difficulty and guarantees ϵ–global optimality for several classes of nonconvex problems. The area of global optimization has received significant attention during the last decade, and the reader interested in global optimization theories, algorithms, applications, and test problems is directed to the books of Horst and Tuy (1990), Neumaier (1990), Floudas and Pardalos (1990), Floudas and Pardalos (1992), Hansen (1992), Horst and Pardalos (1995), and Floudas and Pardalos (1995).

6.3.6 GBD in Continuous and Discrete–Continuous Optimization

We mentioned in remark 1 of the formulation section (i.e., 6.3.1), that problem (6.2) represents a subclass of the problems for which the Generalized Benders Decomposition **GBD** can be applied. This is because in problem (6.2) we considered the $\boldsymbol{y} \in \boldsymbol{Y}$ set to consist of 0–1 variables, while Geoffrion (1972) proposed an analysis for $\boldsymbol{Y}$ being a continuous, discrete or continuous–discrete set.

The main objective in this section is to present the modifications needed to carry on the analysis presented in sections 6.3.1–6.3.5 for continuous $\boldsymbol{Y}$ and discrete-continuous $\boldsymbol{Y}$ set.

The analysis presented for the primal problem (see section 6.3.3.1) remains the same. The analysis though for the master problem changes only in the dual representation of the projection of problem (6.2) (i.e., $v(\boldsymbol{y})$) on the $\boldsymbol{y}$-space. In fact, theorem 3 is satisfied if in addition to the two conditions mentioned in C3 we have that

(iii) For each fixed $\boldsymbol{y}$, $v(\boldsymbol{y})$ is finite, $\boldsymbol{h}(\boldsymbol{x},\boldsymbol{y})$, $\boldsymbol{g}(\boldsymbol{x},\boldsymbol{y})$, and $f(\boldsymbol{x},\boldsymbol{y})$ are continuous on $\boldsymbol{X}$, $\boldsymbol{X}$ is closed and the ε-optimal solution of the primal problem $P(\boldsymbol{y})$ is nonempty and bounded for some $\varepsilon \geq 0$.

Hence, theorem 3 has as assumptions: C1 and C3, which now has (i), (ii), and (iii). The algorithmic procedure discussed in section 6.3.4.2 remains the same, while the theorem for the finite convergence becomes finite ε- convergence and requires additional conditions, which are described in the following theorem:

Theorem 6.3.6 (Finite ε-convergence)
Let

(i) *Y be a nonempty subset of V,*

(ii) *X be a nonempty convex set,*

(iii) *f, g be convex on X for each fixed $y \in Y$,*

(iv) *h be linear on X for each fixed $y \in Y$,*

(v) *f, g, h be continuous on $X \times Y$,*

(vi) *The set of optimal multiplier vectors for the primal problem be nonempty for all $y \in Y$, and uniformly bounded in some neighborhood of each such point.*

Then, for any given $\epsilon > 0$ the **GBD** *terminates in a finite number of iterations.*

Remark 1 Assumption (i) (i.e., $Y \subseteq V$) eliminates the possibility of step 3b, and there are many applications in which $Y \subseteq V$ holds (e.g., variable factor programming). If, however, $Y \nsubseteq V$, then we may need to solve step 3b infinitely many successive times. In such a case, to preserve finite ϵ–convergence, we can modify the procedure so as to finitely truncate any excessively long sequence of successive executions of step 3b and return to step 3a with $\hat{y}$ equal to the extrapolated limit point which is assumed to belong to $Y \cap V$. If we do not make the assumption $Y \subseteq V$, then the key property to seek is that V has a representation in terms of a finite collection of constraints because if this is the case then step 3b can occur at most a finite number of times. Note that if in addition to C1, we have that X represents bounds on the x–variables or X is given by linear constraints, and h, g satisfy the separability condition, then V can be represented in terms of a finite collection of constraints.

Remark 2 Assumption (vi) requires that for all $y \in Y$ there exist optimal multiplier vectors and that these multiplier vectors do not go to infinity, that is they are uniformly bounded in some neighborhood of each such point. Geoffrion (1972) provided the following condition to check the uniform boundedness:

If X is nonempty, compact, convex set and there exists a point $\bar{x} \in X$ such that

$$h(\bar{x}, \bar{y}) = 0,$$

$$g(\bar{x}, \bar{y}) < 0,$$

then the set of optimal multiplier vectors is uniformly bounded in some open neighborhood of $\bar{y}$.

Illustration 6.3.5 This example is taken from Kocis and Grossmann (1988) and can be stated as

$$\begin{aligned} \min_{x,y} \quad & 2x + y \\ \text{s.t.} \quad 1.25 - x^2 - y &\leq 0 \\ x + y &\leq 1.6 \\ x &\geq 0 \\ y &= 0, 1 \end{aligned}$$

Note that the first constraint is concave is x and hence this is a nonconvex problem. Also note that the 0–1 y variable appears linearly and separably from the x variable.

If we use the **v3–GBD** approach, then we introduce one new variable x_1 and one additional equality constraint

$$x_1 - x = 0,$$

and the problem can be written in the following equivalent form:

$$\begin{aligned}
\min_{x,x_1,y} \quad & 2x + y \\
\text{s.t.} \quad 1.25 - x_1 x - y &\leq 0 \\
x_1 - x &= 0 \\
x + y &\leq 1.6 \\
x,\ x_1 &\geq 0 \\
y &= 0,1
\end{aligned}$$

Careful analysis of the set of constraints that define the bounds reveals the following:

(i) For $y = 0$ we have $x^2 \geq 1.25$, and for $y = 1$ we have $x^2 \geq 0.25$ (see first inequality constraint).

(ii) From the second inequality constraint we similarly have that, for $y = 0$, $x \leq 1.6$, while for $y = 1$, $x \leq 0.6$.

Using the aforementioned observations we have a new lower bound for x (and hence x_1);

$$0.5 \leq x \leq 1.6.$$

The set of complicating variables for this example is defined as

$$\mathrm{y} = (x_1, y),$$

and it is a mixed set of a continuous and 0–1 variable. The primal problem takes the form:

$$\begin{aligned}
\min_{x} \quad & 2x + y^k \\
\text{s.t.} \quad 1.25 - x_1^k x - y^k &\leq 0 \\
x_1^k - x &= 0 \\
0.5 \leq x &\leq 1.6
\end{aligned}$$

The relaxed master problem of the **v3–GBD** approach takes the form (assuming feasible primal):

$$\begin{aligned}
\min \quad & \mu_B \\
\text{s.t.} \quad \mu_B &\geq L(x^k, y, \lambda^k, \mu^k) \\
x_1 + y &\leq 1.6 \\
0.5 \leq x_1 &\leq 1.6 \\
y &= 0,1
\end{aligned}$$

where $L(x^k, y, \lambda^k, \mu^k) = 2x^k + y + \lambda^k(x_1 - x^k) + \mu^k(1.25 - x_1 x^k - y)$.

Note that the constraint $x + y \leq 1.6$ has been written as $x_1 + y \leq 1.6$ and since both x_1 and y are complicating variables it is moved directed to the relaxed master problem.

Note also that in this case the primal problem is a linear programming problem, while the relaxed master problem is a mixed-integer linear programming problem. The **v3–GBD** was applied to this problem from several starting points (see Floudas *et al.* (1989)) and the global solution was obtained in two iterations, even though the theoretical conditions for determining the global solution were not satisfied.

6.4 Outer Approximation, OA

6.4.1 Formulation

Duran and Grossmann (1986a; 1986b) proposed an Outer Approximation **OA** algorithm for the following class of MINLP problems:

$$\begin{aligned} & \min_{x,y} \quad c^T y + f(x) \\ & \quad s.t. \quad g(x) + By \leq 0 \\ x \in X &= \{x : x \in \Re^n, A_1 x \leq a_1\} \subseteq \Re^n \\ y \in Y &= \{y : y \in \{0,1\}^q, A_2 y \leq a_2\} \end{aligned} \tag{6.13}$$

under the following conditions:

C1: X is a nonempty, compact, convex set and the functions

$$f : \Re^n \longrightarrow \Re,$$

$$g : \Re^n \longrightarrow \Re^p,$$

are convex in x.

C2: f and g are once continuously differentiable.

C3: A constraint qualification (e.g., Slater's) holds at the solution of every nonlinear programming problem resulting from (6.13) by fixing y.

Remark 1 Note that formulation (6.13) corresponds to a subclass of problem (6.2) that the **GBD** can address. This is due to the inherent assumptions of

(i) Separability in x and y; and

(ii) Linearity in y.

Also, note that problem (6.13) does not feature any nonlinear equality constraints. Hence, the implicit assumption in the **OA** algorithm is that

(iii) Nonlinear equalities can be eliminated algebraically or numerically.

Remark 2 Under the aforementioned assumptions (i) and (ii), problem (6.13) satisfies property (P) of Geoffrion (1972), and hence the **OA** corresponds to a subclass of **v1–GBD** (see sections 6.3.5.1) Furthermore, as we have seen in section 6.3.5.2, assumptions (i) and (ii) make the assumption imposed in **v2–GBD** valid (see remark of section 6.3.5.2) and therefore the **OA** can be considered as equivalent to **v2–GBD** with separability in x and y and linearity in y. Note though that the **v1–GBD** can handle nonlinear equality constraints.

Remark 3 Note that the set of constraints

$$g(x) + By \leq 0,$$

can be written in the following equivalent form:

$$\begin{aligned} g(x) - Cx' &\leq 0, \\ Cx' + By &\leq 0, \end{aligned}$$

by introducing a new set of variables x', and therefore augmenting x to (x, x'), and a new set of constraints. Now, if we define the x variables as (x, x'), and the first constraints as $G(x) \leq 0$, we have

$$\begin{aligned} G(x) &\leq 0, \\ Cx + By &\leq 0, \end{aligned}$$

where the first set of constraints is nonlinear in the x–type variables, while the second set of constraints in linear in both x and y variables. The penalty that we pay with the above transformation is the introduction of the x' and their associated constraints.

6.4.2 Basic Idea

The basic idea in **OA** is similar to the one in **GBD** that is, at each iteration we generate an upper bound and a lower bound on the MINLP solution. The upper bound results from the solution of the problem which is problem (6.13) with fixed y variables (e.g., $y = y^k$). The lower bound results from the solution of the master problem. The master problem is derived using primal information which consists of the solution point x^k of the primal and is based upon an outer approximation (linearization) of the nonlinear objective and constraints around the primal solution x^k. The solution of the master problem, in addition to the lower bound, provides information on the next set of fixed y variables (i.e., $y = y^{k+1}$) to be used in the next primal problem. As the iterations proceed, two sequences of updated upper bounds and lower bounds are generated which are shown to be nonincreasing and nondecreasing respectively. Then, it is shown that these two sequences converge within ϵ in a finite number of iterations.

Remark 1 Note that the distinct feature of **OA** versus **GBD** is that the master problem is formulated based upon primal information and outer linearization.

6.4.3 Theoretical Development

6.4.3.1 The Primal Problem

The primal problem corresponds to fixing the y variables in (6.13) to a 0–1 combination, which is denoted as y^k, and its formulation is

$$\begin{aligned} \min_{x} \quad & c^T y^k + f(x) \\ s.t. \quad & g(x) + By^k \leq 0 \\ & x \in X = \{x : x \in \Re^n, A_1 x \leq a_1\} \end{aligned}$$

Depending on the fixation point y^k, the primal problem can be feasible or infeasible, and these two cases are analyzed in the following:

Case (i): Feasible Primal

If the primal is feasible at iteration k, then its solution provides information on the optimal x^k, $f(x^k)$, and hence the current upper bound $UBD = c^T y^k + f(x^k)$. Using information on x^k, we can subsequently linearize around x^k the convex functions $f(x)$ and $g(x)$ and have the following relationships satisfied:

$$\begin{aligned} f(x) &\geq f(x^k) + \nabla f(x^k)\,(x - x^k), \quad \forall x^k \in X, \\ g(x) &\geq g(x^k) + \nabla g(x^k)\,(x - x^k), \quad \forall x^k \in X, \end{aligned}$$

due to convexity of $f(x)$ and $g(x)$.

Case (ii): Infeasible Primal

If the primal is infeasible at iteration k, then we need to consider the identification of a feasible point by looking at the constraint set:

$$g(x) + By^k \leq 0,$$

and formulating a feasibility problem in a similar way as we did for the **GBD** (see case (ii) of section 6.3.3.1).

For instance, if we make use of the l_1–minimization, we have

$$\begin{aligned} \min_{x \in X} \quad & \sum_{j=1}^{p} a_j \\ s.t. \quad g_j(x) + By^k &\leq a_j, \quad j = 1, 2, \ldots, p \\ a_j &\geq 0 \end{aligned}$$

Its solution will provide the corresponding x^l point based upon we can linearize the constraints:

$$g(x) \geq g(x^l) + \nabla g(x^l)\,(x - x^l), \quad \forall\, x^l,$$

where the right-hand side is a valid linear support.

6.4.3.2 The Master Problem

The derivation of the master problem in the **OA** approach involves the following two key ideas:

(i) Projection of (6.13) onto the y–space; and

(ii) Outer approximation of the objective function and the feasible region.

(i) Projection of (6.13) onto the y–space

Problem (6.13) can be written as

$$\begin{aligned}\min_{y} \quad & \inf_{x} c^T y + f(x) \\ s.t. \quad & g(x) + By \le 0 \\ & x \in X \\ & y \in Y\end{aligned} \tag{6.14}$$

Note that the inner problem is written as infimum with respect to x to cover the case of having unbounded solution for a fixed y. Note also that $c^T y$ can be taken outside of the infimum since it is independent of x.

Let us define $v(y)$:

$$\begin{aligned}v(y) \;=\; & c^T y + \inf_{x} f(x) \\ s.t. \quad & g(x) + By \le 0 \\ & bx \in X\end{aligned} \tag{6.15}$$

Remark 1 $v(y)$ is parametric in the y–variables, and it corresponds to the optimal value of problem (6.13) for fixed y (i.e., the primal problem $P(y^k)$).

Let us also define the set $V$of y's for which exist feasible solutions in the x variables, as

$$V \;=\; \{y : g(x) + By \le 0, \;\; \text{for some } \; x \in X\}, \tag{6.16}$$

Then, problem (6.13) can be written as

$$\begin{aligned}\min_{y} \quad & v(y) \\ s.t. \quad & y \in Y \cap V\end{aligned} \tag{6.17}$$

Remark 2 Problem (6.17) is the projection of (6.13) onto the y–space. The projection needs to satisfy feasibility requirements, and this is represented in (6.17) by imposing $y \in Y \cap V$.

Remark 3 Note that we can replace the infimum with respect to $x \in X$ with the minimum with respect to $x \in X$, since for $y \in Y \cap V$ existence of solution x holds true due to the compactness assumption of X. This excludes the possibility for unbounded solution of the inner problem for fixed $y \in Y \cap V$.

Remark 4 The difficulty with solving (6.17) arises because both V and $v(y)$ are known implicitly. To overcome this difficulty, Duran and Grossmann (1986a) considered outer linearization of $v(y)$ and a particular representation of V.

(ii) Outer Approximation of $v(y)$

The outer approximation of $v(y)$ will be in terms of the intersection of an infinite set of supporting functions. These supporting functions correspond to linearizations of $f(x)$ and $g(x)$ at all $x^k \in X$. Then, the following conditions are satisfied:

$$\begin{aligned} f(x) &\geq f(x^k) + \nabla f(x^l)\,(x - x^k), \quad \forall x^k \in X, \\ g(x) &\geq g(x^k) + \nabla g(x^l)\,(x - x^k), \quad \forall x^k \in X, \end{aligned}$$

due to the assumption of convexity and once continuous differentiability. $\nabla f(x^k)$ represents the n–gradient vector of the $f(x)$ and $\nabla g(x^k)$ is the $(n \times p)$ Jacobian matrix evaluated at $x^k \in X$.

Remark 5 Note that the support functions are linear in x, and as a result $v(y)$ will be a mixed integer linear programming MILP problem.

The constraint qualification assumption, which holds at the solution of every primal problem for fixed $y \in Y \cap V$, coupled with the convexity of $f(x)$ and $g(x)$, imply the following Lemma:

Lemma 6.4.1

$$v(y) = \begin{bmatrix} \min\limits_{x} c^T y + f(x) \\ \text{s.t.} \quad g(x) + By \leq 0 \\ x \in X \end{bmatrix}$$

$$= \begin{bmatrix} \min\limits_{x} & c^T y + f(x^k) + \nabla f(x^k)\,(x - x^k) \\ \text{s.t.} & 0 \geq g(x^k) + \nabla g(x^k)\,(x - x^k) + By \\ & x \in X \end{bmatrix} \quad \forall x^k \in X.$$

Remark 6 It suffices to include those linearizations of the constraints that are active at (x^k, y^k). This implies that fewer constraints are needed in the master problem.

Then, by substituting $v(y)$ of the above Lemma to the projection problem (6.17) we have

$$\begin{aligned} \min_{x} \min_{y} \quad & c^T y + f(x^k) + \nabla f(x^k)\,(x - x^k) \\ \text{s.t.} \quad & 0 \geq g(x^k) + \nabla g(x^k)\,(x - x^k) + By \\ & x \in X \\ & y \in Y \cap V \end{aligned} \tag{6.18}$$

By combining the min operators and introducing a scalar μ_{OA}, problem (6.18) can be written in the following equivalent form:

$$
\begin{aligned}
\min_{x,y,\mu_{OA}} \quad & c^T y + \mu_{OA} && (6.19) \\
\text{s.t.} \quad & \left. \begin{aligned} \mu_{OA} &\geq f(x^k) + \nabla f(x^k)(x - x^k), \forall k \in \mathbf{F} \\ 0 &\geq g(x^k) + \nabla g(x^k)(x - x^k) + By \end{aligned} \right\}, \forall k \in \mathbf{F} \\
& x \in X \\
& y \in Y \cap V
\end{aligned}
$$

where $\mathbf{F} = \{k : x^k \text{ is a feasible solution to the primal } P(y^k)\}$.

Duran and Grossmann (1986a) made the additional assumption that we can replace $y \in Y \cap V$ with $y \in Y$ using as argument that a representation of the constraints $y \in Y \cap V$ is included in the linearizations of problem (6.19) provided that the appropriate integer cuts that exclude the possibility of generation of the same integer combinations are introduced. Subsequently, they defined the master problem of the **OA** as

$$
\begin{aligned}
\min_{x,y,\mu_{OA}} \quad & c^T y + \mu_{OA} && (6.20) \\
\text{s.t.} \quad & \left. \begin{aligned} \mu_{OA} &\geq f(x^k) + \nabla f(x^k)(x - x^k), \forall k \in \mathbf{F} \\ 0 &\geq g(x^k) + \nabla g(x^k)(x - x^k) + By \end{aligned} \right\}, \forall k \in \mathbf{F} \\
& x \in X \\
& y \in Y \\
& \sum_{i \in \mathbf{B}^k} y_i^k - \sum_{i \in \mathbf{NB}^k} y_i^k \leq |\mathbf{B}^k| - 1, \quad k \in \mathbf{F}
\end{aligned}
$$

Remark 7 Note that the master problem (6.20) is a mixed-integer linear programming MILP problem since it has linear objective and constraints, continuous variables (x, μ_{OA}) and 0-1 variables (y). Hence, it can be solved with standard branch and bound algorithms.

Remark 8 The master problem consists of valid linear supports, and hence relaxations of the nonlinear functions, for all points x^k that result from fixing $y = y^k \in Y$ as stated by (6.20). As a result it represents a relaxation of the original MINLP model (6.13), and hence it is a lower bound on its solution, and it is identical to its solution if all supports are included.

Remark 9 It is not efficient to solve the master problem (6.20) directly, since we need to know all feasible x^k points which implies that we have to solve all the primal problems $P(y^k)$, $y \in Y$ (i.e., exhaustive enumeration of 0–1 alternatives). Instead, Duran and Grossmann (1986a) proposed a relaxation of the master problem which will be discussed in the next section.

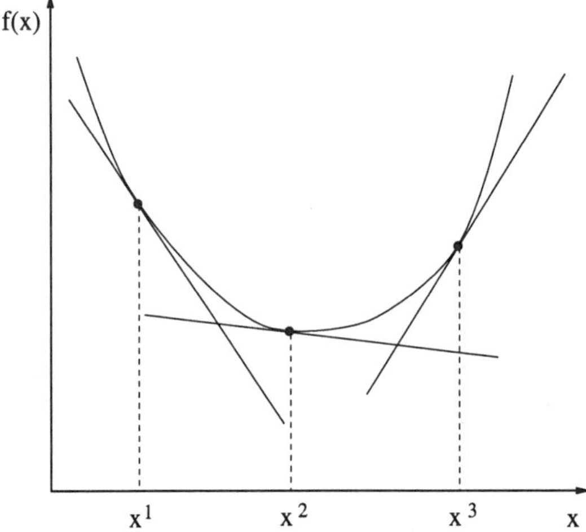

Figure 6.6: Linear support functions.

Remark 10 (Geometrical Interpretation of Master in OA) The master problem of the **OA** can be interpreted geometrically by examining the effect of the linear support function (i.e., outer linearizations) on the objective function and the constraints. Figure 6.6 shows the linear supports of the objective function $f(\boldsymbol{x})$ taken at $\boldsymbol{x}^1, \boldsymbol{x}^2$, and $\boldsymbol{x}^3$.

Note that the linear supports are the tangents of $f(\boldsymbol{x})$ at $\boldsymbol{x}^1, \boldsymbol{x}^2$, and $\boldsymbol{x}^3$ and that they underestimate the objective function. Also, note that accumulation of these linear underestimating supports results in a better approximation of the objective function $f(\boldsymbol{x})$ from the outside (i.e., outer approximation). Notice also that the linear underestimators are valid because $f(\boldsymbol{x})$ is convex in $\boldsymbol{x}$.

Figure 6.7 shows the outer approximation of the feasible region consisting of two linear (i.e., g_3, g_4) and two nonlinear (g_1, g_2) inequalities at a point $\boldsymbol{x}^1$.

u_{11} is the linear support of g_1, while u_{12} is the linear support of g_2, and they both result from linearizing g_1 and g_2 around the point $\boldsymbol{x}^1$. Note that the feasible region defined by the linear supports and g_3, g_4 includes the original feasible region, and hence the relaxation of the nonlinear constraints by taking their linearization corresponds to an overestimation of the feasible region.

Note also that the linear supports provide valid overestimators of the feasible region only because the functions $\boldsymbol{g}(\boldsymbol{x})$ are convex.

With the above in mind, then the master problem of the **OA** can be interpreted geometrically as a relaxation of the original MINLP (in the limit of taking all $\boldsymbol{x}^k$ points it is equivalent) defined as

(i) underestimating the objective function; and

(ii) overestimating the feasible region.

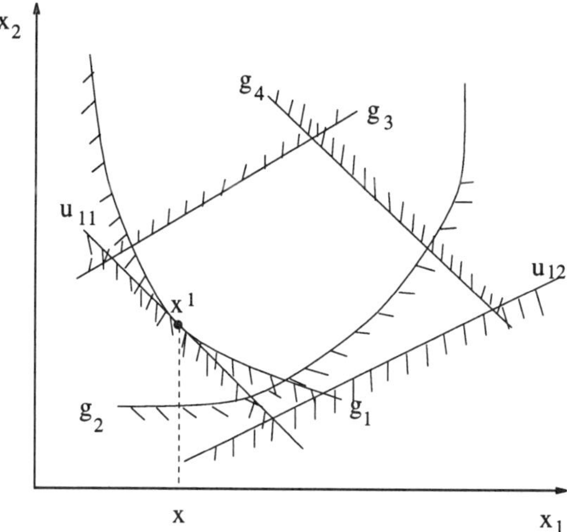

Figure 6.7: Geometric interpretation of outer approximation OA

As such, its solution provides a lower bound on the solution of the original MINLP problem.

Remark 11 Note that at every iteration k we need to add in the master problem of **OA** the linearizations of the active constraints. This implies that the master problem may involve a large number of constraints as the iterations proceed. For instance, if we have $Y = \{0,1\}^q$ and (6.13) has m inequalities, then the master problem should have

$$2^q\,(m+1)$$

constraints so as to be exactly equivalent to (6.13). This provides the motivation for solving the master problem via a relaxation strategy.

6.4.4 Algorithmic Development

The central point in the algorithmic development of **OA** is the solution of the master problem since it is quite straightforward to solve the primal problem which is a convex nonlinear problem.

The natural approach of solving the master problem is relaxation; that is, consider at each iteration the linear supports of the objective and constraints around all previously linearization points. This way, at each iteration a new set of linear support constraints are added which improve the relaxation and therefore the lower bound.

6.4.4.1 Algorithmic Statement of OA

Step 1: Let an initial point $y^1 \in Y$ or $Y \cap V$ if available. Solve the resulting primal problem $P(y^1)$ and obtain an optimal solution x^1. Set the iteration counter $k = 1$. Set the current upper bound $UBD = P(y^1) = v(y^1)$.

Step 2: Solve the relaxed master (**RM**) problem

$$
(\mathbf{RM}) \quad \left[\begin{array}{lll}
\min\limits_{x,y,\mu_{OA}} & c^T y + \mu_{OA} & \\
\text{s.t.} & \left. \begin{array}{rcl} \mu_{OA} & \geq & f(x^k) + \nabla f(x^k)(x - x^k) \\ 0 & \geq & g(x^k) + \nabla g(x^k)(x - x^k) + By \end{array} \right\}, & \forall k \in \mathbf{F} \\
& x \in X & \\
& y \in Y & \\
& \sum\limits_{i \in \mathbf{B}^k} y_i^k - \sum\limits_{i \in \mathbf{NB}^k} y_i^k \leq |\mathbf{B}^k| - 1, & k = 1, 2, \ldots, K - 1
\end{array} \right.
$$

Let (y^{k+1}, μ_{OA}^k) be the optimal solution of the relaxed master problem, where $\mu_{OA}^k + c^T g^{k+1}$ is the new current lower bound on (6.13), $LBD = \mu_{OA}^k + c^T y^{k+1}$ and y^{k+1} is the next point to be considered in the primal problem $P(y^{k+1})$. If $UBD - LBD \leq \epsilon$, then terminate. Otherwise, return to step 1.

If the master problem does not have a feasible solution, then terminate. The optimal solution is given by the current upper and its associated optimal vectors (x, y).

Remark 1 Note that in step 1 instead of selecting a $y^1 \in Y$ or $Y \cap V$ we can solve the continuous relaxation of (6.13) (i.e., treat $0 \leq y \leq 1$) and set the y–variables to their closest integer value. This way, it is still possible that the resulting primal is infeasible which implies that, according to the **OA** algorithm proposed by Duran and Grossmann (1986a), we eliminate this infeasible combination in the relaxed master problem of the first iteration and continue until another y is found that belong to $Y \cap V$. It is clear though at this point that in the relaxed master no information is provided that can be used as linearization point for such a case, apart from the integer cut.

As we have discussed in the outer approximation of $v(y)$, Duran and Grossmann (1986a) made the assumption that $y \in Y$ instead of $y \in Y \cap V$ and introduced the integer cut constraints which they claimed make their assumption valid. Fletcher and Leyffer (1994), however, presented a counter example which can be stated as follows:

$$
\begin{array}{rrcl}
 & \min\limits_{x,\, y} & & -2y - x \\
s.t. & x^2 + y & \leq & 0 \\
 & y & \in & \{-1, 1\}
\end{array}
$$

which has as solution $(x^*, y^*) = (1, -1)$, and objective equal to 1. Let us start at $y^1 = -1$, for which $x^1 = 2$. The master problem according to Duran and Grossmann (1986a) can be written as

$$
\begin{array}{rrcl}
 & \min & & -2y + \mu_{OA} \\
\text{s.t.} & \mu_{OA} & \geq & -x \\
 & 0 & \geq & 1 + 2(x - 1) + y \\
 & y & \in & \{-1, 1\} \cap \{y = -1\}
\end{array}
$$

which has as solution

$$
(y^2, -2y^2 + \mu_{OA}^1) = (1, -2).
$$

Continuing to the next primal $P(y^2)$ we find that

$$\begin{array}{ll} \min\limits_{x} & -2 - x \\ \text{s.t.} & x^2 + 1 \leq 0 \end{array}$$

is infeasible.

Therefore, the integer cut did not succeed in eliminating infeasible combinations.

This example shows clearly that information from infeasible primal problems needs to be included in the master problems and that a representation of the constraint

$$y \in Y \cap V$$

in terms of linear supports has to be incorporated. In other words, we would like to ensure that integer combinations which produce infeasible primal problems are also infeasible in the master problem, and hence they are not generated. Fletcher and Leyffer (1994) showed that for the case of infeasible primal problems the following constraints, which are the linear supports for this case,

$$0 \geq g(x^l) + \nabla g(x^l)(x - x^l) + By, \quad l \in \bar{\mathbf{F}},$$

$$\text{where} \quad \bar{\mathbf{F}} = \{j : P(y^l) \text{ is infeasible and } x^l \text{ solves the feasibility problem}\}$$

need to be incorporated in the master problem. Then, the master problem of **OA** takes the form:

$$\left[\begin{array}{ll} \min\limits_{x,y,\mu_{OA}} & c^T y + \mu_{OA} \\ \text{s.t.} & \left.\begin{array}{l} \mu_{OA} \geq f(x^k) + \nabla f(x^k)(x - x^k), \forall k \in \mathbf{F} \\ \quad 0 \geq g(x^k) + \nabla g(x^k)(x - x^k) + By \end{array}\right\}, \forall k \in \mathbf{F} \\ & 0 \geq g(x^l) + \nabla g(x^l)(x - x^l) + By \Big\}, \forall l \in \bar{\mathbf{F}} \\ & x \in X \\ & y \in Y \end{array}\right. \tag{6.21}$$

Appropriate changes need also to be made in the relaxed master problem described in the algorithmic procedure section. Note also that the correct formulation of the master problem via (6.21) increases the number of constraints to be included significantly.

6.4.4.2 Finite Convergence of OA

For problem (6.13), Duran and Grossmann (1986a) proved finite convergence of the **OA** stated in 6.4.4.1 which is as follows:

> *" If conditions C1, C2, C3 hold and Y is a discrete set, then the* **OA** *terminates in a finite number of iterations."*

Duran and Grossmann (1986a) also proved that a comparison of **OA** with **v2–GBD** yields always:

$$LBD_{\mathbf{OA}} \geq LBD_{\mathbf{v2\text{–}GBD}}$$

where $LBD_{\mathbf{OA}}, LBD_{\mathbf{v2\text{–}GBD}}$ are the lower bounds at corresponding iterations of the **OA** and **v2–GBD**.

Remark 1 The aforementioned property of the bounds is quite important because it implies that the **OA** for convex problems will terminate in fewer iterations that the **v2–GBD**. Note though that this does not necessarily imply that the **OA** will terminate faster, since the relaxed master of **OA** has many constraints as the iterations proceed, while the relaxed master of **v2–GBD** adds only one constraint per iteration.

6.5 Outer Approximation with Equality Relaxation, OA/ER

6.5.1 Formulation

To handle explicitly nonlinear equality constraints of the form $h(x) = 0$, Kocis and Grossmann (1987) proposed the outer approximation with equality relaxation **OA/ER** algorithm for the following class of MINLP problems:

$$
\begin{aligned}
\min_{x,y} \quad & c^T y + f(x) \\
s.t. \quad & h(x) = 0 \\
& g(x) \leq 0 \\
& Cx + By \leq 0 \\
& x \in X = \{x : x \in \Re^n, A_1 x \leq a_1\} \subseteq \Re^n \\
& y \in Y = \{y : y \in \{0,1\}^q, A_2 y \leq a_2\}
\end{aligned}
\tag{6.22}
$$

under the following conditions:

C1: X is a nonempty, compact, convex set, and the functions satisfy the conditions:

$f(x)$ is convex in x,
$g_i(x)\ i \in I_{IN} = \{i : g_i(x) < 0\}$ are convex in x,
$g_i(x)\ i \in I_{EQ} = \{i : g_i(x) = 0\}$ are quasi–convex in x, and
$Th(x)$ are quasi–convex in x,

where T is a diagonal matrix $(m \times m)$ with elements t_{ii}

$$
t_{ii} = \left\{ \begin{array}{rl} -1 & \text{if} \ \ \lambda_i < 0 \\ +1 & \text{if} \ \ \lambda_i > 0, \\ 0 & \text{if} \ \ \lambda_i = 0 \end{array} \right\} \qquad i = 1, 2, \ldots, m
$$

and λ_i are the Lagrange multipliers associated with the m equality constraints.

C2: f, h, and g are continuously differentiable.

C3: A constraint qualification holds at the solution of every nonlinear programming problem resulting from (6.21) by fixing y.

Remark 1 The nonlinear equalities $h(x) = 0$ and the set of linear equalities which are included in $h(x) = 0$, correspond to mass and energy balances and design equations for chemical process systems, and they can be large. Since the nonlinear equality constraints cannot be treated explicitly by the **OA** algorithm, some of the possible alternatives would be to perform:

(i) Algebraic elimination of the nonlinear equalities;

(ii) Numerical elimination of the nonlinear equalities; and

(iii) Relaxation of the nonlinear equalities to inequalities.

Alternative (i) can be applied successfully to certain classes of problems (e.g., design of batch processes Vaselenak *et al.* (1987); synthesis of gas pipelines (e.g., Duran and Grossmann (1986a)). However, if the number of nonlinear equality constraints is large , then the use of algebraic elimination is not a practical alternative.

Alternative (ii) involves the numerical elimination of the nonlinear equality constraints at each iteration of the **OA** algorithm through their linearizations. Note though that these linearizations may cause computational difficulties since they may result in singularities depending on the selection of decision variables. In addition to the potential problem of singularities, the numerical elimination of the nonlinear equality constraints may result in an increase of the nonzero elements, and hence loss of sparsity, as shown by Kocis and Grossmann (1987).

The aforementioned limitations of alternatives (i) and (ii) motivated the investigation of alternative (iii) which forms the basis for the **OA/ER** algorithm.

Remark 2 Note that condition C1 of the **OA/ER** involves additional conditions of quasi–convexity in $\boldsymbol{x}$ for

$$g_i(\boldsymbol{x}), \quad i \in \boldsymbol{I}_{EQ}, \text{ and } \mathrm{T}\boldsymbol{h}(\boldsymbol{x}).$$

Also note that the diagonal matrix T defines the direction for the relaxation of the equalities into inequalities, and it is expected that such a relaxation can only be valid under certain conditions.

6.5.2 Basic Idea

The basic idea in **OA/ER** is to relax the nonlinear equality constraints into inequalities and subsequently apply the **OA** algorithm. The relaxation of the nonlinear equalities is based upon the sign of the Lagrange multipliers associated with them when the primal (problem (6.21) with fixed $\boldsymbol{y}$) is solved. If a multiplier λ_i is positive then the corresponding nonlinear equality $h_i(\boldsymbol{x}) = 0$ is relaxed as $h_i(\boldsymbol{x}) \leq 0$. If a multiplier λ_i is negative, then the nonlinear equality is relaxed as $-h_i(\boldsymbol{x}) \leq 0$. If, however, $\lambda_i = 0$, then the associated nonlinear equality constraint is written as $0 \cdot h_i(\boldsymbol{x}) = 0$, which implies that we can eliminate from consideration this constraint. Having transformed the nonlinear equalities into inequalities, in the sequel we formulate the master problem based on the principles of the **OA** approach discussed in section 6.4.

6.5.3 Theoretical Development

Since the only difference between the **OA** and the **OA/ER** lies on the relaxation of the nonlinear equality constraints into inequalities, we will present in this section the key result of the relaxation and the master problem of the **OA/ER**.

Property 6.5.1
If conditions C1, C3 are satisfied, then for fixed $\boldsymbol{y} = \boldsymbol{y}^k$ *problem (6.22) is equivalent to (6.23):*

$$\min_{\boldsymbol{x}} \quad c^T\boldsymbol{y}^k + f(\boldsymbol{x}) \tag{6.23}$$

$$\begin{aligned} \text{s.t.} \quad & T^k h(x) \leq 0 \\ & g(x) \leq 0 \\ & Cx + By^k \leq \mathbf{d} \\ & x \in X = \{x : x \in \Re^n, A_1 x \leq a_1\} \subseteq \Re^n \end{aligned}$$

where T *is a diagonal matrix* $(m \times m)$ *with elements* t_{ii}^k *defined as*

$$t_{ii}^k = \left\{ \begin{array}{lll} -1 & if & \lambda_i^k < 0 \\ +1 & if & \lambda_i^k > 0, \\ 0 & if & \lambda_i^k = 0 \end{array} \right\} \quad i = 1, 2, \ldots, m$$

Remark 1 Under the aforementioned conditions, the primal problems of (6.22) and (6.23) have unique local solutions which are in fact their respective global solutions since the **KKT** conditions are both necessary and sufficient.

Remark 2 If condition C1 is not satisfied, then unique solutions are not theoretically guaranteed for the NLP primal problems, and as a result the resulting master problems may not provide valid lower bounds. This is due to potential failure of maintaining the equivalence of (6.22) and (6.23). It should be noted that the equivalence between the primal problems of (6.22) and (6.23) needs to be maintained at every iteration. This certainly occurs when the Lagrange multipliers are invariant in sign throughout the iterations k and they satisfy condition C1 at the first iteration.

Remark 3 After the relaxation of the nonlinear equalities we deal with an augmented set of inequalities:

$$\begin{aligned} T^k h(x) &\leq 0, \\ g(x) &\leq 0, \end{aligned}$$

for which the principles of **OA** described in section 6.4 can be applied.

Illustration 6.5.1 To illustrate the relaxation of equalities into inequalities we consider the following simple example taken from Kocis and Grossmann (1987):

$$\begin{aligned} \min \quad & -y + 2x_1 + x_2 \\ \text{s.t.} \quad & x_1 - 2e^{-x_2} = 0 \\ & -x_1 + x_2 + y \leq 0 \\ & 0.5 \leq x_1 \leq 1.4 \\ & y = 0, 1 \end{aligned} \tag{6.24}$$

By fixing $y^1 = 0$ and solving the resulting primal problem, which is nonlinear, we have

$$OBJ = 2.558, \tag{6.25}$$

$$(x_1, x_2) = (0.853, 0.853), \tag{6.26}$$

$$\lambda = -1.619 < 0. \tag{6.27}$$

We have only one equality constraint, and since its Lagrange multiplier λ is negative, then the T^1 matrix (1×1) is

$$t_{11} = -1.$$

Then, the equality is relaxed as:

$$\begin{aligned} T^1 h(x) &\leq 0, \\ -(x_1 - 2\exp(-x_2)) &\leq 0, \\ 2\exp(-x_2) - x_1 &\leq 0. \end{aligned}$$

Remark 4 Note that the relaxed equality

$$2\exp(-x_2) - x_1 \leq 0,$$

is convex in x_1 and x_2, and therefore condition C1 is satisfied.

Remark 5 If instead of the above presented relaxation procedure we had selected the alternative of algebraic elimination of x_1 from the equality constraint, then the resulting MINLP is

$$\begin{aligned} \min \quad & -y + 4\exp(-x_2) + x_2 \\ s.t. \quad & -2\exp -x_2 + x_2 + y \leq 0 \\ & 0.357 \leq x_2 \leq 1.386 \\ & y = 0, 1 \end{aligned} \tag{6.28}$$

By selecting $y = 0$ as in our illustration, we notice that the inequality constraint in (6.28) is in fact nonconvex in x_2. Hence, application application of **OA** to (6.28) cannot guarantee global optimality due to having nonconvexity in x_2.

This simple example clearly demonstrates the potential drawback of the algebraic elimination alternative.

Remark 6 If we had selected x_2 to be algebraically eliminated using the equality constraint, then the resulting MINLP would be

$$\begin{aligned} \min \quad & -y + 2x_1 - 2\ln x_1 + \ln x_2 \\ s.t. \quad & -x_1 - \ln x_1 + \ln 2 + y \leq 0 \\ & 0.5 \leq x_1 \leq 1.4 \\ & y = 0, 1 \end{aligned}$$

Note that in this case the inequality constraint is convex in x_1 and so is the objective function. Hence, application of **OA** can guarantee its global solution.

As a final comment on this example, we should note that when algebraic elimination is to be applied, care must be taken so as to maintain the convexity and in certain cases sparsity characteristics if possible.

6.5.3.1 The Master Problem

The master problem of the **OA/ER** algorithm is essentially the same as problem (6.20) described in section 6.4.3.2, with the difference being that the vector of inequality constraints will be augmented by the addition of the relaxed equalities:

$$T^k h(x) \leq 0.$$

The general form of the relaxed master problem for the **OA/ER** algorithm is

$$\left[\begin{array}{ll}
\min\limits_{x,y,\mu} & c^T y + \mu_{OA} \\
\text{s.t.} & \left.\begin{array}{lll} \mu & \geq & f(x^k) + \nabla f(x^k)\,(x - x^k) \\ 0 & \geq & g(x^k) + \nabla g(x^k)\,(x - x^k) \\ 0 & \geq & T^k \left[h(x^k) + \nabla h(x^k)^T \left(x - x^k\right)\right] \end{array}\right\}, \ \forall k = 1,2,\ldots,K \\
& Cx + By \leq \mathbf{d} \\
& x \in X = \{x : x \in \Re^n, A_1 x \leq a_1\} \subseteq R^n \\
& y \in Y = \{y : y \in \{0,1\}^q, A_2 y \leq a_2\} \\
& \sum\limits_{i \in \mathbf{B}^k} y_i^k - \sum\limits_{i \in \mathbf{NB}^k} y_i^k \leq |B^k| - 1, k = 1,2,\ldots,K \\
& Z_L^{k-1} \leq c^T y + \mu \leq Z_U
\end{array}\right. \tag{6.29}$$

where Z_L^{k-1} is the lower bound at iteration $k-1$, and Z_U is the current upper bound and are used to expedite the solution of (6.29) and to have infeasibility as the termination criterion.

Remark 1 The right-hand sides of the first three sets of constraints in (6.29) can be written in the following form:

$$\begin{array}{rcl}
\nabla f(x^k)^T x - [\nabla f(x^k)^T x^k - f(x^k)] & = & (w^k)^T x - w_o^k \\
\nabla g(x^k)^T x - [\nabla g(x^k)^T x^k - g(x^k)] & = & (S^k) x - s^k \\
\nabla h(x^k)^T x - [\nabla h(x^k)^T x^k] & = & (R^k) x - r^k
\end{array}$$

$$\begin{array}{lrclcrcl}
\text{where} & w^k & = & \nabla f(x^k) & , & w_o^k & = & \nabla f(x^k)^T x^k - f(x^k), \\
& S^k & = & \nabla g(x^k) & , & s^k & = & \nabla g(x^k)^T x^k - g(x^k), \\
& R^k & = & \nabla h(x^k) & , & r^k & = & \nabla h(x^k)^T x^k.
\end{array}$$

Note that $h(x^k) = 0$.

It should be noted that $w^k, w_o^k, S^k, s^k, R^k, r^k$ can be calculated using the above expressions immediately after the primal problem solution since already information is made available at the solution of the primal problem.

Remark 2 The right-hand sides of the first three sets of constraints are the support functions that are represented as outer approximations (or linearizations) at the current solution point $\boldsymbol{x}^k$ of the primal problem. If condition C1 is satisfied then these supports are valid underestimators and as a result the relaxed master problem provides a valid lower bound on the global solution of the MINLP problem.

Remark 3 If linear equality constraints in $\boldsymbol{x}$ exist in the MINLP formulation, then these are treated as a subset of the $\boldsymbol{h}(\boldsymbol{x}) = 0$ with the difference that we do not need to compute their corresponding T matrix but simply incorporate them as linear equality constraints in the relaxed master problem directly.

Remark 4 The relaxed master problem is a mixed integer linear programming MILP problem which can be solved for its global solution with standard branch and bound codes. Note also that if $f(\boldsymbol{x}), \boldsymbol{h}(\boldsymbol{x}), \boldsymbol{g}(\boldsymbol{x})$ are linear in $\boldsymbol{x}$, then we have an MILP problem. As a result, since the relaxed master is also an MILP problem, the **OA/ER** should terminate in 2 iterations.

6.5.4 Algorithmic Development

The **OA/ER** algorithm can be stated as follows:

Step 1: Let an initial point $\boldsymbol{y}^1 \in \boldsymbol{Y}$, or $\boldsymbol{y}^1 \in \boldsymbol{Y} \cap \boldsymbol{V}$ if available. Solve the resulting primal problem $P(\boldsymbol{y}^1)$ and obtain an optimal solution $\boldsymbol{x}^1$ and an optimal multiplier vector λ^1 for the equality constraints $\boldsymbol{h}(\boldsymbol{x}) = \boldsymbol{0}$. Set the current upper bound $UBD = P(\boldsymbol{y}^1) = v(\boldsymbol{y}^1)$.

Step 2: Define the $(m \times m)$ matrix T^k. Calculate $w^k, w_o^k, s^k, S^k, R^k, r^k$.

Step 3: Solve the relaxed master problem **(RM)**:

$$
\begin{aligned}
Z_L^K = \min_{\boldsymbol{x},\boldsymbol{y},\mu} \quad & \boldsymbol{c}^T\boldsymbol{y} + \mu && \textbf{(RM)} \\
\text{s.t.} \quad & \left.\begin{array}{l} \mu \geq (w^k)^T\boldsymbol{x} - w_o^k \\ 0 \geq (S^k)^T\boldsymbol{x} - s^k \\ 0 \geq T^k R^k \boldsymbol{x} - T^k r^k \end{array}\right\}, k = 1, 2, \ldots, K \\
& C\boldsymbol{x} + B\boldsymbol{y} \leq \mathbf{d} \\
& \boldsymbol{x} \in \boldsymbol{X} = \{\boldsymbol{x} : \boldsymbol{x} \in \Re^n, A_1\boldsymbol{x} \leq a_1\} \subseteq R^n \\
& \boldsymbol{y} \in \boldsymbol{Y} = \{\boldsymbol{y} : \boldsymbol{y} \in \{0,1\}^q, A_2\boldsymbol{y} \leq a_2\} \\
& \sum_{i \in \mathbf{B}^k} y_i^k - \sum_{i \in \mathbf{NB}^k} y_i^k \leq |B^k| - 1, \quad k = 1, 2, \ldots, K - 1 \\
& Z_L^{k-1} \leq \boldsymbol{c}^T\boldsymbol{y} + \mu \leq Z_U
\end{aligned}
$$

If the relaxed master problem is feasible then, let $(\boldsymbol{y}^{k+1}, \mu^k)$ be the optimal solution, where $Z_L^k = \boldsymbol{c}^T\boldsymbol{y}^{k+1} + \mu^K$ is the new current lower bound on (6.22), and $\boldsymbol{y}^{k+1}$ the next point to be considered in the primal problem $P(\boldsymbol{y}^{k+1})$.

If $UBD - Z_L^k \leq \epsilon$, then terminate. Otherwise, return to step 1.

If the master problem is infeasible, then terminate.

Remark 1 In the case of infeasible primal problem, we need to solve a feasibility problem. One way of formulating this feasibility problem is the following:

$$\begin{aligned} \min_{\boldsymbol{x}} \quad & \alpha \\ s.t. \quad & \boldsymbol{h}(\boldsymbol{x}) = \boldsymbol{0} \\ & \boldsymbol{g}(\boldsymbol{x}) \leq \boldsymbol{0} \\ & C\boldsymbol{x} + B\boldsymbol{y}^K - \mathbf{d} \leq \alpha \\ & \boldsymbol{x} \in X = \{\boldsymbol{x} : \boldsymbol{x} \in \Re^n, A_1\boldsymbol{x} \leq a_1\} \\ & \alpha \geq 0 \end{aligned} \tag{6.30}$$

Note that in (6.30) we maintain $\boldsymbol{h}(\boldsymbol{x}) = \boldsymbol{0}$ and $\boldsymbol{g}(\boldsymbol{x}) \leq \boldsymbol{0}$ while we allow the $C\boldsymbol{x} + B\boldsymbol{y}^K - \mathbf{d} \leq \boldsymbol{0}$ constraint to be relaxed by α, which we minimize in the objective function. Kocis and Grossmann (1987) suggested the feasibility problem (6.30) and they also proposed the use of the Lagrange multipliers associated with the equality constraints of (6.30) so as to identify the matrix T^k. In case however, that problem (6.30) has no feasible solution, then Kocis and Grossmann (1987) suggested exclusion of this integer combination by an integer cut and solution of the relaxed master problem. This, however, may not be correct using the arguments of Fletcher and Leyffer (1994).

Remark 2 Following similar arguments to those of **OA**, the **OA/ER** algorithm attains finite convergence to the global minimum as long as C1, C2, and C3 are satisfied.

Remark 3 Kocis and Grossmann (1989a) suggested another alternative formulation of the feasibility problem, in which a penalty-type contribution is added to the objective function; that is,

$$\begin{aligned} \min_{\boldsymbol{x}} \quad & c^T\boldsymbol{y} + f(\boldsymbol{x}) + p\,\alpha \\ \text{s.t.} \quad & \boldsymbol{h}(\boldsymbol{x}) = \boldsymbol{0} \\ & \boldsymbol{g}(\boldsymbol{x}) \leq \alpha \\ & C\boldsymbol{x} + B\boldsymbol{y}^K \leq \alpha \\ & \boldsymbol{x} \in X = \{\boldsymbol{x} : \boldsymbol{x} \in \Re^n, A_1\boldsymbol{x} \leq a_1\} \subseteq \Re^n \end{aligned} \tag{6.31}$$

A feasible solution to the primal problem exists when the penalty term is driven to zero. If the primal does not have a feasible solution, then the solution of problem (6.31) corresponds to minimizing the maximum violation of the inequality constraints (nonlinear and linear in $\boldsymbol{x}$). A general analysis of the different types of feasibility problem is presented is section 6.3.3.1.

6.5.5 Illustration

This example is a slightly modified version of the small planning problem considered by Kocis and Grossmann (1987). A representation of alternatives is shown in Figure 6.8 for producing product

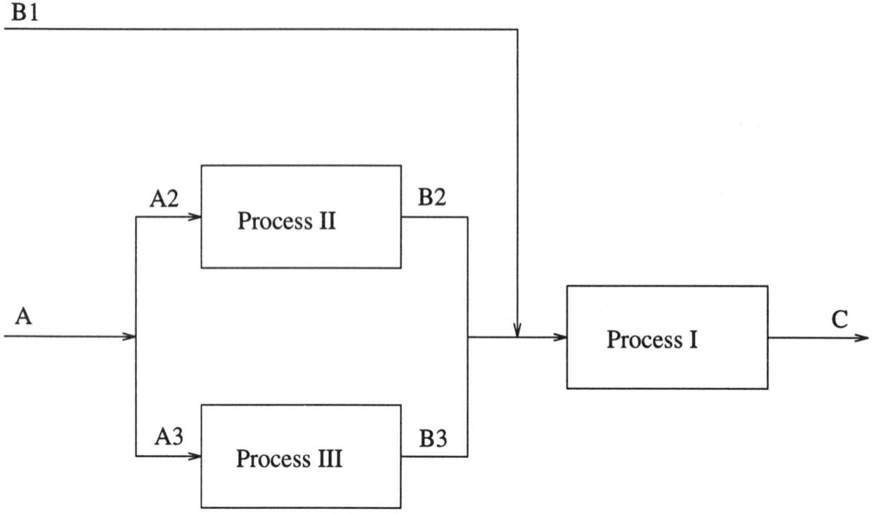

Figure 6.8: Planning problem

C from raw materials A, B via processes I, II, III. Product C can be produced through process I only, process I and III, process I and II. Processes II and III cannot take place simultaneously.

The investment costs of the processes are

$$
\begin{array}{ll}
\text{Process I:} & 3.5y_1 + 2C, \\
\text{Process II:} & 1.0y_2 + 1.0B_2, \\
\text{Process III:} & 1.5y_3 + 1.2B_3,
\end{array}
$$

where y_1, y_2, y_3 are binary variables denoting existence or nonexistence of the processes; B_2, B_3, and C are the flows of products of the processes.

The revenue expressed as a difference of selling the product C minus the cost of raw materials A,B is

$$Revenue = 13C - 1.8\,(A_2 + A_3) - 7B_1,$$

The objective is to minimize the costs minus the revenue, which takes the form:

$$
\begin{aligned}
\text{Objective} &= \text{Costs} - \text{Revenue} \\
&= (3.5y_1 + 2C) + (y_2 + B_2) + (1.5y_3 + 1.2B_3) \\
&\quad -13C + 1.8\,(A_2 + A_3) + 7B_1 \\
&= -11C + 7B_1 + B_2 + 1.2B_3 + 1.8A_2 + 1.8A_3 \\
&\quad +3.5y_1 + y_2 + 1.5y_3.
\end{aligned}
$$

The mass balances for the representation of alternatives shown in Figure 8 are

$$
\begin{array}{ll}
\text{Process I:} & C - 0.9\,(B_1 + B_2 + B_3) = 0, \\
\text{Process II:} & B_2 - \ln(1 + A_2) = 0, \\
\text{Process III:} & B_3 - 1.2\ln(1 + A_3) = 0.
\end{array}
$$

The bounds on the outputs of each process are

$$\text{Process I:} \quad C \leq 1,$$

$$\text{Process II:} \quad B_2 \leq \tfrac{1}{0.9},$$

$$\text{Process III:} \quad B_3 \leq \tfrac{1}{0.9}.$$

Note that imposing the upper bound of 1 on C automatically sets the upper bounds on B_2 and B_3.

The logical constraints for the processes are

$$\begin{array}{ll} \text{Process I:} & C \leq 1y_1, \\ \text{Process II:} & B_2 \leq \tfrac{1}{0.9}y_2, \\ \text{Process III:} & B_3 \leq \tfrac{1}{0.9}y_3. \end{array}$$

We also have the additional integrality constraint:

$$y_2 + y_3 \leq 1.$$

Then, the complete mathematical model can be stated as follows:

$$\begin{array}{rl} \min & -11C + 7B_1 + B_2 + 1.2B_3 + 1.8A_2 \\ & +1.8A_3 + 3.5y_1 + y_2 + 1.5y_3 \\ \text{s.t.} & B_2 - \ln(1 + A_2) = 0 \\ & B_3 - 1.2\ln(1 + A_3) = 0 \\ & C - 0.9\,(B_1 + B_2 + B_3) = 0 \\ & C - 1y_1 \leq 0 \\ & B_2 - \dfrac{1}{0.9}y_2 \leq 0 \\ & B_3 - \dfrac{1}{0.9}y_3 \leq 0 \\ & y_2 + y_3 \leq 1 \\ & C, B_1, B_2, B_3, A_2, A_3 \geq 0 \\ & y_1, y_2, y_3 = 0, 1 \end{array} \tag{6.32}$$

Let us start with $y^1 = (1, 1, 0)$. The primal problem then becomes

$$\begin{array}{rl} \min & -11C + 7B_1 + B_2 + 1.2B_3 + 1.8A_2 \\ & +1.8A_3 + 4.5 \\ \text{s.t.} & B_2 - \ln(1 + A_2) = 0 \\ & B_3 - 1.2\ln(1 + A_3) = 0 \\ & C - 0.9\,(B_1 + B_2 + B_3) = 0 \\ & C - 1 \leq 0 \\ & B_2 - \dfrac{1}{0.9} \leq 0 \\ & B_3 \leq 0 \\ & C, B_1, B_2, B_3, A_2, A_3 \geq 0 \end{array}$$

and has as solution:

$$\begin{aligned} C &= 1, \\ B_1 &= 0, \\ B_2 &= (1/0.9) = 1.11111, \\ B_3 &= 0, \\ A_2 &= 2.037732, \\ A_3 &= 0, \\ Obj &= -1.72097. \end{aligned}$$

The multipliers for the two nonlinear equality constraints are

$$\begin{aligned} \lambda_1 &= 5.46792, \\ \lambda_2 &= 0, \end{aligned}$$

since both λ_1 and λ_2 are nonnegative, then the nonlinear equalities can be relaxed in the form:

$$\begin{aligned} B_2 - \ln(1 + A_2) &\leq 0, \\ B_3 - 1.2\ln(1 + A_3) &\leq 0. \end{aligned}$$

Note that they are both convex and hence condition C1 is satisfied for $T^1 \boldsymbol{h}(\boldsymbol{x}) \leq 0$.

To derive the linearizations around the solution $\boldsymbol{x}^1$ of the primal problem, we only need to consider the two nonlinear relaxed equalities which become

$$\begin{aligned} B_2 - \left[\ln(1 + 2.03773) + \frac{1}{1 + 2.03773}(A_2 - 2.03773)\right] &\leq 0, \\ B_3 - 1.2\left[\ln(1 + 0) + \frac{1}{1 + 0}(A_3 - 0)\right] &\leq 0, \end{aligned}$$

and take the form:

$$\begin{aligned} B_2 - 0.329193A_2 - 0.440303 &\leq 0, \\ B_3 - 1.2A_3 &\leq 0. \end{aligned}$$

Then, the relaxed master problem is of the form:

$$\begin{aligned} \min \quad & 3.5y_1 + y_2 + 1.5y_3 + \mu \\ \text{s.t.} \quad & \mu \geq -11C + 7B_1 + B_2 + 1.2B_3 + 1.8A_2 + 1.8A_3 \\ & 0 \geq B_2 - 0.329193A_2 - 0.440303 \\ & 0 \geq B_3 - 1.2A_3 \\ & C - 0.9\,(B_1 + B_2 + B_3) = 0 \\ & C - 1y_1 \leq 0 \end{aligned}$$

$$
\begin{aligned}
& B_2 - \frac{1}{0.9} y_2 \le 0 \\
& B_3 - \frac{1}{0.9} y_3 \le 0 \\
& y_2 + y_3 \le 1 \\
& y_1 + y_2 - y_3 \le 1 \\
& y_1, y_2, y_3 = 0, 1
\end{aligned}
$$

which has as solution:

$$
\begin{aligned}
y_1 &= 1, \\
y_2 &= 0, \\
y_3 &= 1, \\
C &= 1, \\
B_1 &= 0, \\
B_2 &= 0, \\
B_3 &= (1/0.9) = 1.1111111, \\
A_2 &= 0, \\
A_3 &= 0.925926, \\
\mu &= -8, \\
OBJ &= -3.
\end{aligned}
$$

Thus, after the first iteration we have

$$
\begin{aligned}
UBD &= -1.72097, \\
LBD &= -3,
\end{aligned}
$$

and $y^2 = (1, 0, 1)$.

Solving the primal problem with $y^2 = (1, 0, 1)$ we have the following formulation:

$$
\begin{aligned}
\min \quad & -11C + 7B_1 + B_2 + 1.2B_3 + 1.8A_2 \\
& +1.8A_3 + 5.0 \\
\text{s.t.} \quad & B_2 - \ln(1 + A_2) = 0 \\
& B_3 - 1.2\ln(1 + A_3) = 0 \\
& C - 0.9\,(B_1 + B_2 + B_3) = 0 \\
& C - 1 \le 0 \\
& B_2 - 0 \le 0 \\
& B_3 \le 0 \\
& C, B_1, B_2, B_3, A_2, A_3 \ge 0
\end{aligned}
$$

which has as solution:

$$
\begin{aligned}
C &= 1, \\
B_1 &= 0, \\
B_2 &= 0, \\
B_3 &= (1/0.9) = 1.1111111, \\
A_2 &= 0, \\
A_3 &= 1.5242, \\
OBJ &= -1.9231 \text{ (New Upper Bound)}.
\end{aligned}
$$

The multipliers for the two nonlinear equality constraints are

$$
\begin{aligned}
\lambda_1 &= 0, \\
\lambda_2 &= 3.7863.
\end{aligned}
$$

Since both λ_1 and λ_2 are nonnegative, then we can relax the nonlinear equalities into

$$
\begin{aligned}
B_2 - \ln(1 + A_2) &\leq 0, \\
B_3 - 1.2\ln(1 + A_3) &\leq 0,
\end{aligned}
$$

which are convex and hence satisfy condition C1.

The linearizations around the solution of the primal of the second iteration are

$$
\begin{aligned}
B_2 - A_2 &\leq 0, \\
B_3 - 0.475398A_3 - 0.386507 &\leq 0.
\end{aligned}
$$

Then, the relaxed master problem of the second iteration takes the form:

$$
\begin{aligned}
\min \quad & 3.5y_1 + y_2 + 1.5y_3 + \mu \\
\text{s.t.} \quad & \mu \geq -11C + 7B_1 + B_2 + 1.2B_3 + 1.8A_2 + 1.8A_3 \\
& 0 \geq B_2 - 0.329193A_2 - 0.44303 \\
& 0 \geq B_3 - 1.2A_3 \\
& 0 \geq B_2 - A_2 \\
& 0 \geq B_3 - 0.475398A_3 - 0.386507 \\
& C - 0.9\,(B_1 + B_2 + B_3) = 0 \\
& C - 1y_1 \leq 0 \\
& B_2 - \frac{1}{0.9}y_2 \leq 0 \\
& B_3 - \frac{1}{0.9}y_3 \leq 0 \\
& y_2 + y_3 \leq 1 \\
& y_1 + y_2 - y_3 \leq 1 \\
& y_1 + y_3 - y_2 \leq 1 \\
& y_1, y_2, y_3 = 0, 1 \\
& -3 \leq 3.5y_1 + y_2 + 1.5y_3 + \mu \leq -1.9231
\end{aligned}
$$

which has no feasible solution, and hence termination has been obtained with optimal solution the one of the primal problem of the second iteration

$$OBJ = -1.9231, \quad y = (1, 0, 1).$$

6.6 Outer Approximation with Equality Relaxation and Augmented Penalty, OA/ER/AP

The **OA/ER** algorithm, as we discussed in section 6.5, is based on the assumption of convexity of the functions $f(x)$ and $g(x)$, and quasi–convexity of the relaxed equalities $T^k h(x)$. Under this assumption, noted as condition C1, and conditions C2 and C3 (see section 6.5.1), the **OA/ER** is theoretically guaranteed to identify the global optimum solution. If, however, condition C1 is not met, then the primal NLP problems can be trapped to a local solution, the equivalence of the nonlinear equalities to the relaxed set of the inequalities may not hold and the linearizations of the objective function $f(x)$, the relaxed equalities $T^k h(x)$, and the inequalities $g(x)$ may not underestimate the original functions and hence may cut parts of the feasible region of the original problem. This implies that candidate 0–1 combinations, which belong to the part of the feasible region that is cut by the invalid supports (i.e., linearizations), may be eliminated from further consideration and therefore, only suboptimal solutions will be identified.

Viswanathan and Grossmann (1990) proposed the **OA/ER/AP** algorithm, which is a variant of the **OA/ER** algorithm, with key objective to avoid the limitations imposed by the convexity assumption made in the **OA/ER** algorithm.

6.6.1 Formulation

The **OA/ER/AP** algorithm addresses the same formulation as the **OA/ER** which is

$$\begin{aligned}
\min_{x,y} \quad & c^T y + f(x) \\
\text{s.t.} \quad & h(x) = 0 \\
& g(x) \leq 0 \\
& Cx + By \leq \mathrm{d} \\
& x \in X = \{x : x \in \Re^n, A_1 x \leq a_1\} \subseteq \Re^n \\
& y \in Y = \{y : y \in \{0,1\}^q, A_2 y \leq a_2\}
\end{aligned} \tag{6.33}$$

under the following conditions:

C1: f, h, and g are continuously differentiable.

C2: A constraint qualification holds at the solution of every nonlinear programming problem resulting from (6.33) by fixing y.

Remark 1 Note that conditions C1 and C2 correspond to conditions C2 and C3 respectively of the **OA/ER** (see section 6.5.1). Also, note that since the convexity assumption is not imposed, then even the equivalence of the nonlinear equalities to the relaxed equalities $T^k h(x)$ may no longer be valid.

6.6.2 Basic Idea

Since the convexity assumption is not imposed in the **OA/ER/AP** then (i) the equivalence of $\boldsymbol{h}(\boldsymbol{x})$ and $T^k\boldsymbol{h}(\boldsymbol{x}) \leq \boldsymbol{0}$ may not hold, (ii) the linearizations may not represent valid supports, and (iii) the master problem may not provide a valid lower bound on the solution of (6.33).

The basic idea in **OA/ER/AP** is to address limitations (i), (ii), (iii) by relaxing the linearizations in the master problem, that is by allowing them to be violated and utilizing a penalty-type approach that penalizes these violations of the support functions. The violations of the linearizations are allowed by introducing slack variables, while the penalty of the violations is introduced as an additional set of terms in the objective function that consist of the slack variables multiplied by positive weight factors. This way, because of the relaxation of the constraints, the feasible region is expanded and hence the possibility of cutting off part of the feasible region due to invalid linearizations is reduced.

Remark 1 It is important to note though that such an approach has no theoretical guarantee of not eliminating part of the feasible region, and as a result the global optimum solution determination cannot be guaranteed.

6.6.3 Theoretical Development

Since the main difference between **OA/ER** and **OA/ER/AP** is the formulation of the master problem, we will present in this section the master problem formulation.

6.6.3.1 The Master Problem

The relaxed master problem of the **OA/ER/AP** is based on the relaxed master of the **OA/ER** and is of the form:

$$\left[\begin{array}{l} Z_L^K = \min \quad c^T y + \mu + \sum\limits_k w_k^o s_k^o + \sum\limits_{i,k} w_{i,k}^p p_{i,k} + \sum\limits_{i,k} w_{i,k}^q q_{i,k} \\ \\ \text{s.t.} \\ \qquad \left.\begin{array}{rcl} \mu + s_k^o & \geq & f(\boldsymbol{x}^k) + \nabla f(\boldsymbol{x}^k)(\boldsymbol{x} - \boldsymbol{x}^k) \\ \mathbf{p_k} & \geq & \boldsymbol{g}(\boldsymbol{x}^k) + \nabla \boldsymbol{g}(\boldsymbol{x}^k)(\boldsymbol{x} - \boldsymbol{x}^k) \\ \mathbf{q_k} & \geq & T^k\left[\boldsymbol{h}(\boldsymbol{x}^k) + \nabla \boldsymbol{h}(\boldsymbol{x}^k)(\boldsymbol{x} - \boldsymbol{x}^k)\right] \end{array}\right\} k = 1, 2, \ldots, K \\ \qquad C\boldsymbol{x} + B\boldsymbol{y} \leq \mathbf{d} \\ \qquad \boldsymbol{x} \in X = \{\boldsymbol{x} : \boldsymbol{x} \in \Re^n, A_1\boldsymbol{x} \leq a_1\} \subseteq R^n \\ \qquad \boldsymbol{y} \in Y = \{\boldsymbol{y} : \boldsymbol{y} \in \{0-1\}^q, A_2\boldsymbol{y} \leq a_2\} \\ \qquad \sum\limits_{i \in \mathbf{B}^k} y_i - \sum\limits_{i \in \mathbf{NB}^k} y_i \leq |B^k| - 1, \quad k = 1, 2, \ldots, K-1 \\ \qquad s_k^o, p_{i,k}, q_{i,k} \geq 0, \quad k = 1, 2, \ldots, K \end{array}\right. \qquad (6.34)$$

where $\mathbf{p}_k = \{p_{i,k}\}$ and $\mathbf{q}_k = \{q_{i,k}\}$ are vectors of positive slack variables and s_k^o are positive slack scalars; and $w_o^k, w_{i,k}^p, w_{i,k}^q$ are the weights on the slack variables $s_k^o, p_{i,k}, q_{i,k}$, respectively, that satisfy the following relationships:

$$\begin{aligned} w_k^o &> |\bar{\mu}_k|, \\ w_{i,k}^p &> |\lambda_{i,k}|, \\ w_{i,k}^q &> |\mu_{i,k}|, \end{aligned}$$

where $\bar{\mu}_k, \lambda_{i,k}, \mu_{i,k}$ are the Lagrange multipliers of the primal problem for $y = y^k$ written as

$$\begin{aligned} \min_{x} \quad & c^T y^k + \mu \\ s.t. \quad f(x^k) - \mu &\leq 0 \leftarrow \bar{\mu}_k \\ h(x) &= 0 \leftarrow \lambda_k \\ g(x) &\leq 0 \leftarrow \mu_k \\ Cx + By^k &\leq \mathbf{d} \\ x &\in X = \{x : x \in \Re^n, A_1 x \leq a_1\} \subseteq \Re^n \end{aligned} \tag{6.35}$$

Viswanathan and Grossmann (1990) showed the following :

Property 6.6.1
If (x^k, μ) is a **KKT** *point of (6.35), then it is also a* **KKT** *point of the master problem at $\mathbf{y} = \mathbf{y}^k$.*

6.6.4 Algorithm Development

The **OA/ER/AP** algorithm can be stated as follows:

Step 1: Solve the NLP relaxation of (6.33) (i.e. treat the y variables as continuous with $0 \leq y \leq 1$) to obtain (x^o, y^o). If y^o is a 0–1 combination, terminate. Otherwise, go to step 2.

Step 2: Solve the relaxed master problem (6.34) to identify y^1.

Step 3: Solve the primal problem $P(y^1)$ [i.e., problem (6.35)] to find the upper bound $UBD = P(y^1)$ as well as the Lagrange multipliers.

Step 4: Define the $(m \times m)$ matrix T^k.

Step 5: Solve the relaxed master problem (6.34) to determine y^{k+1} and the lower bound on (6.33) denoted as Z_L^K.

Step 6: Repeat steps 3, 4, and 5 until there is an increase in the optimal value of the feasible primal NLP problems at which point we terminate.

Remark 1 Note that Step 1 of the **OA/ER/AP** algorithm does not require an initial 0–1 combination to be provided but instead it may generate one by solving an NLP relaxation and subsequently solving the relaxed master problem. Such a scheme can be applied in the **GBD**, **OA**, and **OA/ER**.

Remark 2 Note that the algorithm may terminate at step 1 if a 0–1 combination is determined by the NLP relaxation of (6.33). It should be emphasized, however, that since the NLP relaxation does not satisfy any convexity properties in x the obtained solution can be a local solution even though a 0–1 combination for the y variables has been identified.

Remark 3 In the case that the relaxed master problem produces a 0–1 combination y^{k+1} for which the primal problem is infeasible, Viswanathan and Grossmann (1990) suggested the following two alternatives:

Alternative (i): Disregard the infeasible primal solution x^{k+1} and introduce an integer cut that excludes the y^{k+1} combination and solve the relaxed master problem again until you generate a y for which the primal problem is feasible.

Alternative (ii): Add to the relaxed master problem the linearizations around the infeasible continuous point. Note though that to treat the relaxed master problem we need to have information on the Lagrange multipliers. To obtain such information, a feasibility problem needs to be solved and Viswanathan and Grossmann (1990) suggested one formulation of the feasibility problem; that is,

$$\begin{aligned}
\min_{\boldsymbol{x},\alpha} \quad & c^T y^k + f(x) + p\,\alpha \\
\text{s.t.} \quad & h(x) = 0 \\
& g(x) \leq \alpha \\
& Cx + By^k - \mathbf{d} \leq \alpha \\
& x \in X = \{x : x \in \Re^n, A_1 x \leq a_1\} \subseteq Re^n \\
& \alpha \geq 0
\end{aligned} \tag{6.36}$$

where p is a penalty parameter and α is a slack variable applied only to the inequality constraints.

If the solution of (6.36) has $\alpha = 0$, then the primal has a feasible solution. Note that (6.36) does not satisfy any convexity assumptions and as such it can only attain a local solution.

Remark 4 The termination criterion in step 6 is based on obtaining a larger value in the primal problems for consecutive iterations. However, there is no theoretical reason why this should be the criterion for termination since the primal problems by their definition do not need to satisfy any monotonicity property. (i.e., they do not need to be nonincreasing). Based on the above, this termination criterion can be viewed only as a heuristic and may result in premature termination of the **OA/ER/AP**. Therefore, the **OA/ER/AP** can fail to identify the global solution of the MINLP problem (6.33).

6.6.5 Illustration

This problem is a modified version of an example taken from Kocis and Grossmann (1987) and has the following formulation:

$$\begin{aligned} \min \quad & -0.7y + 5(x_1 - 0.5)^2 + 0.8 \\ \text{s.t.} \quad & -\exp(x_1 - 0.2) - x_2 \leq 0 \\ & x_2 + 1.1y \leq -1 \\ & x_1 - 1.2y \leq 0.2 \\ & 0.2 \leq x_1 \leq 1 \\ & -2.22554 \leq x_2 \leq -1 \\ & y = 0, 1 \end{aligned} \tag{6.37}$$

This problem has a convex objective function, but it is nonconvex because of the first constraint which is concave in x_1.

Let us first consider the NLP relaxation of (6.37) which takes the form:

$$\begin{aligned} \min \quad & -0.7y + 5(x_1 - 0.5)^2 + 0.8 \\ \text{s.t.} \quad & -\exp(x_1 - 0.2) - x_2 \leq 0 \\ & x_2 + 1.1y \leq -1 \\ & x_1 - 1.2y \leq 0.2 \\ & 0.2 \leq x_1 \leq 1 \\ & -2.22554 \leq x_2 \leq -1 \\ & 0 \leq y \leq 1 \end{aligned} \tag{6.38}$$

and has as solution

$$\begin{aligned} x_1 &= 0.594, \\ x_2 &= -1.4829, \\ y &= 0.44, \\ OBJ &= 0.53618, \end{aligned}$$

and the first inequality is active with multiplier

$$\mu_1 = 0.636.$$

Linearizing the objective function nonlinear term $(x_1 - 0.5)^2$ and the first nonlinear inequality we have:

$$\begin{aligned} \mu &\geq 5\left[(0.594 - 0.5)^2 + 2(0.544 - 0.5)(x_1 - 0.594)\right] + 0.8, \\ 0 &\geq -\left[\exp(0.594 - 0.2) + \exp(0.544 - 0.2)(x_1 - 0.594)\right] - x_2, \end{aligned}$$

which become

$$\begin{aligned} \mu &\geq 0.94x_1 + 0.28582, \\ 0 &\geq -1.4829x_1 - x_2 - 0.60206. \end{aligned}$$

We now introduce the slack variables

$$s_1^o \text{ and } p_{1,1}$$

for the two linearizations, and set the weights for objective function as

$$\begin{aligned} w_1^o &= 1000, \\ w_{1,1}^o &= 1000 \cdot 0.636 = 636. \end{aligned}$$

Then, the relaxed master problem becomes

$$\begin{aligned} \min \quad & -0.7y + \mu + 1000s_1^o + 636p_{1,1} \\ \text{s.t.} \quad & \mu + s_1^o \geq 0.94x_1 + 0.28582 \\ & p_{1,1} \geq -1.4829x_1 - x_2 - 0.60206 \\ & x_2 + 1.1y \leq -1 \\ & x_1 - 1.2y \leq 0.2 \\ & 0.2 \leq x_1 \leq 1 \\ & -2.22554 \leq x_2 \leq -1 \\ & y = 0, 1 \\ & s_1^o, p_{1,1} \geq 0 \end{aligned} \tag{6.39}$$

The solution of (6.39) is

$$\begin{aligned} y &= 1, \\ x_1 &= 1 \qquad s_1^o = 0, \\ x_2 &= -2.1 \quad p_{1,1} = 0.0159, \\ Z_L^1 &= 10.638. \end{aligned}$$

Solving the primal problem for $y = y^1 = 1$ we have

$$\begin{aligned} \min_{x_1, x_2} \quad & 5(x_1 - 0.5)^2 + 0.1 \\ s.t. \quad & -\exp(x_1 - 0.2) - x_2 \leq 0 \\ & x_2 \leq -2.1 \\ & x_1 \leq 1.4 \\ & 0.2 \leq x_1 \leq 1 \\ & -2.22554 \leq x_2 \leq -1 \end{aligned}$$

and has as solution

$$\begin{aligned} x_1 &= 0.94294, \\ x_2 &= -2.1, \\ OBJ &= 1.07654, \end{aligned}$$

with $\mu_1 = 2.104$.

The linearizations of the objective function and the first constraint are

$$\begin{aligned} \mu &\geq 4.42x_1 - 2.3868, \\ 0 &\geq -2.1x_1 - x_2 - 0.2222. \end{aligned}$$

Introducing s_2^o and $p_{1,2}$ and weights

$$\begin{aligned} w_2^o &= 1000, \\ w_{1,2}^p &= 1000 \cdot 2.104 = 2104, \end{aligned}$$

the relaxed master problem becomes

$$\begin{aligned} Z_L^2 = \min \quad & -0.7y + \mu + 1000s_1^o + 1000s_2^o + 636p_{1,1} + 2104p_{1,2} \\ s.t. \quad \mu + s_1^o &\geq 0.94x_1 + 0.28582 \\ \mu + s_2^o &\geq 4.42x_1 - 2.3868 \\ p_{1,1} &\geq -1.4829x_1 - x_2 - 0.60206 \\ p_{1,2} &\geq -2.1x_1 - x_2 - 0.2222 \\ x_2 + 1.1y &\leq -1 \\ x_1 - 1.2y &\leq 0.2 \\ 0.2 &\leq x_1 \leq 1 \\ -2.22554 &\leq x_2 \leq -1 \\ y &= 0, 1 \\ s_1^o, s_2^o, p_{1,1}, p_{1,2} &\geq 0 \end{aligned}$$

which has solution

$$\begin{aligned} y^2 &= 0, \\ x_1 &= 0.2 \quad s_1^o = s_2^o = 0, \\ x_2 &= -1 \quad p_{1,1} = 0.1015, \ p_{1,2} = 0.4582, \\ Z_L^2 &= 1029.08. \end{aligned}$$

Solving the primal problem for $y = y^2 = 0$ we obtain

$$\begin{aligned} x_1 &= 0.2, \\ OBJ &= 1.25. \end{aligned}$$

Since $P(y^2) > P(y^1)$ we terminate. The optimum solution found is

$$\begin{aligned} y &= 1, \\ x_1 &= 0.94194, \\ x_2 &= -2.1, \\ OBJ &= 1.07654. \end{aligned}$$

6.7 Generalized Outer Approximation, GOA

6.7.1 Formulation

Fletcher and Leyffer (1994) generalized the **OA** approach proposed by Duran and Grossmann (1986a), to the class of optimization problems stated as

$$\begin{aligned} \min_{x,y} \quad & f(x,y) \\ s.t. \quad & g(x,y) \leq 0 \\ & x \in X \subseteq \Re^n \\ & y \in Y = \{0,1\}^q \end{aligned} \tag{6.40}$$

under the following conditions:

C1: X is a nonempty, compact, convex set and the functions

$$\begin{aligned} f &: \Re^n \times \Re^q \longrightarrow \Re, \\ g &: \Re^n \times \Re^q \longrightarrow \Re^p, \end{aligned}$$

are convex.

C2: f and g are continuously differentiable.

C3: A constraint qualification (e.g., Slater's) holds at the solution of every nonlinear programming problem resulting from (6.40) by fixing y.

Remark 1 Problem (6.40) allows for nonlinearities in the 0–1 variables to be treated directly and as such it includes the class of pure 0–1 nonlinear programming problems. Note also that Fletcher and Leyffer (1994) assumed y being integer while here we present y as 0-1 variables based on remark 1 of section 6.1.

Remark 2 Note that in problem (6.40) no assumption of separability of x and y or linearity in y is made. From this point of view, problem (6.40) under conditions C1–C3 generalizes problem (6.13) under its respective assumptions C1, C2, C3.

6.7.2 Basic Idea

The basic idea in **GOA** is similar to the one in **OA**, with the key differences being the (i) treatment of infeasibilities, (ii) new formulation of the master problem that considers the infeasibilities explicitly, and (iii) unified treatment of exact penalty functions.

6.7.3 Theoretical Development

6.7.3.1 Treatment of Infeasibilities in the Primal Problems

If the primal problem $P(y^k)$ is infeasible, then the following general feasibility problem was proposed by Fletcher and Leyffer (1994):

$$\begin{aligned} \min_{x} \quad & \sum_{i \in I'} w_i g_i^+(x, y^k) \\ s.t. \quad & g_i(x, y^k) \leq 0, \quad i \in I \\ & x \in X \end{aligned} \tag{6.41}$$

where $y = y^k$ is fixed y, w_i are the weights, $g_i^+(x, y^k) = \max[0, g_i(x, y^k)]$, I is the set of feasible inequality constraints, and I' is the set of infeasible inequality constraints.

Remark 1 Note that the general feasibility problem (6.41) has in the objective function a summation over only the infeasible inequality constraints. Note also that a similar approach was presented in section 10.3.3.1. Fletcher and Leyffer (1994) proved the following important property for problem (6.41).

Lemma 6.7.1 If the primal problem $P(y^k)$ is infeasible so that x^k solves problem (6.41) and has

$$\sum_{i \in I'} w_i g_i^+(x^k, y^k) > 0,$$

then, $y = y^k$ is infeasible in the constraints

$$\left. \begin{aligned} 0 \geq g_i(x^k, y^k) + \nabla g_i(x^k, y^k)^T \begin{pmatrix} x - x^k \\ y - y^k \end{pmatrix}, \quad \forall i \in I' \\ 0 \geq g_i(x^k, y^k) + \nabla g_i(x^k, y^k)^T \begin{pmatrix} x - x^k \\ y - y^k \end{pmatrix}, \quad \forall i \in I \end{aligned} \right\} \forall\, x \in X.$$

Remark 2 Note that the aforementioned constraints are the linearizations of the nonlinear inequality constraints around the point (x^k, y^k). Then, the above property states that infeasibility in the primal problem $P(y^k)$ implies that the linearization cuts of infeasible and feasible inequalities around (x^k, y^k) are violated.

6.7.3.2 The Master Problem

Fletcher and Leyffer (1994) claimed that the master problem proposed by Duran and Grossmann (1986a) does not account for infeasibilities in the primal problems, and derived the master problem for the **GOA** based on the same ideas of

(i) Projection of (6.40) onto the y–space; and

(ii) Outer approximation of the objective function and the feasible region, but in addition utilizing the key property discussed in section 6.7.3.1 for treating correctly the infeasible primal problems case. This is denoted as representation of V via outer approximation.

(i) Projection of (6.40) onto the y–space

Problem (6.40) can be written as

$$\begin{aligned} \min_{y} \min_{x} \quad & f(x,y) \\ s.t. \quad & g(x,y) \leq 0 \\ & x \in X \\ & y \in Y \end{aligned} \tag{6.42}$$

Note that the inner problem is written as minimum with respect to x since the infimum is attained due to the compactness assumption of the set X.

Let us define $v(y)$:

$$\begin{aligned} v(y) = \min_{x} \quad & f(x,y) \\ s.t. \quad g(x,y) & \leq 0 \\ x & \in X \end{aligned} \tag{6.43}$$

Let us also define the set V of y's for which there exist feasible solution in the x variables as

$$V = \{y : g(x,y) \leq 0 \text{ for some } x \in X\}\,. \tag{6.44}$$

Then, problem (6.40) can be written as

$$\begin{aligned} \min_{y} \quad & v(y) \\ \text{s.t.} \quad & y \in Y \cup V \end{aligned} \tag{6.45}$$

Remark 1 Problem (6.45) is the projection of (6.40) onto the y–space. The projection has to meet feasibility requirements which are represented in (6.45) by imposing $y \in Y \cap V$.

Remark 2 $v(y)$ and V are only known implicitly and to overcome this difficulty Fletcher and Leyffer (1994) proposed their representation in (6.45) through outer approximation.

(ii) Outer Approximation of $v(y)$

Since a constraint qualification (condition 3) holds at the solution of every primal problem $P(y^k)$ for every $y^k \in Y \cap V$, then the projection problem (6.45) has the same solution as the problem:

$$\left[\begin{array}{cc} \min\limits_{y} & \left[\begin{array}{ll} \min\limits_{x} & f(x^k,y^k) + \nabla f(x^k,y^k)^T \begin{pmatrix} x - x^k \\ 0 \end{pmatrix} \\ \text{s.t.} & g(x^k,y^k) + \nabla g(x^k,y^k)^T \begin{pmatrix} x - x^k \\ 0 \end{pmatrix} \leq 0 \\ & x \in X \end{array}\right] \\ & y \in Y \cap V \end{array}\right] \tag{6.46}$$

Remark 3 Note that the inner problem in (6.46) is $v(y)$ with linearized objective and constraints around x^k. The equivalence in solution between (6.45) and (6.46) is true because of the convexity condition and the constraint qualification.

Remark 4 It should be emphasized that we need to include only the linearizations of the inequality constraints that are active at the solution of the primal problem $P(y^k)$. This is important because it reduces the number of linearizations that need to included in the master problem.

Remark 5 The convexity assumption (condition C1) implies that (x^k, y^k) is feasible in the inner problem of (6.46) for all $k \in \mathbf{F}$:

$$\mathbf{F} = \{k : y^k \text{ is a feasible solution to } P(y^k)\}\,.$$

By introducing a scalar variable μ_{GOA}, the master problem can be formulated as

$$\left[\begin{array}{lll} \min\limits_{x,y,\mu_{\mathrm{GOA}}} & \mu_{\mathrm{GOA}} & \\ \text{s.t.} & & \\ & \left.\begin{array}{rcl} \mu_{\mathrm{GOA}} & \geq & f(x^k,y^k) + \nabla f(x^k,y^k)^T \begin{pmatrix} x - x^k \\ y - y^k \end{pmatrix} \\ 0 & \geq & g(x^k,y^k) + \nabla g(x^k,y^k)^T \begin{pmatrix} x - x^k \\ y - y^k \end{pmatrix} \end{array}\right\} k \in \mathbf{F} & \\ & x \in X & \\ & y \in Y \cap V & \end{array}\right] \tag{6.47}$$

Remark 6 Note that in (6.47) we still need to find a representation of the set V. In other words, we have to ensure that the 0–1 assignments which produce infeasible primal subproblems are also infeasible in the master problem (6.47).

(iii) Representation of V

Let us define as the set $\bar{\mathbf{F}}$ the 0–1 assignment for which the primal problem $P(y^k)$ is infeasible; that is,

$$\bar{\mathbf{F}} = \{k : P(y^k) \text{ is infeasible and } x_k \text{ solves the feasibility problem (6.41)}\}$$

Then, it follows from the property of section 6.7.3.1 that the constraints

$$0 \geq g(x^k, y^k) + \nabla g(x^k, y^k)^T \begin{pmatrix} x - x^k \\ y - y^k \end{pmatrix}, \quad \forall k \in \bar{\mathbf{F}} \tag{6.48}$$

exclude the 0–1 assignments for which the primal problem $P(y^k)$ is infeasible.

Remark 7 The constraints (6.48) define the set of $y \in Y \cap V$, and hence we can now formulate the master problem correctly. Note that we replace the set $y \in Y \cap V$ with constraints (6.48) that are the outer approximations at the points y^k for which the primal is infeasible and the feasibility problem (6.41) has as solution x^k.

The master problem of **GOA** now takes the form:

$$\left[\begin{array}{ll} \min\limits_{x,y,\mu_{GOA}} & \mu_{GOA} \\ \text{s.t.} & \\ & \left.\begin{array}{lcl} \mu_{GOA} & \geq & f(x^k, y^k) + \nabla f(x^k, y^k)^T \begin{pmatrix} x - x^k \\ y - y^k \end{pmatrix} \\ 0 & \geq & g(x^k, y^k) + \nabla g(x^k, y^k)^T \begin{pmatrix} x - x^k \\ y - y^k \end{pmatrix} \end{array}\right\} \forall\, k \in \mathbf{F} \\ & \left. 0 \geq g(x^k, y^k) + \nabla g(x^k, y^k)^T \begin{pmatrix} x - x^k \\ y - y^k \end{pmatrix} \right\} \forall k \in \bar{\mathbf{F}} \\ & x \in X \\ & y \in Y \end{array}\right. \tag{6.49}$$

Remark 8 Under conditions C1, C2, and C3, problem (6.49) is equivalent to problem (6.40). Problem (6.49) is a MILP problem.

Remark 9 Problem (6.49) will be treated by relaxation in an iterative way, since it is not practical to solve it directly on the ground that all primal problems $P(y^k)$ need to be solved first.

Remark 10 If $Y = \{0,1\}^q$ and we have p inequality constraints, then the master problem (6.49) will have

$$2^q\,(p+1)$$

constraints. Hence, the solution of (6.49) via relaxation in an iterative framework is preferred.

6.7.4 Algorithmic Development

The **GOA** algorithm can be stated as follows:

Step 1: Let an initial point $y^1 \in Y$ or $y^1 \in Y \cap V$ if available. Solve the resulting primal $P(y^1)$ or the feasibility problem (6.41) for $y = y^1$ if $P(y^1)$ is infeasible, and let the solution be x^1. Set the iteration counter $k = 1$. Set the current upper $UBD = P(y^1) = v(y^1)$ (if primal is feasible)

Step 2: Solve the relaxed master problem **(RM)**,

$$
(\mathbf{RM}) \left[
\begin{array}{ll}
\min\limits_{x,y,\mu_{\mathbf{GOA}}} & \mu_{\mathbf{GOA}} \\
\text{s.t.} & \\
& \left.\begin{array}{l}
\mu_{\mathbf{GOA}} \geq f(x^k,y^k) + \nabla f(x^k,y^k)^T \begin{pmatrix} x - x^k \\ y - y^k \end{pmatrix} \\
0 \quad \geq g(x^k,y^k) + \nabla g(x^k,y^k)^T \begin{pmatrix} x - x^k \\ y - y^k \end{pmatrix}
\end{array}\right\} \forall\, k \in \mathbf{F}^i \\
& 0 \geq g(x^k,y^k) + \nabla g(x^k,y^k)^T \begin{pmatrix} x - x^k \\ y - y^k \end{pmatrix} \Bigg\}, \forall k \in \bar{\mathbf{F}}^i \\
& x \in X \\
& y \in Y \\
& \sum\limits_{i \in \mathbf{B}^k} y_i^k - \sum\limits_{i \in \mathbf{NB}^k} y_i^k \leq |\mathbf{B}^k| - 1
\end{array}
\right.
$$

where $\mathbf{F}^i = \{k | k \leq i : \exists \text{ a feasible solution } x^k \text{ to } P(y^k)\}$,
$\bar{\mathbf{F}}^i = \{k | k \leq i : P(y^k) \text{ is infeasible and } x^k \text{ solves problem (6.41) for } y = y^k\}$.

Let $(y^{k+1}, \mu^k_{\mathbf{GOA}})$ be the optimal solution of the relaxed master problem **(RM)**, where the new current lower bound is

$$LBD = \mu^k_{\mathbf{GOA}}$$

and y^{k+1} is the next 0–1 assignment point to be considered in the primal problem $P(y^{k+1})$.

If $UBD - LBD \leq \epsilon$, then terminate. Otherwise, go to step 1.

If the relaxed master problem does not have a feasible solution, then terminate with optimal solution being the one given by the current upper bound.

Remark 1 Note that in the relaxed master problem we need to include the linearizations of only the active constraints at the feasible primal problems.

Remark 2 Under conditions C1, C2, and C3, and $|Y| < \infty$, the **GOA** terminates in a finite number of steps and determines the global solution of problem (6.40).

6.7.5 Worst–Case Analysis of GOA

Fletcher and Leyffer (1994) studied the worst-case performance of **GOA** in an attempt to present the potential limitations that the outer approximation algorithm of Duran and Grossmann (1986a) may exhibit despite the encouraging experience obtained from application to engineering problems.

They constructed the following example for which all integer feasible points need to be examined before finding the solution, that is, complete enumeration is required.

$$\begin{array}{ll} \min\limits_{y} & f(y) = (y - \epsilon)^2 \\ \text{s.t.} & y \in \{0, \epsilon, \ldots, 1/2, 1\} \end{array}$$

in which $\epsilon = 2^{-p}$ for some $p > 1$.

Starting from $y^1 = 0$, which is adjacent value to the solution $y^* = \epsilon$, the next iterate is $y^2 = 1$ which is an extreme feasible point. Then, the **GOA** works its way back to the solution $y^* = \epsilon$ by visiting each remaining integer assignment

$$y^i = 2^{-i+1}, \quad i = 3, 4, \ldots, p+1.$$

6.7.6 Generalized Outer Approximation with Exact Penalty, GOA/EP

6.7.6.1 The Primal Problem

In **GOA/EP**, we do not distinguish between feasible and infeasible primal problem, but instead formulate the following primal problem suggested by Fletcher and Leyffer (1994) which is based on an exact penalty function:

$$\left[\begin{array}{ll} \min\limits_{x} & \Phi(x, y^k) = f(x, y^k) + \sigma\, ||g(x, y^k)^+|| \\ \text{s.t.} & x \in X \end{array} \right. \tag{6.50}$$

where X includes only linear constraints on x;
$g(x, y^k)^+ = \max\left(0, g(x, y^k)\right)$;
$||\cdot||$ is a norm in $\Re^p$; and
σ is a sufficiently large penalty parameter.

Remark 1 If σ is sufficiently large and condition C3 is valid, then problems (6.50) and $P(y^k)$ have the same solution.

6.7.6.2 The Master Problem

Following a similar development as in the **GOA**, we obtain the following formulation for the master problem of **GOA/EP**:

$$\left[\begin{array}{l} \min\limits_{x, y, \mu} \quad \mu \\ \text{s.t.} \\ \quad \left. \begin{array}{rcl} \mu & \geq & f(x^k, y^k) + \nabla f(x^k, y^k)^T \begin{pmatrix} x - x^k \\ y - y^k \end{pmatrix} \\ 0 & \geq & g(x^k, y^k) + \nabla g(x^k, y^k)^T \begin{pmatrix} x - x^k \\ y - y^k \end{pmatrix} \end{array} \right\} \forall\, k \in \mathbf{FP} \\ \quad x \in X \\ \quad y \in Y \end{array} \right. \tag{6.51}$$

where **FP** $= \{k : x^k$ is an optimal solution to (6.50)$\}$.

Remark 1 Note that in the master problem (6.51), we do not need to consider linearizations around infeasible primal points since we consider problem (6.50).

Remark 2 Under conditions C1, C2, and C3, problem (6.51) is equivalent to (6.40) in the sense that (x^*, y^*) solves (6.40) if and only if it solves (6.51).

Remark 3 Replacing **(RM)** by the relaxation of (6.51) we obtain the **GOA/EP** algorithm.

6.7.6.3 Finite Convergence of GOA/EP

Fletcher and Leyffer (1994) proved that: If conditions C1 and C2 hold and $|Y| < \infty$, then the **GOA/EP** algorithm terminates in a finite number of iterations.

Remark 1 Note that the condition C3 is not required for finite convergence of the **GOA/EP**. This is because in exact penalty functions a constraint qualification does not form part of the first-order necessary conditions. However, note that C3 is needed to ensure that the solution (x^*, y^*) of the **GOA/EP** algorithm also solves (6.40).

6.8 Comparison of GBD and OA-based Algorithms

In sections 6.3–6.7 we discussed the generalized benders decomposition **GBD** and the outer approximation based algorithms (i.e., **OA, OA/ER, OA/ER/AP, GOA**), and we identified a number of similarities as well as key differences between the two classes of MINLP algorithms.

The main similarity between the **GBD** and **OA** variants is the generation of two bounding sequences: (i) an upper bounding sequence which is nonincreasing if we consider the updated upper bounds, and (ii) a lower bounding sequence which is nondecreasing. Hence, both classes of algorithms are based on a decomposition of the original MINLP model into subproblems providing at each iteration an upper and a lower bound on the sought solution and subsequently showing under certain conditions that the two sequences become ϵ–close in a finite number of iterations.

In addition to the aforementioned similarity though, there exist several key differences which are the subject of discussion in this section. These differences can be classified as being due to the

(i) Formulation,

(ii) Equalities,

(iii) Nonlinearities in y and $x - y$,

(iv) Primal Problem,

(v) Master Problem,

(vi) Quality of Lower Bounds

and will be discussed in the following.

6.8.1 Formulation

The general formulations of MINLP models that can be addressed via **GBD** and **OA** are:

$$\left[\begin{array}{rrl} & & \textbf{GBD} \\ \min\limits_{x,y} & f(x,y) & \\ \text{s.t.} & h(x,y) & = 0 \\ & g(x,y) & \leq 0 \\ & x & \in X \\ & y & \in Y \end{array}\right. \qquad \left[\begin{array}{rrl} & & \textbf{OA} \\ \min\limits_{x,y} & c^T y + f(x) & \\ \text{s.t.} & g(x) & \leq 0 \\ & Cx + By & \leq 0 \\ & x & \in X \subseteq \Re^n \\ & y & \in Y = \{0,1\}^q \end{array}\right.$$

In the MINLP model formulation that **GBD** can address we note that

(i) No separability/linearity assumption on the vector of y variables is made;

(ii) The sets X, Y need not be continuous and integer, respectively, but they can be mixed continuous–integer;

(iii) Nonlinear equality constraint can be treated explicitly;

(iv) A key motivation is to partition the variables into $\boldsymbol{x}$ and $\boldsymbol{y}$ in such a fashion that we exploit special structure existing in the model.

The **OA** algorithm, on the other hand, addresses a class of problems in which additional assumptions need to be made; that is,

(i) The $\boldsymbol{y}$–variables participate separably and linearly;

(ii) The $\boldsymbol{x}$–variables are continuous, while the $\boldsymbol{y}$–variables are integer;

(iii) Nonlinear equalities cannot be treated explicitly but only through algebraic or numerical elimination.

Note that the key feature of **GBD**, in constrast to **OA**, is that it allows for exploiting existing special structure in the $\boldsymbol{x}$–domain, $\boldsymbol{y}$–domain or in the $\boldsymbol{x}$–$\boldsymbol{y}$ domain. At the same time it allows for invoking special structure by appropriate definition of the $\boldsymbol{x}$ and $\boldsymbol{y}$ vectors and classification of the set of constraints. Examples of invoking special structure are the work of (a) Floudas *et al.* (1989), and Aggarwal and Floudas (1990b), who invoked convexity or linearity in $\boldsymbol{x}$ for fixed $\boldsymbol{y}$ and vice versa; (b) Floudas and Ciric (1989), who invoked the property of total unimodularity in the decomposed primal subproblem which allowed to be solved as a continuous problem and hence avoided the combinatorial nature of the original problem; and (c) Paules and Floudas (1992), who invoked staircase structure in the inner level subproblems for problem of stochastic programming in Process Synthesis.

6.8.2 Nonlinear Equality Constraints

The **GBD** algorithm can address nonlinear equality constraints explicitly without the need for algebraic or numerical elimination as in the case for **OA**.

The **OA/ER**, however, can also treat nonlinear equality constraints $\boldsymbol{h}(\boldsymbol{x}) = \boldsymbol{0}$ by transforming them into inequalities using the direction matrix T^k:

$$T^k \boldsymbol{h}(\boldsymbol{x}) \leq \boldsymbol{0},$$

$$T^k = \{t_{ii}\} = \begin{cases} -1 & \text{if} \quad \lambda_i^k < 0, \\ +1 & \text{if} \quad \lambda_i^k > 0, \\ 0 & \text{if} \quad \lambda_i^k = 0. \end{cases}$$

Note though that the additional condition of Quasi–convexity of $T^k \boldsymbol{h}(\boldsymbol{x})$ needs to be imposed so as to obtain equivalence between the original model and the one with the relaxed set of equalities into inequalities.

Therefore, the condition of quasi-convexity of $T^k \boldsymbol{h}(\boldsymbol{x})$ has to be checked at every iteration in the **OA/ER** algorithm, which is not needed in the **GBD** algorithm.

6.8.3 Nonlinearities in y and Joint $x - y$

The **GBD** algorithm can address models with nonlinearities in the $\boldsymbol{y}$ variables or in the joint $\boldsymbol{x}$–$\boldsymbol{y}$ domain explicitly, that is without necessarily introducing additional variables and/or constraints.

The **OA, OA/ER, OA/ER/AP** algorithms, cannot treat nonlinearities in $\boldsymbol{y}$ and joint $\boldsymbol{x}$–$\boldsymbol{y}$ explicitly since they are based on the assumptions of separability of $\boldsymbol{x}$ and $\boldsymbol{y}$ and linearity in the $\boldsymbol{y}$ vector of variables. They can however treat nonlinearities of the forms:

(i) $\psi(y)$ nonlinear

(ii) $\phi(x, y)$ nonlinear

by reformulation of the model that involves:

(a) Addition of new variables, and

(b) Addition of new constraints.

For instance, by introducing a new vector of continuous variables $\boldsymbol{x}^1$ and a new set of constraints $\boldsymbol{x}^1 - \boldsymbol{y} = \boldsymbol{0}$, the nonlinearities of forms (i) and (ii) can be written as

$$\begin{aligned} \phi(y) &= \phi(x^1), \\ \phi(x, y) &= \phi(x, x^1), \\ x^1 - y &= 0. \end{aligned}$$

Note that the new constraints $\boldsymbol{x}^1 - \boldsymbol{y} = \boldsymbol{0}$ are linear in $\boldsymbol{y}$ and hence satisfy the condition of separability/linearity of the **OA, OA/ER**, and **OA/ER/AP** algorithms.

Note also, however, that if we want to transform general nonlinear equality constraints of the form:

$$h(x, y) = 0,$$

then we will have

$$\begin{aligned} h(x, x^1) &= 0, \\ x^1 - y &= 0, \end{aligned}$$

and we subsequently need to check the equivalence conditions of

$$T^k h(x, x^1) \le 0,$$

that require quasi-convexity of $T^k h(x, x^1) \le 0$.

Therefore, there is a penalty associated with the aforementioned transformations in the **OA, OA/ER**, and **OA/ER/AP** algorithms so as to treat general nonlinearities. It is, however, the **GOA** approach that represents an alternative general way of treating only inequality constraints.

Remark 1 Note that the aforementioned transformations are similar to the ones used in **v2–GBD** of Floudas *et al.* (1989) even though they are more restrictive since they treat only the 0–1 $\boldsymbol{y}$ variables.

6.8.4 The Primal Problem

The primal problems of **GBD** and **OA**-based algorithms are similar but not necessarily identical. Their formulations for **GBD**, and **OA/ER**, **OA/ER/AP** are

$$\left[\begin{array}{llll} & \textbf{GBD} & & \\ \min\limits_{x} & f(x,y^k) & & \\ \text{s.t.} & h(x,y^k) & = & 0 \\ & g(x,y^k) & \leq & 0 \\ & x & \in & X \end{array}\right] \qquad \left[\begin{array}{llll} & & \textbf{OA/ER} & \\ \min\limits_{x} & c^T y^k + f(x) & & \\ & \text{s.t. } h(x) & \leq & 0 \\ & g(x) & \leq & 0 \\ & Cx + By^k & \leq & d \\ & x & \in & X \end{array}\right]$$

Note that in **GBD**, the primal problem can be a linear programming LP or nonlinear programming NLP problem depending on the selection of the y vector of variables which can be a mixed set of continuous integer variables.

In the **OA** and its variants, however, the primal problem will be a nonlinear programming NLP problem if any of the $f(x), h(x), g(x)$ is nonlinear.

Illustration 6.8.1 Let us consider the example:

$$\begin{aligned} \min_{x,y} \quad & y + x \\ \text{s.t.} \quad & x \exp(y) \leq 10 \\ & 0 \leq x \leq 10 \\ & y = 0, 1 \end{aligned}$$

In the **GBD** which can treat joint $x - y$ nonlinearities explicitly, the primal problem is

$$\begin{aligned} \min_{x} \quad & y^k + x \\ \text{s.t.} \quad & x \exp y^k \leq 10 \\ & 0 \leq x \leq 10 \end{aligned}$$

which is a linear programming LP problem in x.

In the **OA**, however, we need to introduce a new variable x^1 and a new constraint

$$x^1 - y = 0,$$

and transform the problem into

$$\begin{aligned} \min_{x,y,x^1} \quad & y + x \\ \text{s.t.} \quad & x \exp(x^1) \leq 10 \\ & x^1 - y = 0 \\ & 0 \leq x \leq 10 \\ & 0 \leq x^1 \leq 1 \\ & y = 0, 1 \end{aligned}$$

The primal problem then becomes

$$\begin{aligned}
\min_{x,x^1} \quad & y^k + x \\
\text{s.t.} \quad & x \exp(x^1) \leq 10 \\
& x^1 - y^k = 0 \\
& 0 \leq x \leq 10 \\
& 0 \leq x^1 \leq 1
\end{aligned}$$

which is a nonlinear programming NLP problem in x and x^1. Note though that x^1 can be eliminated from $x^1 - y^k = 0$.

6.8.5 The Master Problem

The master problem in **GBD** and its variants **v1**, **v2**, and **v3–GBD** involve one additional constraint at each iteration. This additional constraint is the support function of the Lagrange function (see section 6.3). As a result, the master problem in **GBD** and its variants correspond to a small size problem even when a large number of iterations has been reached.

The master problem in **OA** and it variants **OA/ER**, **OA/ER/AP** involves linearizations of the nonlinear objective function, the vector of transformed nonlinear equalities and the original nonlinear inequalities around the optimum solution $\boldsymbol{x}^k$ of the primal problem at each iteration. As a result, a large number of constraints are added at each iteration to the master problem. Therefore, if convergence has not been reached in a few iterations the effort of solving the master problem, which is a mixed-integer linear programming MILP problem, increases.

Since the master problem in **OA** and its variants has many more constraints than the master problem in **GBD** and its variants, the lower bound provided by **OA** is expected to be better than the lower bound provided by the **GBD**. To be fair, however, in the comparison between **GBD** and **OA** we need to consider the variant of **GBD** that satisfied the conditions of **OA**, namely, separability and linearity of the $\boldsymbol{y}$ variables as well as the convexity condition and the constraint qualification, instead of the general **GBD** algorithm. The appropriate variant of **GBD** for comparison is the **v2–GBD** under the conditions of separability and linearity of the $\boldsymbol{y}$ vector. Duran and Grossmann (1986a) showed that

$$(LBD)^k_{\mathbf{OA}} \geq (LBD)^k_{\mathbf{v2\text{–}GBD}};$$

that is, under convexity, conditions separability/linearity conditions; and the constraint qualification condition, the lower bound provided by the **OA** algorithm at each iteration k is tighter than the lower bound provided by **v2–GBD** algorithm. This implies that convergence to the solution of the MINLP model will be achieved in fewer iterations via the **OA** instead of the **v2–GBD** algorithm. This, however, does not necessarily imply that the total computational effort is smaller in the **OA** than the **v2–GBD** algorithm, simply because the master problem of the **OA** involves many constraints which correspond to linearizations around $\boldsymbol{x}^k$, while the **v2–GBD** features one constraint per iteration. Hence, there is a trade-off between the quality of the lower bounds and the total computational effort.

Illustration 6.8.2 We consider here the illustrative example used in section 6.6.5 of the **OA/ER**, and apply the **v2–GBD** on model (6.32) starting from the same starting point $\boldsymbol{y}^1 = (1, 1, 0)$.

The solution of the primal problem is

$$\begin{aligned} C &= 1, \\ B_1 &= 0, \\ B_2 &= (1/0.9) = 1.1111, \\ B_3 &= 0, \\ A_2 &= 2.03773, \\ A_3 &= 0, \\ OBJ &= -1.72097, \end{aligned}$$

with Lagrange multipliers

$$\begin{aligned} \lambda_1^1 &= 5.46792, \\ \lambda_2^1 &= 0, \\ \lambda_3^1 &= 7.77777, \end{aligned}$$

$$\begin{aligned} \mu_1^1 &= 3.2222, \\ \mu_2^1 &= 0.53209, \\ \mu_3^1 &= 5.8. \end{aligned}$$

Then the Lagrange function becomes

$$\begin{aligned} L(\boldsymbol{x}^1, \boldsymbol{y}, \lambda^1, \mu^1) &= -6.22097 + 3.5y_1 + y_2 + 1.5y_3 \\ &\quad +\lambda_1^1 (B_2^1 - \ln(1 + A_2^1)) \\ &\quad +\lambda_2^1 (B_3^1 - 1.2 \ln(1 + A_2^1)) \\ &\quad +\lambda_3^1 (C^1 - 0.9(B_1^1 + B_2^1 + B_3)) \\ &\quad +\mu_1^1 (C^1 - y^1) \\ &\quad +\mu_2^1 (B_2^1 - (1/0.9)y_2) \\ &\quad +\mu_3^1 (B_3^1 - (1/0.9)y_3) \\ &= -6.22097 + 3.5y_1 + y_2 + 1.5y_3 \\ &\quad +3.2222(1 - y_1) \\ &\quad +0.53209(1.1111 - 1.1111y_2) \\ &\quad +4.3(0 - 1.1111y_3) \\ &= -0.2778y_1 + 0.4088y_2 - 4.9444y_3 - 2.4076. \end{aligned}$$

Then the master problem for the **v2–GBD** is

$$\begin{aligned} \min \quad & \mu_B \\ \text{s.t.} \quad & \mu_B \geq 0.2778y_1 + 0.4088y_2 - 4.9444y_3 - 2.4076 \\ & y_2 + y_3 \leq 1 \\ & y_1, y_2, y_3 = 0, 1 \end{aligned}$$

and has solution:

$$\begin{aligned} y_1 &= 0, \\ y_2 &= 0, \\ y_3 &= 1, \\ \mu_B &= -7.352. \end{aligned}$$

Then after one iteration of the **v2–GBD** we have

$$\begin{aligned} UBD &= -1.72097, \\ LBD &= -7.352. \end{aligned}$$

Note that after one iteration

$$(LBD)_{\mathbf{OA}} = -3 > (LBD)_{v2-\mathbf{GBD}} = -7.352.$$

6.8.6 Lower Bounds

In the previous section we showed that if convexity holds along with quasi-convexity of $T^k \boldsymbol{h}(\boldsymbol{x}) \leq \boldsymbol{0}$ and the constraint qualification, then the **OA/ER** terminates in fewer iterations than the **v2–GBD** since the lower bound of the **OA/ER** is better than the one of the **v2–GBD**.

It should be noted, however, that we need to check whether $T^k \boldsymbol{h}(\boldsymbol{x}) \leq \boldsymbol{0}$ is quasi-convex at each iteration. If the quasi-convexity condition is not satisfied, then the obtained lower bound by **OA/ER** may not be a valid one; that is, it may be above the global solution of the MINLP model. This may happen due to the potential invalid linearizations which may cut off part of the feasible region.

For the case of **OA/ER/AP**, the termination criterion corresponds to a heuristic and hence the correct optimum solution may not be obtained.

If the convexity conditions are satisfied along with the constraint qualification then the **GBD** variants and the **OA** attain their global optimum.

6.9 Generalized Cross Decomposition, GCD

6.9.1 Formulation

Holmberg (1990) generalized the approach proposed by Van Roy (1983) to the class of optimization problems stated as

$$\begin{aligned} \min_{x,y} \quad & f(x,y) \\ \text{s.t.} \quad & g_1(x,y) \leq 0 \\ & g_2(x,y) \leq 0 \\ & x \in X \subseteq \Re^n \\ & y \in Y = \{0,1\}^q \end{aligned} \tag{6.52}$$

under the following conditions:

C1: The functions

$$\begin{aligned} f &: \Re^n \times \Re^q \longrightarrow \Re, \\ g_1 &: \Re^n \times \Re^q \longrightarrow \Re^{p_1}, \\ g_2 &: \Re^n \times \Re^q \longrightarrow \Re^{p_2}, \end{aligned}$$

are proper convex functions for each fixed $y \in Y = \{0,1\}^q$ (e.g., f is proper convex if $\text{dom}(f) = \{x \in \Re^n | f < +\infty\}$ is not empty and $f > -\infty, \ \forall x \in \text{dom}(f)$).

C2: X is a nonempty, compact, convex set and the functions f, g_1, and g_2 are bounded and Lipschitzian on (X, Y).

C3: the optimization with respect to x of the Lagrange function can be performed independently of y.

Remark 1 Formulation (6.52) corresponds to a subclass of the problems which the generalized cross decomposition **GCD** of Holmberg (1990) can be applied. Holmberg (1990) studied the more general case of $y \subseteq \Re^q$ and defined the vector y in a similar way that Geoffrion (1972) did for the generalized benders decomposition **GBD**.

Remark 2 Condition **C3** is the property **P** of Geoffrion (1972) which was discussed in the section of **v1–GBD**. Since the primal problem can be feasible or infeasible and the Lagrange functions generated have a different form, condition **C3** can be interpreted as follows:

Case I - Feasible Primal

Condition **C3** is that there exist functions q_1 and q_3 such that

$$L(x,y,\mu_1,\mu_2) = f(x,y) + \mu_1^T g_1(x,y) + \mu_2^T g_2(x,y) = q_1\left(q_3(x,\mu_1,\mu_2), y, \mu_1, \mu_2\right)$$

$$\forall x \in X, \forall y \in Y, \forall \mu_1 \geq 0, \mu_2 \geq 0,$$

where q_3 is a scalar function and q_1 is increasing in its first argument.

Case II - Infeasible Primal

Condition **C3** is that there exist functions q_2 and q_4 such that

$$\bar{L}(x, y, \bar{\mu_1}, \bar{\mu_2}) = \bar{\mu_1}^T g_1(x, y) + \bar{\mu_2}^T g_2(x, y) = q_2(q_4(x, \bar{\mu_1}, \bar{\mu_2}), y, \bar{\mu_1}, \bar{\mu_2})$$

$$\forall x \in X, \forall y \in Y, \forall (\bar{\mu_1}, \bar{\mu_2}) \in \Lambda,$$

$$\text{where} \quad \Lambda = \left\{ \bar{\mu_1} \in \Re^{p_1},\ \bar{\mu_2} \in \Re^{p_2} \ : \ \bar{\mu_1} \geq 0,\ \bar{\mu_2} \geq 0,\ \sum_{i=1}^{p_1+p_2} \mu_i = 1 \right\},$$

q_4 is a scalar function , and

q_2 is increasing in its first argument.

Remark 3 Note that due to conditions **C1** and **C2** we have that

(i) q_1 is proper convex, bounded, and Lipschitzian on (X, Y) for fixed $\mu_1 \geq 0,\ \mu_2 \geq 0$;

(ii) q_1 is proper convex, bounded, and Lipschitzian on (X, Y) for fixed $(\bar{\mu_1}, \bar{\mu_2}) \in \Lambda$.

6.9.2 Basic Idea

The Generalized Cross Decomposition **GCD** consists of two phases, that is the primal and dual subproblem phase (i.e., phase I) and the master problem phase (i.e., phase II), and appropriate convergence tests. In phase I, the primal subproblem provides an upper bound on the sought solution of (6.52) and Lagrange multipliers μ_1^k, μ_2^k for the dual subproblem. The dual subproblem provides a lower bound on the solution of (6.52) and supplies y^k for the primal subproblem. Both the primal and the dual subproblem generate cuts for the master problem in phase II. At each iteration of the **GCD**, a primal and dual subproblem are solved, and a primal convergence test is applied on y^k, while a dual convergence test is applied on μ_1^k, μ_2^k. If any convergence test fails, then we enter phase II that features the solution of a master problem and return subsequently to phase I. Figure 6.9 depicts the essential characteristics of **GCD**.

The key idea in **GCD** is to make extensive use of phase I (i.e., primal, dual subproblems) and limit as much as possible the use of phase II (i.e., master problem) by the application of appropriate convergence tests. This is because the master problem is known to be a more difficult and cpu time consuming problem, than the primal and dual subproblems of phase I.

6.9.3 Theoretical Development

This section presents the theoretical development of the Generalized Cross Decomposition, **GCD**. Phase I is discussed first with the analysis of the primal and dual subproblems. Phase II is presented subsequently for the derivation of the problem while the convergence tests are discussed last.

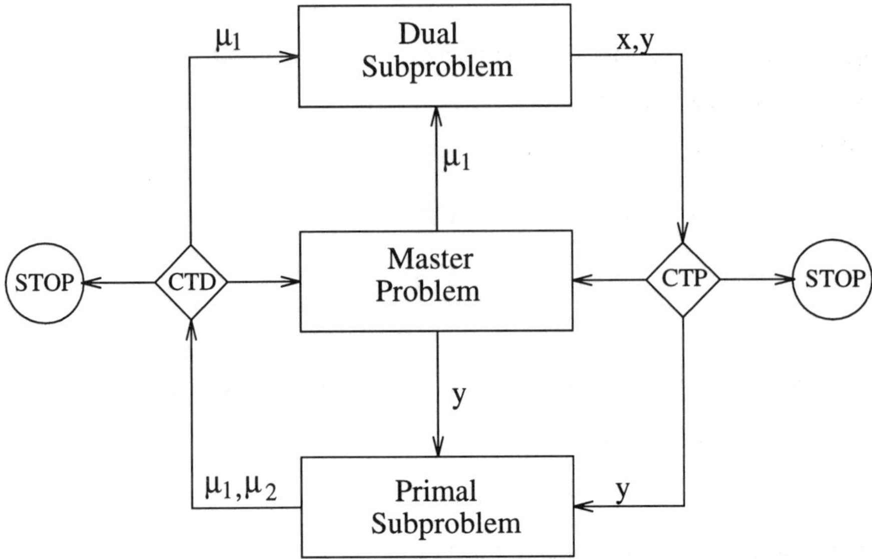

Figure 6.9: Pictorial representation of generalized cross decomposition

6.9.3.1 Phase I - Primal and Dual Subproblems

The primal subproblem results from fixing the $\boldsymbol{y}$–vector of variables in (6.52) to a particular 0–1 combination denoted as $\boldsymbol{y}^k$ and takes the form

$$\left[\begin{array}{lrcl} \min\limits_{\boldsymbol{x}} & f(\boldsymbol{x},\boldsymbol{y}^k) & & \\ s.t. & g_1(\boldsymbol{x},\boldsymbol{y}^k) & = & \mathbf{0} \\ & g_2(\boldsymbol{x},\boldsymbol{y}^k) & \leq & \mathbf{0} \\ & \boldsymbol{x} & \in & \boldsymbol{X} \subseteq \Re^n \end{array}\right. \qquad (P(\boldsymbol{y}^k))$$

Remark 1 The primal subproblem $P(\boldsymbol{y}^k)$ is convex in $\boldsymbol{x}$, due to condition **C1**.

Fixing the $\boldsymbol{y}$ variables to $\boldsymbol{y}^k$ may result in a feasible or infeasible primal subproblem. Hence, we will discuss the two cases separately in the following:

Case (i): Feasible Primal Subproblem

A feasible solution to $P(\boldsymbol{y}^k)$ consists of $\boldsymbol{x}^k$, $P(\boldsymbol{y}^k)$ which is the upper bound, and the Lagrange multipliers μ_1^k, μ_2^k. The Lagrange function then takes the form:

$$L(\boldsymbol{x},\boldsymbol{y},\mu_1^k,\mu_2^k) = f(\boldsymbol{x},\boldsymbol{y}) + {\mu_1^k}^T g_1(\boldsymbol{x},\boldsymbol{y}) + {\mu_2^k}^T g_2(\boldsymbol{x},\boldsymbol{y})$$

Case (ii): Infeasible Primal Subproblem

If the primal subproblem is infeasible, then the following feasibility problem is formulated:

$$\left[\begin{array}{ll} \min\limits_{x} & \alpha \\ s.t. & g_1(x,y^k) \leq \alpha \\ & g_2(x,y^k) \leq \alpha \\ & x \in X \subseteq \Re^n \end{array}\right. \qquad (FP(y^k))$$

The solution of $FP(y^k)$, which is convex, provides the Lagrange multipliers $\bar{\mu}_1{}^l, \bar{\mu}_2{}^l$ for the inequality constraints. Then, the Lagrange function takes the form:

$$\bar{L}(x,y,\bar{\mu}_1{}^l,\bar{\mu}_2{}^l) = \bar{\mu}_1{}^{lT} g_1(x,y) + \bar{\mu}_2{}^{lT} g_2(x,y),$$

Remark 2 Note that $FP(y^k)$ does not provide an upper bound for (6.52).

Having obtained μ_1^k (i.e., in case of feasible primal) from the solution of $P(y^k)$, the dual subproblem, for the case of feasible primal, takes the following form:

$$\left[\begin{array}{ll} \min\limits_{x,y} & f(x,y) + \mu_1^{kT} g_1(x,y) \\ s.t. & g_2(x,y) \leq 0 \\ & x \in X \subseteq \Re^n \\ & y \in Y = \{0-1\}^q \end{array}\right. \qquad (D(\mu_1^k))$$

Remark 3 The dual subproblem $D(\mu_1^k)$ can be nonconvex in the joint $x - y$ space. As a result, its solution may not correspond to a valid lower bound on (6.52).

Remark 4 Note that the objective function in $D(\mu_1^k)$ corresponds to a Lagrange relaxation.

If the primal subproblem is infeasible, then the dual subproblem takes the form

$$\left[\begin{array}{ll} \min\limits_{x,y} & \bar{\mu}_1^{lT} g_1(x,y) \\ s.t. & g_2(x,y) \leq 0 \\ & x \in X \subseteq \Re^n \\ & y \in Y = \{0-1\}^q \end{array}\right. \qquad (FD(\bar{\mu}_1^{lT}))$$

where $\bar{\mu}_1^l$ are the Lagrange multipliers of $g_1(x,y^k) \leq 0$ for the feasibility problem $FP(y^k)$.

Remark 5 Note that the solution of $FD(\bar{\mu}_1^l)$ does not provide any bound on the (6.52), and it can only provide a dual cut that will eliminate $\bar{\mu}_1^l$ from further consideration.

6.9.3.2 Phase II - Master Problem

The master problem can be formulated by using primal or dual information. As a result, there are two types of master problems: the primal master and the Lagrange Relaxation master problem. In the sequel, we will discuss the derivation of each master problem.

Primal Master Problem

The derivation of the primal master problem follows the same steps as the derivation of the master problem in Generalized Benders Decomposition **GBD**. The final form of the primal master problem is:

$$\left[\begin{array}{llll} \min\limits_{y,\mu_c} & \mu_c & & \\ s.t. & \mu_c & \geq & \inf\limits_{x\in X} f(x,y) + \mu_1^T g_1(x,y) + \mu_2^T g_2(x,y), \;\; \forall(\mu_1,\mu_2) \geq 0 \\ & 0 & \geq & \inf\limits_{x\in X} \bar{\mu}_1^T g_1(x,y) + \bar{\mu}_2^T g_2(x,y), \;\; \forall(\bar{\mu}_1,\bar{\mu}_2) \in \Lambda \\ & y & \in & Y \end{array}\right. \tag{PM}$$

Remark 1 The primal master, **(PM)**, has an infinite number of cuts that correspond to each nonnegative pair of Lagrange multipliers (μ_1, μ_2) and to each $(\bar{\mu}_1, \bar{\mu}_2) \in \Lambda$. Each cut involves an optimization problem (i.e. minimization with respect to $x \in X$ of $L(x, y, \mu_1, \mu_2)$ or $\bar{L}(x, y, \bar{\mu}_1, \bar{\mu}_2)$ which is parametric in $y \in Y$ and in theory should be solved for all $y \in Y$.

Remark 2 Using condition **C3**, which was analyzed in the formulation section, the cuts that correspond to feasible primal problems can be written as

$$\begin{array}{rcl} \inf\limits_{x\in X} L(x,y,\mu_1,\mu_2) & = & \inf\limits_{x\in X} q_1(q_3(x,\mu_1,\mu_2),y,\mu_1,\mu_2) \\ & & \forall\, y,\;\; \forall(\mu_1,\mu_2) \;\geq\; 0 \end{array} \tag{6.53}$$

Since q_1 is proper, convex, bounded, and Lipschitzian on X, and X is compact and convex, the infimum in (6-50) is attained, and hence can be replaced by minimum. Similarly the cuts that correspond to infeasible primal problems can be written in terms of q_2, q_4 as follows:

$$\begin{array}{rcl} \inf\limits_{x\in X} \bar{L}(x,y,\bar{\mu}_1,\bar{\mu}_2) & = & \inf\limits_{x\in X} q_2(q_4(x,\bar{\mu}_1,\bar{\mu}_2),y,\bar{\mu}_1,\bar{\mu}_2) \\ & & \forall\, y,\;\; \forall(\bar{\mu}_1,\bar{\mu}_2) \;\geq\; 0 \end{array} \tag{6.54}$$

Since q_2 is proper, convex, bounded, and Lipschitzian on X, and X is compact and convex, the infimum can be replaced by the minimum with respect to x.

Remark 3 Since q_1 and q_2 are increasing in their first arguments, q_3 and q_4, respectively, the minimization of q_1 and q_2 with respect to $x \in X$ can be performed in $q_3(x, \mu_1, \mu_2)$ and $q_4(x, \bar{\mu}_1, \bar{\mu}_2)$ instead, that is,

$$\min_{x \in X} q_1(q_3(x, \mu_1, \mu_2), y, \mu_1, \mu_2) = q_1(\min_{x \in X} q_3(x, \mu_1, \mu_2), y, \mu_1, \mu_2) \quad \forall\, y,\ \forall\, (\mu_1, \mu_2) \geq 0, \tag{6.55}$$

$$\min_{x \in X} q_2(q_4(x, \bar{\mu}_1, \bar{\mu}_2), y, \bar{\mu}_1, \bar{\mu}_2) = q_2(\min_{x \in X} q_4(x, \bar{\mu}_1, \bar{\mu}_2), y, \bar{\mu}_1, \bar{\mu}_2) \quad \forall\, y,\ \forall (\bar{\mu}_1, \bar{\mu}_2) \geq 0. \tag{6.56}$$

Note that y does not affect the minimization with respect to $x \in X$, and hence the minimizations

$$\min_{x \in X} \quad q_3(x, \mu_1, \mu_2) \quad \text{and}$$
$$\min_{x \in X} \quad q_4(x, \mu_1, \mu_2)$$

need to be performed only once and will be valid $\forall\, y \in Y$. Then, the primal master problem takes the form:

$$\left[\begin{array}{llll} \min\limits_{y, \mu_c} & \mu_c & & \\ s.t. & \mu_c & \geq & q_1\left(\min\limits_{x \in X} q_3(x, \mu_1, \mu_2)\right) \quad \forall\, (\mu_1, \mu_2) \geq 0 \\ & 0 & \geq & q_2\left(\min\limits_{x \in X} q_4(x, \bar{\mu}_1, \bar{\mu}_2), y, \bar{\mu}_1, \bar{\mu}_2)\right) \quad \forall (\bar{\mu}_1, \bar{\mu}_2) \in \Lambda \\ & y & \in & Y \end{array}\right. \tag{6.57}$$

Remark 4 Formulation 6.57 still contains an infinite number of cuts. Selecting a finite number of cuts can be done by fixing (μ_1, μ_2) and $(\bar{\mu}_1, \bar{\mu}_2)$ to (μ_1^k, μ_2^k), $k \in K$ and $(\bar{\mu}_1^l, \bar{\mu}_2^l)$, $l \in L$, respectively, where k and l and the indices corresponding to the iterations performed. Then we have the following relaxed primal master problem:

$$\left[\begin{array}{llll} \min\limits_{y, \mu_c} & \mu_c & & \\ s.t. & \mu_c & \geq & q_1\left(\min\limits_{x \in X} q_3(x, \mu_1^k, \mu_2^k)\right) \quad \forall\, k = 1, 2, \ldots, K \\ & 0 & \geq & q_2\left(\min\limits_{x \in X} q_4(x, \bar{\mu}_1^l, \bar{\mu}_2^l), y, \bar{\mu}_1^l, \bar{\mu}_2^l)\right) \quad \forall\, l = 1, 2, \ldots L \\ & y & \in & Y \end{array}\right. \tag{6.58}$$

Remark 5 For each (μ_1^k, μ_2^k), the minimization of q_3 with respect to $x \in X$; that is,

$$\min_{x \in X} q_3(x, \mu_1^k, \mu_2^k)$$

can be done explicitly, and let us denote its minimizers as x^k. Similarly, for each $(\bar{\mu}_1^l, \bar{\mu}_2^l)$ the minimization of q_4 with respect to $x \in X$; that is,

$$\min_{x \in X} q_4(x, \bar{\mu}_1^l, \bar{\mu}_2^l)$$

can be carried out independently and let us denote its minimizers as $\bar{x}^l$. Then we have

$$\min_{x \in X} q_3(x, \mu_1^k, \mu_2^k) = q_3(x^k, \mu_1^k, \mu_2^k), \quad k = 1, 2, \ldots, K \tag{6.59}$$

$$\min_{x \in X} q_4(x, \bar{\mu}_1^l, \bar{\mu}_2^l) = q_4(x^k, \bar{\mu}_1^l, \bar{\mu}_2^l), \quad l = 1, 2, \ldots, L \tag{6.60}$$

and q_1, q_2 take the form:

$$q_1(q_3(x^k, \mu_1^k, \mu_2^k), y, \mu_1^k, \mu_1^k), \quad k = 1, \ldots, K,$$

$$q_2(q_4(x^k, \bar{\mu}_1^l, \bar{\mu}_2^l), y, \bar{\mu}_1^l, \bar{\mu}_2^l), \quad l = 1, 2, \ldots, L.$$

Denoting the cuts of (6.59) and (6.60) as $q_1^k(y)$, and $q_2^l(y)$, respectively, the relaxed primal master problem **(RPM)** (6.59) takes the following final form:

$$\left[\begin{array}{lrcl} \min\limits_{y, \mu_c} & \mu_c & & \\ s.t. & \mu_c & \geq & q_1^k(y), \quad k = 1, 2, \ldots, K \\ & 0 & \geq & q_2^l(y), \quad l = 1, 2, \ldots, L \\ & y & \in & Y \end{array}\right. \quad \textbf{(RPM)}$$

$$\begin{array}{rrcl} \text{where} & q_1^k(y) & = & q_1(q_3(x^k, \mu_1^k, \mu_2^k), y, \mu_1^k, \mu_2^k), \\ & q_2^l(y) & = & q_2(q_4(\bar{x}^l, \bar{\mu}_1^l, \bar{\mu}_2^l), y, \bar{\mu}_1^l, \bar{\mu}_2^l). \end{array}$$

Remark 6 The **(RPM)** problem represents a lower bound on the optimal solution of (6.52).

Lagrange Relaxation Master Problem

The derivation of the Lagrange relaxation master problem employs Lagrangian duality and considers the dualization of the $g_1(x,y) \leq 0$ constraints only. The dual takes the following form:

$$\max_{\mu_1 \geq 0} \left[\begin{array}{ll} \min\limits_{x \in X, y \in Y} & f(x,y) + \mu_1^T g_1(x,y) \\ s.t. & g_2(x,y) \leq 0 \\ & x \in X \\ & y \in Y \end{array}\right] \tag{6.61}$$

Remark 7 Note that the inner problem in (6.61) is parametric in μ_1, and for fixed value of $\mu_1 = \mu_1^k$ it corresponds to the dual subproblem $D(\mu_1^k)$ presented in phase I.

If we denote the solution of the dual subproblem $D(\mu_1^k)$ as (x^k, y^k), assuming feasibility, and define

$$h^k(\mu_1) = f(x^k, y^k) + \mu_1^T g_1(x^k, y^k), \quad k = 1, 2, \ldots, K,$$

then, the relaxed Lagrange relaxation master **RLRM** problem becomes

$$\left[\begin{array}{lll} \max\limits_{\mu_1, \mu_c'} & \mu_c' & \\ \text{s.t.} & \mu_c' \leq h^k(\mu_1), & k = 1, 2, \ldots, K \\ & \mu_1 \geq 0 & \end{array}\right. \qquad \textbf{(RLRM)}$$

$$\text{where} \quad h^k(\mu_1) = f(x^k, y^k) + \mu_1^T g_1(x^k, y^k).$$

Remark 8 Solution of **RLRM** will provide a valid upper bound in (6.52) only if (6.52) is convex. If it is not convex then a duality gap may occur and **RLRM** can only be used as a heuristic. Note also that **RLRM** is an upper bound on 6.61.

Remark 9 If the primal subproblem is infeasible, then the dual subproblem takes the form of $FD(\bar{\mu}_1^l)$ (see phase I).

In such cases, we introduce

$$h^l(\bar{\mu}_1) = \bar{\mu}_1^{l^T} g_1(x^k, y^k), \quad l = 1, \ldots, L$$

and $0 \leq h^l(\bar{\mu}_1), \quad l = 1, \ldots, L$ in the **RLRM**.

Remark 10 The solution of the dual subproblem $D(\mu_1^k)$ is a valid lower bound on the Lagrangean dual (6.61). The solution of the dual subproblem $D(\mu_1^k)$ will also be a valid lower bound on (6.52) if the $D(\mu_1^k)$ is convex in the joint $x - y$ space.

6.9.3.3 Convergence Tests

The convergence tests of the **GCD** make use of the notions of (i) upper bound improvement, (ii) lower bound improvement, and (iii) cut improvement. An upper bound improvement corresponds to a decrease in the upper bound UBD obtained by the primal subproblem $P(y^k)$. A lower bound improvement corresponds to an increase in the lower bound LBD obtained by the dual subproblem $D(\mu_1^k)$. A cut improvement corresponds to generating a new cut which becomes active and hence is not dominated by the cuts generated in previous iterations. If the cut is generated in the relaxed primal master problem **(RPM)** it is denoted as a primal cut improvement. If the cut is generated in the relaxed Lagrange relaxation master problem then the improvement is classified as Lagrange relaxation cut improvement.

The basic idea in the convergence tests CT is to provide answers to the following three questions:

Q1: Can y^k provide an upper bound improvement [i.e., can the solution of $P(y^k)$ be strictly less than the current UBD] ?

Q2: Can y^k provide a lower bound improvement [i.e., can the solution of $D(\mu_1^k)$ be strictly greater than the current LBD] ?

Q3: Can we obtain a Lagrange relaxation cut improvement for $\bar{\mu}_1^k$ (i.e., for unbounded solutions) ?

The convergence tests CT that provide the answers to the aforementioned questions are formulated as

CTP: If $q_1^k(y^c) < UBD$ for $k = 1, \ldots, K$, and $q_2^l(y^c) \leq 0$ for $l = 1, \ldots, L$, where y^c is the current y, then y^c will provide an upper bound improvement. If not, use a master problem.

CTD: If $h^k(\mu_1^c) > LBD$ for $k = 1, \ldots, K$, where μ_1^c is the current μ_1, then μ_1^c will provide a lower bound improvement. If not, use a master problem.

CTDU: If $h^l(\bar{\mu}_1^c) > 0$ for $l = 1, \ldots, L$, where $\bar{\mu}_1^c$ is the current $\bar{\mu}_1$, then $\bar{\mu}_1^c$ will provide a cut improvement. If not, use a master problem.

Remark 1 The first condition of the CTP test (i.e., $q_1^k(y^c) < UBD$ for $k = 1, \ldots, K$) and the CTD test correspond to feasible problems and as a result are denoted as "value convergence" tests. The second condition of the CTP test (i.e., $q_2^l(y^c) \leq 0$ for $l = 1, \ldots, L$) and the CTDU test correspond to feasibility problems and are denoted as "feasibility convergence" tests.

Holmberg (1990) proved the following theorem and lemma for finite termination after applying the convergence tests CT:

Theorem 6.9.1
The convergence tests CTP and CTD are necessary for bound improvement and sufficient for bound improvement or cut improvement. The convergence test CTDU is sufficient for cut improvement.

Lemma 6.9.1 For model (6.52) in which Y is a finite discrete set the convergence test CT will fail after a finite number of iterations, and hence the generalized cross decomposition **GBD** algorithms will solve (6.52) exactly in a finite number of steps.

Remark 2 Note that the convergence tests CT contain a primal part (i.e., CTP) and a dual part (i.e., CTD). As a result, it suffices if one of them fails after a finite number of steps. It is in fact the CTP test which makes certain that the convergence tests CT will fail after a finite number of iterations. Note also that the number of optimal solutions to $D(\mu_1^k)$ for different μ_1^k is infinite which implies that the lower bound improvements can occur an infinite number of times without having the CTD test failing. Similarly, the cut improvement can be repeated infinitely and as a result the CTDU test can provide an infinite number of cut improvements without failing. Based on the above discussion, to attain finite convergence (i.e., have the convergence tests CT failing after a finite number of steps) we have to rely on the CTP test which is based on the primal master problem. Therefore, we have to make certain that the primal master problem is used, and we do not exclusively use the Lagrange relaxation master problem.

6.9.4 Algorithmic Development

Figure 6.9 presented the generic algorithmic steps of the generalized cross decomposition **GCD** algorithm, while in the previous section we discussed the primal and dual subproblems, the relaxed primal master problem, the relaxed Lagrange relaxation master problem, and the convergence tests.

Prior to presenting the detailed algorithmic statement of **GCD**, we introduce the following definitions:

P^k**:** the optimal objective value of the primal subproblem $P(y^k)$. This is a valid upper bound on (6.52).

UBD the updated lowest objective value of $P(y^k)$'s, that is $UBD = \min_k P(y^k)$. This is the tightest current upper bound.

$(\mu_C)^k$**:** the optimal objective function value of the relaxed primal master problem **RPM**, with K cuts. This is a valid lower bound on (6.52).

D^k the optimal objective value of the dual subproblem $D(\mu_1^k)$. This is a lower bound on the Lagrangian dual and a valid lower bound on (6.52) only if $D(\mu_1^k)$ is convex in $x - y$.

LBD**:** the updated highest value of the lower bound. If $D(\mu_1^k)$ are convex in $x - y$, then

$$LBD = \max_k \{D^k, (\mu_C)^k\}.$$

If $D(\mu_1^k)$ are not convex in $x - y$, then

$$LBD = \max_k \{(\mu_C)^k\},$$

$(\mu'_C)^k$**:** the optimal objective function value of the relaxed Lagrange relaxation master problem **RLRM**. This is a valid upper bound on the Lagrangean dual problem.

Remark 1 Note that P^k does not have a monotonic behavior; that is, it may fluctuate. UBD, however, is monotonically nonincreasing. Note also that D^k does not satisfy any monotonicity and hence it may also fluctuate. The $\max_k\{D^k\}$ however is monotonically nondecreasing.

Remark 2 The LBD is monotonically nondecreasing, and $(\mu'_C)^k$ is monotonically nonincreasing.

Remark 3 The upper bounding sequence $\{P^k\}$, and the lower bounding sequence $\{\mu_C^k\}$ will converge to the optimal value of (6.52). This corresponds to the Generalized Benders Decomposition **GBD** part of the Generalized Cross Decomposition.

Remark 4 The upper bounding sequence $\{\mu_C'^k\}$ and the lower bounding sequence $\{D^k\}$ will converge to the optimal value of the Lagrange dual (6.58). This corresponds to the Dantzig-Wolfe part of the Generalized Cross Decomposition. The optimal value of the Lagrange dual (6.58) is in general less than the optimal value of (6.52) due to potential existence of a duality if the problem is nonconvex.

Remark 5 If $D(\mu_1^k)$ are convex then the **GCD** utilizes the best lower bound of the **GBD** and Dantzig-Wolfe.

6.9.4.1 Algorithmic Statement of GCD

Assuming the the problem (6.52) has a finite optimal solution value, we can now state the complete **GCD** algorithm. Figure 6.10 depicts the algorithmic steps of **GCD**.

Step 1: Set the iteration counters $k = 1$ for feasible primal subproblems, and $l = 1$ for infeasible primal subproblem. Set the $UBD = +\infty$ and the $LBD = -\infty$, and select the convergence tolerance $\epsilon \geq 0$. Choose an initial point y^1.

Step 2: Apply the CTP test for $y = \hat{y}$ (i.e., for the current y except the y^1 of the starting point). If the CTP test is passed, then go to step 3. Otherwise, go to step 4.

Step 3: Solve the primal problem for $y = \hat{y}$, that is the problem $P(\hat{y})$. We have two cases: feasible and infeasible primal subproblems:

Step 3A: Feasible Primal $P(\hat{y})$

The primal subproblem has an optimal objective value $\hat{P}$, a solution $\hat{x}$, and multiplier vectors $\hat{\mu}_1, \hat{\mu}_2$. Update the upper bound $UBD = \min_{\hat{y}}\{\hat{P}\}$. If $|UBD - LBD| \leq \epsilon$, then terminate. Otherwise, go to step 5A.

Step 3B: Infeasible Primal $P(\hat{y})$

The primal subproblem does not have a feasible solution for $y = \hat{y}$. Solve the feasibility problem $FP(\hat{y})$ to determine the multiplier vectors $\hat{\bar{\mu}}_1, \hat{\bar{\mu}}_2$. Go to step 5B.

Step 4: Solve a relaxed master problem. We distinguish two cases, the relaxed primal master and the relaxed Lagrange relaxation master problem:

Step 4A: Relaxed Primal Master (RPM)

Let $(\hat{y}, \hat{\mu}_C)$ be an optimal solution of the relaxed primal master problem. $\hat{\mu}_C$ is a lower bound on (6.52); that is, the current lower bound is $LBD = \hat{\mu}_C$. If $|UBD - LBD| \leq \epsilon$, then terminate. Otherwise, go to step 3.

Step 4B: Relaxed Lagrange Relaxation Master (RLRM)

Let $(\hat{\mu}_1, \hat{\mu}'_C)$ be an optimal solution of the relaxed Lagrange relaxation master problem. $\hat{\mu}'_C$ is a valid upper bound on (6.52) only if (6.52) is convex. Go to step 6.

Step 5: Apply the CTD test for $\mu_1 = \hat{\mu}_1$ of the CTDU test for $\mu_1 = \hat{\bar{\mu}}_1$.

step 5A: CTD test for $\mu_1 = \hat{\mu}_1$

If the CTD test is passed, go to step 6. Otherwise, go to step 8 or go to step 8 or step 4B.

Step 5B: CTDU test for $\hat{\bar{\mu}}_1 = \bar{\mu}_1$

If the CTD test is passed, go to step 7. Otherwise, go to step 8 or step 4B.

Step 6: Solve the dual subproblem for $\mu_1 = \hat{\mu}_1$; that is, the problem $D(\hat{\mu}_1)$. Let $(\hat{x}, \hat{y}, \hat{D})$ be its solution. Update the lower bound by $LBD = \max\{LBD, \hat{D}\}$ by assuming that $D(\hat{\mu}_1)$ is convex in $\boldsymbol{x} - \boldsymbol{y}$. If $|UBD - LBD| \leq \epsilon$, then terminate. Otherwise, set $k = k + 1, \boldsymbol{y}^{k+1} = \hat{\boldsymbol{y}}$, and return to step 2.

Step 7: Solve the dual subproblem $FD(\bar{\mu}_1)$ which results if the primal subproblem is infeasible. Let $(\hat{\boldsymbol{x}}, \hat{\boldsymbol{y}})$ be the solution of $FD(\bar{\mu}_1)$. Set $l = l + 1$, $\boldsymbol{y}^{l+1} = \hat{\boldsymbol{y}}$ and return to step 2.

Step 8: Solve the relaxed primal master (RPM). Let $(\hat{\boldsymbol{y}}, \hat{\mu}_C)$ be its optimal solution. Update the lower bound, that is $LBD = \hat{\mu}_C$. If $|UBD - LBD| \leq \epsilon$, then terminate. Otherwise, set $k = k + 1, \boldsymbol{y}^{k+1} = \hat{\boldsymbol{y}}$, and return to step 3.

Remark 1 Note that the **GCD** is based on the idea that it is desirable to solve as few master problems as possible since these are the time consuming problems. Therefore, if the convergence tests CTP, CTD, CTDU are passed at each iteration, then we generate cuts and improved bounds on the sought solution of (6.52). This is denoted as the "subproblem" phase. If a sufficiently large number of cuts are generated in the subproblem phase then we may need to solve a master problem only a few times prior to obtaining the optimal solution.

Remark 2 In the initial iterations the master problems will have only a few cuts and as a result the primal and dual subproblems are expected to provide tighter bounds. As the iterations increase however the number of cuts in the master problems increases, and it is expected that the master problem will provide tighter bounds. On the above grounds, it is a good idea to delay the use of master problems as much as possible (i.e., depending on the outcome of the convergence tests) and start with only the primal and dual subproblem phase first. It is important to note that the rate of convergence in the generalized cross decomposition **GCD** depends on how balanced the primal and dual subproblems are.

Since we aim at avoiding the master problems if possible then we have to create balanced primal and dual subproblems on the grounds that the output of the primal subproblem is used as the input to the dual subproblem and vice versa.

Remark 3 From Figure 6.10, note that

(i) If the tests CTD, CTDU are not passed at all iterations and the relaxed primal master problem is used only, then the **GCD** reduces to the generalized benders decomposition **GBD**.

(ii) If the test CTP is not used at all iterations and the relaxed Lagrange master problem is used only, then the **GCD** reduces to the Dantzig-Wolfe decomposition.

A natural choice of master problem is to use the relaxed primal master if the CTP test is not passed, and to use the relaxed Lagrange relaxation master if the CTD or CTDU test is not passed. Note also that it is not necessary to use both master problems.

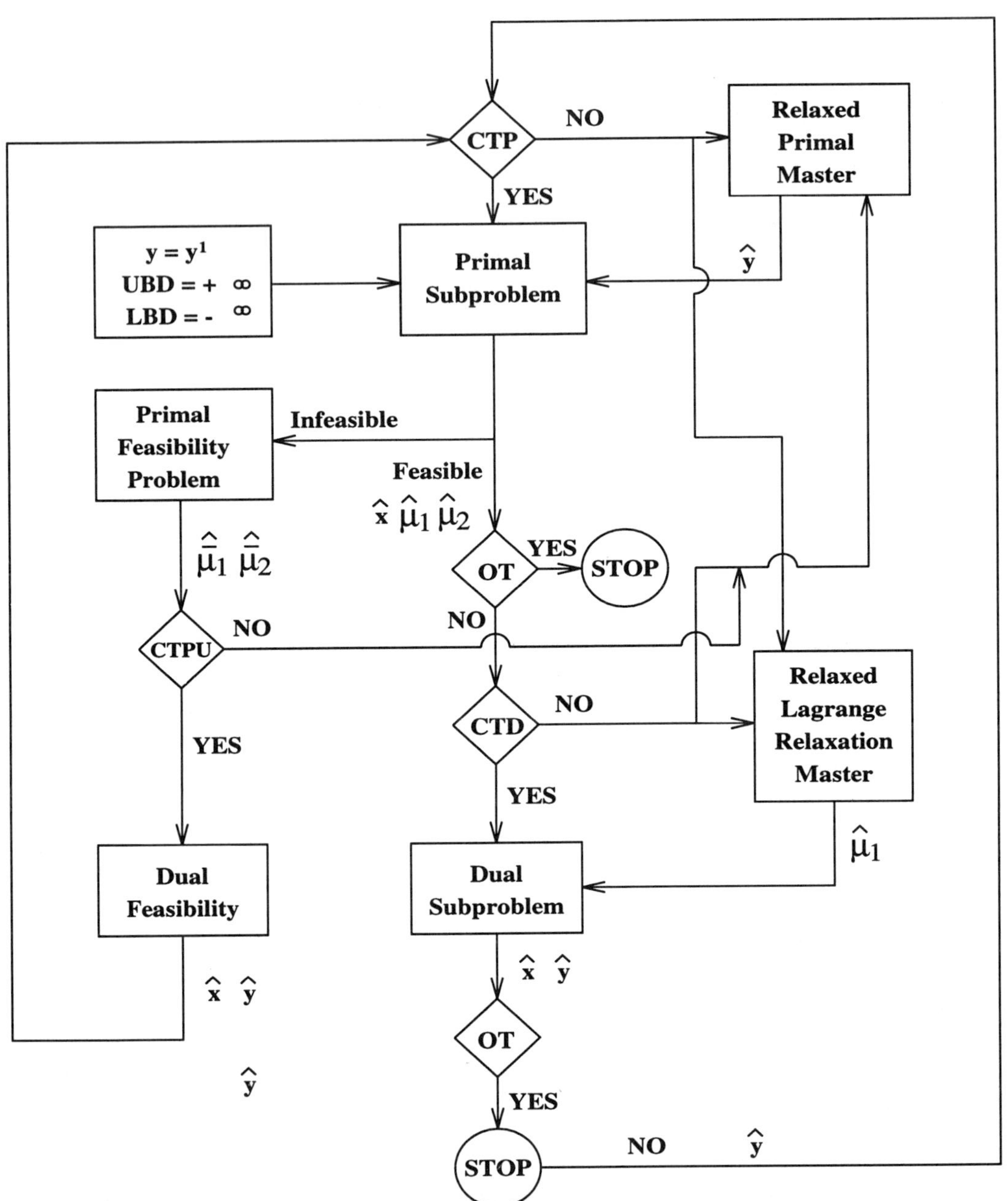

Figure 6.10: Flow diagram of generalized cross decomposition.

Remark 4 The **GCD** algorithm can start with either the primal subproblem, as shown in Figure 6.10, or with the dual subproblem, depending on whether a good primal or dual initial point is available.

6.9.4.2 Finite Convergence of GCD

For formulation (6.52), Holmberg (1990) proved finite convergence of the **GCD** algorithm stated in the previous section which is as follows:

Theorem 6.9.2
(Finite Convergence)

If conditions C1, C2, C3 hold and $\boldsymbol{Y}$ *is a finite discrete set, then the Generalized Cross Decomposition* **GCD** *will solve (6.52) exactly in a finite number of steps.*

6.9.5 GCD under Separability

Under the separability assumption we have that

$$\begin{aligned} f(x,y) &= f_1(x) + f_2(y), \\ g_1(x,y) &= g_{11}(x) + g_{12}(y), \\ g_2(x,y) &= g_{21}(x) + g_{22}(y). \end{aligned}$$

Then problem (6.52) takes the form:

$$\begin{aligned} \min_{x,y} \quad & f_1(x) + f_2(y) \\ s.t. \quad & g_{11}(x) + g_{12}(y) \leq 0 \\ & g_{21}(x) + g_{22}(y) \leq 0 \\ & x \in X \subseteq \Re^n \\ & y \in Y = \{0,1\}^q \end{aligned}$$

Remark 1 The separability assumption results in condition $C3$ being satisfied. In fact, condition $C3$ can be satisfied with a weaker assumption than separability, and this was discussed in the previous sections.

In the sequel, we will discuss the effect of the separability assumption on the primal subproblem, dual subproblem and the primal and Lagrange relaxation master problems.

Primal Subproblem

The primal subproblem takes the form:

$$\begin{aligned} \min_{x} \quad & f_1(x) + f_2(y^k) \\ s.t. \quad & g_{11}(x) + g_{12}(y^k) \leq 0 \\ & g_{21}(x) + g_{22}(y^k) \leq 0 \\ & x \in X \end{aligned}$$

and the analysis of feasible, infeasible primal is along the same lines of the **GCD** presented in an earlier section.

Remark 2 Condition C1 of the **GCD** with the separability assumption becomes that $f_1(x), g_{11}(x)$, $g_{21}(x)$ are convex in x. As a result the solution of the primal subproblem $P(y^k)$ would correspond to a global optimum solution. Also note that due to condition C2, the primal subproblem is stable.

Dual Subproblem

The dual subproblem with the separability assumption becomes

$$\left.\begin{array}{ll} \min\limits_{x,y} & f_1(x) + f_2(y) + \mu_1^{k^T} \left[g_{11}(x) + g_{12}(y)\right] \\ s.t. & g_{21}(x) + g_{22}(y) \leq 0 \\ & x \in X \subseteq \Re^n \\ & y \in Y = \{0,1\}^q \end{array}\right\} D(\mu_1^k)$$

Remark 3 If, in addition to separability of x and y, we assume that the y variables participate linearly; that is,

$$\begin{aligned} f_2(y) &= c^T y, \\ g_{12}(y) &= B_1 y, \\ g_{22}(y) &= B_2 y, \end{aligned}$$

then, the dual subproblem $D(\mu_1^k)$ becomes a convex MINLP problem which can be solved for its global solution with either Generalized Benders Decomposition, or the Outer Approximation algorithm. As a result, in this case the dual subproblems will provide a valid lower bound on (6.52). Also note that if we linearize $f_1(x), g_{11}(x), g_{21}(x)$ at a point x^k, then the dual subproblem becomes a mixed-integer linear programming MILP problem which can be solved with standard branch and bound codes (e.g., CPLEX) for its global solution. More important though, the solution of this MILP problem is a valid lower bound on the solution of (6.52).

The form of this MILP model for the dual subproblem is

$$\left.\begin{array}{ll} \min\limits_{x,y} & f_1^{lin}(x) + c^T y + \mu_1^{k^T} \left[g_{11}^{lin}(x) + B_1(y)\right] \\ s.t. & g_{21}^{lin}(x) + B_2(y) \leq 0 \\ & x \in X \subseteq \Re^n \\ & y \in Y = \{0,1\}^q \end{array}\right\} D^{lin}(\mu_1^k)$$

Remark 4 The above form of the dual subproblem will provide a weaker lower bound than the one resulting by solving the convex MINLP for its global solution This is because $D^{lin}(\mu_1^k)$ would correspond to having only 1 iteration of the **OA** algorithm for the convex MINLP.

Primal Master Problem

Under the separability assumption we have

$$\begin{aligned}
\min_{x \in X} L(x,y,\mu_1,\mu_2) &= \min_{x \in X} f_1(x) + f_2(y) + \mu_1^T \left[g_{11}(x) + g_{12}(y)\right] + \mu_2^T \left[g_{21}(x) + g_{22}(y)\right] \\
&= f_2(y) + \mu_1^T g_{12}(y) + \mu_2^T g_{22}(y) + \min_{x \in X} \left[f_1(x) + \mu_1^T g_{11}(x) + \mu_1^T g_{21}(x)\right]
\end{aligned}$$

and

$$\begin{aligned}
\min_{x \in X} \bar{L}(x,y,\bar{\mu}_1,\bar{\mu}_2) &= \min_{x \in X} \bar{\mu}_1^T \left[g_{11}(x) + g_{12}(y)\right] + \bar{\mu}_2^T \left[g_{21}(x) + g_{22}(y)\right] \\
&= \bar{\mu}_1^T g_{12}(y) + \bar{\mu}_2^T g_{22}(y) + \min_{x \in X} \left[\bar{\mu}_1^T g_{11}(x) + \bar{\mu}_1^T g_{21}(x)\right].
\end{aligned}$$

As a result, q_3 and q_4 take the form:

$$\begin{aligned}
q_3 &= f_1(x) + \mu_1^T g_{11}(x) + \mu_2^T g_{21}(x), \\
q_4 &= \bar{\mu}_1^T g_{11}(x) + \bar{\mu}_2^T g_{21}(x).
\end{aligned}$$

The relaxed primal master (RPM) problem is of the same form discussed in an earlier section with the additional definitions of q_3 and q_4. Its solution represents a valid lower bound on (6.52). Also, note that if we assume linearity in the y-space, then the (RPM) problem becomes a MILP problem.

Lagrange Relaxation Master Problem

Under the assumption of (i) separability of x and y, and (ii) linearity in y-space, the dual subproblem can be solved for its global solution or a valid relaxation of it (i.e., convex MINLP linearized at x) can provide a valid lower bound on (6.52).

As a result, the relaxed Lagrange master problem assuming feasibility will be of the form:

$$\left.\begin{aligned}
\min_{\mu_1,\mu_C'} \quad & \mu_C' \\
s.t. \quad & \mu_C' \leq h^k(\mu_1) \quad k = 1,2,\ldots,K \\
& \mu_1 \geq 0
\end{aligned}\right\} \textbf{(RLRM)}$$

where $h^k(\mu_1) = f_1(x^k) + c^T y^k + \mu_1^T \left[g_{11}(x^k) + g_{12}(y^k)\right]$ with the important difference that its solution represents a valid upper bound on (6.52).

Remark 5 The assumptions of $x - y$ separability and linearity in y result in $D(\mu_1^k)$ and (RLRM) providing valid lower and upper bounds on (6.52). Hence, the algorithmic steps of **GCD** can be simplified in this case. More importantly, however, the **GCD** will attain a global optimum solution of (6.52) under the conditions of separability in $x - y$ and linearity in y.

Illustration

This example was used as an illustration in v2-**GBD** and is of the form:

$$\begin{aligned}
\min \quad & y_1 + y_2 + y_3 + 5x^2 \\
s.t. \quad & 3x - y_1 - y_2 \leq 0 \\
& -x + 0.1y_2 + 0.25y_3 \leq 0 \\
& -y_1 - y_2 - y_3 \leq -2 \\
& -y_1 - y_2 - 2(y_3 - 1) \leq 0 \\
& 0.2 \leq x \leq 1 \\
& y_1, y_2, y_3 = 0, 1
\end{aligned}$$

Note that the separability of $x - y$ and linearity of y assumptions are satisfied. We will partition the constraints so as to have the first two inequalities in $g_1(x, y)$ and the third and fourth inequality, which are only functions of y variables in $g_2(x, y)$. As a result we have

$$\begin{aligned}
f_1(x) &= 5x^2, \\
f_2(y) &= y_1 + y_2 + y_3, \\
g_{11}(x) &= (3x, -x), \\
gg_{11}(y) &= (-y_1 - y_2, 0.1y_2 + 0.25y_3), \\
g_{21}(x) &= (0, 0), \\
g_{22}(y) &= [-y_1 - y_2 - y_3 + 2, -y_1 - y_2 - 2(y_3 - 1)].
\end{aligned}$$

Iteration 1:

Set $y^1 = (y_1, y_2, y_3) = (1, 1, 1)$ and $UBD = \infty$, $LBD = -\infty$.

The primal subproblem becomes

$$\begin{aligned}
\min \quad & 3 + 5x^2 \\
s.t. \quad & 3x - 2 \leq 0 \\
& -x + 0.35 \leq 0 \\
& 0.2 \leq x \leq 1
\end{aligned}$$

This problem is feasible and has as solution:

$$\begin{aligned}
x^1 &= 0.35\,,\ UBD = 3.6125, \\
\mu_{1,1}^1 &= 0, \\
\mu_{1,2}^1 &= 3.5,
\end{aligned}$$

where $\mu_{1,1}^1$, $\mu_{1,2}^1$ are the Lagrange multipliers for the two constraints $g_1(x, y)$.

Applying the CTD test we have

$$\begin{aligned}
h^1(\mu_{1,1}, \mu_{1,2}) &= f_1(x^1) + f_2(y^1) + \mu_{1,1}\left(3x^1 - y_1^1 - y_2^1\right) + \mu_{1,2}\left(-x^1 + 0.1y_2^1 + 0.25y_3^1\right) \\
&= 0.6125 + 3 + \mu_{1,1}(-0.95) + \mu_{1,2}(0) \\
&= 3.6125 - \mu_{1,1} \cdot 0.95 \\
&= 3.6125 - 0 \cdot 0.95 = 3.6125 > -\infty
\end{aligned}$$

and hence the CTD test is passed.

For $\mu_{1,1} = 0$, $\mu_{2,2} = 3.5$ we solve the dual subproblem, which takes the form:

$$\begin{aligned}
\min_{x,y} \quad & y_1 + y_2 + y_3 + 5x^2 + 3.5\left[-x + 0.1y_2 + 0.25y_3\right] \\
s.t. \quad & -y_1 - y_2 - y_3 + 2 \leq 0 \\
& -y_1 - y_2 - 2(y_3 - 1) \leq 0 \\
& 0.2 \leq x \leq 1 \\
& y_1, y_2, y_3 = 0, 1
\end{aligned}$$

Note that the dual subproblem is a convex MINLP problem with a convex term $(5x^2)$ in the objective function. We linearize $5x^2$ around $x^1 = 0.35$ resulting in

$$5[(0.35)^2 + 2 \cdot (0.35)(x - 0.35)] = 0.35x - 0.6125.$$

The linearized dual subproblem $D^{lin}(\mu_1^k)$ becomes

$$\begin{aligned}
\min \quad & y_1 + 1.35y_2 + 1.875y_3 - 0.6125 \\
s.t. \quad & -y_1 - y_2 - y_3 + 2 \leq 0 \\
& -y_1 - y_2 - 2(y_3 - 1) \leq 0 \\
& y_1, y_2, y_3 = 0, 1
\end{aligned}$$

and has as solutions:

$$y^2 = (1, 1, 0)\,, \; LBD = 1.7375.$$

Iteration 2:

Applying the CTP test for $y = y^2$ we have

$$\begin{aligned}
q_3 &= 5(0.35)^2 + 0 \cdot g_{11}(x) + 3.5(-0.35) = -0.6125, \\
q_1 &= y_1 + y_2 + y_3 + 3.5(0.1y_2 + 0.25y_3) - 0.6125, \\
q_1(y^2) &= 1 + 1 + 0 + 3.5(0.1 \cdot 1 + 0.25 \cdot 0) - 0.6125 \\
&= 1.7375 \; < UBD = 3.6125.
\end{aligned}$$

Hence, the CTP test is passed and we continue with the primal subproblem. For $y = y^2 = (1, 1, 0)$ we solve the primal subproblem and obtain:

$$\begin{aligned}
x^2 &= 0.2, \\
\mu_{1,1}^2 &= 0, \\
\mu_{1,2}^2 &= 0,
\end{aligned}$$

and the objective function is 2.2. The new upper bound becomes $UBD = 2.2$. Applying the CTD test we have

$$\begin{aligned}
h^2(\mu_{1,1}, \mu_{1,2}) &= f_1(x^2) + f_2(y^2) \\
&= 2.2 > LBD = 1.7375,
\end{aligned}$$

and therefore the CTD test is passed.

For $\mu^2_{1,1} = \mu^2_{1,2} = 0$ we solve the dual subproblem which becomes

$$\begin{array}{ll}\min & y_1 + y_2 + y_3 + 5x^2 \\ s.t. & -y_1 - y_2 - y_3 + 2 \leq 0 \\ & -y_1 - y_2 - 2(y_3 - 1) \leq 0 \\ & 0.2 \leq x \leq 1 \\ & y_1, y_2, y_3 = 0, 1\end{array}$$

We linearize $5x^2$ around $x = 0.2$ and we have

$$5x^2 = 5[0.4x - 0.04] = 2x - 0.2.$$

Then, the linearized dual subproblem becomes

$$\begin{array}{ll}\min & y_1 + y_2 + y_3 + 2x - 0.2 \\ s.t. & -y_1 - y_2 - y_3 + 2 \leq 0 \\ & -y_1 - y_2 - 2(y_3 - 1) \leq 0 \\ & 0.2 \leq x \leq 1 \\ & y_1, y_2, y_3 = 0, 1\end{array}$$

which has a solution:

$$\begin{aligned}x &= 0.2, \\ y_1 &= 1, \\ y_2 &= 1, \\ y_3 &= 0,\end{aligned}$$

and objective value equal to 2.2. Hence $LBD = 2.2$. At this point we have

$$\begin{aligned}UBD &= 2.2, \\ LBD &= 2.2,\end{aligned}$$

and hence we terminate.

6.9.6 GCD in Continuous and Discrete–Continuous Optimization

In remark 1 of the formulation section of the **GCD** we mentioned that problem (6.52) is a subclass of problems for which the Generalized Cross Decomposition can be applied. This is due to having $\boldsymbol{Y} = \{0, 1\}^q$ in (6.52) instead of the general case of the set ***Y*** being a continuous, discrete, or discrete-continuous nonempty, compact set. The main objective in this section is to discuss the modifications in the analysis of the **GCD** for the cases of the ***Y*** set being continuous or discrete-continuous.

The analysis for the primal subproblem, the dual subproblem, the primal master problem, and the Lagrange relaxation master problem remains the same. The only difference is that if the ***Y*** set is continuous, then the dual subproblem and the primal master problem are nonlinear continuous optimization problems in x and y, respectively.

The convergence tests, however, need to be modified on the grounds that it is possible to have an infinite number of both primal and dual improvements if Y is continuous, and hence not attain termination in a finite number of steps. To circumvent this difficulty, Holmberg (1990) defined the following stronger ϵ-improvements:

Definition 6.9.1 (ϵ-bound improvement): ϵ-bound improvement is defined as an improvement by at least ϵ of the upper bound or the lower bound.

Definition 6.9.2 (ϵ-cut improvement): ϵ-cut improvement is defined as a generation of a new cut that is at least ϵ better than all known cuts at some iteration.

Holmberg (1990) proposed the ϵ-convergence tests, (CT–ϵ) as follows:

CTP-ϵ: If $q_1^k(y^c) \leq UBD - \epsilon$ for $k = 1, 2, \ldots, K$ and $q_1^k(y^c) \leq -\epsilon$ for $k = 1, 2, \ldots, L$, where y^c is the current y, then y^c will provide an upper bound improvement. If not, use a master problem.

CTD-ϵ: If $h^k(\mu_1^c) \geq LBD + \epsilon$ for $k = 1, 2, \ldots, K$, where μ_1^c is the current μ_1, then μ_1^c will provide a lower bound improvement. If not, use a master problem.

CTDU-ϵ: If $h^l(\bar{\mu}_1^c) \geq \epsilon$ for $l = 1, 2, \ldots, L$, where $\bar{\mu}_1^c$ is the current $\bar{\mu}^1$ will provide a cut-improvement. If not, use a master problem.

Remark 1 The first condition of CTP-ϵ and the CTD-ϵ are classified as "ϵ-value convergence" tests since they correspond to feasible problems. The second condition of CTP-ϵ and the CTDU-ϵ are denoted as "ϵ-feasibility convergence" tests since they correspond to feasibility problems.

Remark 2 The ϵ-convergence test are sufficient for ϵ-improvement but they are not necessary. An additional condition, which is an inverse Lipschitz assumption, needs to be introduced so as to prove the necessity. This additional condition states that for points of a certain distance apart the value of the feasibility cut should differ by at least some amount. This is stated in the following theorem of Holmberg (1990).

Theorem 6.9.3
The ϵ-value convergence tests of CTP-ϵ, the feasibility tests of CTP, and the ϵ-convergence tests of CTD-ϵ are necessary for ϵ-bound improvement. The ϵ-convergence tests are sufficient for one of the following:

(i) *ϵ-bound improvement;*

(ii) *ϵ-cut improvement;*

(iii) *ϵ_1-bound improvement and ϵ_2-cut improvement such that $\epsilon_1 + \epsilon_2 = \epsilon$.*

With the above theorem as a basis Holmberg (1990) proved the finiteness of the convergence tests which is stated as

Lemma 6.9.2 The ϵ-convergence tests will fail after a finite number of steps.

Remark 3 Note that in proving finiteness we need to use the relaxed primal master problem. Also note that it suffices that CTP-ϵ fails for finiteness. However, we cannot show that the CTD-ϵ test will fail after a finite number of steps.

Remark 4 If the primal subproblem has a feasible solution for every $y \in Y$, then the **GCD** algorithm will attain finite ϵ-convergence (i.e. $UBD - LBD \leq \epsilon$) in a finite number of steps for any given $\epsilon > 0$. Obviously, in this case the ϵ-feasibility convergence tests are not needed.

Remark 5 If $Y \not\subseteq V$, then we may have asymptotic convergence of the feasibility cuts, that is, the solution of the feasibility problems gets closer and closer to zero which corresponds to feasible solution, but never actually becomes feasible. One way to circumvent this difficulty is by defining an ϵ-feasible solution which corresponds to using a linear penalty function that transforms the feasibility cuts to value cuts. Holmberg (1990) showed that the **GCD** equipped with ϵ-convergence tests does not have weaker asymptotic convergence than the Generalized Benders Decomposition **GBD**.

Summary and Further Reading

This chapter introduces the fundamentals of mixed-integer nonlinear optimization. Section 6.1 presents the motivation and the application areas of MINLP models. Section 6.2 presents the mathematical description of MINLP problems, discusses the challenges and computational complexity of MINLP models, and provides an overview of the existing MINLP algorithms.

Section 6.3 presents the Generalized Benders Decomposition **GBD** approach. Section 6.3.1 and section 6.3.2 present the formulation and the basic idea of **GBD**, respectively. Section 6.3.3 discusses the theoretical development for the primal and master problem along with the geometrical interpretation of the master problem. Section 6.3.4 discusses a relaxation approach for solving the master problem, presents a general algorithmic strategy for **GBD** and discusses its finite convergence. Section 6.3.5 presents three variants of the **GBD**, illustrates them with simple examples, discusses their relations and the effect of imposing the additional conditions of separability in $x - y$ and linearity in y. Section 6.3.6 discusses the **GBD** for continuous and discrete- continuous optimization, presents the finite ϵ-convergence theorem, and illustrates the approach via a simple example. Further reading on **GBD** can be found in the suggested references as well as in Geoffrion and Graves (1974), McDaniel and Devine (1977), Rardin and Unger (1976), Magnanti and Wong (1981), Wolsey (1981), Rouhani *et al.* (1985), Lazimy (1982, 1985), Flippo (1990), Flippo and Rinnoy Kan (1990,1993), and Adams and Sherali (1993).

Section 6.4 discusses the Outer Approximation **OA** approach. Sections 6.4.1 and 6.4.2 present the formulation, conditions, and the basic idea of **OA**. Section 6.4.3 presents the development of the primal and master problem, as well as the geometrical interpretation of the master problem. Section 6.4.4 presents the **OA** algorithm and its finite ϵ-convergence.

Section 6.5 presents the Outer Approximation with Equality Relaxation **OA/ER** for handling nonlinear equality constraints. Sections 6.5.1 and 6.5.2 discuss the formulation, assumptions and the basic idea. Section 6.5.3 discusses the equality relaxation, and illustrates it with a simple example, and presents the formulation of the master problem. Section 6.5.4 discusses the **OA/ER** algorithm and illustrates it with a small planning problem.

Section 6.6 discusses the Outer Approximation with Equality Relaxation and Augmented Penalty **OA/ER/AP** approach. In Sections 6.6.1 and 6.6.2 the formulation and basic idea are presented, while in section 6.6.3 the master problem is derived. Section 6.6.4 presents the **OA/ER/AP** algorithm and illustrates it with a nonconvex example problem. The reader is referred to the suggested references in sections 6.4,6.5 and 6.6 for further reading in the outer approximation based algorithms.

Section 6.7 presents the Generalized Outer Approximation **GOA** approach. After a brief discussion on the problem formulation, the primal and master subproblem formulations are developed, and the **GOA** algorithm is stated in section 6.7.4. In Section 6.7.5, the worst case analysis of **GOA** is discussed, while in section 6.7.6 the Generalized Outer Approximation with exact Penalty **GOA/EP** and its finite convergence are discussed.

Section 6.8 compares the **GBD** approach and the **OA**based approaches with regard to the formulation, handling of nonlinear equality constraints, nonlinearities in y and joint $x - y$, the primal problem, the master problem, and the quality of the lower bounds.

Section 6.9 presents the Generalized Cross Decomposition **GCD** approach. Section 6.9.1 discusses the formulation, and Section 6.9.2 present the basic idea of the **GCD**. In Section 6.9.3, the theoretical development of the primal subproblem, dual subproblem, the primal master problem and the Lagrange relaxation master problem is discussed along with the convergence tests and their finite termination. Section 6.9.4 presents the detailed algorithmic steps, the relation between the **GCD**, the Dantzig-Wolfe and Generalized Benders Decomposition, and the finite convergence of **GCD**. Section 6.9.5 discusses the **GCD** under the additional assumptions of separability in $\boldsymbol{x}$ and $\boldsymbol{y}$, and linearity in $\boldsymbol{y}$, along with an illustration of the **GCD**. Section 6.9.6 discusses the modifications needed in **GCD** for continuous and discrete-continuous $\boldsymbol{Y}$ set , and presents the corresponding convergence results. Further reading in this subject can be found in Van Roy (1986), Holmberg (1991), Holmberg (1992), and Vlahos (1991).

Problems on Mixed Integer Nonlinear Optimization

1. Consider the design of multiproduct batch plants. The plant consists of M processing stages manufacturing fixed amount Q_i of N products. For each stage j, the number of parallel units N_j and and their sizes V_j have to be determined along with the batch sizes B_i and cycle times T_{Li} for each product i. The data for this problem is given in Table 1. The MINLP formulation for this problem is given below:

$$
\begin{aligned}
z &= \min \sum_{j=1}^{M} \alpha_j N_j V_j^{\beta_j} \\
\text{subject to} & \\
V_j &\geq S_{ij} B_i \quad i = 1, N \quad j = 1, M \\
N_j T_{Li} &\geq t_{ij} \quad i = 1, N \quad j = 1, M \\
\sum_{i=1}^{N} \frac{Q_i T_{Li}}{B_i} &\leq H \\
1 \leq N_j &\leq N_j^U \quad j = 1, M \\
V_j^L \leq V_j &\leq V_j^U \quad j = 1, M \\
T_{Li}^L \leq T_{Li} &\leq T_{Li}^U \quad i = 1, N \\
B_i^L \leq B_i &\leq B_i^U \quad i = 1, N
\end{aligned}
$$

where α_j and β_j are cost coefficients. The lower and upper bounds for V_j and N_j are given in the table while valid bounds for T_{Li} and B_i can be determined from the given data.

No. of stages, M			6			
No. of products, N			5			
Cost coefficient			$\alpha_j = \$250$	$\beta_j = 0.6$	$j = 1, M$	
Bounds on volumes			$V_j^L = 300\text{lt}$	$V_j^U = 3000\text{lt}$	$j = 1, M$	
Max. No. of parallel units				$N_j^U = 4$	$j = 1, M$	
Horizon time			H = 6000h			
Q_i: production rate of i (kg)						
	A	B	C	D	E	
	250,000	150,000	180,000	160,000	120,000	
S_{ij}: size factor for product i in stage j (l kg^{-1})						
Product	1	2	3	4	5	6
A	7.9	2.0	5.2	4.9	6.1	4.2
B	0.7	0.8	0.9	3.4	2.1	2.5
C	0.7	2.6	1.6	3.6	3.2	2.9
D	4.7	2.3	1.6	2.7	1.2	2.5
E	1.2	3.6	2.4	4.5	1.6	2.1
t_{ij}: processing time for product i in stage j (h)						
A	6.4	4.7	8.3	3.9	2.1	1.2
B	6.8	6.4	6.5	4.4	2.3	3.2
C	1.0	6.3	5.4	11.9	5.7	6.2
D	3.2	3.0	3.5	3.3	2.8	3.4
E	2.1	2.5	4.2	3.6	3.7	2.2

Apply the v2-GBD and OA with the starting point of $N_j^0 = (3,3,4,4,3,3)$.

2. Consider the following problem:

$$\begin{aligned} \min z \;&=\; 5y_1 + 6y_2 + 8y_3 + 10x_1 - 7x_6 - 18\ln(x_2+1) \\ &-\; 19.2\ln(x_1 - x_2 + 1) + 10 \end{aligned}$$

$$\begin{aligned} \text{s.t. } 0.8\ln(x_2+1) + 0.96\ln(x_1-x_2+1) - 0.8x_6 &\geq 0 \\ x_2 - x_1 &\leq 0 \\ x_2 - Uy_1 &\leq 0 \\ x_1 - x_2 - Uy_2 &\leq 0 \\ \ln(x_2+1) + 1.2\ln(x_1-x_2+1) - x_6 - Uy_3 &\geq -2 \\ y_1 + y_2 &\leq 1 \\ y \in \{0,1\}^3, a \leq x \leq b, x = (x_j{:}j = 1,2,6) &\in \Re^3 \\ a^T = (0,0,0), b^T = (2,2,1), U &= 2 \end{aligned}$$

With initial point $y^0 = (1,0,1)$ obtain its solution by applying:

(i) v2-GBD,

(ii) OA,

(iii) GCD.

Is its solution a global optimum? Why?

3. Consider the following problem:

$$\begin{aligned} \min z &= 5y_1 + 8y_2 + 6y_3 + 10y_4 + 6y_5 \\ &- 10x_3 - 15x_5 - 15x_9 + 15x_{11} + 5x_{13} - 20x_{16} \\ &+ \exp(x_3) + \exp(x_5/1.2) - 60\ln(x_{11} + x_{13} + 1) + 140 \end{aligned}$$

$$\begin{aligned} \text{s.t.} \quad -\ln(x_{11} + x_{13} + 1) &\leq 0 \\ -x_3 - x_5 - 2x_9 + x_{11} + 2x_{16} &\leq 0 \\ -x_3 - x_5 - 0.75x_9 + x_{11} + 2x_{16} &\leq 0 \\ x_9 - x_{16} &\leq 0 \\ 2x_9 - x_{11} - 2x_{16} &\leq 0 \\ -0.5x_{11} + x_{13} &\leq 0 \\ 0.2x_{11} - x_{13} &\leq 0 \\ \exp(x_3) - Uy_1 &\leq 1 \\ \exp(x_5/1.2) - Uy_2 &\leq 1 \\ 1.25x_9 - Uy_3 &\leq 0 \\ x_{11} + x_{13} - Uy_4 &\leq 0 \\ -2x_9 + 2x_{16} - Uy_5 &\leq 0 \\ y_1 + y_2 &= 1 \\ y_4 + y_5 &\leq 1 \\ y \in \{0,1\}^5, a \leq x \leq b, x = (x_j{:}j = 3,5,9,11,13,16) &\in \Re^6 \\ a^T = (0,0,0,0,0,0), b^T = (2,2,2,-,-,3), U &= 10 \end{aligned}$$

with an initial guess of $y^0 = (1,0,0,0,0)$.

Obtain its solution by applying:

(i) v2-GBD,

(ii) OA,

(iii) GCD.

Is its solution a global optimum? Why?

4. Consider the following problem:

$$\begin{aligned} \min z &= 5y_1 + 8y_2 + 6y_3 + 10y_4 + 6y_5 + 7y_6 + 4y_7 + 5y_8 \\ &- 10x_3 - 15x_5 + 15x_{10} + 80x_{17} + 25x_{19} + 35x_{21} - 40x_9 \\ &+ 15x_{14} - 35x_{25} + \exp(x_3) + \exp(x_5/1.2) - 65\ln(x_{10} + x_{17} + 1) \\ &- 90\ln(x_{19} + 1) - 80\ln(x_{21} + 1) + 120 \end{aligned}$$

$$
\begin{aligned}
\text{s.t.} - 1.5\ln(x_{19}+1) - \ln(x_{21}+1) - x_{14} &\le 0\\
-\ln(x_{10}+x_{17}+1) &\le 0\\
-x_3 - x_5 + x_{10} + 2x_{17} + 0.8x_{19} + 0.8x_{21} - 0.5x_9 - x_{14} - 2x_{25} &\le 0\\
-x_3 - x_5 + 2x_{17} + 0.8x_{19} + 0.8x_{21} - 2x_9 - x_{14} - 2x_{25} &\le 0\\
-2x_{17} - 0.8x_{19} - 0.8x_{21} + 2x_9 + x_{14} + 2x_{25} &\le 0\\
-0.8x_{19} - 0.8x_{21} + x_{14} &\le 0\\
-x_{17} + x_9 + x_{25} &\le 0\\
-0.4x_{19} - 0.4x_{21} + 1.5x_{14} &\le 0\\
0.16x_{19} + 0.16x_{21} - 1.2x_{14} &\le 0\\
x_{10} - 0.8x_{17} &\le 0\\
-x_{10} + 0.4x_{17} &\le 0\\
\exp(x_3) - Uy_1 &\le 1\\
\exp(x_5/1.2) - Uy_2 &\le 1\\
x_9 - Uy_3 &\le 0\\
0.8x_{19} + 0.8x_{21} - Uy_4 &\le 0\\
2x_{17} - 2x_9 - 2x_{25} - Uy_5 &\le 0\\
x_{19} - Uy_6 &\le 0\\
x_{21} - Uy_7 &\le 0\\
x_{10} + x_{17} - Uy_8 &\le 0\\
y_1 + y_2 &= 1\\
y_4 + y_5 &\le 1\\
-y_4 + y_6 + y_7 &= 0\\
y_3 - y_8 &\le 0\\
y \in \{0,1\}^8, a \le x \le b, x = (x_j{:}j = 3,5,10,17,19,21,9,14,25) &\in \Re^9\\
a^T = (0,0,0,0,0,0,0,0,0), b^T = (2,2,1,2,2,2,2,1,3), U &= 10
\end{aligned}
$$

Obtain its solution by applying:

(i) v2-GBD,

(ii) OA,

(iii) GCD.

Is its solution a global optimum? Why?

5. Consider the bilinear programming problem:

$$
\begin{aligned}
\max_{x,y} \quad & -x_1 + x_2 + y_1 + (x_1 - x_2)(y_1 - y_2)\\
s.t. \quad & x_1 + 4x_2 \le 8\\
& 4x_1 + x_2 \le 12\\
& 3x_1 + 4x_2 \le 12
\end{aligned}
$$

$$
\begin{aligned}
&2y_1 + y_2 \le 8\\
&y_1 + 2y_2 \le 8\\
&y_1 + y_2 \le 5\\
&x_1, x_2, y_1, y_2 \ge 0
\end{aligned}
$$

Project on the y variables and apply v2-GBD and GCD from the following starting points:

(i) $(y_1, y_2) = (4, 0)$,
(ii) $(y_1, y_2) = (1, 1)$,
(iii) $(y_1, y_2) = (3, 2)$.

Explain why different solutions may be reached depending on the starting point.

6. Consider the indefinite quadratic problem:

$$
\begin{aligned}
\max_x \quad & -x_1^2 + x_2^2 + 6x_1 - 4x_2\\
s.t. \quad & x_2 - 2x_1 \le 2\\
& 2x_1 - x_2 \le 8\\
& -x_1 - 2x_2 - x_3 \le -4\\
& x_2 \le 4\\
& x_1, x_2 \ge 0
\end{aligned}
$$

Apply the v2-GBD and GCD by using the transformation:

$$
x_1^2 = x_1 \cdot y, \quad x_1 - y = 0,
$$

and projecting on the y variable. Consider the following starting points for y: $-1, 1, 1$, and 0. Is the obtained solution a global optimum?

7. Consider the following problem:

$$
\begin{aligned}
\min\ & 37.293239x_1 + 0.8356891x_1x_5 + 5.3578547x_3^2 - 40792.141\\
s.t.\ & -0.0022053x_3x_5 + 0.0056858x_2x_5 + 0.0006262x_1x_4 - 6.665593 \le 0\\
& 0.0022053x_3x_5 - 0.0056858x_2x_5 - 0.0006262x_1x_4 - 85.334407 \le 0\\
& 0.0071317x_2x_5 + 0.0021813x_3^2 + 0.0029955x_1x_2 - 29.48751 \le 0\\
& -0.0071317x_2x_5 - 0.0021813x_3^2 - 0.0029955x_1x_2 + 9.48751 \le 0\\
& 0.0047026x_3x_5 + 0.0019085x_3x_4 + 0.0012547x_1x_3 - 15.699039 \le 0\\
& -0.0047026x_3x_5 - 0.0019085x_3x_4 - 0.0012547x_1x_3 + 10.699039 \le 0\\
& 78 \le x_1 \le 102\\
& 33 \le x_2 \le 45\\
& 27 \le x_3 \le 45\\
& 27 \le x_4 \le 45\\
& 27 \le x_5 \le 45
\end{aligned}
$$

Introduce two new variables y_1, y_2:

$$y_1 - x_1 = 0, \quad y_2 - x_3 = 0,$$

and apply the v3-GBD and GCD by projecting on y_1, y_2 and x_5.

8. Consider the heat exchanger network problem:

$$
\begin{aligned}
OBJ = \quad \min \quad & 1200 \left[\frac{800}{2.5\left(2/3\sqrt{(320-t_2)(300-t_1)} + \frac{(320-t_2)+(300-t_1)}{6}\right)} \right]^{0.6} \\
+ \quad & 1200 \left[\frac{1000}{0.2\left(2/3\sqrt{(340-t_4)(300-t_3)} + \frac{(340-t_4)+(300-t_3)}{6}\right)} \right]^{0.6}
\end{aligned}
$$

subject to

$$
\begin{aligned}
f_1 + f_2 &= 10.0 \\
f_1 + f_6 &= f_3 \\
f_2 + f_5 &= f_4 \\
f_5 + f_7 &= f_3 \\
f_6 + f_8 &= f_4 \\
100f_1 + t_4 f_6 &= t_1 f_3 \\
100f_2 + t_2 f_5 &= t_3 f_4 \\
f_3(t_2 - t_1) &= 800 \\
f_4(t_4 - t_3) &= 1000 \\
t_1 &\leq 290 \\
t_2 &\leq 310 \\
t_3 &\leq 290 \\
t_4 &\leq 330 \\
f_1, f_2, f_3, f_4, f_5, f_6, f_7, f_8 &\geq 0
\end{aligned}
$$

The objective function is convex. Apply the v3-GBD and the GCD by projecting on the f variables.

9. Consider the problem:

$$
\begin{aligned}
z = \quad & \min_{x,y} 2x_1 + 3x_2 + 1.5y_1 + 2y_2 - 0.5y_3 \\
s.t. \quad & (x_1)^2 + y_1 = 1.25 \\
& (x_2)^{1.5} + 1.5y_2 = 3.0 \\
& x_1 + y_1 \leq 1.6 \\
& 1.333x_2 + y_2 \leq 3.0 \\
& -y_1 - y_2 + y_3 \leq 0 \\
& x_1, x_2 \geq 0 \\
& y_1, y_2, y_3 \in \{0, 1\}
\end{aligned}
$$

(i) Apply the OA/ER/AP by projecting on the y_1, y_2, y_3.

(ii) Apply the v3-GBD after writing:

$$\begin{aligned} x_1^2 - x_1\omega_1 &= 0, \\ x_1 - \omega_1 &= 0, \\ x_2^{1.5} - \omega_2\omega_5 &= 0, \\ \omega_2 - \omega_3 &= 0, \\ \omega_3 - \omega_4 &= 0, \\ \omega_5 - \omega_3\omega_4 &= 0, \end{aligned}$$

and projecting on $\omega_1, \omega_2, \omega_4$.Consider two starting points:

$$\begin{aligned} (\omega_1, \omega_2, \omega_4) &= (0.5, 1.145, 1.145), \\ (\omega_1, \omega_2, \omega_4) &= (1.0, 1.260, 1.357). \end{aligned}$$

10. Consider the problem:

$$\begin{aligned} \min \quad & 7.5y_1 + 5.5y_2 + 7v_1 + 6v_2 + 5x \\ s.t. \quad & z_1 = 0.9[1 - \exp(-0.5v_1)]x_1 \\ & z_2 = 0.8[1 - \exp(-0.4v_2)]x_2 \\ & x_1 + x_2 - x = 0 \\ & z_1 + z_2 = 10 \\ & v_1 \leq 10y_1 \\ & v_2 \leq 10y_2 \\ & x_1 \leq 20y_1 \\ & x_2 \leq 20y_2 \\ & y_1 + y_2 = 1 \\ & x_1, x_2, z_1, z_2, v_1, v_2 \geq 0 \\ & y_1, y_2 \in \{0, 1\}^2 \end{aligned}$$

(i) Apply the OA/ER/AP, and

(ii) Apply the v2-GBD.

11. Consider the distillation sequencing problem:

$$\max 35P1A + 30P2B - 10F1 - 8F2 - F4A - F4B - 4F5A - 4F5B - 2YF - 50YD$$

$$\begin{aligned} s.t. \quad F3A &= 0.55F1 + 0.50F2 \\ F3B &= 0.45F1 + 0.50F2 \\ F4A &= E4 \cdot F3A \\ F4B &= E4 \cdot F3B \end{aligned}$$

$$
\begin{aligned}
F5A &= E5 \cdot F3A \\
F5B &= E5 \cdot F3B \\
F6A &= E6 \cdot F3A \\
F6B &= E6 \cdot F3B \\
F7A &= F3A - F4A - F5A - F6A \\
F7B &= F3B - F4B - F5B - F6B \\
F8A &= 0.85 \cdot F4A \\
F8B &= 0.20 \cdot F4B \\
F9A &= 0.15 \cdot F4A \\
F9B &= 0.80 \cdot F4B \\
F10A &= 0.975 \cdot F5A \\
F10B &= 0.050 \cdot F5B \\
F11A &= 0.025 \cdot F5A \\
F11B &= 0.950 \cdot F5B \\
P1A &= F8A + F10A + F6A \\
P1B &= F8B + F10B + F6B \\
P2A &= F9A + F11A + F7A \\
P2B &= F9B + F11B + F7B \\
F4A + F4B &\geq 2.5 \cdot YF \\
F4A + F4B &\leq 25 \cdot YF \\
F5A + F5B &\geq 2.5 \cdot YD \\
F5A + F5B &\leq 25 \cdot YD \\
P1A &\geq 4 \cdot P1B \\
P2B &\geq 3 \cdot P2A \\
P1A + P1B &\leq 15 \\
P2A + P2B &\leq 18 \\
E4, E5, E6 &\leq 1.0 \\
F1, F2 &\leq 25.0 \\
YD, YF &= 0, 1
\end{aligned}
$$

(i) Apply the v2-GBD and OA/ER/AP by projecting on YD, YF, and

(ii) Apply the v3-GBD by projecting on a continuous set of variables, and YD, YF. Is the obtained solution a global optimum?

12. Consider the following *Mixed-Integer Quadratic Program*:

$$
\begin{aligned}
\max_{x,y} \quad & c^T x + d^T y + \frac{1}{2} x^T Q x + x^T R y + \frac{1}{2} y^T S y \\
\text{s.t.} \quad & Ax + Ey \leq b \\
& x \in Z_+^{n_1}
\end{aligned}
$$

$$y \in \Re^{n_2}$$
$$y \geq 0$$

where

$$B = \begin{pmatrix} Q & R \\ R^T & S \end{pmatrix}$$

is a *symmetric, negative semi-definite matrix.* Show that the dual problem is:

$$\begin{aligned} \min \quad & u^T b + (c - A^T u + Qv + Rw)^T \bar{x} - \frac{1}{2} v^T Q v - v^T R w - \frac{1}{2} w^T S w \\ s.t. \quad & E^T u - R^T v - S w \geq d \\ & u \geq 0 \end{aligned}$$

where $\bar{x}$ is the projected set of integer variables.

Part 3

Applications in Process Synthesis

Chapter 7 Process Synthesis

This chapter provides an introduction to Process Synthesis. Sections 7.1 and 7.2 discuss the components of a chemical process system and define the process synthesis problem. Section 7.3 presents the different approaches in the area of process synthesis. Section 7.4 focuses on the optimization approach and discusses modeling issues. Finally, Section 7.5 outlines application areas which are the subject of discussion in chapters 8, 9 and 10.

7.1 Introduction

Process Synthesis, an important research area within chemical process design, has triggered during the last three decades a significant amount of academic research work and industrial interest. Extensive reviews exist for the process synthesis area as well as for special classes of problems (*e.g.*, separation systems, heat recovery systems) and for particular approaches (*e.g.*, insights-based approach, optimization approach) applied to process synthesis problems. These are summarized in the following:

Overall Process Synthesis: Hendry *et al.* (1973)
Hlavacek (1978)
Westerberg (1980)
Stephanopoulos (1981)
Nishida *et al.* (1981)
Westerberg (1989)
Gundersen (1991)

Heat Exchanger Network Synthesis: Gundersen and Naess (1988)

Separation System Synthesis: Westerberg (1985)
Smith and Linnhoff (1988)

Optimization in Process Synthesis: Grossmann (1985) , (1989) , (1990)

Floquet *et al.* (1988)
Grossmann *et al.* (1987)
Floudas and Grossmann (1994)

Prior to providing the definition of the process synthesis problem we will describe first the overall process system and its important subsystems. This description can be addressed to the overall process system or individual subsystems and will be discussed in the subsequent sections.

7.1.1 The Overall Process System

An overall process system can be represented as an integrated system that consists of three main interactive components (see Figure 7.1):

(i) Chemical plant,

(ii) Heat recovery system,

(iii) Utility system.

In the chemical plant, the transformation of the feed streams (*e.g.*, raw materials) into desired products and possible by-products takes place. In the heat recovery system, the hot and cold process streams of the chemical plant exchange heat so as to reduce the hot and cold utility requirements. In the utility plant, the required utilities (*e.g.*, electricity and power to drive process units) are provided to the chemical plant while hot utilities (*e.g.*, steam at different pressure levels) are provided to the heat recovery system. Figure 7.1 shows the interactions among the three main components which are the hot and cold process streams for (i) and (ii), the electricity and power demands for (i) and (iii), and the hot utilities (*e.g.*, fuel, steam at different pressure levels, hot water) and cold utilities (*e.g.*, cooling water, refrigerants) for (ii) and (iii).

The chemical plant consists of the following three main interactive subsystems (see Figure 7.2):

(a) Reactor system,

(b) Separation system,

(c) Recycle system.

The reactor system may consist of a number of reactors which can be continuous stirred tank reactors, plug flow reactors, or any representation between the two above extremes, and they may operate isothermally, adiabatically or nonisothermally. The separation system depending on the reactor system effluent may involve only liquid separation, only vapor separation or both liquid and vapor separation schemes. The liquid separation scheme may include flash units, distillation columns or trains of distillation columns, extraction units, or crystallization units. If distillation is employed, then we may have simple sharp columns, nonsharp columns, or even single complex distillation columns and complex column sequences. Also, depending on the reactor effluent characteristics, extractive distillation, azeotropic distillation, or reactive distillation may be employed. The vapor separation scheme may involve absorption columns, adsorption units,

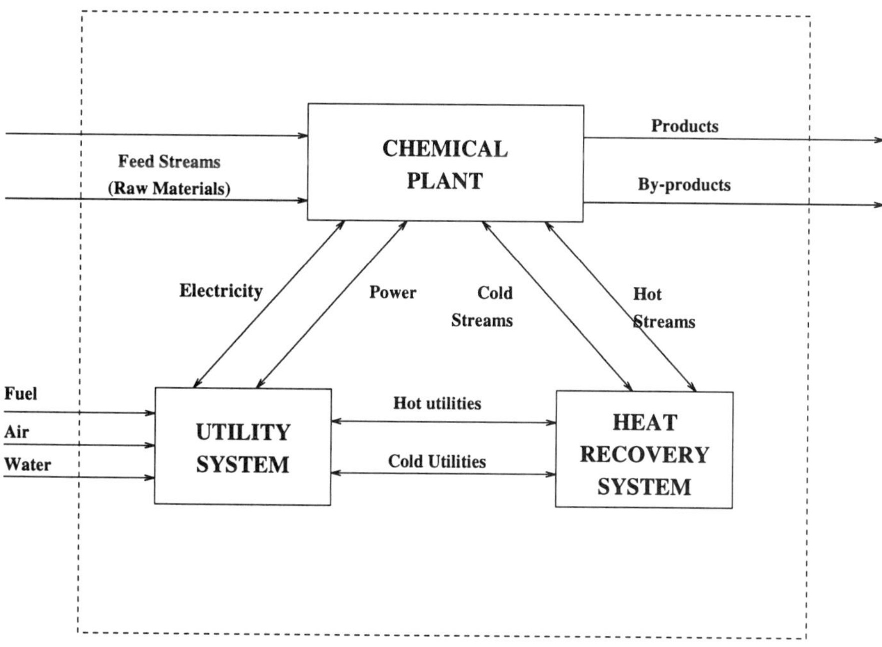

Figure 7.1: An overall process system

membrane separators, or condensation devices. The recycle system may feature a number of gas-recycles and liquid recycles from the separation system to the reactor system and has as typical units compressors and pumps for the gas and liquid recycles, respectively. In the presence of impurities, the gas recycle may involve a purge stream.

It is important to highlight the interactions taking place among the three aforementioned sub-systems which are also indicated in Figure 7.2. The feed streams can be directly fed to the reactor system, or they can be directed to the separation system first for feed purification and subsequently directed to the separation system where valuable components are recovered and are directed back to the reactor system via the recycle system. Note that if several reactors are employed in the reactor system (*e.g.*, reactors 1 and 2) then alternatives of reactor 1-separator- recycle-reactor 2-separator-recycle may take place in addition to the other alternatives of reactor system-separation system-recycle system.

The heat recovery system has as units heat exchangers, and its primary objective is to reduce energy requirements expressed as hot and cold utilities by appropriately exchanging heat between the hot and cold process streams.

The utility system has as typical units steam and gas turbines, electric motors, electric generators, fired or waste heat boilers, steam headers at different pressures and auxiliary units (*e.g.*, vacuum condenser, water treater, deaerator), and provides the required electricity, power, and utilities.

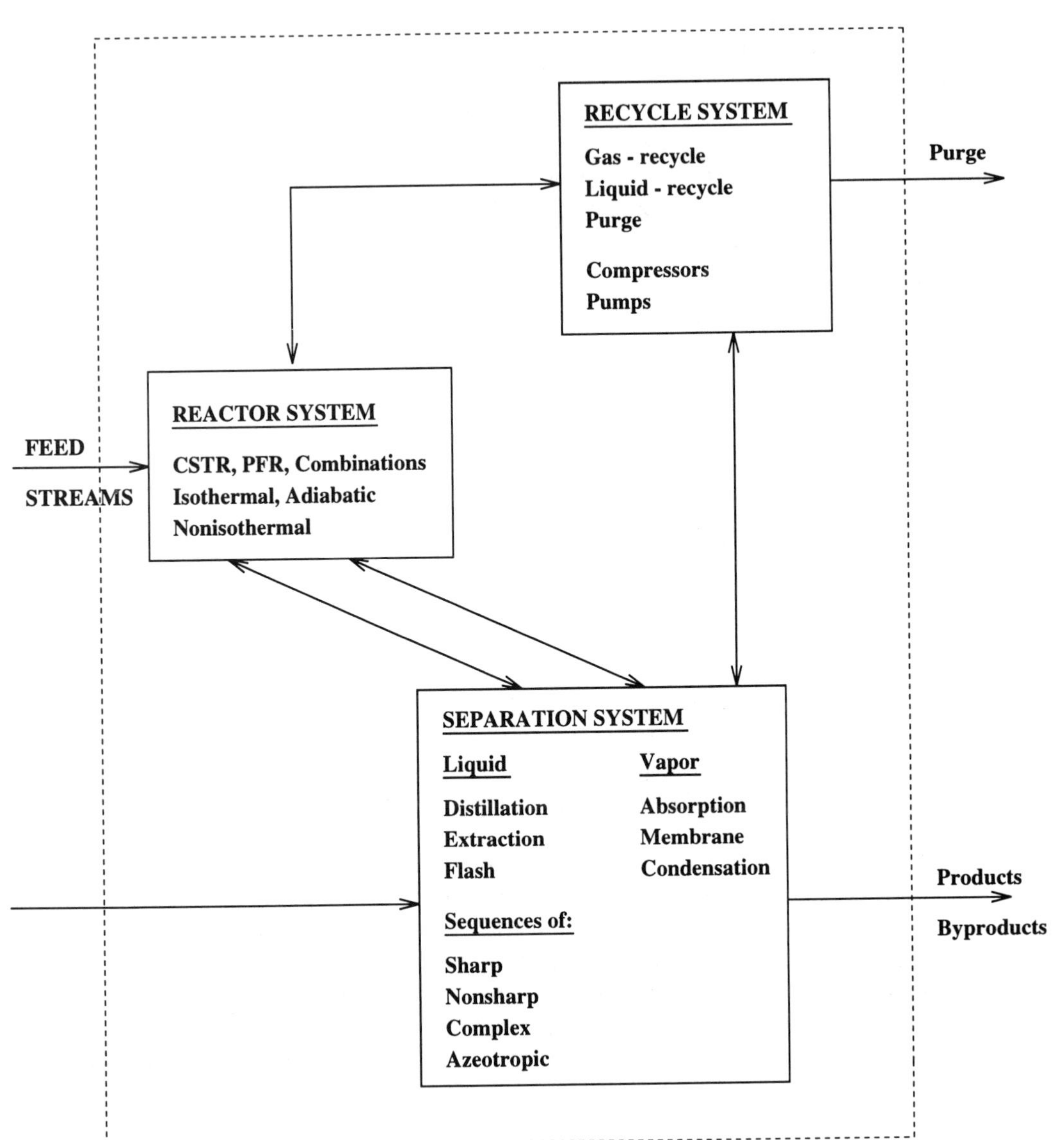

Figure 7.2: The chemical plant

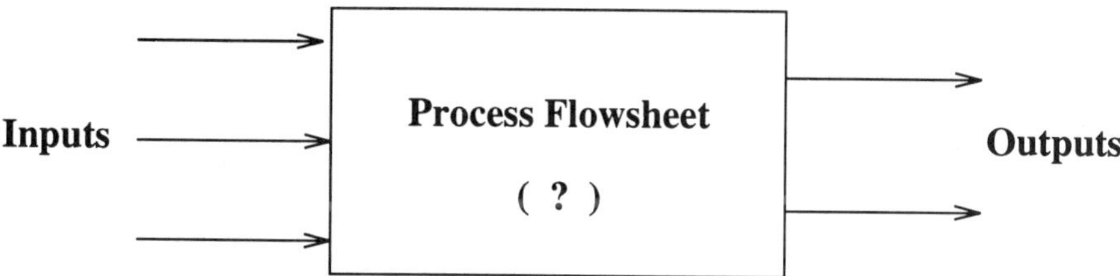

Figure 7.3: A schematic of process synthesis problem

Having described the overall process system, its three main interactive components (*i.e.*, the chemical plant, the heat recovery system and the utility system), as well as the three subsystems of the chemical plant (*i.e.*, the reactor system, the separation system, and the recycle system) we can now define the process synthesis problem.

7.2 Definition

Given are the specifications of the inputs (*e.g.*, feed streams) that may correspond to raw materials, and the specifications of the outputs (e.g., products, by-products) in either an overall process system or one of its subsystems (see Figure 7.3).

The primary objective of process synthesis is to develop systematically process flowsheet(s) that transform the available raw materials into the desired products and which meet the specified performance criteria of

(a) Maximum profit or minimum cost,

(b) Energy efficiency,

(c) Good operability with respect to

- (i) Flexibility,
- (ii) Controllability,
- (iii) Reliability,
- (iv) Safety,
- (v) Environmental regulations.

Remark 1 The unknown process flowsheet box in Figure 7.3 can be an overall process system, one of its three main systems (see Figure 7.1), one of the subsystems of the chemical plant (see Figure 7.2), or any combination of the above.

To determine the optimal process flowsheet(s) with respect to the imposed performance criteria, we have to provide answers to the following key questions:

Q1: Which process units should be used in the process flowsheet?

Q2: How should the process units be interconnected?

Q3: What are the optimal operating conditions and sizes of the selected process units?

The answers to questions Q1 and Q2 provide information on the topology/structure of the process flowsheet since they correspond to the selection of process units and their interconnections, while the answer to question Q3 provides information on the optimal values of the design parameters and operating conditions.

Note also that Q1 involves discrete decision making while Q2 and Q3 make selections among continuous alternatives. As a result, the determination of the optimal process flowsheet(s) can be viewed conceptually as a problem in the mixed discrete-continuous optimization domain.

The performance criteria are multiple ranging from economic to thermodynamic to operability considerations. As a result, the process synthesis problem can be classified mathematically as a

Multi-objective mixed discrete-continuous optimization problem.

Note, however, that most of the research work in process synthesis has been based on the single-objective simplification with the notable exception of the recent work of Luyben and Floudas (1994a, b) on the interaction of design/synthesis and control. As a result, in the rest of this book we will focus on process synthesis approaches that are single-objective or decompose the original problem so as to become a single-objective problem.

The nature of the multi-objective process synthesis problem is depicted in Figure 7.4 where it is shown that the optimal structure/topology of the process and the operating conditions of the process units result from the trade-offs of (Gundersen 1991):

(i) Capital Investment,

(ii) Energy,

(iii) Raw Materials Utilization,

(iv) Operability.

7.2.1 Difficulties/Challenges in Process Synthesis

The mixed discrete-continuous nonlinear nature of the process synthesis problem implies two major challenges which are due to its:

(i) Combinatorial nature, and

(ii) Nonlinear (nonconvex) characteristics.

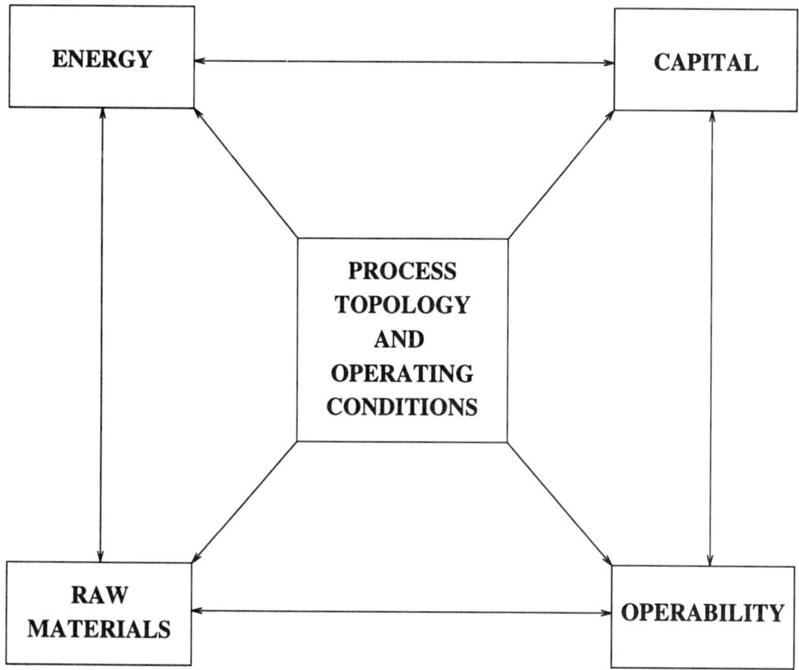

Figure 7.4: Trade-offs in process synthesis

By assigning, for instance, binary variables to represent the existence or not of process units in a total process system, the potential matches of hot and cold streams so as to reduce the utility consumption or the existence of trays in a distillation-based separation system, the resulting total number of binary variables can grow to 1000 or more which implies that we have to deal with a large combinatorial problem (*i.e.*, 2^{1000} process flowsheets).

Due to the interconnections of process units, the nonlinear models that describe each process unit, and the required search for the optimal operating conditions and sizing of each unit, the resulting model is highly nonlinear and contains nonconvexities. As a result, we have to deal with the issue of whether the obtained solution of the mixed integer nonlinear programming MINLP model is a local solution or a global solution, and subsequently to explore algorithms that allow the determination of the global solution.

Therefore, we need to address the following two issues:

Q1: How can we deal with the large combinatorial problem effectively?

Q2: How can we cope with the highly nonlinear model with respect to the quality of its solution (*i.e.*, local vs. global)?

To address these issues a number of approaches have been proposed in the last three decades which are discussed in the next section.

7.3 Approaches in Process Synthesis

To meet the objectives of the process synthesis problem, three main approaches have been proposed, which are based on

(i) Heuristics and evolutionary search,

(ii) Targets, physical insights, and

(iii) Optimization.

In (i), rules of thumb based on engineering experience are applied so as to generate good initial process flowsheets which are subsequently improved via a set of evolutionary rules applied in a systematic fashion. The principle advantage of (i) is that *near* optimal process flowsheets can be developed quickly at the expense, however, of not being able to evaluate the quality of the solution. An additional drawback of (i) is the potential conflicting rules which imply arbitrary weighting schemes for the considered alternatives.

In (ii), the physical and thermodynamic principles are utilized so as to obtain targets for the optimal process flowsheet(s). These targets can correspond to upper or lower bounds on the best possible process flowsheet(s) and provide key information for the improvement of existing processes. This approach is very powerful in reducing the huge combinatorial problem in cases where the bounds are close to the best solution and hence reduce the search space for the best alternatives. Note, however, that (ii) makes the assumption that process flowsheets derived based on targets (*e.g.*, energy utilization) are *near* optimal from the economic point of view (*e.g.*, total annualized cost) which is only true if the target (*e.g.*, energy) is the dominant cost item. As a result, the main drawback of (ii) is that capital cost considerations are not taken into account appropriately but only through artificial decomposition and hence provide only guidelines.

In (iii), an explicit or implicit representation of a number of process flowsheets that are of interest is proposed first and subsequently formulated as an optimization model. The solution of the resulting optimization problem extracts from the proposed set of process structures the one that best meets the performance criterion. The main advantages of (iii) are that (a) it represents a systematic framework that takes into consideration explicitly all types of interactions (*i.e.*, capital and operating costs) simultaneously, and (b) it generates automatically the best process flowsheet(s) for further investigation. Its main drawback, however, has to do with our capability to efficiently and rigorously solve the resulting large-scale mathematical models which call for theoretical and algorithmic advances on the optimization approaches. It should also be noted that (iii) can find the best process structure(s) only out of these included in the representation of alternatives.

Remark 1 Note that the borderlines between the three main approaches are not necessarily distinct. For instance, the targets in (ii) can be viewed as heuristics or rules that simplify the combinatorial problem and allow for its decomposition into smaller, more tractable problems (see chapter on heat exchanger network synthesis via decomposition approaches). The optimization approach (iii) can formulate thermodynamic targets, or targets on the attainable region of reaction mechanisms as optimization models, and can either utilize them so as to decompose the large-scale problem or follow a simultaneous approach that treats the full-scale mathematical model. The first alternative can be considered as target (heuristic)-optimization while the second as exclusively within the optimization approach.

In the following chapters we will discuss the optimization approach as applied to several classes of process synthesis problems, and we will present optimization models for the targets (*i.e.*, decomposition of process synthesis tasks) as well as for the overall nondecomposed mathematical representation of the alternatives. Prior to doing this, however, we will discuss the basic elements of the optimization approach (iii) in the next section.

7.4 Optimization Approach in Process Synthesis

7.4.1 Outline

The optimization approach in process synthesis consists of the following main steps:

Step 1: Representation of Alternatives A *superstructure* is postulated in which all process alternative structures of interest are embedded and hence are candidates for feasible or optimal process flowsheet(s). The superstructure features a number of different process units and their interconnections.

Step 2: Mathematical Model The superstructure is modeled mathematically as the following general formulation:

$$\left\{\begin{array}{ll} \min\limits_{x,y} & f(x,y) \\ \text{s.t.} & h(x,y) = 0 \\ & g(x,y) \leq 0 \\ & x \in X \leq R^n \\ & y \in Y = \{0,1\}^l \end{array}\right. \tag{7.1}$$

where

x is a vector of n continuous variables that may represent flow rates, temperatures, pressures, compositions of the process streams and sizes of the process units;

y is a vector of l $0-1$ variables that denote the potential existence (*i.e.*, $y_i = 1$) or not (*i.e.*, $y_i = 0$) of a process unit i in the optimal process flowsheet(s);

$f(x,y)$ is a single objective function which represents the performance criterion;

$h(x,y)$ are the m equality constraints that may denote total mass and component balances, energy balances, equilibrium relationships which constitute the process constraints;

$g(x, y)$ are the p inequality constraints which correspond to design specifications, restrictions, feasibility constraints, and logical constraints;

Step 3: Algorithmic Development: The mathematical model (7.1), which describes all process alternatives embedded in the superstructure, is studied with respect to the theoretical properties that the objective function and constraints may satisfy. Subsequently, efficient algorithms are utilized for its solution which extracts from the superstructure the optimal process structure(s).

Remark 1 All three aforementioned steps are very important each one for a different reason. The representation problem (*i.e.*, step 1) is crucial on the grounds of determining a superstructure which on the one hand is rich enough to allow all alternatives to be included and on the other hand is clever enough to eliminate undesirable structures. The modeling problem (*i.e.*, step 2) is of great importance since the effective evaluation of the embedded process alternatives depends to a high degree on how well modeled the superstructure is. The algorithmic development problem (*i.e.*, step 3) depends heavily on the mathematical model and its properties and is important since it represents the strategy based upon which the best alternatives are identified quickly and without the need to enumerate all possible structures.

We will discuss each of the three main steps of the optimization approach in the following sections.

7.4.2 Representation of Alternatives

The development of the appropriate superstructure is of major importance since the optimal process flowsheet(s) sought will be as good as the postulated representation of alternative structures.

Conceptually, the representation of alternative process flowsheet(s) is based on elementary graph theory ideas. By representing: each unit of the superstructure as a node, each input and output as a node, the interconnections among the process units as two-way arcs, the interconnections between the inputs and the process units as one-way arcs, the interconnections between the process units and the outputs as one-way arcs, and the interconnections between the inputs and the outputs as one-way arcs, then we have a bipartite planar graph that represents all options of the superstructure.

As an illustration let us consider the representation of alternative flowsheets that have one input, two outputs, and three process units. The complete superstructure is shown in Figure 7.5.

Remark 1 Note that given the process units, inputs, and outputs such a graph representation includes all possible alterative interconnections among the units, the inputs, and the outputs. Note also that the one-way arcs directly from the inputs to the outputs correspond to by-pass streams of the process units. Arcs originating from a unit and returning to the same unit without going through any other node have not been incorporated for simplicity of the presentation.

The superstructures of overall process systems or their important subsystems will be discussed in detail in the following chapters.

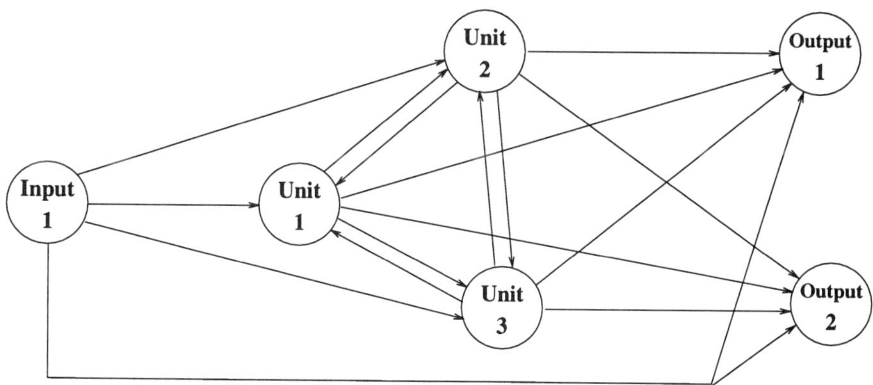

Figure 7.5: Illustration of superstructure

7.4.3 Mathematical Model of Superstructure

The general mathematical model of the superstructure presented in step 2 of the outline, and indicated as (7.1), has a mixed set of $0 - 1$ and continuous variables and as a result is a mixed-integer optimization model. If any of the objective function and constraints is nonlinear, then (7.1) is classified as mixed- integer nonlinear programming MINLP problem.

If the objective function and constraints are linear then (7.1) becomes a MILP problem. If in addition we fix the y variables to either 0 or 1, then the resulting formulation of (7.1) will be a nonlinear programming NLP problem or a linear programming LP problem depending on whether the objective function and constraints are nonlinear or linear.

The mixed-integer nature of the general mathematical model (7.1) implies a number of difficulties that are associated with the modeling aspects. Nemhauser and Wolsey (1988) emphasized that *in integer programming, formulating a good model is of crucial importance to solving it.* It is not uncommon that one formulation of a given problem is almost unsolvable while another formulation may be much easier to solve.

In the sequel, we will discuss a few simple guidelines in modeling MINLP problems:

7.4.3.1 Modeling with 0-1 Variables only

Some typical examples of modeling with 0-1 variables exclusively include:

(i) Select among process units $i \in PU$:

$$\sum_{i \in PU} y_i = 1 : \quad \text{Select only one unit,}$$

$$\sum_{i \in PU} y_i \leq 1 : \quad \text{Select at most one unit,}$$

$$\sum_{i \in PU} y_i \geq 1 : \quad \text{Select at least one unit.}$$

(ii) Select unit i only if unit j is selected

$$y_j - y_i \leq 0.$$

If $y_j = 1$, then $y_i = 1$. If $y_j = 0$, then y_i can be 0 or 1. This set of conditions is useful when there exist several sequences of process units out of which one sequence is to be selected (*e.g.*, sequences of distillation columns).

(iii) Propositional logic expressions

Let us consider a set of clauses $P_i, i = 1, 2, \ldots, r$ which are related via the logical operators:

OR	denoted as $\vee$
AND	denoted as $\wedge$
EXCLUSIVE OR (EOR)	denoted as $\oplus$
NEGATION	denoted as $\neg$

Then a proposition is any logical expression that is based on a set of clauses P_i which are related via the aforementioned logical operators.

Remark 1 The additional logical operator of IMPLICATION can be constructed from the aforementioned logical operators. For instance, clause P_1 implies clause P_2 (*i.e.*, $P_1 \Rightarrow P_2$) is equivalent to

$$\neg P_1 \vee P_2.$$

The propositional logic expressions can be expressed in an equivalent mathematical representation by associating a binary variable y_i with each clause P_i. The clause P_i being true, corresponds to $y_i = 1$, while the clause P_i being false, corresponds to $y_i = 0$. Note that $(\neg P_i)$ is represented by $(1 - y_i)$. Examples of basic equivalence relations between propositions and linear constraints in the binary variables include the following (Williams (1988)):

Proposition	**Mathematical Representation**
1. $P_1 \vee P_2 \vee P_3$	$y_1 + y_2 + y_3 \geq 1$
2. $P_1 \wedge P_2 \wedge P_3$	$y_1 \geq 1$ $y_2 \geq 1$ $y_3 \geq 1$
3. $\neg P_1 \vee P_2$ Implication: $(P_1 \Rightarrow P_2)$	$1 - y_1 + y_2 \geq 1$ or $y_1 - y_2 \leq 0$
4. $(\neg P_1 \vee P_2) \wedge (\neg P_2 \vee P_1)$ Equivalence: P_1 if and only if P_2	$\left\{ \begin{array}{c} y_1 - y_2 \leq 0 \\ y_2 - y_1 \leq 0 \end{array} \right\}$ or $y_1 = y_2$
5. $P_1 \oplus P_2 \oplus P_3$ Exactly one of the variables	$y_1 + y_2 + y_3 = 1$
6. $P_1 \vee P_2 \Rightarrow P_3$ This is equivalent to $(\neg P_1 \vee P_3) \wedge (\neg P_2 \vee P_3)$	$y_1 - y_3 \leq 0$ $y_2 - y_3 \leq 0$

Remark 2 Note that using the above basic equivalence relations we can systematically convert any arbitrary propositional logic expression into a set of linear equality and inequality constraints. The basic idea in an approach that obtains such an equivalence is to reduce the logical expression into its equivalent *conjunctive normal form* which has the form:

$$Q_1 \wedge Q_2 \wedge \ldots \wedge Q_n,$$

where Q_i, $i = 1, 2, \ldots, n$ are composite clauses of the P_i clauses written as disjunctions of the P_i's.

Note that each of the Q_i's $i = 1, 2, \ldots, n$ must be true independently (*i.e.*, each Q_i correspond to one constraint; see proposition 2). Also, note that since each Q_i can be written as a disjunction of the P_i's then it can be expressed via proposition 1 as a linear constraint.

Procedure to Obtain a Conjunctive Normal Form

Clocksin and Mellish (1981) proposed a systematic procedure that results in a conjunctive normal form of a propositional logic expression. This procedure involves the following three steps:

Step 1: Replace the *Implication* by its equivalent disjunction:

$$P_1 \Rightarrow P_2 \Longleftrightarrow \neg P_1 \wedge P_2.$$

Step 2: Apply DeMorgan's Theorem to put the negation inward:

$$\begin{aligned} \neg(P_1 \wedge P_2) &\Longleftrightarrow \neg P_1 \vee \neg P_2, \\ \neg(P_1 \vee P_2) &\Longleftrightarrow \neg P_1 \wedge \neg P_2. \end{aligned}$$

Step 3: Distribute the logical operator OR over the logical operator AND recursively using the equivalence of

$$(P_1 \wedge P_2) \vee P_3 \Longleftrightarrow (P_1 \vee P_3) \wedge (P_2 \vee P_3).$$

Illustration 7.4.1 This example is taken from Raman and Grossmann (1991), who first introduced the propositional logic expressions equivalence in chemical engineering. It involves the proposition:

$$(P_1 \wedge P_2) \vee P_3 \Rightarrow P_4 \vee P_5.$$

We apply here the procedure of Clocksin and Mellish (1981).

Step 1: The proposition is equivalent to

$$\neg[(P_1 \wedge P_2) \vee P_3] \vee (P_4 \vee P_5).$$

Step 2: Put the negation inward in two steps:

$$[\neg(P_1 \wedge P_2) \wedge \neg P_3] \vee (P_4 \vee P_5),$$
$$[(\neg P_1 \vee \neg P_2) \wedge \neg P_3] \vee (P_4 \vee P_5).$$

Step 3: Apply the operators OR over AND recursively:

$$[(\neg P_1 \vee \neg P_2) \vee (P_4 \vee P_5)] \wedge [\neg P_3 \vee (P_4 \vee P_5)],$$

which is written as

$$[\neg P_1 \vee \neg P_2 \vee P_4 \vee P_5] \wedge [\neg P_3 \vee P_4 \vee P_5].$$

This is the desired conjunctive normal form:

$$Q_1 \wedge Q_2,$$

where

$$\begin{aligned} Q_1 &= \neg P_1 \vee \neg P_2 \vee P_4 \vee P_5, \\ Q_2 &= \neg P_3 \vee P_4 \vee P_5. \end{aligned}$$

We assign y_1 to P_1, y_2 to P_2, y_3 to P_3, y_4 to P_4, y_5 to P_5. Then, the composite clauses Q_1 and Q_2 can be written as linear constraints of the form (see proposition 1)

$$\begin{aligned} Q_1 &\Longleftrightarrow 1 - y_1 + 1 - y_2 + y_4 + y_5 \geq 1 \text{ or } y_1 + y_2 - y_4 - y_5 \leq 1, \\ Q_2 &\Longleftrightarrow 1 - y_3 + y_4 + y_5 \geq 1 \text{ or } y_3 - y_4 - y_5 \leq 0. \end{aligned}$$

Since the original proposition has been equivalently transformed to the proposition

$$Q_1 \wedge Q_2,$$

with the above-stated Q_1 and Q_2 composite clauses, then the original proposition is equivalently written in mathematical form as the two independent linear constraints (see proposition 2):

$$\begin{aligned} y_1 + y_2 - y_4 - y_5 &\leq 1, \\ y_3 - y_4 - y_5 &\leq 0. \end{aligned}$$

7.4.3.2 Modeling with Continuous and Linear 0-1 Variables

(i) Activation and Deactivation of Continuous Variables

The 0-1 y variables can be used to activate or de-activate the continuous x variables, as well as inequality or equality constraints. For instance, if a process unit i does not exist (*i.e.*, $y_i = 0$), then the inlet flowrate to the unit i, F_i should be zero, while if the process unit i exists (*i.e.*, $y_i = 1$), then the inlet flowrate should be within given lower and upper bounds. This can be expressed as

$$F_i^L y_i \leq F_i \leq F_i^U y_i,$$

where F_i^L, F_i^U are the lower and upper bounds on the flowrate F_i. For $y_i = 0$, we have $0 \leq F_i \leq 0$ and hence $F_i = 0$. For $y_i = 1$, though, we have $F_i^L \leq F_i \leq F_i^U$.

(ii) Activation and Relaxation of Constraints

Activation and deactivation of inequality and equality constraints can be obtained in a similar way. For instance, let us consider the model of a process unit i that consists of one inequality $g(x) \leq 0$ and one equality $h(x) = 0$. If the process unit i does not exist (*i.e.*, $y_i = 0$), then both the equality and inequality should be relaxed. If, however, the process unit i exists (*i.e.*, $y_i = 1$), then the inequality and equality constraints should be activated. This can be expressed by introducing positive slack variables for the equality and inequality constraint and writing the model as

$$\begin{aligned} h(x) + s_1^+ - s_1^- &= 0, \\ g(x) &\leq s_2, \\ s_1^+ + s_1^- &\leq U_1 \cdot (1 - y_i), \\ s_2 &\leq U_2 \cdot (1 - y_i), \\ s_1^+, s_1^-, s_2 &\geq 0. \end{aligned}$$

Note that if $y_i = 1$, then $s_1^+ = s_1^- = s_2 = 0$, while if $y_i = 0$, then the slack variables are allowed to vary so as to relax the model constraints. Note also that s_2 can be eliminated by writing

$$g(x) - U_2(1 - y_i) \leq 0.$$

Note that care should be given to identifying the tightest upper bounds U_1 and U_2 in order to avoid a poor linear programming relaxation. The same argument holds true also for the F_i^L and F_i^U on the inlet flow rate F_i of unit i.

(iii) Nodes with Several Inputs

It should be noted that when several m input streams with flowrates F_j are directed to the same process unit i, then there are two basic alternatives of writing the logical expressions. *Alternative 1* is of the form:

$$\begin{aligned} \sum_{j=1}^{m} F_j - U y_i &\leq 0, \\ F_j &\geq 0 \quad j = 1, 2, \ldots, m, \end{aligned}$$

while *alternative 2* is

$$\begin{aligned} F_j - U y_i \leq 0 &\quad j = 1, 2, \ldots, m, \\ f_j \geq 0 &\quad j = 1, 2, \ldots, m. \end{aligned}$$

Alternative 1 has a single logical constraint while *alternative 2* has m logical constraints.

(iv) Logical Constraints in Multiperiod Problems

As another interesting example, let us consider the logical constraints in a multiperiod problem where z_i denotes the selection of a unit i and y_i^t denotes the operation or not of unit i at time period t. The logical constraints can be written in two alternative forms. *Alternative 1* takes the form:

$$\sum_{t=1}^{T} y_i^t - T \cdot z_i \leq 0.$$

Note that if process unit i is not selected (*i.e.*, $z_i = 0$), then $y_i^t = 0$ for all periods of operation. If, however, unit i is selected (*i.e.*, $z_i = 1$) then any of the T periods of operation is possible.

Alternative 2 becomes

$$y_i^t - z_i \leq 0 \quad t = 1, 2, \ldots, T,$$

and consists of T linear inequalities instead of one as shown in *alternative 1*.

Remark 1 It is interesting to note though that *alternative 2* provides a better way of modeling on the grounds that it provides a tighter continuous LP relaxation than *alternative 1*. Furthermore the continuous relaxation of *alternative 2* corresponds to the convex hull of 0-1 solutions. Note that the disaggregation of constraints in *alternative 2* results in a tighter continuous relaxation at the expense of having a potentially large set of constraints. It is not clear what is generally better with respect to the modeling aspects. This is due to the existing trade-off between the number of imposed constraints and the tightness of the continuous relaxation.

(v) Either-Or Constraints

$$\textbf{Either } f_1(x) \leq 0 \quad \textbf{or} \quad f_2(x) \leq 0.$$

This logical expression can be written in the following equivalent mathematical form by introducing a 0-1 variable y_i which takes the value of 1 if $f_1(x) \leq 0$ and 0 if $f_2(x) \leq 0$:

$$\begin{aligned} f_1(x) - U(1 - y_1) &\leq 0, \\ f_2(x) - U y_1 &\leq 0, \end{aligned}$$

where U is an upper positive bound.

Note that if $y_1 = 1$ we have

$$\begin{aligned} f_1(x) &\leq 0, \\ f_2(x) &\leq U, \end{aligned}$$

while if $y_1 = 0$ we have

$$\begin{aligned} f_1(x) &\leq U, \\ f_2(x) &\leq 0. \end{aligned}$$

(vi) Constraint Functions in Logical expressions

The following three illustrations of constraint functions are taken from Raman and Grossmann (1992).

Illustration 7.4.2 The first example that we consider here is stated as

$$\text{If } f(x) \leq 0, \text{ then } g(x) \geq 0.$$

We introduce y_1 associated with $f(x) \leq 0$ and y_2 associated with $g(x) \geq 0$, where both y_1, y_2 are binary variables. Then, we introduce the following inequality constraints that define the types of association:

$$\begin{aligned} f(x) \leq 0: \quad & L_1 y_1 + \epsilon && \leq f(x) \leq U_1(1 - y_1), \\ g(x) \geq 0: \quad & L_2(1 - y_2) && \leq g(x) \leq U_2 y_2 - \epsilon, \end{aligned}$$

where L_1, U_1 are the lower (negative) and upper bounds on $f(x)$;
L_2, U_2 are the lower (negative) and upper bounds on $g(x)$; and
ϵ is a small positive constant.

Analyzing the constraints for $f(x) \leq 0$, we see that

$$\begin{aligned} &\text{If } y_1 = 1, \text{ then} \quad && L_1 + \epsilon \leq f(x) \leq 0 \ (L_1 \text{ has to be negative}). \\ &\text{If } y_1 = 0, \text{ then} \quad && \epsilon \leq f(x) \leq U_1. \end{aligned}$$

Analyzing the constraints for $g(x) \geq 0$, we see that

$$\begin{aligned} &\text{If } y_2 = 1, \text{ then} \quad && 0 \leq g(x) \leq U_2 - \epsilon. \\ &\text{If } y_2 = 0, \text{ then} \quad && L_2 \leq g(x) \leq -\epsilon \ (L_2 \text{ has to be negative}). \end{aligned}$$

If we define the clauses

$$\begin{aligned} P_1 &: \ f(x) \leq 0, \\ P_2 &: \ g(x) \geq 0, \end{aligned}$$

then this example corresponds to the proposition

$$P_1 \Rightarrow P_2,$$

which is equivalent to

$$\neg P_1 \vee P_2,$$

which can be written as (see proposition 3)

$$1 - y_1 + y_2 \geq 1 \text{ or } y_1 - y_2 \leq 0.$$

Then, the equivalent representation in mathematical form is

$$\begin{aligned} & & y_1 - y_2 & \leq 0, \\ L_1 y_1 + \epsilon &\leq & f(x) & \leq U_1(1 - y_1), \\ L_2(1 - y_2) &\leq & g(x) & \leq U_2 y_2 - \epsilon. \end{aligned}$$

Illustration 7.4.3 The second example involves the logical expression:

$$\text{If } f(x) \le 0 \text{ and } h(x) = 0, \text{ then } g(x) \ge 0.$$

Define the three clauses:

$$\begin{aligned} P_1 &: f(x) \le 0, \\ P_2 &: h(x) = 0, \\ P_3 &: g(x) \ge 0, \end{aligned}$$

and then the proposition is

$$P_1 \wedge P_2 \Rightarrow P_3,$$

which is equivalent to

$$\neg(P_1 \wedge P_2) \vee P_3,$$

and becomes

$$\neg P_1 \vee \neg P_2 \vee P_3.$$

Assigning the 0-1 variables y_1, y_2, y_3 to P_1, P_2, P_3 respectively, then the above proposition can be written as

$$1 - y_1 + 1 - y_2 + y_3 \ge 1 \quad \text{or} \quad y_1 + y_2 - y_3 \le 1.$$

We have discussed in the previous example that $f(x) \le 0$ and $g(x) \ge 0$ can be expressed as

$$\begin{aligned} L_1 y_1 + \epsilon &\le f(x) \le U_1(1 - y_1), \\ L_2(1 - y_3) &\le g(x) \le U_3 y_3 - \epsilon, \end{aligned}$$

while in the activation and relaxation of constraints part (ii) we discussed how the equality constraint $h(x) = 0$ can be written as

$$\begin{aligned} & h(x) + s_1^+ - s_1^- = 0, \\ & s_1^+ + s_1^- \le U_2(1 - y_2), \\ & s_1^+, s_1^- \ge 0. \end{aligned}$$

Then the equivalent mathematical representation of the logical expression:

$$\begin{aligned} & y_1 + y_2 - y_3 \le 1, \\ L_1 y_1 + \epsilon & \le f(x) \le U_1(1 - y_1), \\ L_2(1 - y_3) & \le g(x) \le U_3 y_3 - \epsilon, \\ & h(x) + s_1^+ - s_1^- = 0, \\ & s_1^+ + s_1^- \le U_2(1 - y_2), \\ & s_1^+, s_1^- \ge 0. \end{aligned}$$

Illustration 7.4.4 This example illustrates the modeling of nondifferentiable functions with 0-1 variables. Let us consider the non differentiable function $\phi(x)$ defined as

$$\phi(x) = \max[0, f(x)],$$

This definition states that

$$\begin{aligned} \phi(x) &= f(x) \quad \text{if } f(x) \geq 0, \\ \phi(x) &= 0 \quad \text{if } f(x) < 0, \end{aligned}$$

and the nondifferentiability is at $f(x) = 0$. If we define the 0-1 variable y_1 associated with $f(x) \geq 0$:

$$L_1(1 - y_1) \leq f(x) \leq U_1 y_1,$$

then, if $y_1 = 1$ we have $0 \leq f(x) \leq U_1$, and if $y_1 = 0$ we have $L_1 \leq f(x) \leq 0$ where L_1, U_1 are the lower (≤ 0) and upper bounds on $f(x)$, respectively. In addition, we have to impose the conditions that

$$\begin{aligned} \phi(x) = f(x) \quad &\text{if} \quad f(x) \geq 0, \\ \phi(x) = 0 \quad &\text{if} \quad f(x) < 0. \end{aligned}$$

One way of obtaining these conditions is by defining $\phi(x) - f(x)$:

$$\begin{aligned} L_2(1 - y_1) &\leq \phi(x) - f(x) \leq U_2(1 - y_1), \\ L_3 y_1 &\leq \phi(x) \leq U_3 y_1, \end{aligned}$$

where L_2, U_2 are lower and upper bounds on $\phi(x) - f(x)$; and
L_3, U_3 are lower and upper bounds on $\phi(x)$.

Note that, if $y_1 = 1$ we have

$$\begin{aligned} 0 &\leq \phi(x) - f(x) \leq 0 \Rightarrow \phi(x) = f(x), \\ L_3 &\leq \phi(x) \leq U_3. \end{aligned}$$

If $Y_1 = 0$, we have

$$\begin{aligned} L_2 &\leq \phi(x) - f(x) \leq U_2, \\ 0 &\leq \phi(x) \leq 0 \Rightarrow \phi(x) = 0. \end{aligned}$$

Then the equivalent representation is

$$\begin{aligned} L_1(1 - y_1) \leq & \quad f(x) & \leq U_1 y_1, \\ L_2(1 - y_1) \leq & \quad \phi(x) - f(x) & \leq U_2(1 - y_1), \\ L_3 y_1 \leq & \quad \phi(x) & \leq U_3 y_1. \end{aligned}$$

Note also, that we can simplify it further since

$$\phi(x) \geq 0 \text{ and } \phi(x) - f(x) \geq 0,$$

which implies that $L_2 = L_3 = 0$. Then we have

$$\begin{aligned} L_1(1-y_1) \leq \quad & f(x) & \leq U_1 y_1, \\ 0 \leq \quad & \phi(x) - f(x) & \leq U_2(1-y_1), \\ 0 \leq \quad & \phi(x) & \leq U_3 y_1. \end{aligned}$$

Another alternative way of writing the conditions

$$\begin{aligned} \phi(x) - f(x) &= 0 \text{ if } f(x) \geq 0, \\ \phi(x) &= 0 \text{ if } f(x) < 0, \end{aligned}$$

is by introducing slacks $s_1^+, s_1^-, s_2^+, s_2^-$:

$$\begin{aligned} \phi(x) - f(x) + s_1^+ - s_1^- &= 0, \\ \phi(x) + s_2^+ - s_2^- &= 0, \\ s_1^+ + s_1^- - U_2(1-y_1) &\leq 0, \\ s_2^+ + s_2^- - U_3 y_1 &\leq 0, \\ s_1^+, s_1^-, s_2^+, s_2^- &\geq 0. \end{aligned}$$

Since $\phi(x) - f(x) \geq 0$ and $\phi(x) \geq 0$, then we only need s_1^-, s_2^- and hence we have

$$\begin{aligned} \phi(x) - f(x) - s_1^- &= 0, \\ \phi(x) - s_2^- &= 0, \\ s_1^- - U_2(1-y_1) &\leq 0, \\ s_1^- - U_3 y_1 &\leq 0, \\ s_1^-, s_2^- &\geq 0. \end{aligned}$$

Then the equivalent mathematical representation of

$$\phi(x) = \max[0, f(x)]$$

is

$$\begin{aligned} L_1(1-y_1) \leq f(x) &\leq U_1 y_1, \\ \phi(x) - f(x) - s_1^- &= 0, \\ \phi(x) - s_2^- &= 0, \\ s_1^- - U_2(1-y_1) &\leq 0, \\ s_2^- - U_3 y_1 &\leq 0, \\ s_1^-, s_2^- &\geq 0. \end{aligned}$$

Also, note that we can further eliminate s_1^-, s_2^- from the equality constraints and finally have

$$\begin{aligned} L_1(1-y_1) \leq \quad & f(x) & \leq U_1 y_1, \\ 0 \leq \quad & \phi(x) - f(x) - U_2(1-y_1) & \leq 0, \\ 0 \leq \quad & \phi(x) - U_3 y_1 & \leq 0, \end{aligned}$$

which is the same final form as the one shown by using alternative 1.

7.4.3.3 Modeling with Bilinear Products of Continuous and 0-1 Variables

Illustration 7.4.5 This example comes from control systems with time delays and is taken from Psarris and Floudas (1990) (see model P3) where bilinear products of the form $x_{ij} \cdot y_{ij}$ exist in inequality constraints of the form:

$$\sum_{j=1}^{n} x_{ij}y_{ij} - x_{ij} \leq 0 \quad i = 1, 2, \ldots, n,$$

and the objective function is of the form:

$$\min \sum_{i} \sum_{j} x_{ij}$$

where x_{ij} are continuous variables and y_{ij} are 0-1 variables. Lower and upper bounds on the x_{ij} variables L and U are also provided (*i.e.* $L \leq x_{ij} \leq U$).

The modeling approach applied in Psarris and Floudas (1990) used the idea proposed by Petersen (1971) and which was extended by Glover (1975). The basic idea is to introduce new variables h_{ij} for each bilinear product:

$$h_{ij} = x_{ij}y_{ij} \ \forall (ij)$$

and introduce four additional constraints for each (ij):

$$\begin{aligned} x_{ij} - U(1 - y_{ij}) &\leq h_{ij} \leq x_{ij} - L(1 - y_{ij}), \\ Ly_{ij} &\leq h_{ij} \leq Uy_{ij}, \end{aligned}$$

which are linear in x_{ij}, y_{ij}, and h_{ij}.

Note that if $y_{ij} = 1$, then the above constraints become

$$\begin{aligned} x_{ij} &\leq h_{ij} \leq x_{ij}, \\ L &\leq h_{ij} \leq U, \end{aligned}$$

and the first two constraints imply that $h_{ij} = x_{ij}$ while the second two constraints are relaxed.

If $y_{ij} = 0$, then we have

$$\begin{aligned} x_{ij} - U &\leq h_{ij} \leq x_{ij} - L, \\ 0 &\leq h_{ij} \leq 0, \end{aligned}$$

and the second two constraints imply that $h_{ij} = 0$, while the first two constraints are relaxed since $x_{ij} - U \leq 0$ and $x_{ij} - L \geq 0$.

Remark 1 If $L = 0$, then the four additional linear constraints become:

$$\begin{aligned} x_{ij} - U(1 - y_{ij}) &\leq h_{ij} \leq x_{ij}, \\ 0 &\leq h_{ij} \leq Uy_{ij}. \end{aligned}$$

Note that the nonnegativity constraint of h_{ij} can be omitted since $x_{ij} \geq 0$ and hence in this case we only need to introduce three constraints and one variable for each (ij).

Remark 2 In this particular example, since we minimize the sum of x_{ij}, and the minimization of the objective function will make the value of h_{ij} equal to x_{ij} if $y_{ij} = 1$ and h_{ij} equal to zero if $y_{ij} = 0$, then we only need 2 linear constraints which are

$$\begin{aligned} x_{ij} - U(1 - y_{ij}) &\leq h_{ij}, \\ L y_{ij} &\leq h_{ij}, \end{aligned}$$

as pointed out by Psarris and Floudas (1990). If in addition $L = 0$ (*i.e.*, x_{ij} nonnegative), then we need to introduce only 1 constraint and 1 variable per (ij):

$$x_{ij} - U(1 - y_{ij}) \leq h_{ij}.$$

Illustration 7.4.6 This example is taken from Voudouris and Grossmann (1992) , and corresponds to multiple choice structures that arise in discrete design problems of batch processes. The model has bilinear inequality constraints of the form:

$$\alpha_{ij} \sum_{s=1}^{N(i)} d_{is} x_j y_{is} - \beta_{ij} w_j \leq 0 \; j \in J(i), \;\; i = 1, 2, \ldots, n,$$

where x_j are continuous variables with bound L_j, U_j; y_{is} are 0-1 variables, and the following constraint holds:

$$\sum_{s=1}^{N(i)} y_{is} = 1.$$

To remove the bilinear terms $x_j y_{is}$, Voudouris and Grossmann (1992) proposed the introduction of new variables h_{ijs}:

$$\begin{aligned} x_j = \sum_{s=1}^{N(i)} h_{ijs} & \qquad j \in J(i), \; i = 1, 2, \ldots, n, \\ L_j y_{is} \leq h_{ijs} \leq U_j y_{is} & \qquad j \in J(i), \; i = 1, 2, \ldots, n, \;\; s = 1, 2, \ldots, N(i). \end{aligned}$$

Then the inequality constraints become

$$\alpha_{ij} \sum_{s=1}^{N(i)} d_{is} h_{ijs} - \beta_{ij} w_j \leq 0 \; j \in J(i), \;\; i = 1, 2, \ldots, n$$

with the previously defined constraints of h_{ijs}.

Remark 3 An alternative way of modeling this problem is via the approach followed in illustration 1 and the observation made in Remark 2. Torres (1991) showed that when the bilinear terms participate only in inequalities, then it suffices to consider only the following constraints:

$$\left. \begin{aligned} L_j y_{is} &\leq h_{ijs} \\ x_j - U_j(1 - y_{is}) &\leq h_{ijs} \end{aligned} \right\} \quad \begin{aligned} & i = 1, 2, \ldots, n, \\ & j \in J(i), \\ & s = 1, 2, \ldots, N(i). \end{aligned}$$

Note that this set of constraints are fewer than the ones proposed by Voudouris and Grossmann (1992), but they may produce a weaker relaxation.

Illustration 7.4.7 This example is taken from Glover (1975) and is as follows

We have a continuous variable x, and a 0-1 variable y:

$$L_0(w) \leq x \leq U_0(w) \text{ when } y = 0,$$
$$L_1(w) \leq x \leq U_1(w) \text{ when } y = 1.$$

where $L_0(w), L_1(w)$ are lower bounds of x which are linear functions of other continuous variables w; and $U_0(w), U_1(w)$ are upper bounds of x which are linear functions of other continuous variables w.

A natural way to model this problem is via the following two inequalities:

$$L_0(w) + [L_1(w) - L_0(w)]y \leq x \leq U_0(w) + [U_1(w) - U_0(w)]y.$$

Note that if $y = 1$, we have

$$L_1(w) \leq x \leq U_1(w),$$

while if $y = 0$, we have

$$L_0(w) \leq x \leq U_0(w),$$

which is the desired result.

However, since L_0, L_1, U_0, U_1 are linear functions of other continuous variables w, we have nonlinear products of continuous times 0-1 variables.

To remove this limitation, Glover (1975) proposed a clever approach that is based upon Petersen's (1971) linearization idea.

First we define the constants $\underline{U}_0\bar{U}_0, \underline{U}_1, \bar{U}_1, \underline{L}_0, \bar{L}_0, \underline{L}_1, \bar{L}_1$:

$$\begin{aligned}
\underline{L}_0 &\leq L_0(w) \leq \bar{L}_0, \\
\underline{U}_0 &\leq U_0(w) \leq \bar{U}_0, \\
\underline{L}_1 &\leq L_1(w) \leq \bar{L}_1, \\
\underline{U}_1 &\leq U_1(w) \leq \bar{U}_1.
\end{aligned}$$

Then the appropriate set of linear inequalities is

$$\begin{aligned}
L_0(w) + (\underline{L}_1 - \bar{L}_0)y &\leq x \leq U_0(w) + (\bar{U}_1 - \underline{U}_0)y, \\
L_1(w) + (\underline{L}_0 - \bar{L}_1)(1-y) &\leq x \leq U_1(w) + (\bar{U}_0 - \underline{U}_1)(1-y).
\end{aligned}$$

Note that if $y = 0$, we have

$$\begin{aligned}
L_0(w) &\leq x \leq U_0(w), \\
L_1(w) + \underline{L}_0 - \bar{L}_1 &\leq x \leq U_1(w) + \bar{U}_0 - \underline{U}_1.
\end{aligned}$$

where the first constraints are the ones desired while the second constraints become redundant.

If $y = 1$, we have

$$\begin{aligned}
L_0(w) + \underline{L}_1 - \bar{L}_0 &\leq x \leq U_0(w) + \bar{U}_1 - \underline{U}_0, \\
L_1(w) &\leq x \leq U_1(w),
\end{aligned}$$

where the second constraints are the desirable and the first constraints are redundant.

7.4.3.4 Modeling Nonlinearities of Continuous Variables

Modeling nonlinear expressions of the continuous variables is of paramount importance on the grounds that appropriate modeling techniques can result in formulations that are convex and hence their global solutions can be attained. Therefore, the modeling task of nonlinearities of continuous variables will not only affect the efficiency of the algorithms but it may also affect drastically the quality of the solution obtained in terms of attaining a local versus a global optimum.

With the above in mind, we will focus here on

(i) Separable concave functions,

(ii) Convexification procedures, and

(iii) Nonconvexities.

(i) Modeling separable concave functions

Suppose we have a separable concave function

$$C(x_1, x_2, x_3) = C_1(x_1) + C_2(x_2) + C_3(x_3)$$

where $C_1(x_1)$ denotes the cost of a process unit 1 with capacity x_1;
$C_2(x_2)$ is the cost of process unit 2 with capacity x_2;
$C_3(x_3)$ is the cost of process unit 3 with capacity x_3; and
$C_1(x_1), C_2(x_2), C_3(x_3)$ are concave functions in x_1, x_2, and x_3, respectively.

Figure 7.6 shows the concave cost function $C_1(x_1)$ as a function of x_1.

An alternative way of modeling $C_1(x_1)$ is via a set of piece- wise linear functions that correspond to fixed charge cost models of the form:

$$\begin{cases} \alpha + \beta \cdot x & \text{if} \quad L < x \le U, \\ 0 & \text{if} \quad x = 0. \end{cases}$$

Let us consider a piecewise linear approximation of $C(x_1)$ specified by the points:

$$\begin{aligned} A &: [\gamma_1, C(\gamma_1)], \\ B &: [\gamma_2, C(\gamma_2)], \\ C &: [\gamma_3, C(\gamma_3)], \\ D &: [\gamma_4, C(\gamma_4)]. \end{aligned}$$

For any $\gamma_1 \le x_1 \le \gamma_4$, we can write

$$\begin{aligned} & x_1 = \lambda_1\gamma_1 + \lambda_2\gamma_2 + \lambda_3\gamma_3 + \lambda_4\gamma_4, \\ & \lambda_1 + \lambda_2 + \lambda_3 + \lambda_4 = 1, \\ & \lambda_1, \lambda_2, \lambda_3, \lambda_4 \ge 0. \end{aligned}$$

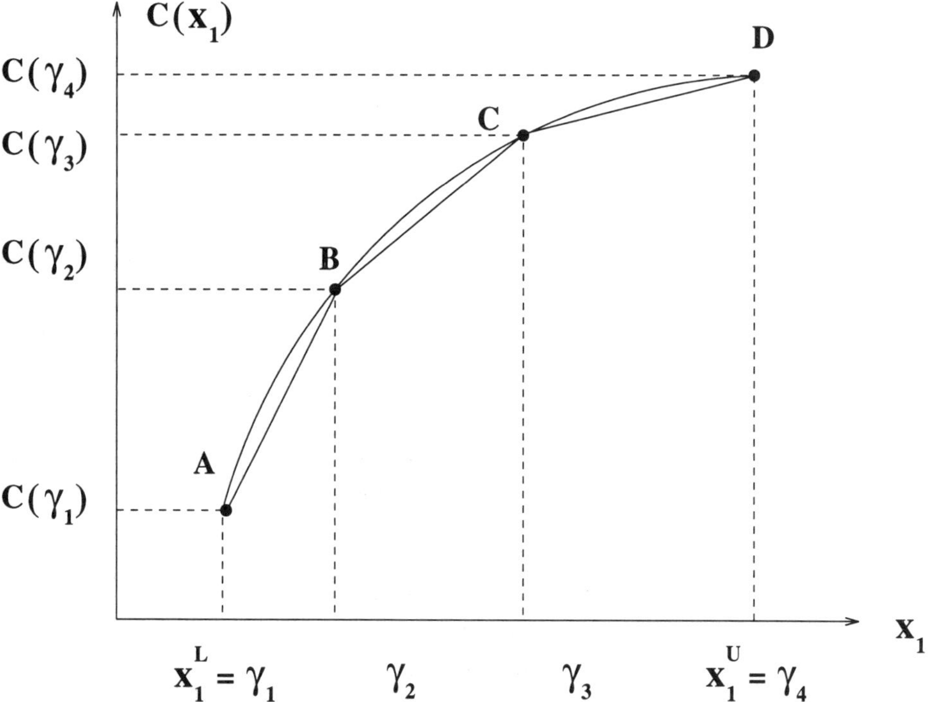

Figure 7.6: Concave function and a piecewise linear approximation

Hence, if $\gamma_2 \leq x_1 \leq \gamma_3$, then

$$\begin{aligned} x_1 &= \lambda_2\gamma_2 + \lambda_3\gamma_3, \\ \lambda_2 &+ \lambda_3 = 1, \\ \lambda_2&, \lambda_3 \geq 0, \end{aligned}$$

and we have for the cost function that

$$C(x_1) = \lambda_2 C(\gamma_2) + \lambda_3 C(\gamma_3).$$

Generalizing, we can write that

$$C(x_1) = \lambda_1 C(\gamma_1) + \lambda_2 C(\gamma_2) + \lambda_3 C(\gamma_3) + \lambda_4 C(\gamma_4),$$

$$\begin{aligned} \lambda_1 + \lambda_2 + \lambda_3 + \lambda_4 &= 1, \\ \lambda_1, \lambda_2, \lambda_3, \lambda_4 &\geq 0. \end{aligned}$$

Condition on λ's

In the piecewise linear approximation of the concave function shown in Figure 7.6, there are three linear sections (*i.e.*, $\gamma_1 \leq x_1 \leq \gamma_2, \gamma_2 \leq x_1 \leq \gamma_3$, and $\gamma_3 \leq x_1 \leq \gamma_4$) and four points. As a result, the condition on the λ's that must be imposed is that

At most two of the λ's are strictly positive.

This condition results from having x_1 as an intermediate point in any of these linear sections or as one of the four points. This condition on λ's can be modeled by introducing one binary variable for each section:

$$\begin{aligned} &y_1 \quad \text{for} \quad \gamma_1 \leq x_1 \leq \gamma_2, \\ &y_2 \quad \text{for} \quad \gamma_2 \leq x_1 \leq \gamma_3, \\ &y_3 \quad \text{for} \quad \gamma_3 \leq x_1 \leq \gamma_4, \end{aligned}$$

such that

$$y_1 \begin{cases} 1 & \text{if } \gamma_1 \leq x_1 \leq \gamma_2 \\ 0 & \text{otherwise} \end{cases}$$

$$y_2 \begin{cases} 1 & \text{if } \gamma_2 \leq x_1 \leq \gamma_3 \\ 0 & \text{otherwise} \end{cases}$$

$$y_3 \begin{cases} 1 & \text{if } \gamma_3 \leq x_1 \leq \gamma_4 \\ 0 & \text{otherwise} \end{cases}$$

Note that we must have

$$y_1 + y_2 + y_3 = 1,$$

which indicates that only one section will be selected.

We also need to introduce the following constraints:

$$\begin{aligned}
\lambda_1 &\leq y_1, \\
\lambda_2 &\leq y_1 + y_2, \\
\lambda_3 &\leq y_2 + y_3, \\
\lambda_4 &\leq y_3.
\end{aligned}$$

Note that if $y_1 = 1$, then $y_2 = y_3 = 0$ and only λ_1 and λ_2 can be positive. If $y_2 = 1$, then $y_1 = y_3 = 0$ which implies that only λ_2 and λ_3 can be positive. If $y_3 = 1$, then $y_1 = y_2 = 0$ and hence only λ_3 and λ_4 can be positive.

Then the piecewise linear approximation of the approximation of the concave function $C(x_1)$ can be formulated as

$$\begin{cases}
C(x_1) = \lambda_1 C(\gamma_1) + \lambda_2 C(\gamma_2) + \lambda_3 C(\gamma_3) + \lambda_4 C(\gamma_4) \\
x_1 = \lambda_1\gamma_1 + \lambda_2\,\gamma_2 + \lambda_3\gamma_3 + \lambda_4\gamma_4 \\
\lambda_1 + \lambda_2 + \lambda_3 + \lambda_4 = 1 \\
\lambda_1 \leq y_1 \\
\lambda_2 \leq y_1 + y_2 \\
\lambda_3 \leq y_2 + y_3 \\
\lambda_4 \leq y_3 \\
y_1 + y_2 + y_3 = 1 \\
\lambda_1, \lambda_2, \lambda_3, \lambda_4 \geq 0 \\
y_1, y_2, y_3 = 0 - 1
\end{cases}$$

Remark 1 Note that $\gamma_1, \gamma_2, \gamma_3, \gamma_4$ are known points. Hence, the unknown continuous variables are $x_1, \lambda_1, \lambda_2, \lambda_3, \lambda_4$, and the binary variables are y_1, y_2, y_3. As a result of this three-piece linear approximation of the concave function $C(x_1)$ we have increased the number of continuous variables by four, introduced three binary variables, two linear constraints on the continuous variables, four logical constraints and one constraint on the binary variables only.

Remark 2 We can generalize the formulation of a piecewise linear approximation of a concave function by considering K points:

$$[\gamma_i, C(\gamma_i)]\ i = 1, 2, \ldots, K$$

The general formulation is

$$\begin{cases} C(x_1) = \sum_{i=1}^{K} \lambda_i C(\gamma_i) \\ x_1 = \sum_{i=1}^{K} \lambda_i \gamma_i \\ \sum_{i=1}^{K} \lambda_i = 1 \\ \lambda_1 \leq y_1 \\ \lambda_i \leq y_{i-1} + y_i & i = 2, 3, \ldots, (K-1) \\ \lambda_K \leq y_{K-1} \\ \sum_{i=1}^{K-1} y_i = 1 \\ \lambda_i \geq 0 & i = 1, 2, \ldots, K \\ y_i = \{0-1\}^{K-1} \end{cases}$$

Remark 3 An alternative way of modeling the concave function $C(x_1)$ as a piecewise linear approximation with three segments as shown in Figure 7.6, is the following:

First we obtain expressions of $C(x_1)$ in each of the linear segments, which are

$$\begin{aligned} C(x_1) &= \alpha_1 + \beta_1 \cdot x_1 \quad \text{for} \quad \gamma_1 \leq x_1 \leq \gamma_2, \\ C(x_2) &= \alpha_2 + \beta_2 \cdot x_1 \quad \text{for} \quad \gamma_2 \leq x_1 \leq \gamma_3, \\ C(x_1) &= \alpha_3 + \beta_3 \cdot x_1 \quad \text{for} \quad \gamma_3 \leq x_1 \leq \gamma_4, \end{aligned}$$

where

$$\begin{aligned} \alpha_1 &= C(\gamma_1) - \beta_1 \cdot \gamma_1 \quad , \quad \beta_1 = \frac{C(\gamma_2) - C(\gamma_1)}{\gamma_2 - \gamma_1}, \\ \alpha_2 &= C(\gamma_2) - \beta_2 \cdot \gamma_2 \quad , \quad \beta_2 = \frac{C(\gamma_3) - C(\gamma_2)}{\gamma_3 - \gamma_2}, \\ \alpha_3 &= C(\gamma_3) - \beta_3 \cdot \gamma_3 \quad , \quad \beta_3 = \frac{C(\gamma_4) - C(\gamma_3)}{\gamma_4 - \gamma_3}. \end{aligned}$$

Then we introduce the binary variables y_1, y_2, y_3 each one associated with segment 1, 2, and 3 respectively, and the continuous variables x_1^1, x_1^2, x_1^3 and we can hence write

$$\begin{cases} C(x_1) = (\alpha_1 \cdot y_1 + \beta_1 x_1^1) + (\alpha_2 \cdot y_2 + \beta_2 x_1^2) + (\alpha_3 \cdot y_3 + \beta_3 x_1^3) \\ x_1 = x_1^1 + x_1^2 + x_1^3 \\ \gamma_1 \cdot y_1 \leq x_1^1 \leq \gamma_2 \cdot y_1 \\ \gamma_2 \cdot y_2 \leq x_1^2 \leq \gamma_3 \cdot y_2 \\ \gamma_3 \cdot y_3 \leq x_1^3 \leq \gamma_4 \cdot y_3 \\ y_1 + y_2 + y_3 = 1 \\ y_1, y_2, y_3 = 0 - 1 \end{cases}$$

If $y_1 = 1$, then $y_2 = y_3 = 0$, which implies that $x_1^2 = x_1^3 = 0$ and hence $x_1 = x_1^1$.
If $y_2 = 1$, then $y_1 = y_3 = 0$, which implies that $x_1^1 = x_1^3 = 0$ and hence $x_1 = x_1^2$.
If $y_3 = 1$, then $y_1 = y_2 = 0$, which implies that $x_1^1 = x_1^2 = 0$ and hence $x_1 = x_1^3$.

Comparing the above formulation with the one presented earlier we notice that we introduce three continuous variables instead of four, one constraint on the continuous variables instead of two, and six logical constraints instead of four.

This type of modeling approach can also be generalized to K points, and hence $(K-1)$ linear segments and results in:

$$\left\{\begin{array}{rcll} C(x_1) & = & \sum_{i=1}^{K-1}(\alpha_i y_i + \beta_i x_1^i) & \\ x_1 & = & \sum_{i=1}^{K-1} x_1^i & \\ \gamma_i \cdot y_i & \leq & x_1^i \leq \gamma_{i+1} \cdot y_i & i = 1, 2, \ldots, (K-1) \\ \sum_{i=1}^{K_1} y_i & = & 1 & \\ y_i & \in & \{0-1\}^{K-1} & \end{array}\right.$$

(ii) Convexification

Another approach toward the modeling of nonlinear functions of continuous variables is to identify classes of nonlinear functions that can be transformed so as to result in convex nonlinear functions. An important class of nonlinear functions that can be convexified arises in geometric programming and is denoted as posynomials. Posynomials are of the form:

$$f(x) = \sum_j c_j \prod_j x_i^{\alpha_i},$$

where $c_j \geq 0$, and α_i can be any arbitrary rational number, and $x_i \geq 0$.

By introducing new variables z_i and the following logarithmic transformation:

$$z_i = \ln x_i \text{ or } (x_i = e^{z_i}),$$

the posynomial can be transformed into:

$$\begin{aligned} f(z) &= \sum_j c_j \prod_j e^{\alpha_i z_i} \\ &= \sum_j c_j e^{\sum_i \alpha_i z_i}. \end{aligned}$$

The sum $w = \sum_i \alpha_i z_i$ is a linear function in z_i and hence convex. The function e^w is convex and so is the function e^w with w being a linear function in z_i. Note that this expression is convex since it is a sum of positive convex functions (*i.e.*, exponentials).

Remark 4 The above is true if $c_j \geq 0$ for all j. In case the coefficients change sign then we have a signomial which cannot be convexified, with the logarithmic transformation.

Illustration 7.4.8 This example has a function that is the sum of a bilinear term, a concave term, and a fractional term with positive coefficients and variables:

$$f(x_1, x_2, x_3) = x_1 x_2 + x_1^{0.6} + \frac{x_1}{x_3}.$$

Defining

$$\begin{aligned} z_1 &= \ln x_1, \\ z_2 &= \ln x_2, \\ z_3 &= \ln x_3, \end{aligned}$$

we have

$$\begin{aligned} f(z_1, z_2, z_3) &= e^{z_1} e^{z_2} + e^{0.6 z_1} + e^{z_1 - z_3} \\ &= e^{z_1 + z_2} + e^{0.6 z_1} + e^{z_1 - z_3}, \end{aligned}$$

which is convex in z_1, z_2, z_3 since it is a sum of convex functions.

Illustration 7.4.9 This example illustrates the case where convexification is not possible, and has a function $f(x)$ of positive $x \geq 0$:

$$f(x) = x_1 \cdot x_2 - x_3 \cdot x_4.$$

Defining

$$\begin{aligned} z_1 &= \ln x_1, \\ z_2 &= \ln x_2, \\ z_3 &= \ln x_3, \\ z_4 &= \ln x_4, \end{aligned}$$

we have

$$f(z_1, z_2, z_3, z_4) = e^{z_1 + z_2} - e^{z_3 + z_4},$$

which is a difference of two convex functions and is a non-convex function.

Illustration 7.4.10 This example is taken from Grossmann and Sargent (1979), Kocis and Grossmann (1988) and represents the design of multiproduct batch plants. The plant consists of M processing stages manufacturing fixed amount of Q_i of N products. At each stage j, the number of parallel units N_j and their sizes V_j are determined along with the batch sizes $_i$ and cycle times

T_{Li} for each product i. The model is:

$$
\left\{
\begin{array}{lll}
\min & \sum_{j=1}^{M} \alpha_j N_j V_j^{B_j} & \\
\text{s.t.} & \left.\begin{array}{rl} V_j \geq & S_{ij}B_i \\ N_j T_{Li} \geq & t_{ij} \end{array}\right\} & i = 1,2,\ldots,N;\ j = 1,2,\ldots,M \\
& \sum_{i=1}^{N} \frac{Q_i T_{Li}}{B_i} \leq \ H & \\
& \left.\begin{array}{rcl} 1 \leq & N_j & \leq N_j^U \\ V_j^L \leq & V_j & \leq V_j^U \end{array}\right\} & j = 1,2,\ldots,M \\
& \left.\begin{array}{c} T_{Li}^L \leq T_{Li} \leq T_{Li}^U \\ B_i^L \leq B_i \leq B_i^U \end{array}\right\} & i = 1,2,\ldots,N \\
& N_j \text{ integer} &
\end{array}
\right.
$$

where α_i, β_j positive constants (*e.g.* $\alpha_j = 200, \beta_j = 0.6$);
S_{ij} are the given size factors for product i per stage j;
Q_i are given production rates;
t_{ij} are given processing times for product i in stage j; and
H is the given horizon time.

Note that the first set of constraints are linear while the rest are nonlinear. Also note that in the objective function we have sum of products of integers times concave functions, while the second set of constraints has bilinear terms of integer and continuous variables. The third set of constraints has sum of fractional terms.

Defining

$$
\begin{aligned}
v_j &= \ln(V_j), \\
n_j &= \ln(N_j), \\
b_i &= \ln(B_i), \\
t_{Li} &= \ln(T_{Li}),
\end{aligned}
$$

and substituting (*i.e.*, eliminating V_j, n_j, B_i, T_{Li}) we have

$$
\left\{
\begin{array}{lll}
& \min \sum_{j=1}^{M} \alpha_j e^{(n_j + \beta_j v_j)} & \\
\text{s.t.} & \left.\begin{array}{rcl} v_j & \geq & \ln[S_{ij}] + b_i \\ n_j + t_{Li} & \geq & \ln[t_{ij}] \end{array}\right\} & i = 1,2,\ldots,N;\ j = 1,2,\ldots,M \\
& \sum_{i=1}^{N} Q_i e^{(t_{Li} - b_i)} \leq H & \\
& \left.\begin{array}{c} 0 \leq n_j \leq \ln[N_j^U] \\ \ln[V_j^L] \leq v_j \leq \ln[V_j^U] \end{array}\right\} & j = 1,2,\ldots,M \\
& \left.\begin{array}{c} \ln[T_{Li}^L] \leq t_{Li} \leq \ln[T_{Li}^U] \\ \ln[B_i^L] \leq b_i \leq \ln[B_i^U] \end{array}\right\} & i = 1,2,\ldots,N
\end{array}
\right.
$$

Note that in this formulation we have eliminated the nonconvexities. The nonlinearities appear in the form of exponential functions in the objective and the third constraint.

Remark 5 Another application of the logarithmic transformation for convexification is presented in Floudas and Paules (1988) for the synthesis of heat-integrated sharp distillation sequences.

(iii) Nonconvexities

If nonlinear expressions that can not be convexified exist in the mathematical model, then we have to resort to global optimization approaches so as to obtain the global solution with theoretical guarantee. The research area of global optimization has received significant attention in the last ten years in both the optimization and chemical engineering community. As a result, there exist now algorithms that can attain the global solution for several classes of nonlinear programming problems. The recent books of Horst and Tuy (1990), Floudas and Pardalos (1990), Floudas and Pardalos (1992), Horst and Pardalos (1995), and Floudas and Pardalos (1995) provide an exposition to the recent advances.

Finally, summarizing the section 7.4.3 on *Mathematical Model of Superstructure*, we should try to derive a model that

(a) has a tight continuous relaxation,

(b) has as linear structure as possible, and

(c) has as few nonconvexities as possible.

7.4.4 Algorithmic Development

Parts I and II of this book presented the fundamentals of nonlinear and mixed integer nonlinear optimization. In addition, in Part II we discussed a number of important algorithms for dealing with MINLP problems. These algorithms are based on the decomposition of the original MINLP model into the primal and master problem and the generation of two upper and lower bounding sequences that converge within ϵ in a finite number of iterations. These algorithms are immediately applicable to the Process Synthesis models that we discussed in the previous section, and their application will be illustrated in the following chapters.

During the last decade we have experienced a significant algorithmic development of MINLP algorithms, as well as extensive computational experience with applications in chemical engineering. It is interesting to note that most of the activity was initiated in chemical engineering mainly due to the natural formulation of process synthesis problems as MINLP models. Two main schools of thought have emerged: the one at Carnegie- Mellon University by Grossmann and co-workers and the other at Princeton University by Floudas and co-workers. The former school focused on outer approximation-based algorithms while the latter concentrated on variants of the generalized benders decomposition approach with key objective the exploitation of existing special structure or creation of special structure via clever decomposition schemes.

As a result, there now exist automated implementations of MINLP algorithms which are identified as **DICOPT++** (Viswanathan and Grossmann, 1990), **APROS** (Paules and Floudas, 1989), as well as the library **OASIS** (Floudas, 1990). These automated implementations make use of the modeling systems **GAMS** (Brooke *et al.*, 1988) which has emerged as a major modeling tool in which problems can be specified in algebraic form and automatically interfaced with linear, nonlinear and mixed integer linear solvers.

A recent development of automated implementations of MINLP algorithms, which does not rely on the GAMS modeling system, is **MINOPT** (Rojnuckarin and Floudas, 1994).

MINOPT (Mixed Integer Nonlinear OPTimizer) is written entirely in C and solves MINLP problems by a variety of algorithms that include: (i) the Generalized Benders Decomposition **GBD**, (ii) the Outer Approximation with Equality Relaxation **OA/ER**, (iii) the Outer Approximation with Equality Relaxation and Augmented Penalty **OA/ER/AP**, and (iv) the Generalized Cross Decomposition **GCD**.

The primary objectives in the design of the **MINOPT** system have been: (i) to provide an efficient and user-friendly interface for solving MINLP problems, (ii) to allow for the other user-specified subroutines such as process simulation packages, and (iii) to provide an expandable platform that can incorporate easily additional MINLP algorithms.

MINOPT has a number of important features that include:

- Front-end parser
- Extensive options
- User specified subroutines
- Callable as subroutine

The front-end parser allows user-friendly interaction and permits the inclusion of nonlinear constraints without recompilation. It employs a set notation which allows for writing the mathematical model in compact expressions.

The extensive options allow the user to tune the performance of each algorithm. Some of the frequently used options are (i) the incorporation of integer cuts, (ii) the solution of continuous relaxation problems, (iii) alternative feasibility problems, (iv) an automatic initialization procedure, (v) a tightening of bounds procedure, and (vi) solvers' parameter changes.

The user specified subroutines allow for connections to various other programs such as process simulators and ordinary differential equation solvers. Currently, **MINOPT** is connected to the DASOLV (Jarvis and Pantelides, 1992) integrator, and can solve MINLP models with differential and algebraic constraints.

7.5 Application Areas

In the following chapters we will discuss the application of the optimization MILP and MINLP approaches to Process Synthesis problems arising in central chemical engineering process systems, sub-systems, and overall process systems. These include

- Heat exchanger network synthesis

- Separation system synthesis
- Reactor system synthesis
- Reactor-separator-recycle synthesis

Summary and Further Reading

This chapter presents an introduction to the area of process synthesis. Section 7.1 discusses the overall process system and its components. Section 7.2 defines the process synthesis problems and discusses the challenges associated with it due to its combinatorial and nonconvex nature. Section 7.3 presents the different approaches for addressing process synthesis problems. Section 7.4 focuses on the optimization approach in process synthesis. An outline of the key steps in such an approach is presented first in section 7.4.1 with a discussion on the representation of alternatives following in section 7.4.2. Section 7.4.3 addresses the development of mathematical models for the representation of alternatives. The modeling topics of (i) 0-1 variables only, (ii) continuous and linear binary variables, (iii) bilinear products of continuous and binary variables, as well as (iv) modeling nonlinearities of continuous variables are discussed and illustrated with several examples. Section 7.4.4 discusses algorithmic development issues while section 7.5 briefly outlines a number of important application areas. Further reading on recent advances in the area of process synthesis can be found in Floudas and Grossmann (1994).

Chapter 8 Heat Exchanger Network Synthesis

This chapter focuses on heat exchanger network synthesis approaches based on optimization methods. Sections 8.1 and 8.2 provide the motivation and problem definition of the HEN synthesis problem. Section 8.3 discusses the targets of minimum utility cost and minimum number of matches. Section 8.4 presents synthesis approaches based on decomposition, while section 8.5 discusses simultaneous approaches.

8.1 Introduction

Heat exchanger network HEN synthesis is one of the most studied synthesis/design problems in chemical engineering. This is attributed to the importance of determining energy costs and improving the energy recovery in chemical processes. The comprehensive review of Gundersen and Naess (1988) cited over 200 publications while a substantial annual volume of studies has been performed in the last few years.

The HEN synthesis problem, in addition to its great economic importance features a number of key difficulties that are associated with handling:

(i) The potentially explosive combinatorial problem for identifying the best pairs of hot and cold streams (i.e., matches) so as to enhance energy recovery;

(ii) Forbidden, required, and restricted matches;

(iii) The optimal selection of the HEN structure;

(iv) Fixed and variable target temperatures;

(v) Temperature dependent physical and transport properties;

(vi) Different types of streams (e.g., liquid, vapor, and liquid-vapor); and

(vii) Different types of heat exchangers (e.g., counter-current, noncounter-current, multi-stream), mixed materials of construction, and different pressure ratings.

It is interesting to note that the extensive research efforts during the last three decades toward addressing these aforementioned difficulties/issues exhibit variations in their objectives and types of approaches which are apparently cyclical. The first approaches during the 1960s and early 1970s treated the HEN synthesis problem as a single task (i.e., no decomposition into sub-tasks). The work of Hwa (1965) who proposed a simplified superstructure which he denoted as composite configuration that was subsequently optimized via separable programming was a key contribution in the early studies, as well as the tree searching algorithms of Pho and Lapidus (1973). Limitations on the theoretical and algorithmic aspects of optimization techniques were, however, the bottleneck in expanding the applicability of the mathematical approaches at that time. As a result, most of the research efforts focused on identifying targets based on the minimum utility cost, the minimum number of matches for vertical and non-vertical heat transfer. The work of Hohmann (1971) which received little recognition in the 1970s and the work of Linnhoff and Flower (1978a; 1978b) introduced the concept of the pinch point. As a result, the research emphasis shifted from single task approaches to multitask procedures that represent simple techniques for decomposing the original problem into subproblems which are the aforementioned targets. Furthermore, HEN synthesis approaches based on decomposition into the following three separate tasks:

(i) Minimum utility cost

(ii) Minimum number of matches

(iii) Minimum investment cost network configurations

were developed via thermodynamic, heuristic, and optimization approaches.

Target (i) without constraints on the matches was addressed via the feasibility table (Hohmann, 1971), the problem table analysis (Linnhoff and Flower, 1978a), and the $T - Q$ diagram (Umeda *et al.*, 1979). For unconstrained and constrained matches rigorous mathematical models were developed, namely the LP transportation model of Cerda *et al.* (1983), its improvement of the LP transshipment model of Papoulias and Grossmann (1983) and the modified transshipment model of Viswanathan and Evans (1987).

Target (ii) was addressed rigorously by Cerda and Westerberg (1983) as a Mixed Integer Linear Programming MILP transportation model and by Papoulias and Grossmann (1983) as an MILP transshipment model. Both models determine the minimum number of matches given the minimum utility cost.

Target (iii) determines the minimum investment cost network configuration given that targets (i) and (ii) are performed sequentially (Floudas *et al.*, 1986).

The principle advantage of HEN synthesis approaches based on decomposition into targets (i.e., subtasks) is that they involve simpler subproblems that can be treated in a much easier fashion than the original single-task problem. It is, however, this identical advantage that introduces a number of limitations. The main limitation of a sequential synthesis method is that the trade-offs between the utility cost, the number of matches, the area requirement and the minimum investment cost are not taken into account appropriately. More specifically, early decisions on the level of energy recovery to be achieved by the **(HEN)**, and decisions to decompose the original problem into subproblems based on the location of the pinch point(s) may result in HEN designs that are suboptimal networks. As a result of these limitations of the sequential HEN synthesis approaches, researchers in the late 1980s and early 1990s focused on simultaneous optimization approaches that attempt to treat the synthesis problem as a single-task problem. This way, all trade-offs are taken into consideration properly, and advances in theoretical and algorithmic aspects of optimization have allowed for important strides to be achieved via the optimization based synthesis approaches. Floudas and Ciric (1989) proposed a mixed integer nonlinear programming MINLP model for treating simultaneously targets (ii) and (iii) and investigated issues related to attaining the best solution of target (iii). Ciric and Floudas (1990) showed that the assumption of decomposing based on the pinch point is rather artificial and applied the simultaneous approach of targets (ii) and (iii) to pseudo-pinch HEN synthesis problems. Ciric and Floudas (1991) posed the HEN synthesis problem as a single optimization problem in which all decisions on utility consumption, matches, and network configuration are treated simultaneously and demonstrated the benefit of a simultaneous approach versus a sequential approach. Yee and Grossmann (1990) proposed an MINLP model of a simplified superstructure based on the assumption of isothermal mixing for the simultaneous optimization of HENs. Their assumption of isothermal mixing results in a linear set of constraints which make the MINLP model robust at the expense of excluding a number of structural alternatives. The same type of superstructure was employed by Yee *et al.* (1990a) for targeting of energy and area simultaneously and by Yee *et al.* (1990b) for coupling the process with the heat recovery system. Dolan *et al.* (1989; 1990) instead of focusing on MINLP modeling and algorithmic development, applied the simulated annealing algorithms for the simultaneous synthesis of HENs.

8.2 Problem Statement

The HEN synthesis problem can be stated as follows:

> Given are a set of hot process streams, HP, to be cooled, and a set of cold process streams, CP, to be heated. Each hot and cold process stream has a specified heat capacity flowrate while their inlet and outlet temperatures can be specified exactly or given as inequalities. A set of hot utilities, HU, and a set of cold utilities, CU, along with their corresponding temperatures are also provided.
>
> Then, the objective is to determine the heat exchanger network of the least total annualized cost.
>
> Key elements that characterize such a HEN are the:
>
> (i) Hot and cold utilities required,
>
> (ii) Stream matches and the number of units,
>
> (iii) Heat loads of each heat exchanger,
>
> (iv) Network configuration with flowrates and temperatures of all streams, and
>
> (v) Areas of heat exchangers.

8.2.1 Definition of Temperature Approaches

Three temperature approaches are utilized at different stages of the HEN synthesis approaches. These are essentially parameters which, when they are fixed, then restrict the search space and hence simplify the problem. If, however, these three parameters are not fixed but they are treated as optimization variables explicitly, then we impose no restrictions on the search space. These parameters are defined as

$HRAT$, Heat recovery approach temperature, is used to quantify utility consumption levels.

$TIAT$, Temperature interval approach temperature, is used to partition the temperature range into intervals.

ΔT_{min} **or** $EMAT$, Minimum approach temperature, specifies the minimum temperature difference between two streams exchanging heat within an exchanger.

The three temperature approaches satisfy the relationship:

$$HRAT \geq TIAT \geq \Delta T_{min}.$$

The utility consumption levels are parametrized by $HRAT$, while the $TIAT$ controls the residual flow out of the temperature intervals Ciric and Floudas (1990). In general, $HRAT$ is allowed to vary, while $TIAT$ may be set equal to $HRAT$ (i.e., strict-pinch case) or may be fixed to a constant value strictly less than $HRAT$ (i.e., pseudo-pinch case). ΔT_{min} is set greater than or equal to a small positive value, ϵ, which in the limit can be zero.

8.3 Targets for HEN Synthesis

8.3.1 Minimum Utility Cost

8.3.1.1 Problem Statement

The minimum utility cost target problem of HENs can be stated as:

> Given a $HRAT$, determine the minimum utility cost of a heat exchanger network without prior knowledge of the HEN configuration.

This is a very important target since it corresponds to the maximum energy recovery that can be attained in a feasible HEN for a fixed $HRAT$. This target allows for the elimination of several HEN structures which are not energy efficient and leads to near-optimal solutions as long as the energy is the dominant cost item compared to the investment cost.

Remark 1 The above-mentioned target can be stated as the minimum utility consumption problem. We have stated it as minimum utility cost so as to distinguish the effect of multiple hot and cold utilities. This way, by assigning a different cost for each type of utility (e.g,. fuel, steam at different pressure levels, hot water, cooling water, refrigerants) we can explicitly determine the effect on the cost of each utility. Note though that the mathematical models for utility consumption and utility cost provide identical solutions for the utility loads due to their linear nature, as we will see in a later section.

8.3.1.2 Concept

The key concept which allows for the determination of the minimum utility cost prior to knowing the HEN structure is the *pinch point.* A pinch point limits the maximum energy integration and hence is regarded as a thermodynamic bottleneck that prevents further energy integration whenever the feasibility constraints of heat exchange are violated. To illustrate the concept of pinch point(s) we will consider the following example.

Illustration 8.3.1 (Pinch point)

This example is taken from Floudas and Ciric (1989) and consists of two hot process streams, two cold process streams, and has $HRAT = 30°C$. The inlet, outlet temperatures, and the flowrate heat capacities are shown in Table 8.1. If we consider a first-law analysis for this example and simply calculate the heat available in the hot process streams and the heat required by the cold process streams, we have

$$\begin{array}{rclcrl} Q^{H1} &=& 5\,(95-75) &=& 100 & \text{kW}, \\ Q^{H2} &=& 50\,(80-75) &=& 250 & \text{kW}, \\ Q^{C1} &=& 10\,(30-90) &=& -600 & \text{kW}, \\ Q^{C2} &=& 12.5\,(60-70) &=& -125 & \text{kW}. \end{array}$$

Therefore we have $Q^{H1} + Q^{H2} = 350$ kW available from the hot process streams while there is a demand of 725 kW for the cold process streams. Based on this analysis then, we may say that we need to purchase $(725 - 350) = 375$ kW of hot utilities.

We will see, however, that this is not a correct analysis if a pinch point exists. In other words, the implicit assumption in the first law analysis is that we always have feasible heat exchange for the given $HRAT = 30°C$. To illustrate the potential bottleneck, let us consider the graphical representation of the hot and cold process streams in the $T - Q$ diagram shown in Figure 8.1.

The piecewise linear representation denoted as ABC is the *hot composite curve* while the representation denoted as $DEFG$ is the *cold composite curve*. It is instructive to outline the basic

Stream	$T^{in}(°C)$	$T^{out}(°C)$	$FC_p(\text{kW}°C^{-1})$
$H1$	95	75	5
$H2$	80	75	50
$C1$	30	90	10
$C2$	60	70	12.5

Table 8.1: Data for illustration of pinch point

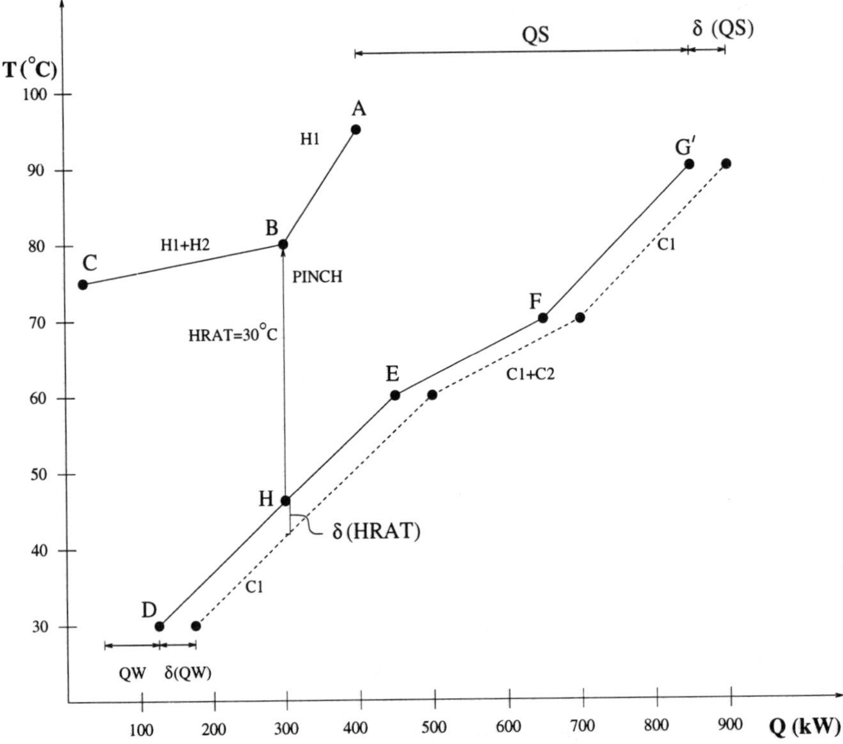

Figure 8.1: $T - Q$ illustration of pinch

steps of constructing the hot and cold composite curves in a $T - Q$ diagram given the data of the illustrative example.

Hot composite curve

Point A: It can be any point of $T = 95°C$ denoting the inlet temperature of the hottest process stream (i.e., stream $H1$ in our example).

Point B: It has $T = 80°C$ which corresponds to the inlet temperature of $H2$ (i.e., second hottest stream).

Segment AB: Only $H1$ exists in this segment and the heat load of $H1$ in this segment is $5(95 - 80) = 75$ kW. Note also that the slope of the line AB is $\frac{1}{F_{H1}} = \frac{1}{5}$.

Point C: It has $T = 75°C$ which is the outlet temperature of both $H1$ and $H2$.

Segment BC: Both $H1$ and $H2$ exist in this segment, and their joint heat loads are $(5 + 50)(80 - 75) = 275$ kW. Hence, point C is defined as $(Q^B - 275, 75)$. Note also that the slope of the line BC is $\frac{1}{F_{H1}+F_{H2}} = \frac{1}{55}$.

Cold composite curve

Point D: It has $T = 30°C$ which corresponds to the inlet temperature of the coldest process stream (i.e., $C1$ in our example).

Point E: It has $T = 60°C$ which denotes the inlet temperature of $C2$ (i.e., second coldest stream).

Segment DE: Only $C1$ exists in this segment and its heat load is $10(60 - 30) = 300$ kW. Hence, point E is defined as $(Q^D + 300, 60)$. Note also that the slope of the line DE is $\frac{1}{F_{C1}} = \frac{1}{10}$.

Point F: It has $T = 70°C$, which denotes the outlet temperature of $C2$.

Segment EF: Both $C1$ and $C2$ exist in this segment, and their joint heat load is $(10 + 12.5)(70 - 60) = 225$ kW. Hence point F is defined as $(Q^E + 225, 70)$. Note also that the slope of line EF is $\frac{1}{F_{C1}+F_{C2}} = \frac{1}{22.5}$.

Point G: It has $T = 90°C$ which is the outlet temperature of cold stream $C1$.

Segment FG: Only $C1$ exists in this segment, and its heat load is $10(90 - 70) = 200$ kW. Hence, point G is defined as $(Q^F + 200, 90)$. Also, note that the slope of the line FG is $\frac{1}{F_{C1}} = \frac{1}{10}$, which implies that the line segments DE and FG are parallel.

Heat Recovery Approach Temperature, $HRAT$

The requirement for energy recovery is that the $HRAT = 30°C$. Therefore, we have to move the *hot and cold composite curves* parallel to the T axis as close to each other up to the point at which the perpendicular distance of any two points of the composite curves becomes equal to $HRAT = 30°C$.

As shown in Figure 8.1, this takes place between the points B and H. In fact, point H is defined as the intersection of $T^B - 30$ and the segment DE as the cold composite curve approaches the hot one. Therefore, the pinch point occurs between the temperatures:

$$80 - 50°C.$$

Remark 1 Since we cannot bring the two composite curves closer, the pinch point represents the bottleneck for further heat recovery. In fact, it partitions the temperature range into two subnetworks, one above the pinch and one below the pinch. Heat flow cannot cross the pinch since there will be violations in the heat exchange driving forces. As a result, we need a hot utility at the subnetwork above the pinch and a cold utility at the subnetwork below the pinch. In other words, having identified the pinch point, we can now apply the first law analysis to each subnetwork separately and determine the hot and cold utility requirements. These can be read from the $T - Q$ diagram since they correspond to the horizontal segments AG and CD, respectively. Hence, for our example we have:

$$\begin{aligned} QS &= 450 \text{ kW}, \\ QW &= 75 \text{ kW}. \end{aligned}$$

Note also that the introduction of hot stream $H2$ is responsible for the pinch point (i.e., the inlet temperature of $H2$).

Remark 2 The first law analysis applied to the overall network provides $QS = 375$ kW and $QW = 0$, which is incorrect since it does not take into account the thermodynamic feasibility imposed by $HRAT = 30°C$. Note that the correct utilities $QS = 450$ kW, $QW = 75$ kW satisfy the overall energy balance $450 - 75 = 375$ kW.

Remark 3 If we increase the $HRAT$ by $\delta(HRAT)$, then as we see in Figure 8.1 the required amount of hot and cold utilities increase, since the cold composite curve moves perpendicularly down by $\delta(HRAT)$. Then, we have

$$\begin{aligned} QS &+ \delta(QS), \\ QW &+ \delta(QW), \end{aligned}$$

the required hot and cold utilities.

8.3.1.3 Partitioning of Temperature Intervals

In this section, we will discuss the partitioning of the temperature range into temperature intervals (TI) assuming that $TIAT = HRAT$. In section 8.5.2 we will discuss the effect of partitioning with $TIAT \leq HRAT$ upon the residual heat flows from one interval to the next.

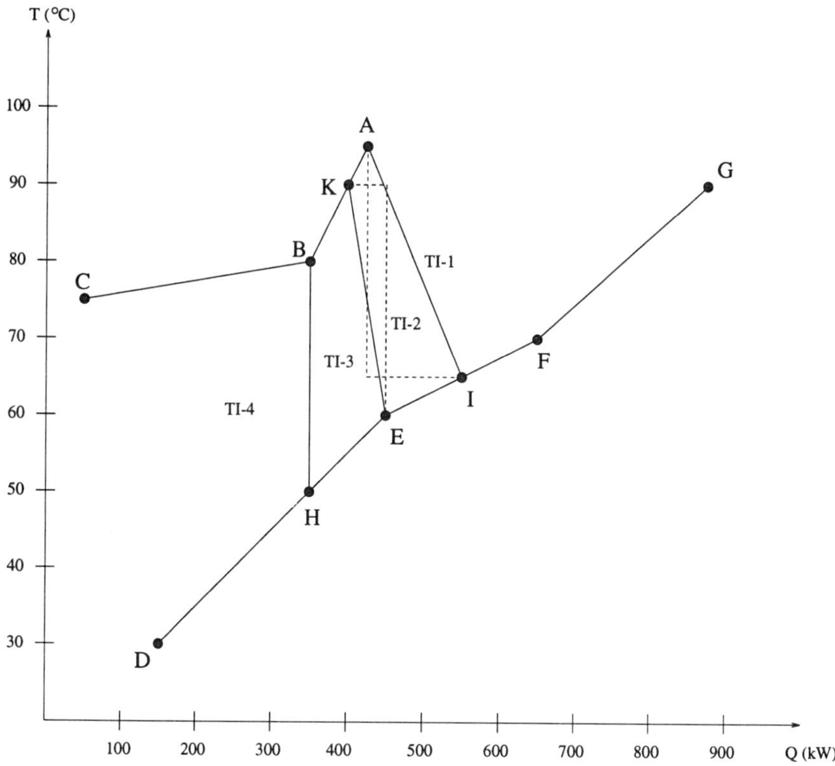

Figure 8.2: Temperature intervals partitioning

Figure 8.2 shows the temperature intervals for the illustration of section 8.3.1.2 assuming that $TIAT$ is equal to $HRAT$. Note that in this example there exist four temperature intervals:

$$\begin{aligned} TI-1 &: AIFG, \\ TI-2 &: AKEI, \\ TI-3 &: KBHE, \\ TI-4 &: CBHD, \end{aligned}$$

with the pinch point located between $TI-3$ and $TI-4$ on the line BH.

The guidelines of the temperature interval partitioning are (see also Cerda *et al.* (1983), Linnhoff and Flower (1978a), Floudas and Ciric (1989)):

(i) For each of the inlets of the hot process streams (i.e., points A, B in our example) determine the corresponding points in the cold composite curve defined as $T^H - TIAT = T^H - HRAT$. In the illustrative example, point I is defined from point A and point H is defined from point B. Then, draw the lines between A and I, and B and H.

(ii) For each of the inlets of the cold process streams (i.e., point D, E in our example) determine the corresponding points in the hot composite curve defined as $T^C + TIAT = T^C + HRAT$. In the illustrative example, point K is defined from point E, but point D does not have a corresponding point since there is no intersection with the hot composite curve. Then, draw the line between E and K.

Remark 1 The above guidelines assume that we have one hot and one cold utility each located at the top and bottom respectively of the temperature range. In the case, however, where there exist multiple hot and cold utilities then we first identify the hottest hot utility and the coldest cold utility and define the rest of the utilities as intermediate hot and intermediate cold utilities. Subsequently, we treat the intermediate hot and cold utilities as hot and cold process streams for purposes of the partitioning into temperature intervals. In other words, we have to apply the two aforementioned guidelines for both the hot and cold process streams and intermediate hot and cold utilities. Another pictorial representation of the four temperature intervals is shown in Figure 8.3 along with the hot and cold stream data.

Remark 2 Note that hot stream $H1$ does not participate in $TI-1$, while cold streams $C1$ and $C2$ do participate. As a result, hot utilities are required.

Remark 3 A hot process stream cannot transfer heat to a cold process stream that exists in a higher TI because of driving force violations. For instance, hot stream $H1$ cannot transfer heat to cold streams $C1$, $C2$ at $TI-1$. Similarly, hot stream $H2$ cannot transfer heat to cold stream $C2$ at all. Also, hot stream $H2$ can only transfer heat to cold stream $C2$ at $TI-4$.

Remark 4 Heat available at a higher TI can be transferred to a cold stream at a lower interval via residual heat flows from one interval to the next, which are shown in Figure 8.3 as R_1, R_2, and R_3.

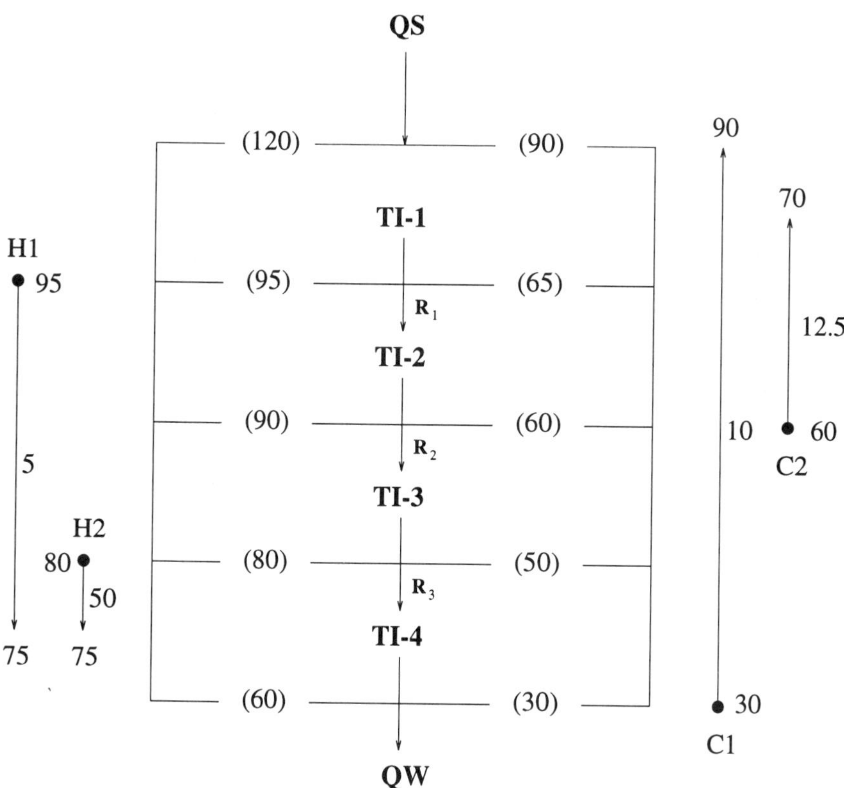

Figure 8.3: Temperature intervals of illustration

Remark 5 There exist upper bounds on the amounts of heat that can be transferred from a hot stream to a cold stream. These upper bounds correspond to the minimum of

(i) maximum heat available by the hot stream, and

(ii) maximum heat needed by the cold stream.

For instance, in $TI - 2$, the upper bound on the heat that can be transferred from $H1$ to $C2$ is

$$\begin{aligned} Q_{H1-C2} &\leq \min\{5(95-90), 12.5(65-60)\} \\ &= \min\{25, 62.5\} \;=\; 25. \end{aligned}$$

The upper bound on the heat that can be transferred from $H1$ to $C1$ in $TI - 3$ is

$$\begin{aligned} Q_{H1-C1} &\leq \min\{5(95-80), 10(60-50)\} \\ &= \min\{75, 100\} = 75. \end{aligned}$$

The key observation in calculating this upper bound is that the available heat of $H1$ is not only the amount of heat at $TI - 3$, which is $5(90 - 80) = 50$, but also the available heat at $TI - 2$ for $H1$ which can be transferred via the heat residual R_2.

Remark 6 If a phase change takes place during the temperature variation of a process stream, then the temperature partitioning should be modified so as to contain a component of the temperature variation in the first phase, a component for the phase change, and a component for the temperature variation in the second phase, in the form of a piece-wise linear approximation. The three components are treated subsequently as if they are separate steams, and the same analysis, that is described in the following sections, applies.

8.3.1.4 LP Transshipment Model

The target of minimum utility cost in HENs can be formulated as a linear programming LP transshipment model which corresponds to a well known model in operations research (e.g., network problems). The transshipment model is used to determine the optimum network for transporting a commodity (e.g., a product) from sources (e.g., plants) to intermediate nodes (e.g., warehouses) and subsequently to destinations (e.g., markets).

Papoulias and Grossmann (1983) drew the analogy between the transshipment model and the HEN, which is shown in Table 8.2. Using this analogy, heat is considered as a commodity which is transferred form the hot process streams and hot utilities to the cold process streams and cold utilities via the temperature intervals. The partitioning procedure discussed in the previous section allows only for feasible transfer of heat in each temperature interval (see also the remarks of section 8.3.1.3).

Figure 8.4 shows pictorially the analogy between the transshipment model and the heat exchanger network. The nodes on the left indicate the sources while the nodes on the right denote the destinations. The intermediate nodes, shown as boxes, are the warehouses. The simple arrows denote the heat flow from sources to warehouses and from the warehouses to destinations, while the highlighted arrows denote the heat flow from one warehouse to the one immediately below.

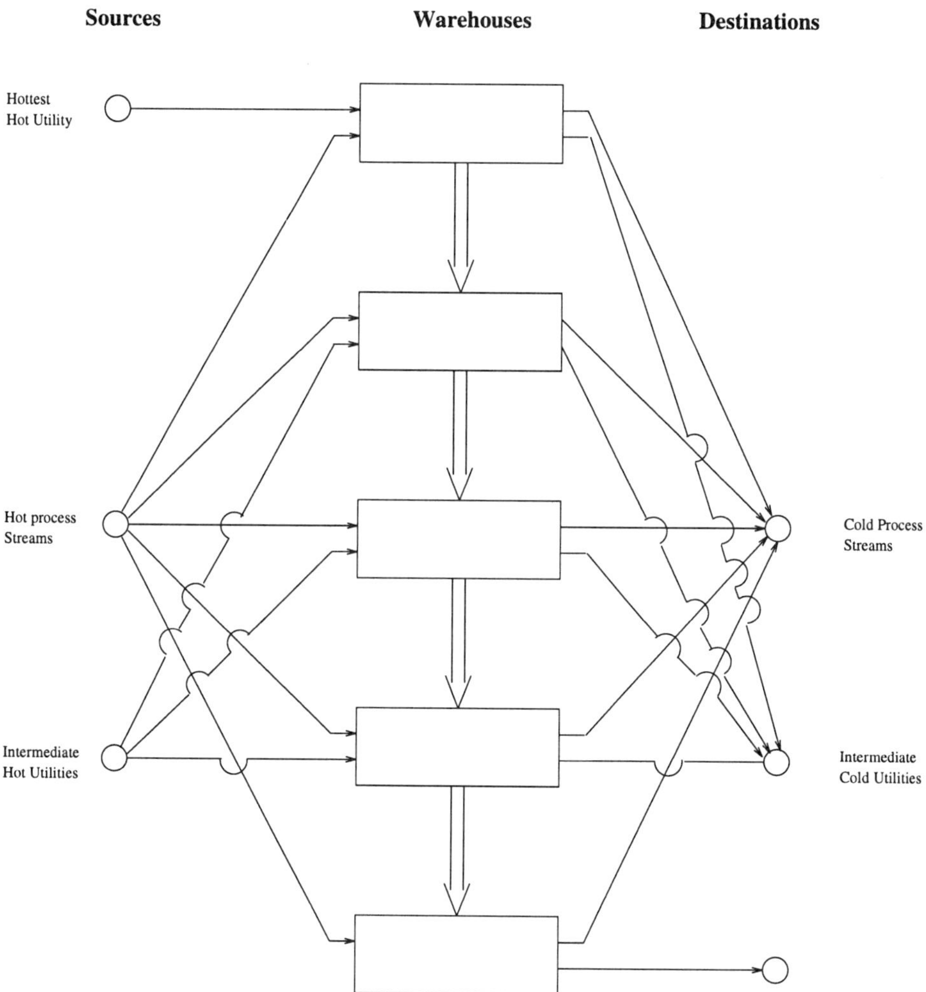

Figure 8.4: Analogy between the transshipment model and the HEN

Transshipment model		HEN
commodity	=	heat
intermediated nodes (i.e. warehouses)	=	temperature intervals
sources	=	hot process streams and hot utilities
destinations	=	cold process streams and cold utilities

Table 8.2: Analogy of transshipment model in HEN synthesis

Remark 1 Note that in Figure 8.4 we have classified the sources as consisting of

Sources:
- Hottest hot utility
- Intermediate hot utilities
- Hot process streams

while the destinations are

Destinations:
- Coldest hot utility
- Intermediate cold utilities
- Cold process streams

This classification is done so as to be consistent with the partitioning procedure presented in the previous section and to treat the general case of multiple hot and cold utilities.

Remark 2 Note that in the top temperature interval, there is no heat residual entering. The only heat flows entering are those of the hottest hot utility and of the hot process streams. Similarly in the bottom temperature interval there is no heat residual exiting. The only heat flows exiting are those of the cold utility and the cold process streams.

Having presented the pictorial representation of the transshipment model we can now state the basic idea for the minimum utility cost calculation.

Basic Idea of the LP Transshipment Model

The basic idea of the minimum utility cost calculation via the transshipment representation is to (i) introduce variables for all potential heat flows (i.e., sources to warehouses, warehouses to destinations, warehouses to warehouses), (ii) write the overall energy balances around each warehouse, and (iii) write the mathematical model that minimizes the utility cost subject to the energy balance constraints.

To write the mathematical model we need to first define the following indices and sets:

$$TI = \{k \mid k = 1, 2, \ldots, K\},$$

$$\begin{aligned} HP_k &= \{i \mid \text{hot process stream } i \text{ is present in interval } k\}, \\ CP_k &= \{j \mid \text{cold process stream } j \text{ is present in interval } k\}, \\ HU_k &= \{i \mid \text{hot utility } i \text{ is present in interval } k\}, \\ CU_k &= \{j \mid \text{cold utility } j \text{ is present in interval } k\}, \\ i &: \text{hot process stream/utility}, \\ j &: \text{cold process stream/utility}, \\ k &: \text{temperature interval}. \end{aligned}$$

We also need to introduce the following variables:

$$\begin{aligned} QS_{ik} &: \text{heat load of hot utility } i \text{ entering temperature interval } k, \\ QW_{jk} &: \text{heat load of cold utility } j \text{ entering temperature interval } k, \\ R_k &: \text{heat residual load out of temperature interval } k, \\ Q^H_{ik} &: \text{heat load of hot process stream } i \text{ entering temperature interval } k, \\ Q^C_{jk} &: \text{heat load of hot process stream } j \text{ entering temperature interval } k, \end{aligned}$$

representing all potential heat flows.

Remark 3 Note that Q^H_{ik} and Q^C_{ik} are not actual variables since they are specified form the data and the temperature partitioning. The heat load, Q^H_{ik} is given by

$$Q^H_{ik} = F_i \left(C_p\right)_{ik} \Delta T_{ik},$$

where F_i is the flowrate of hot stream i, $(C_p)_{ik}$ is the heat capacity of hot stream i in interval k, and ΔT_{ik} is the temperature change of hot stream i in interval k. Note that for constant flowrate heat capacities $F_i(C_p)_{ik}$ is given by the data of the problem, while ΔT_{ik} can be obtained from the data and the temperature partitioning.

Similarly, we have

$$Q^C_{jk} = F_j \left(C_p\right)_{jk} \Delta T_{jk},$$

where F_j, $(C_p)_{jk}$, ΔT_{jk} are defined for the cold streams.

Then the pictorial representation of the kth temperature interval along with the associated variables (i.e., hot and cold utility loads, heat residual loads) and the parameters (i.e., heat loads of hot and cold streams) is shown in Figure 8.5.

Remark 4 Note that for the top and bottom temperature interval (i.e. for $k = 1$ and $k = K$) we have $R_0 = R_K = 0$ (i.e., no heat residuals entering or exiting).

Remark 5 Heat enters the k'th temperature interval from

(i) Hot process streams and hot utilities whose temperature range includes part of, or the whole temperature interval, and

(ii) The previous $(k-1)$th temperature interval that is at a higher temperature via the heat residual R_{k-1}. This residual heat flow cannot be utilized in the $(k-1)$th interval (i.e., it is in excess).

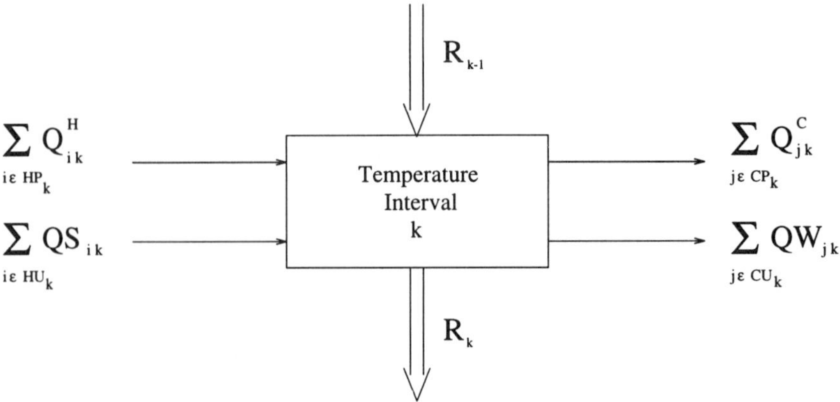

Figure 8.5: Heat flow pattern of kth temperature interval

Note also that heat exits the kth temperature interval and it is directed to:

(i) Cold process streams and cold utilities whose temperature range includes part of the whole temperature interval, and

(ii) The next temperature interval that is at a lower temperature via the heat residual R_k. This heat load is in excess and cannot be utilized in the kth interval.

Remark 6 All the variables defining the different heat flow loads are nonnegative, and therefore we have

$$\left.\begin{aligned} R_k &\geq 0 \\ QS_{ik} &\geq 0 \\ QW_{jk} &\geq 0 \end{aligned}\right\} k \in TI.$$

Remark 7 From Figure 8.5, the overall energy balance around temperature interval k can be simply written as

$$R_k - R_{k-1} + \sum_{j \in CU_k} QW_{jk} - \sum_{i \in HU_k} QS_{ik} = \sum_{j \in HP_k} Q_{ik}^H - \sum_{j \in CP_k} Q_{jk}^C,$$

which is a linear equality constraint.

Remark 8 The minimum utility cost criterion can now be formulated as

$$\sum_{i \in HU} C_i \sum_{k \in TI} QS_{ik} + \sum_{j \in CU} C_j \sum_{k \in TI} QW_{jk},$$

where C_i and C_j are the costs of hot utility i and cold utility$j$, respectively.

Note that the minimum utility cost criterion is linear in the variables QS_{ik} and QW_{jk}.

Then the transshipment model for minimum utility cost is

$$\left.\begin{array}{ll}
\min & \sum\limits_{i\in HU} C_i \sum\limits_{k\in TI} QS_{ik} + \sum\limits_{j\in CU} C_j \sum\limits_{k\in TI} QW_{jk} \\
s.t. & R_k - R_{k-1} + \sum\limits_{j\in CU_k} QW_{jk} - \sum\limits_{i\in HU_k} QS_{ik} = \\
 & \quad = \sum\limits_{i\in HP_k} Q^H_{ik} - \sum\limits_{j\in CP_k} Q^C_{jk} \quad , \quad k \in TI \\
 & QS_{ik}, QW_{jk}, R_k \geq 0 \quad k \in TI, i \in HU, j \in CU \\
 & R_0 = R_K = 0
\end{array}\right\} \textbf{(P1)}$$

Remark 9 Due to the linearity of the objective function and constraints, **P1** is a linear programming LP problem. Its solution will provide the optimal values for the hot and cold utility loads, as well as the values for the heat residuals R_k. The occurrence of any pinch point(s) can be found by simply checking which of the heat residual flows R_k are equal to zero.

Remark 10 Note that if the hot and cold utilities take place at a fixed temperature which is usually the case then we only have QS_{ik} defined for one interval (i.e., the one which hot utility i is entering). As a result, we do not need to define QS_{ik} for all intervals. Similarly, we introduce QW_{ik} only for the intervals at which the cold utility j exits. If this occurs at one interval only, then we have to introduce one variable for each hot, and one for each cold utility.

Remark 11 Cerda *et al.* (1983) first proposed the transportation model for the calculation of the minimum utility cost, and subsequently Papoulias and Grossmann (1983) presented the transshipment model **P1** which requires fewer variables and constraints than the transportation model.

Illustration 8.3.2 (LP Transshipment Model)

To illustrate the application of model **P1** for the calculation of the minimum utility cost, we will consider the example used in section 8.3.1.2, with data provided in Table 8.1 of that section. This example features constant flow rate heat capacities, one hot and one cold utility being steam and cooling water, respectively. Here, we assume $C_i = C_j = 1$, that is we meet the minimum utility consumption.

The temperature interval partitioning along with the transshipment representation is shown in Figure 8.6. Note that in Figure 8.6, we also indicate the heat loads provided by the hot process streams at each temperature intervals as well as the heat loads needed by the cold process streams at each temperature interval. Note also that the optimization variables are QS, QW, R_1, R_2, and R_3.

The energy balances around each temperature interval are written as follows:

$$\begin{array}{lcrcl}
TI-1 & : & R_1 - QS & = & 0 - 250 - 52.5 & = & -312.5, \\
TI-2 & : & R_2 - R_1 & = & 25 - 50 - 62.5 & = & -87.5, \\
TI-3 & : & R_3 - R_2 & = & 50 - 100 & = & -50, \\
TI-4 & : & QW - R_3 & = & 25 + 250 - 200 & = & 75.
\end{array}$$

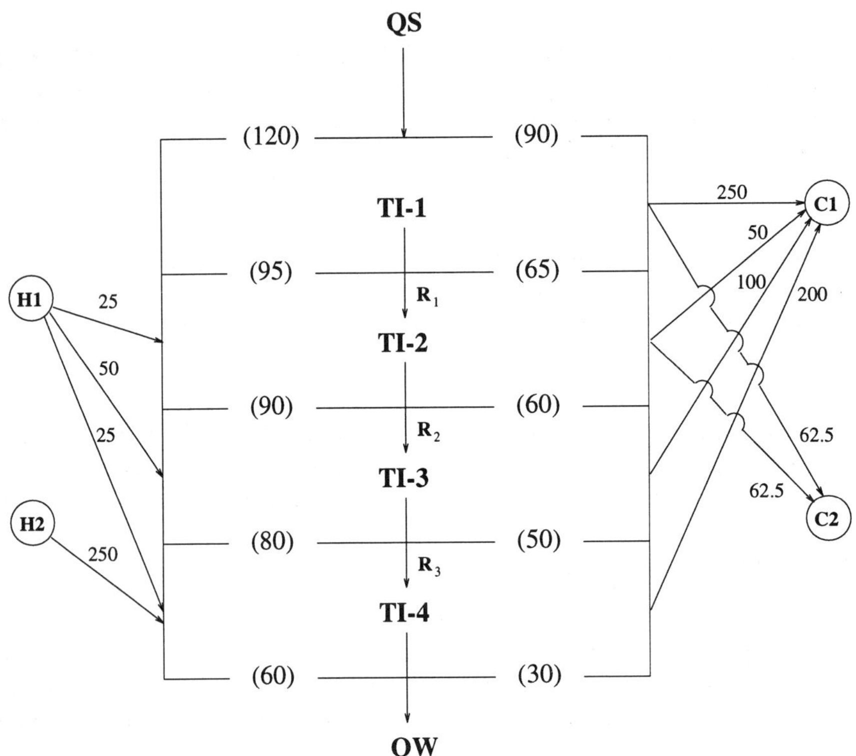

Figure 8.6: Transshipment representation for illustration 8.3.2

Then, the LP transshipment model for minimum utility consumption takes the form:

$$\begin{aligned} \min \quad & QS + QW \\ s.t. \quad & R_1 - QS = -312.5 \\ & R_2 - R_1 = -87.5 \\ & R_3 - R_2 = -50 \\ & QW - R_3 = 75 \\ & QS, QW, R_1, R_2, R_3 \geq 0 \end{aligned}$$

This model features four equalities, five variables and has linear objective function and constraints. Its solution obtained via GAMS/MINOS is:

$$\begin{aligned} QS &= 450, \\ QW &= 75, \\ R_1 &= 137.5, \\ R_2 &= 50, \\ R_3 &= 0. \end{aligned}$$

Since $R_3 = 0$, there is a pinch point between $TI - 3$ and $TI - 4$. Hence, the problem can be decomposed into two independent subnetworks, one above the pinch and one below the pinch point.

Remark 12 When we have one hot and one cold utility, it is possible to solve the LP transshipment model **(P1)** by hand. This can be done by solving the energy balances of $TI - 1$ for R_1, $TI - 2$ for R_2, $TI - 3$ for R_3, and $TI - 4$ for QW which become

$$\begin{aligned} R_1 &= QS - 312.5 & , & \\ R_2 &= R_1 - 87.5 &= & \; QS - 400, \\ R_3 &= R_2 - 50 &= & \; QS - 450, \\ QW &= R_3 + 75 &= & \; QS - 375. \end{aligned}$$

Since R_1, R_2, R_3, $R_4 \geq 0$ we have

$$\begin{aligned} QS &\geq 312.5, \\ QS &\geq 400, \\ QS &\geq 450, \\ QS &\geq 375. \end{aligned}$$

The objective function to be minimized becomes

$$QS + QW = 2 \cdot QS - 375.$$

Then, we seek the minimum QS that satisfies all the above four inequalities. This is

$$QS = 450,$$

Stream	$T^{in}(°C)$	$T^{out}(°C)$	$FC_p(\text{kW}°C^{-1})$
$H1$	95	75	5
$H2$	80	75	50
$C1$	30	90	10
$C2$	60	70	12.5
HP steam	$300°C$	at $ 70 / kWyr	
Hot water	$70°C$	at $ 30 / kWyr	
CW	$20°C$	at $ 20 / kWyr	

Table 8.3: Data for illustration 8.3.3

and hence,

$$\begin{aligned} QW &= 75, \\ R_1 &= 137.5, \\ R_2 &= 50, \\ R_3 &= 0. \end{aligned}$$

Illustration 8.3.3 (Multiple Utilities)

This example is a modified version of illustration 8.3.2, and its data are shown in Table 8.3. Since we have two hot utilities and HP steam is the hottest hot utility we treat the intermediate hot utility (i.e., hot water) as a hot stream for the partitioning into temperature intervals. Then the pictorial representation of the transshipment model becomes (see Figure 8.7):

The energy balances around the TI's are

$$\begin{array}{llll} TI-1 : & R_1 - QS & = 0 - 250 - 52.5 & = -312.5, \\ TI-2 : & R_2 - R_1 & = 25 - 50 - 62.5 & = -87.5, \\ TI-3 : & R_3 - R_2 & = 50 - 100 & = -50, \\ TI-4 : & R_4 - R_3 & = 25 + 250 - 200 & = 175, \\ TI-5 : & QW - R_3 - QS_2 & = -100 & \end{array}$$

Then, the minimum utility cost model **P1** is

$$\begin{array}{ll} \min & 70 \cdot QS_1 + 30 \cdot QS_2 + 20 \cdot QW \\ s.t. & R_1 - QS_1 = -312.5 \\ & R_2 - R_1 = -87.5 \\ & R_3 - R_2 = -50 \\ & R_4 - R_3 = 175 \\ & QW - R_4 - QS_2 = -100 \\ & QS_1, QS_2, QW, R_1, R_2, R_3 \geq 0 \end{array}$$

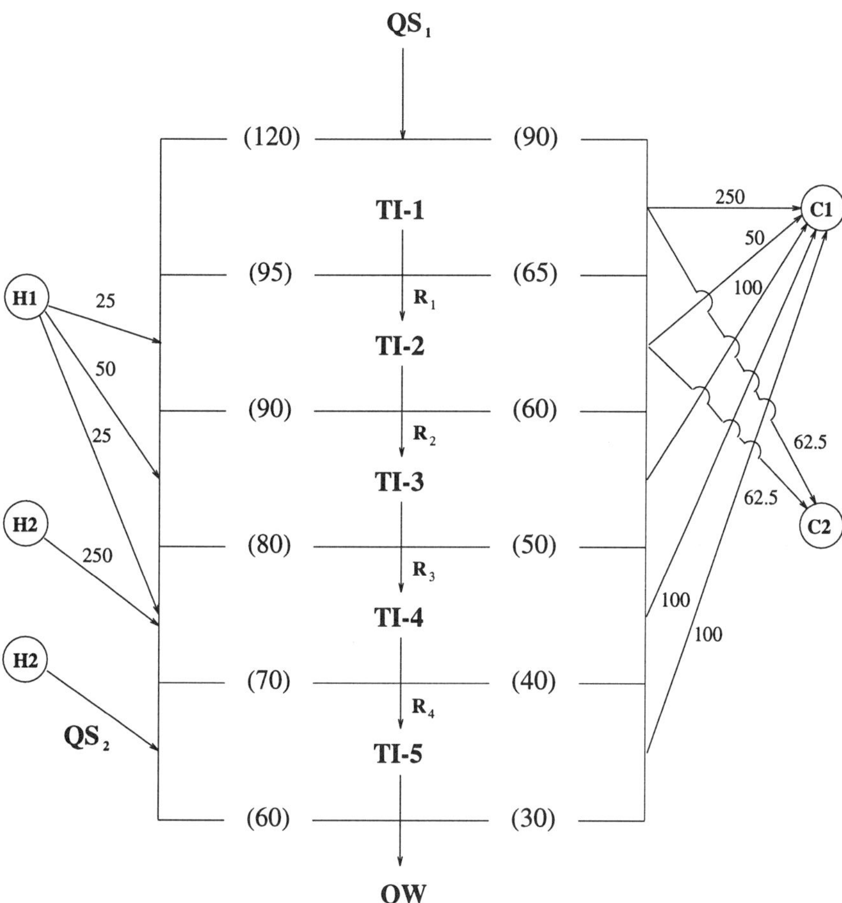

Figure 8.7: Transshipment representation for illustration 8.3.3

and features five equalities, seven variables and linear objective function and constraints. Its solution obtained with GAMS/MINOS is

$$\begin{aligned}
QS_1 &= 450.0,\\
QS_2 &= 0.0,\\
QS_W &= 75.0,\\
R_1 &= 137.5,\\
R_2 &= 50.0,\\
R_3 &= 0.0,\\
R_4 &= 175.0.
\end{aligned}$$

8.3.2 Minimum Number of Matches

In the previous section we discussed the minimum utility cost target and its formulation as an LP transshipment model. The solution of the LP transshipment model provides:

(i) The required loads of hot and cold utilities, and

(ii) The location of pinch point(s) if any.

The pinch point(s) decompose the overall HEN problem into subproblems corresponding to above and below the pinch, if a single pinch point exists, or above the first pinch, between consecutive pinch point(s) and below the bottom pinch point, if multiple pinch points exist. These subproblems are denoted as *subnetworks*. It is the existence of pinch points that allows these subnetworks to be treated independently since we have made the implicit assumption that heat cannot cross the pinch point.

A useful target postulated so as to distinguish among the many HENs that satisfy the minimum utility cost is the minimum number of matches problem which is stated in the following section.

8.3.2.1 Problem Statement

Given the information provided from the minimum utility cost target (i.e., loads of hot and cold utilities, location of pinch points, and hence subnetworks), determine for each subnetwork the minimum number of matches (i.e., pairs of hot and cold process streams, pairs of hot utilities and cold process streams, pairs of cold utilities and hot process streams, and pairs of hot-hot or cold-cold process streams exchanging heat), as well as the heat load of each match.

Remark 1 The implicit assumption in postulating this target is that HENs that satisfy it are usually close to an optimal or near optimal total annualized cost solution. Such an assumption is made since we do not take into account explicitly the investment cost of the heat exchangers but instead lump everything into the minimum number of matches target.

8.3.2.2 MILP Transshipment Model

Papoulias and Grossmann (1983) proposed an MILP transshipment model for the formulation of the minimum number of matches target. This model is applied to each subnetwork of the HEN problem.

Basic Idea of MILP Transshipment Model

The basic idea in the transshipment model for the minimum number of matches target is to model explicitly the potential heat exchange between all pairs of streams (excluding hot utilities to cold utilities) with respect to

(i) Existence of each match,

(ii) Amount of heat load of each match, and

(iii) Amount of heat residual of each hot process stream/utility.

The potential existence of each match is modeled via the binary variables y_{ij}:

$$y_{ij} = \begin{cases} 1 \text{ , if match } (ij) \text{ takes place} \\ 0 \text{ , otherwise} \end{cases}$$

where $i \in HP \cup HU, j \in CP \cup CU$, and we exclude matches $(i, j) \quad i \in HU, j \in CU$.

The amount of heat load of each match (ij) is modeled through the introduction of continuous variables Q_{ij} and Q_{ijk}:

$$Q_{ij} = \sum_k Q_{ijk},$$

where Q_{ijk} is the heat exchanged in match (ij) at interval k and Q_{ij} is the heat load of match (ij) over all intervals of the subnetwork under consideration.

The heat residual of each hot process stream/utility exiting each temperature interval k is modeled via the continuous variables, $R_{i,k}$ and R_k:

$$R_k = \sum_i R_{i,k},$$

where $R_{i,k}$ is the heat residual of hot process stream/utility i out of temperature interval k, and R_k is the total residual heat exiting interval k.

The pictorial representation of a temperature interval k is shown in Figure 8.8, where we have a hot process stream $H1$, a hot utility $S1$ potentially exchanging heat with a cold process stream $C1$; that is, we have two potential matches $(H1, C1)$ and $(S1 - C1)$.

Remark 1 Note that even though the amount of hot utility $S1$, which enters at an upper TI, is known from the minimum utility cost calculation, the heat load $Q_{S1,C1,k}$ is a variable. Also, the residuals $R_{S1,k-1}$, $R_{S1,k}$ are unknown and hence are variables. Known fixed quantities are the

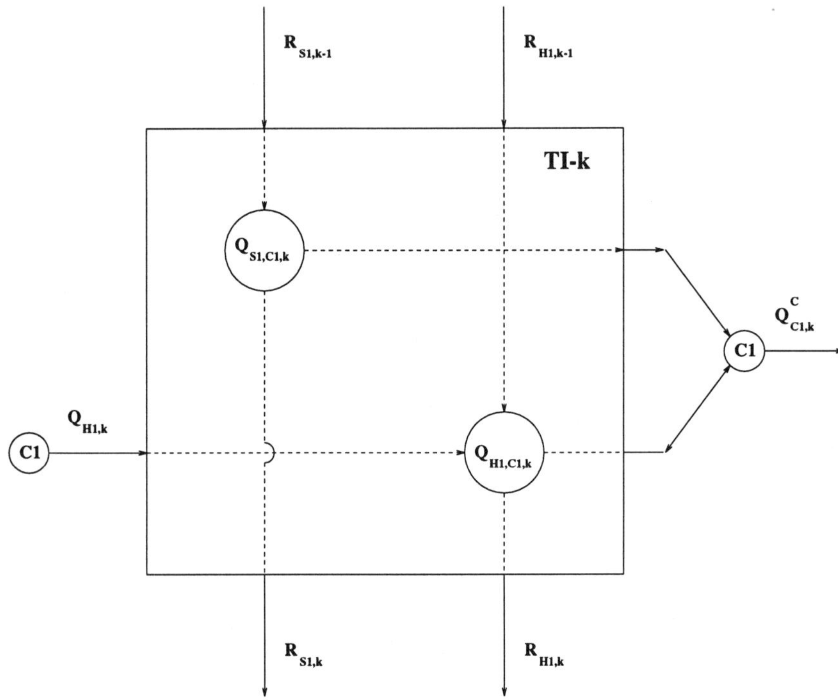

Figure 8.8: Graphical representation of $TI - k$ for the MILP transshipment model

$Q^H_{H1,k}$ and $Q^C_{C1,k}$ which represent the heat available from hot stream $H1$ and the heat needed by cold stream $C1$, respectively, at interval k. The residuals $R_{H1,k-1}$, $R_{H1,k}$, and the $Q_{H1,C1,k}$ are variables.

Having introduced the appropriate binary and continuous variables (i.e., y_{ij}, Q_{ijk}, Q_{ij}, R_{ik}, R_k), the basic building blocks of the model for the minimum number of matches target are

(i) Energy balances for each hot process stream and at each temperature interval k,

(ii) Energy balances for each hot utility and at each temperature interval k,

(iii) Energy balances for each cold process stream and at each temperature interval k,

(iv) Energy balances for each cold utility and at each temperature interval k,

(v) Definitions of total residual flows at each interval,

(vi) Definitions of heat loads of each match,

(vii) Relations between heat loads Q_{ij} and binary variables y_{ij},

(viii) Nonnegativity constraints on continuous variables,

(ix) Zero top and bottom residual constraints,

(x) Integrality conditions on y_{ij}.

The objective function involves the sum of binary variables representing all potential matches. Then, the mathematical model for the minimum number of matches target can be stated as follows:

$$\min \sum_{i \in HP \cup HU} \sum_{j \in CP \cup CU} y_{ij}$$

s.t.

$$\left.\begin{aligned}
R_{i,k} - R_{i,k-1} + \sum_{j \in CP_k \cup CU_k} Q_{ijk} &= Q^H_{ik} \quad , \quad i \in HP_k \\
R_{i,k} - R_{i,k-1} + \sum_{j \in CP_k} Q_{ijk} &= Q^H_{ik} \quad , \quad i \in HU_k \\
\sum_{i \in HP_k \cup HU_k} Q_{ijk} &= Q^C_{jk} \quad , \quad j \in CP_k \\
\sum_{i \in HP_k} Q_{ijk} &= Q^C_{jk} \quad , \quad j \in CU_k \\
R_k - \sum_{i \in HP_k \cup HU_k} R_{ik} &= 0
\end{aligned}\right\} k \in TI \qquad \textbf{(P2)}$$

$$\left.\begin{aligned}
&Q_{i,j} = \sum_{k \in TI} Q_{ijk} \\
&L_{ij} y_{ij} \le Q_{ij} \le \mathcal{U}_{ij} y_{ij} \\
&Q_{ijk} \ge 0, k \in TI; R_{ik} \ge 0, k \in TI \\
&R_0 = R_K = 0 \\
&y_{ij} = 0 - 1
\end{aligned}\right\} i \in HP \cup HU \; , \; j \in CP \cup CU$$

Remark 2 Note that we exclude from the potential matches the pairs of hot utilities to cold utilities. This is done by simply not including y_{ij} variables for such in both the objective function and constraints.

Remark 3 The objective function is a linear sum of all y_{ij}'s and simply minimizes the number of potential matches. The energy balances and the definition constraints are linear constraints in the residuals and heat loads. The relations between the continuous and binary variables are also linear since L_{ij} and U_{ij} are parameters corresponding to lower and upper bounds, respectively, on the heat exchange of each match (ij). It is important to understand the key role of these constraints which is to make certain that if a match does not exist, then its heat load should be zero while if the match takes place, then its heat load should be between the provided bounds. We can observe that:

$$\begin{array}{lll} \text{If} \quad y_{ij} = 1 & \text{, then} & L_{ij} \leq Q_{ij} \leq U_{ij}, \\ \text{If} \quad y_{ij} = 0 & \text{, then} & 0 \leq Q_{ij} \leq 0 \quad \text{and hence } Q_{ij} = 0. \end{array}$$

The appropriate definition of the lower and upper bounds can have a profound effect on the computational effort of solving the model **P2**. In fact, the tighter the bounds, the less effort is required, even though the same solution can be obtained for arbitrarily large U_{ij}. Finally, the nonnegativity and the top and bottom residual constraints are also linear. The variables are a mixed set of continuous and binary variables. Therefore, **P2** corresponds to a mixed-integer linear programming MILP transshipment model.

The solution of **P2** provides the minimum matches and their heat loads for each subnetwork. Floudas *et al.* (1986) (see Appendix A of reference) showed that the matches correspond to feasible heat exchanger units and hence, this target is equivalent to the target of minimum number of units.

Remark 4 Since **P2** is an MILP, its global solution can be attained with standard branch and bound algorithms which are implemented in commercially available packages (e.g., LINDO, ZOOM, SCICONIC, OSL, CPLEX).

It should be noted, however, that the MILP model **P2** can have several global solutions (i.e., solutions with the same objective function), and it is very straightforward to generate all such solutions for further examination by incorporating the appropriate integer cuts:

$$\sum_{i \in B} y_{ij} - \sum_{i \in NB} y_{ij} \leq \mid B \mid -1,$$

and resolving the MILP model **P2** . Note also that since we made the implicit assumption of near-optimality, it is important to generate all global solutions that have minimum number of matches and subsequently analyze them with respect to actual investment cost.

Remark 5 (Calculation of Lower and Upper Bounds, L_{ij} and U_{ij})

To calculate the tightest bounds on the heat loads of each potential match, within the considered subnetwork, we will distinguish the following cases:

Case 1 – No restrictions on matches:

In this case, the upper bound U_{ij} of potential match (ij) is given by the minimum of the heat loads of streams i and j; that is,

$$U_{ij} = \min \left\{ \sum_{k \in TI} Q^H_{ik}, \sum_{k \in TI} Q^C_{jc} \right\}.$$

The lower bound L_{ij} is used primarily so as to avoid small heat exchangers and in this case of no restriction we will have $L_{ij} = 0$.

Case 2 – Required Matches:

In this case, if one match (ij) must take place for instance, then we set

$$y_{ij} = 1,$$

and hence eliminate one binary variable from the model. If no restriction are imposed on the heat loads, then the upper and lower bounds are the same as in case 1.

Case 3 – Forbidden Matches:

In this case, if one match (ij) must not take place for instance, then this is equivalent to setting:

$$y_{ij} = 0,$$

and the upper and lower bounds should be

$$U_{ij} = L_{ij} = 0,$$

which will make the heat loads Q_{ij} zero. Note, however, that from the modeling point of view it is better to eliminate the binary variable y_{ij} as well as the continuous variables Q_{ijk} and their associated constraints from the model **P2** from the very beginning. This will result in fewer variables and constraints.

Case 4 – Restricted Heat Loads on Matches:

In this case, the upper and lower bounds of a match (ij) will be those provided:

$$\begin{aligned} U_{ij} &= U^R_{ij}, \\ L_{ij} &= L^R_{ij}. \end{aligned}$$

Case 5 – Preferred Matches:

In this case, the objective function should reflect the preferences via assigning weights to each match. Then, the objective function will become

$$\min \sum_{i \in HP \cup HU} \sum_{j \in CP \cup CU} \omega_{ij} y_{ij},$$

where ω_{ij} are the weights assigned to each match (ij). The calculation of upper and lower bounds on the heat loads, however, is not affected by such preferences.

Remark 6 Cerda and Westerberg (1983) developed an MILP model based on the transportation formulation which could also handle all cases of restricted matches. Furthermore, instead of solving the MILP transportation model via an available solver, they proposed several LP relaxations that can avoid the associated combinatorial problems. The drawback, though, of this model is that it requires more variables and constraints. Viswanathan and Evans (1987) proposed a modified transportation model for which they used the *out-of-kilter* algorithm to deal with constraints.

Remark 7 (Minimum Utility Cost with Restricted Matches)

The MILP transshipment model **P2** can be modified so as to correspond to a minimum utility cost calculation with restricted matches. The key modification steps are

(i) Write the **P2** model for the overall HEN and not for each subnetwork since we do not know the location of pinch point(s),

(ii) Separate the restricted from the non-restricted matches:

$$\begin{aligned} (ij) &\in RM \quad : \quad \text{restricted match} \\ (ij) &\in NRM \quad : \quad \text{non-restricted match} \end{aligned}$$

and set

$$y_{ij} = 1 \quad , \quad (ij) \in NRM,$$

since these potential matches have no restrictions on the heat transfer with lower and upper bounds:

$$\begin{aligned} L_{ij} &= 0, \\ U_{ij} &= \min\left\{\sum_k Q^H_{ik} \quad , \quad \sum_k Q^C_{jk}\right\}. \end{aligned}$$

This way we eliminate the binary variables for the nonrestricted matches and we simply provide valid upper and lower bounds.

For the restricted matches (i.e., required, forbidden, restricted heat loads) we impose the appropriate constraints as discussed in cases 2,3 and 4.

(iii) The objective function becomes

$$\min \sum_{i \in HU} C_i \sum_{k \in TI} Q^H_{ik} + \sum_{j \in CU} C_j \sum_{k \in TI} Q^C_{jk},$$

since the hot and cold utility loads are treated as variables explicitly.

Note that with the above modifications, **P2** becomes a linear programming problem since the restricted matches also fix the binary variables $y_{ij}(ij) \in RM$, with variables, Q^H_{ik} , $i \in HU$, Q^C_{jk} , $j \in CU$, Q_{ijk} , R_{ik}. Then the formulation of minimum utility cost with restricted matches becomes

$$\begin{array}{ll}
\min & \sum_{i\in HU} C_i \sum_{k\in TI} Q^H_{ik} + \sum_{j\in CU} C_j \sum_{k\in TI} Q^C_{jk} \\
s.t. & \\
& \left.\begin{array}{rcll}
R_{i,k} - R_{i,k-1} + \sum_{j\in CP_k\cup CU_k} Q_{ijk} &=& Q^H_{ik} \quad , & i\in HP_k \\
R_{i,k} - R_{i,k-1} + \sum_{j\in CP_k} Q_{ijk} &=& Q^H_{ik} \quad , & i\in HU_k \\
\sum_{i\in HP_k\cup HU_k} Q_{ijk} &=& Q^C_{jk} \quad , & j\in CP_k \\
\sum_{i\in HP_k} Q_{ijk} &=& Q^C_{jk} \quad , & j\in CU_k \\
R_k - \sum_{i\in HP_k\cup HU_k} R_{ik} &=& 0 &
\end{array}\right\} k\in TI \\
& \left.\begin{array}{l}
Q_{i,j} = \sum_{k\in TI} Q_{ijk} \\
0 \le Q_{i,j} \le \mathcal{U}_{ij}
\end{array}\right\} (i,j) \in NRM \\
& \left.\begin{array}{l}
Q_{i,j} = \sum_{k\in TI} Q_{i,j,k} \\
Q_{ij} = 0 \text{ (forbidden)} \\
\mathcal{L}^R_{ij} \le Q_{ij} \le \mathcal{U}^R_{ij} \text{ (restricted)} \\
R_0 = R_{K-1} = 0
\end{array}\right\} (i,j) \in RM
\end{array} \qquad \textbf{(P3)}$$

Note that the solution of **P3** will in addition provide information on the heat loads Q_{ij} of the potential matches.

Illustration 8.3.4 (MILP Transshipment Model)

This example, which corresponds to the motivating example of Gundersen and Grossmann (1990) below the pinch point, features three hot process streams, two cold streams, and a cold utility with data shown in Table 8.4.

The pictorial transshipment representation is shown in Figure 8.9. Note that there are two temperature intervals, and there exist nine potential matches:

$$\begin{array}{c}
H1 - C1, H1 - C2, H1 - CW, \\
H2 - C1, H2 - C2, H2 - CW, \\
H3 - C1, H3 - C2, H3 - CW.
\end{array}$$

In the top temperature interval only the matches between the hot process streams and the cold process streams can take place. In the bottom interval we have the matches of $H1$, $H2$, $H3$ with $C1$ and CW only since $C2$ does not participate in this TI. As a result we need to introduce 12 instead of 18 continuous variables Q_{ijk}. We need to introduce nine binary variables for the aforementioned potential matches. The MILP transshipment model **P2** is:

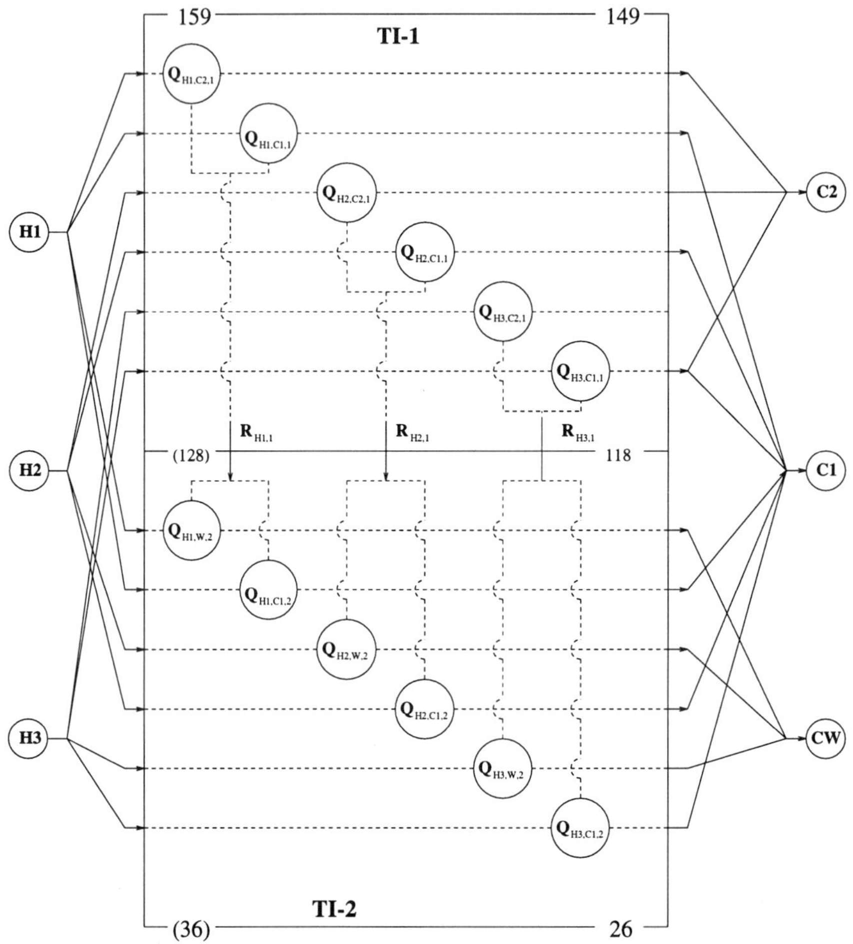

Figure 8.9: Transshipment graphical representation for minimum number of matches target

Stream	$T^{in}(°C)$	$T^{out}(°C)$	$FC_p(\text{kW}°C^{-1})$
$H1$	159	77	228.5
$H2$	159	88	20.4
$H3$	159	90	53.8
$C1$	26	127	93.3
$C2$	118	149	196.1
CW	15	30	
$QW_{min} = 8395.2\text{kW}$			
$HRAT = 10°C$			

Table 8.4: Data for illustration of the MILP transshipment

$$\begin{aligned} \min \quad & y_{H1,C1} + y_{H1,C2} + y_{H1,W} + y_{H2,C1} + y_{H2,C2} + \\ & y_{H2,W} + y_{H3,C1} + y_{H3,C2} + y_{H3,W} \\ s.t. \quad & \end{aligned}$$

$$\left.\begin{array}{lll} R_{H1,1} + & Q_{H1,C1,1} + Q_{H1,C2,1} & = 7083.5 \\ R_{H2,1} + & Q_{H2,C1,1} + Q_{H2,C2,1} & = 632.4 \\ R_{H3,1} + & Q_{H3,C1,1} + Q_{H3,C2,1} & = 1667.8 \\ & Q_{H1,C1,1} + Q_{H2,C1,1} + Q_{H3,C1,1} & = 839.7 \\ & Q_{H1,C2,1} + Q_{H2,C2,1} + Q_{H3,C2,1} & = 6079.1 \\ R_1 - R_{H1,1} - R_{H2,1} - R_{H3,1} & & = 0 \end{array}\right] TI-1$$

$$\left.\begin{array}{lll} -R_{H1,1} + & Q_{H1,C1,2} + Q_{H1,W,2} & = 11653.5 \\ -R_{H2,1} + & Q_{H2,C1,2} + Q_{H2,W,2} & = 816 \\ -R_{H3,1} + & Q_{H3,C1,2} + Q_{H3,W,2} & = 2044.4 \\ & Q_{H1,C1,2} + Q_{H2,C1,2} + Q_{H3,C1,2} & = 8583.6 \\ & Q_{H1,W,2} + Q_{H2,W,2} + Q_{H3,W,2} & = 8395.2 \\ R_1 - R_{H1,1} - R_{H2,1} - R_{H3,1} & & = 0 \end{array}\right] TI-2$$

$$\left.\begin{array}{lll} Q_{H1,C1} & = & Q_{H1,C1,1} + Q_{H1,C1,2} \\ Q_{H1,C2} & = & Q_{H1,C2,1} \\ Q_{H1,W} & = & Q_{H1,W,2} \\ Q_{H2,C1} & = & Q_{H2,C1,1} + Q_{H2,C1,2} \\ Q_{H2,C2} & = & Q_{H2,C2,1} \\ Q_{H2,W} & = & Q_{H2,W,2} \\ Q_{H3,C1} & = & Q_{H3,C1,1} + Q_{H3,C1,2} \\ Q_{H3,C2} & = & Q_{H3,C2,1} \\ Q_{H3,W} & = & Q_{H3,W,2} \end{array}\right] \text{Definitions of heat loads}$$

$$\left.\begin{array}{lll} Q_{H1,C1} & \leq & U_{H1,C1} \cdot y_{H1,C1} \\ Q_{H1,C2} & \leq & U_{H1,C2} \cdot y_{H1,C2} \\ Q_{H1,W} & \leq & U_{H1,W} \cdot y_{H1,W} \\ Q_{H2,C1} & \leq & U_{H2,C1} \cdot y_{H2,C1} \\ Q_{H2,C2} & \leq & U_{H2,C2} \cdot y_{H2,C2} \\ Q_{H2,W} & \leq & U_{H2,W} \cdot y_{H2,W} \\ Q_{H3,C1} & \leq & U_{H3,C1} \cdot y_{H3,C1} \\ Q_{H3,C2} & \leq & U_{H3,C2} \cdot y_{H3,C2} \\ Q_{H3,W} & \leq & U_{H3,W} \cdot y_{H3,W} \end{array}\right] \text{Logical constraints}$$

$$\begin{array}{rll} Q_{H1,C1,1}, Q_{H1,C1,2}, Q_{H1,C2,1}, Q_{H1,W,2} & \geq & 0 \\ Q_{H2,C1,1}, Q_{H2,C1,2}, Q_{H2,C2,1}, Q_{H2,W,2} & \geq & 0 \\ Q_{H3,C1,1}, Q_{H3,C1,2}, Q_{H3,C2,1}, Q_{H3,W,2} & \geq & 0 \\ R_{H1,1}, R_{H2,1}, R_{H3,1} & \geq & 0 \end{array}$$

$$y_{H1,C1}, y_{H1,C2}, y_{H1,W}, y_{H2,C1}, y_{H2,C2}, y_{H2,W}, y_{H3,C1}, y_{H3,C2}, y_{H3,W} = 0-1$$

where

$$\begin{array}{lll} U_{H1,C1} & = & \min\{7083.5 + 11653.5, 839.7 + 8583.6\} = 9423.3 \\ U_{H1,C2} & = & \min\{7083.5, 6079.1\} = 6079.1 \\ U_{H1,W} & = & \min\{7083.5 + 11653.5, 8395.2\} = 8395.2 \\ U_{H2,C1} & = & \min\{632.4 + 816, 839.7 + 8583.6\} = 1448.4 \\ U_{H2,C2} & = & \min\{632.4, 6079.1\} = 632.4 \\ U_{H2,W} & = & \min\{632.4 + 816, 8395.2\} = 1448.4 \\ U_{H3,C1} & = & \min\{1667.8 + 2044.4, 839.7 + 8583.6\} = 3712.2 \\ U_{H3,C2} & = & \min\{1677.8, 6079.1\} = 1667.8 \\ U_{H3,W} & = & \min\{1677.8 + 2044.4, 8395.2\} = 3712.2 \end{array}$$

Solving this MILP transshipment model with GAMS/CPLEX and applying integer cuts, the following four global solutions are obtained each with five matches, but different distributions of heat loads:

Solution 1:

Match(ij)	Q_{ij}
$H1-C1$	9423.3
$H1-C2$	6079.1
$H1-W$	3234.6
$H2-W$	1448.4
$H3-W$	3712.2

Solution 2:

Match(ij)	Q_{ij}
$H1 - C1$	7974.9
$H1 - C2$	6079.1
$H1 - W$	4683.0
$H2 - C1$	1448.4
$H3 - W$	3712.2

Solution 3:

Match(ij)	Q_{ij}
$H1 - C1$	5711.1
$H1 - C2$	6079.1
$H1 - W$	6946.8
$H2 - W$	1448.4
$H3 - C1$	3712.2

Solution 4:

Match(ij)	Q_{ij}
$H1 - C1$	4262.7
$H1 - C2$	6079.1
$H1 - W$	8395.2
$H2 - C1$	1448.4
$H3 - C1$	3712.2

Remark 8 Note that all four solutions are equivalent with respect to the minimum number of matches target (i.e., all have fixed matches), and therefore we need to introduce additional criteria to distinguish which solution is more preferable. Since the four solutions feature different heat load distributions, one approach to distinguish them is to somehow determine the minimum investment cost network for each and rank them with respect to their investment cost. This of course assumes that we have a methodology that given the heat loads and the matches provides the minimum investment cost network. We will discuss such an approach in section 8.4.

Another approach is the modify the MILP transshipment model **P2** so as to have preferences among multiple global solutions of model **P2** according to their potential of vertical heat transfer between the composite curves. Such an approach was proposed by Gundersen and Grossmann (1990) as a good heuristic and will be discussed in section 8.3.3.

8.3.2.3 How to Handle Hot to Hot and Cold to Cold Matches

A number of research groups suggested that in certain cases it is desirable to allow for matches between hot-to-hot, and cold-to-cold process streams (Grimes *et al.*, 1982; Viswanathan and Evans, 1987; Dolan *et al.*, 1989). It is worth noting that the MILP transshipment model of Papoulias and

Stream	$T^{in}(^\circ F)$	$T^{out}(^\circ F)$	$FC_p(\text{kW}^\circ F^{-1})$
$H1$	320	200	16.6668
$H2$	480	280	20
$C1$	140	320	14.4501
$C2$	240	500	11.53
S	540	540	
CW	100	180	
$HRAT = 18^\circ F$			
Match $H1 - C1$ is not allowed			
Cold-to-cold match between $C2 - C1$ may take place			

Table 8.5: Data of illustration 8.3.5

Grossmann (1983) does not take into account such alternatives. In this section, we will show that hot-hot, cold-cold matches can be incorporated within the MILP transshipment framework.

The basic ideas in such an approach are

(i) For a cold-to-cold match, one cold stream must behave as a hot stream first with unknown outlet temperature which is equal to the real inlet temperature of the same cold stream. This implies that we have to introduce one additional hot stream with inlet temperature the inlet temperature of the cold stream and unknown outlet temperature. At the same time, the real inlet temperature of the cold stream is unknown and hence the heat load of the cold stream will be a piecewise linear function of its real inlet unknown temperature.

(ii) For a hot-to-hot match, one hot stream must behave as a cold stream first with unknown outlet temperature which is equal to the real inlet temperature of the same hot stream. Similarly to (i), we have to introduce an additional cold stream and express the heat load of the hot stream as a piecewise linear function of its real unknown inlet temperature.

(iii) To identify the piecewise linear functions, we introduce the binary variables and utilize the modeling approach of piecewise linear approximation discussed in the Process Synthesis chapter.

We will illustrate this approach with the $4SP1$ problem presented next.

Illustration 8.3.5 (Cold-to-Cold and Forbidden Matches)

This example consists of two hot, two cold streams, one hot and one cold utility and has a forbidden match between hot stream $H1$ and cold stream $C1$. Its data is shown in Table 8.5.

Since the $C2 - C1$ match is allowed, we introduce a third hot stream $H3$ which has as inlet temperature the inlet temperature of $C2$ and unknown outlet temperature which is equal to the real

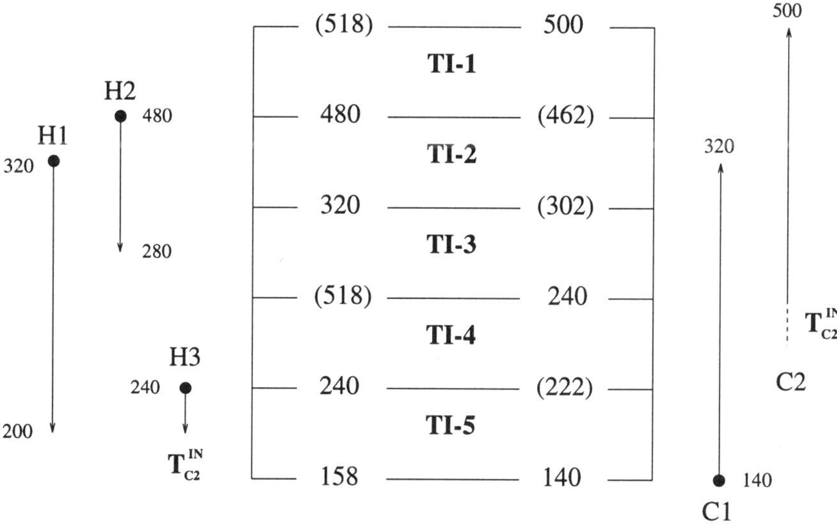

Figure 8.10: Temperature partitioning of cold-to-cold illustration

inlet temperature of $C2$. This is depicted in the temperature partitioning shown in Figure 8.10. The outlet temperature of $H3$ should be equal to the inlet temperature of $C2$ since $H3$ corresponds to the part of $C2$ which acts as hot stream in the $C2 - C1$ potential match. Note that the heat load of $H3$ is

$$Q^{H}_{H3,5} = 11.53\,(240 - T^{IN}_{C2}),$$

which is a linear function of T^{IN}_{C2}.

The heat load however of C_2 is a piecewise linear function of $T^{IN}_{C_2}$ since depending on its value it can be in temperature interval 4 or temperature interval 5. To obtain such a piecewise linear functionality, we introduce two binary variables y_4, y_5:

$$\begin{aligned} y_4 &= \begin{cases} 1, \text{if } T^{IN}_{C_2} \text{ is in } TI-4 \\ 0, \text{otherwise} \end{cases} \\ y_5 &= \begin{cases} 1, \text{if } T^{IN}_{C_2} \text{ is in } TI-5 \\ 0, \text{otherwise} \end{cases} \\ y_4 + y_5 &= 1 \end{aligned}$$

and two continuous variables $T^{IN,4}_{C_2}$, $T^{IN,5}_{C_2}$:

$$T^{IN,4}_{C_2} + T^{IN,5}_{C_2} = T^{IN}_{C_2},$$

with

$$\begin{aligned} 222 \leq T^{IN,4}_{C_2} &\leq 240, \\ 158 \leq T^{IN,5}_{C_2} &\leq 222. \end{aligned}$$

Note that $T_{C_2}^{IN,5}$ cannot be lower than 158 if the $C2 - C1$ match takes place.

The heat loads of $C2$ in $TI - 4$ and $TI - 5$ are functions at the $T_{C_2}^{IN}$:

$$\begin{aligned} Q_{C2,4}^{C}\left(T_{C_2}^{IN}\right) &= a_4 \cdot y_4 + b_4 \cdot T_{C_2}^{IN,4}, \\ Q_{C2,5}^{C}\left(T_{C_2}^{IN}\right) &= a_5 \cdot y_5 + b_5 \cdot T_{C_2}^{IN,5}, \end{aligned}$$

where a_4, b_4, a_5, b_5 are

$$\begin{aligned} b_4 &= \frac{Q_{C2,4}^{C}(240) - Q_{C2,4}^{C}(222)}{240 - 222} = -11.53, \\ b_5 &= \frac{Q_{C2,5}^{C}(222) - Q_{C2,5}^{C}(158)}{222 - 158} = -11.53, \\ a_5 &= Q_{C2,5}^{C}(158) - b_5 \cdot 158 \\ &= 11.53 \cdot 222, \\ a_4 &= Q_{C2,4}^{C}(222) - b_4 \cdot 222 \\ &= 11.53 \cdot 240. \end{aligned}$$

Then

$$\begin{aligned} Q_{C2,4}^{C} &= 11.53 \left[240 \cdot y_4 - T_{C_2}^{IN,4}\right], \\ Q_{C2,5}^{C} &= 11.53 \left[222 \cdot y_5 - T_{C_2}^{IN,5}\right]. \end{aligned}$$

Note that these expressions are piecewise linear in $T_{C_2}^{IN,4}$ and $T_{C_2}^{IN,5}$, and hence they will not change the nature of the transshipment model formulations for

(i) Minimum utility cost with restricted matches,

(ii) Minimum number of matches.

Formulations **P2** and **P3** can now be applied in a straightforward fashion for (i) and (ii) respectively. The differences in these formulations are that

(i) $T_{C_2}^{IN}, T_{C_2}^{IN,1}, T_{C_2}^{IN,2}$ are variables, and

(ii) y_4, y_5 are additional binary variables.

8.3.3 Minimum Number of Matches for Vertical Heat Transfer

As we discussed on Section 8.3.2.2, the MILP transshipment model may have several global solutions which all exhibit the same number of matches. To establish which solution is more preferable, Gundersen and Grossmann (1990) proposed as criterion the *vertical heat transfer* from the hot composite to the cold composite curve with key objective the minimization of the total heat transfer area.

The vertical heat transfer between the hot and cold composite curves utilizes as a means of representation the partitioning of enthalpy (Q) into enthalpy intervals (EI). The partitioning into enthalpy intervals has a number of similarities with the partitioning of temperature intervals presented in section 8.3.1.3, but it has at the same time a number of key differences outlined next.

8.3.3.1 Partitioning of Enthalpy Intervals

In this section we will discuss first the partitioning of enthalpy (Q) into enthalpy intervals (EI) assuming that $TIAT = EMAT = HRAT$ and later on we will discuss the case of $TIAT = EMAT < HRAT$.

Case I: $EMAT = TIAT = HRAT$

The basic idea in this case is to consider all kink points of both the hot and cold composite curves and draw vertical lines at these kink points. These vertical lines define the enthalpy intervals. Note that the list of kink points includes supply and *targets* of hot and cold streams. To illustrate such an (EI) partitioning let us consider the example used in section 8.3.1.2 for which the temperature interval partitioning is depicted in Figure 8.2 of section 8.3.1.3. The (EI) partition for this example is shown in Figure 8.11. Note that in this example there are six EIs.

$$
\begin{array}{lcll}
EI-1 & : & F'FGG & \text{– heat transfer from steam to } C1, \\
EI-2 & : & E'EFF' & \text{– heat transfer from steam to } C1+C2, \\
EI-3 & : & AA'EE' & \text{– heat transfer from steam to } C1, \\
EI-4 & : & BB'A'A & \text{– heat transfer from } H1 \text{ to } C1, \\
EI-5 & : & D'DB'B & \text{– heat transfer from } H1+H2 \text{ to } C1, \\
EI-6 & : & CC'DD' & \text{– heat transfer from } H1+H2 \text{ to Water.}
\end{array}
$$

In the first three EIs we have heat transfer from stream to cold process streams. In the $EI-6$ we have heat transfer from hot process streams to cooling water. In $EI-4$, $EI-5$ we have heat transfer from hot process streams to cold process streams.

An alternative representation of the six EI's is shown in Figure 8.12.

Case II: $EMAT = TIAT < HRAT$

In such a case, we create the *balanced* hot and cold composite curves. These correspond to eliminating the intervals that have utilities and hence end up only with interval $EI-4$ and $EI-5$. We need to create the *balanced* curves since the hot and cold utility loads are calculated for an $HRAT$ and the partitioning into EI's will be based on $EMAT = TIAT < HRAT$.

The basic idea in (EI) partitioning in this case is to identify first the points that are:

(i) $T^H_{supply} - EMAT$ in the cold composite curve, and

(ii) $T^C_{supply} + EMAT$ in the hot composite curve.

and add them to the list of kink points which involves the endpoints. Then, we follow the same approach as in case I.

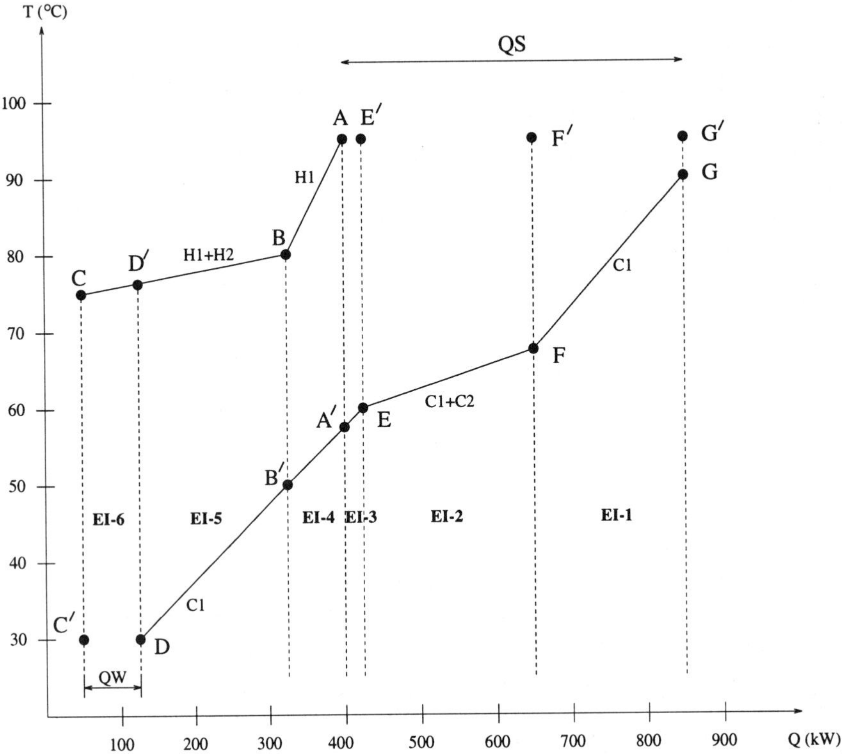

Figure 8.11: Enthalpy interval partitioning for case I

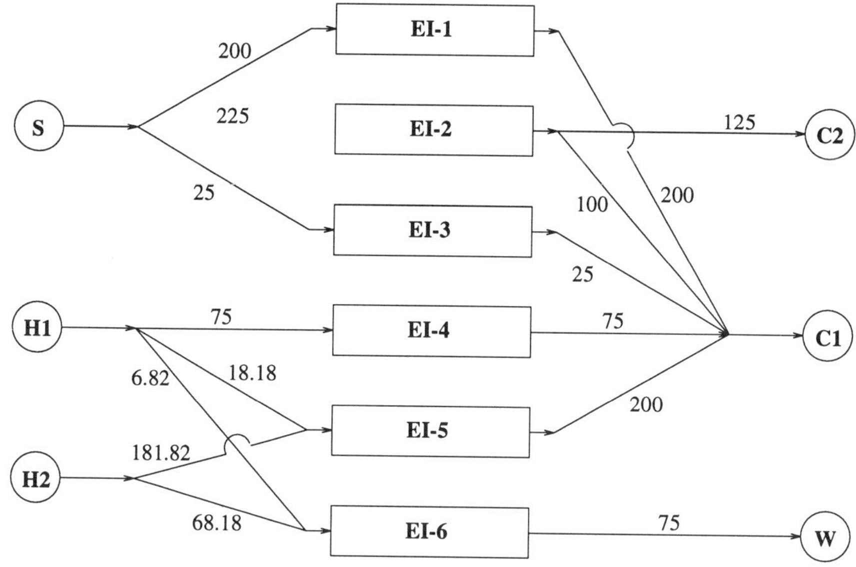

Figure 8.12: Representation of enthalpy intervals for case I

We illustrate this case in Figure 8.13 for $EMAT = TIAT = 25°C$ which is less than $HRAT = 30°C$. Note that the balanced hot and cold composite curves are

Balanced hot composite curve: $D'BA$;

Balanced cold composite curve: $DB'A'$.

Also note that the new point H is created due to

$$T^B - 25,$$

which is the temperature of the kink point B at the hot composite curve minus the $EMAT = TIAT$, which corresponds to BB'' in Figure 8.13. As a result, we have three enthalpy intervals:

$$\begin{aligned} EI-1 &: AA'HH', \\ EI-2 &: H'HB'B, \\ EI-3 &: BB'DD'. \end{aligned}$$

An alternative representation like the one shown in Figure 8.12 for case I can similarly be developed for case II.

8.3.3.2 Basic Idea

The basic idea in discriminating among multiple global solutions of the MILP transshipment model of Papoulias and Grossmann (1983) is to use as criterion the verticality of heat transfer,

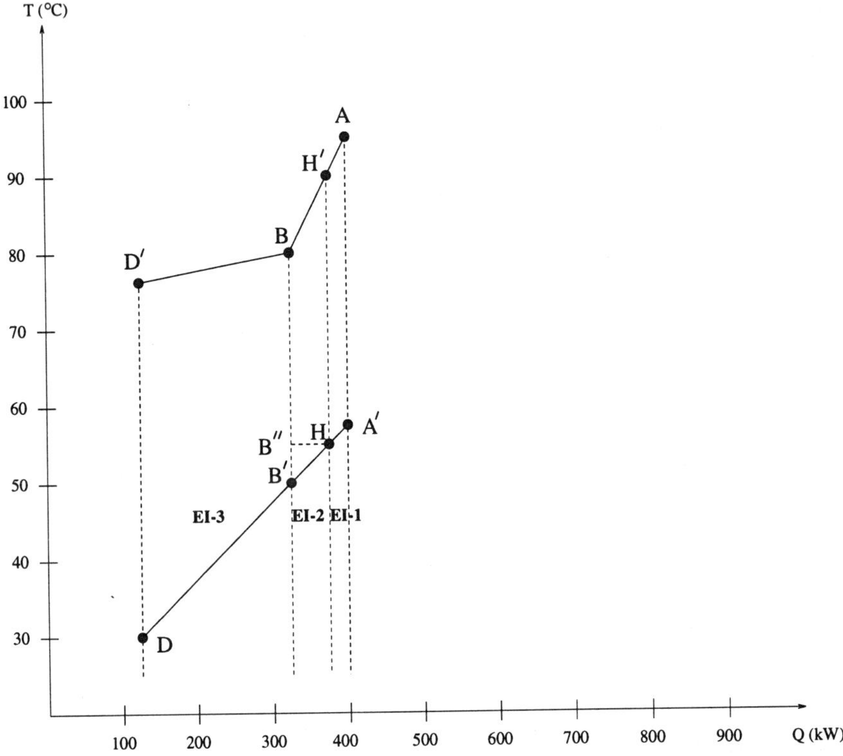

Figure 8.13: Enthalpy interval partitioning for case II

determine from the (EI) partitioning the heat loads of vertical heat transfer, and compare them with the heat loads of the matches. If the heat loads of the matches are larger than the maximum loads of the vertical heat transfer, then the objective function is augmented with a penalty term which attempts to drive the heat loads of the matches down to the maximum vertical heat transfer loads. In other words, if nonvertical heat transfer occurs, then the objective function is penalized.

To formulate the modified transshipment model based on the above idea we introduce the variables S_{ij} which should satisfy the constraints

$$\begin{aligned} S_{ij} &\geq Q_{ij} - Q_{ij}^V, \\ S_{ij} &\geq 0, \end{aligned}$$

where Q_{ij}^V are parameters that denote the maximum vertical heat transfer in a match (ij), and which can be calculated from the (EI) partitioning as follows:

$$Q_{ij}^V = \sum_{l \in EI} \min \left\{ Q_{il}^H, Q_{jl}^C \right\},$$

with $EI = \{l :$ is an enthalpy interval $i, l = 1, 2, \ldots L\}$.

The penalty type term that is added to the objective function takes the form:

$$\epsilon \cdot \sum_{i \in HP \cup HU} \sum_{j \in CP \cup CU} S_{ij},$$

where the weight factor ϵ is given by

$$\epsilon = \frac{1}{\left[\sum_{i \in HP \cup HU} Q_i \right]}.$$

8.3.3.3 Vertical MILP Transshipment Model

The modified MILP transshipment model that favors vertical heat transfer is of the following form:

$$\begin{aligned} \min \quad & \sum_{i \in HP \cup HU} \sum_{j \in CP \cup CU} y_{ij} + \epsilon \sum_{i \in HP \cup HU} \sum_{j \in CP \cup CU} S_{ij} \\ s.t. \quad & \\ & \text{constraints of } \mathbf{P2} \\ & \left. \begin{aligned} S_{ij} &\geq Q_{ij} - Q_{ij}^V \\ S_{ij} &\geq 0 \end{aligned} \right\} \begin{aligned} & i \in HP \cup HU \\ & j \in CP \cup CU \end{aligned} \end{aligned} \qquad \textbf{(P4)}$$

Remark 1 Note that the weight factor ϵ has been selected to be the inverse of the total heat exchanged. As a result, the penalty term that corresponds to the nonvertical heat transfer (also called *criss-cross* heat transfer) is divided by the total heat exchanged, and hence it cannot be more than one. This implies that such an objective function will identify the minimum number of matches and at the same time select the combination of heat loads of the matches that correspond to the most vertical heat transfer.

Remark 2 Model **P4** is an MILP which can solved with commercially available solvers. It has more constraints and variables when compared to **P2** due to the introduction of the S_{ij}'s and the aforementioned constraints. These additional variables and constraints, however, do not increase significantly the computational effort for its solution. Note however that its solution is a *lower bound* on the nonvertical heat transfer since it does not take the flow rates of the streams into consideration.

Remark 3 Model **P4** will provide good results only when the heat transfer coefficients are equal or close in values, since in this case the vertical heat transfer results in minimum heat transfer area. If however, the heat transfer coefficients are different, then nonvertical heat transfer can result in less heat transfer area. Therefore, for such cases the vertical MILP model **P4** is not applicable since it will discriminate among multiple global solutions of **P2** with the wrong criterion.

Another implicit assumption in model **P4**, in addition to favoring vertical heat transfer, is that the minimum total heat transfer area goes hand in hand with the minimum total investment cost solution. It should be emphasized that there exist cases where this is not true.

Remark 4 Model **P4** is applied to each subnetwork, that is after decomposition based on the location of the pinch point(s). If, however, we apply model **P4** to overall networks without decomposing them into subnetworks, then the quality of *lower bound* on the nonvertical heat transfer becomes worse. This is due to the fact that the additional variables and constraints have been applied for the overall heat transfer in each match (ij). As a result they do not provide any direction/penalty for local differences, that is, differences between heat exchange loads versus maximum vertical heat transfer loads at each temperature interval $k \in TI$. This deficiency can be remedied by introducing the variables $S_{ik}, k \in TI$ and the parameters Q_{ik}^V corresponding to each temperature interval k, along with the constraints:

$$S_{ik} \geq \sum_{j \in CP_k \cup CU_K} Q_{ijk}^H - Q_{ik}^V,$$

with appropriate definition of Q_{ik}^V (see Gundersen and Grossmann (1990)).

Remark 5 If we fix the number of matches (e.g., equal to U), and ask the question about which heat load distribution is more vertical, then we can address this issue via the following model:

$$\begin{array}{lll} \min & \sum_{i \in HP \cup HU} \sum_{j \in CP \cup CU} S_{ij} + \delta \sum_{i \in HP \cup HU} \sum_{j \in CP \cup CU} y_{ij} & \\ s.t. & \text{constraints of } \mathbf{(P4)} & \mathbf{(P5)} \\ & \sum_{i \in HP \cup HU} \sum_{j \in CP \cup CU} y_{ij} = U & \end{array}$$

where δ takes a small positive value. Based on remark 4, model **P5** can also be written with penalty terms applied to each temperature interval.

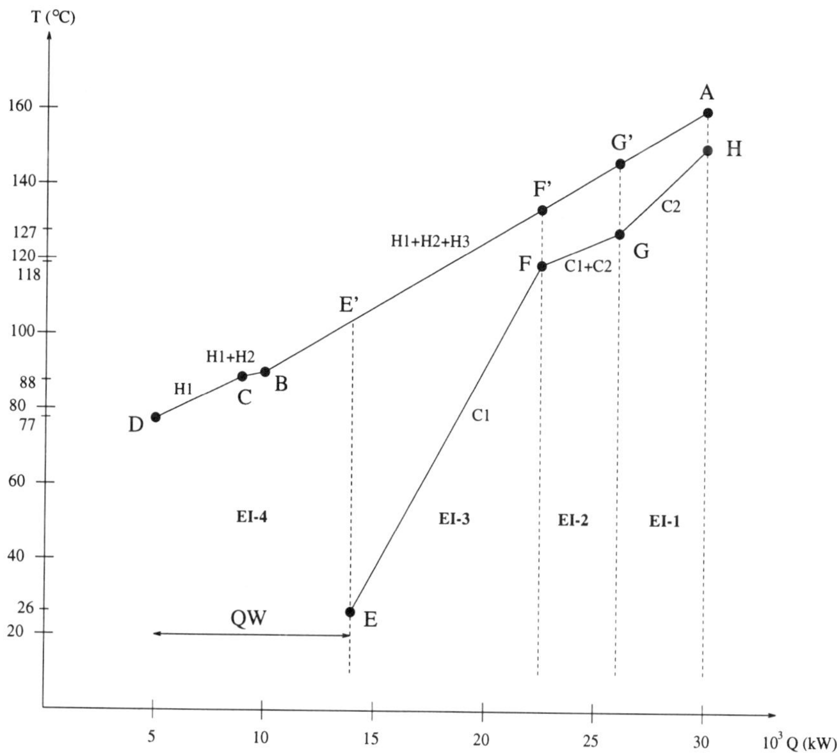

Figure 8.14: Enthalpy intervals in **T-Q** form

Illustration 8.3.6 (Vertical MILP Transshipment Model P4)

This example is the same example used as illustration of the MILP transshipment model **P2** in section 8.3.2.2. The constraints of **P2** were presented, and hence we will focus here on the:

(i) Enthalpy interval partitioning

(ii) Calculation of Q_{ij}^V,

(iii) Additional constraints, and

(iv) Penalty term of the objective function.

The (EI) partitioning shown in Figure 8.14 is based upon $EMAT = TIAT = HRAT = 10°C$.

$$\begin{aligned}
EI-1 &: AHG'G'' \\
EI-2 &: G'GFF' \\
EI-3 &: F'IEE' \\
EI-4 &: E'ED
\end{aligned}$$

We can calculate the heat loads available from the hot streams and the heat loads needed by the cold streams at each EI, as follows:

$$EI-1: \quad \left[\begin{array}{l} Q^{C}_{C2,1} = 196.1(149-127) = 4314.2 \\ Q^{H}_{H1,1} + Q^{H}_{H2,1} + Q^{H}_{H3,1} = (228.5+20.4+53.8)(159-T_{G'}) \\ \text{By equating the outputs we have} \\ 159 - T_{G'} = 14.2524 \\ T_{G'} = 144.7476 \\ \text{Then we obtain} \\ Q^{H}_{H1,1} \;=\; 228.5(159-T_{G'}) \;=\; 3256.6723 \\ Q^{H}_{H2,1} \;=\; 20.4(159-T_{G'}) \;=\; 290.7488 \\ Q^{H}_{H3,1} \;=\; 53.8(159-T_{G'}) \;=\; 762.5031 \end{array}\right.$$

$$EI-2: \quad \left[\begin{array}{l} Q^{C}_{C1,2} = 93.3(127-118) = 839.7 \\ Q^{C}_{C2,2} = 196.1(127-118) = 1764.9 \\ Q^{H}_{H1,2} + Q^{H}_{H2,2} + Q^{H}_{H3,2} = (228.5+20.4+53.8)(T_{G'}-T_{F'}) \\ \text{By equating the sum of the first two with the third line we have:} \\ T_{G'} - T_{F'} = 8.60456 \\ T_{F'} = 136.14304 \\ \text{Then, we obtain} \\ Q^{H}_{H1,2} \;=\; 228.5\;\;(T_{G'}-T_{F'}) \;=\; 1966.1417 \\ Q^{H}_{H2,2} \;=\; 20.4\;\;(T_{G'}-T_{F'}) \;=\; 175.533 \\ Q^{H}_{H3,2} \;=\; 53.8\;\;(T_{G'}-T_{F'}) \;=\; 462.9253 \end{array}\right.$$

$$EI-3: \left[\begin{array}{l} Q^C_{C1,3} = 93.3(118-26) = 8583.6 \\ Q^H_{H1,3} + Q^H_{H2,3} + Q^H_{H3,3} = (228.5+20.4+53.8)(T_{F'} - T_{E'}) \\ \text{By equating the above two lines we have:} \\ T_{F'} - T_{E'} = 28.3567 \\ T_{E'} = 107.78634 \\ \text{Then, we obtain} \\ Q^H_{H1,3} \ = \ 228.5 \ \ (T_{F'} - T_{E'}) \ = \ 6479.5262 \\ Q^H_{H2,3} \ = \ 20.4 \ \ (T_{F'} - T_{E'}) \ = \ 578.4785 \\ Q^H_{H3,3} \ = \ 53.8 \ \ (T_{F'} - T_{E'}) \ = \ 1525.5952 \end{array}\right.$$

$$EI-4: \left[\begin{array}{lcccl} QW & = & 8395.2 & & \\ Q^H_{H1,4} & = & 228.5 & (T_{E'} - 77) & = 7034.6787 \\ Q^H_{H2,4} & = & 20.4 & (T_{E'} - 88) & = 403.6414 \\ Q^H_{H3,4} & = & 53.8 & (T_{E'} - 90) & = 956.9051 \end{array}\right.$$

With the above data, we can now calculate the maximum vertical heat transfer of each match (ij).

$$\begin{array}{rcl} H1-C1 & : & Q^V_{H1,C1} = \min\{3252.6723, 0\} + \min\{1966.1417, 839.7\} + \\ & & \qquad \min\{6479.5262, 8583.6\} + \min\{7034.6784, 0\} \\ & & \qquad = 7319.2262, \\ H1-C2 & : & Q^V_{H1,C2} = 5021.5723, \\ H1-W & : & Q^V_{H1,W} = 7034.6787, \\ H2-C1 & : & Q^V_{H2,C1} = 754.0115, \\ H2-C2 & : & Q^V_{H2,C2} = 466.2818, \\ H2-W & : & Q^V_{H2,W} = 403.6414, \\ H3-C1 & : & Q^V_{H3,C1} = 1988.5205, \\ H3-C2 & : & Q^V_{H3,C2} = 1225.4284, \\ H3-W & : & Q^V_{H3,W} = 956.9051. \end{array}$$

The weight factor ϵ is calculated as

$$\begin{aligned} \epsilon &= \frac{1}{228.5(159-77) + 20.4(159-88) + 53.8(159-90)} \\ &= 4.18452 \cdot 10^{-5}. \end{aligned}$$

The additional constraints are then

$$\begin{aligned} S_{H1,C1} &\geq Q_{H1,C1} - Q^V_{H1,C1}, \\ S_{H1,C2} &\geq Q_{H1,C2} - Q^V_{H1,C2}, \\ S_{H1,W} &\geq Q_{H1,W} - Q^V_{H1,W}, \end{aligned}$$

$$
\begin{aligned}
S_{H2,C1} &\geq Q_{H2,C1} - Q^V_{H2,C1},\\
S_{H2,C2} &\geq Q_{H2,C2} - Q^V_{H2,C2},\\
S_{H2,W} &\geq Q_{H2,W} - Q^V_{H2,W},\\
S_{H3,C1} &\geq Q_{H3,C1} - Q^V_{H3,C1},\\
S_{H3,C2} &\geq Q_{H3,C2} - Q^V_{H3,C2},\\
S_{H3,W} &\geq Q_{H3,W} - Q^V_{H3,W},
\end{aligned}
$$

and the nonnegativity constraints on the S_{ij} variables. Solving the vertical MILP problem resulted in the following ranking of the four solutions:

Solution	$\sum\sum S_{ij}$
3	3494
4	4096
2	4863
1	6787

8.4 Decomposition Based HEN Synthesis Approaches

In section 8.3.1.4 we discussed the LP transshipment model for the minimum utility cost target. For a given $HRAT$, the solution of the LP transshipment model provides

(i) The minimum required heat loads of hot and cold utilities, and

(ii) The location of all pinch point(s), if any.

Based on (ii), we can decompose the overall HEN problem into subnetworks (i.e., above the top pinch, between top pinch and intermediate pinch, between intermediate pinch points, between bottom pinch and intermediate pinch, and below the bottom pinch) which furthermore can be treated independently.

In section 8.3.2.2 we discussed the MILP transshipment model for the minimum number of matches target. For a given minimum utility cost solution (i.e., given $HRAT$'s, QS_i, QW_j, location of pinch points and hence subnetworks), the solution of the MILP transshipment model, which is applied to each subnetwork provides:

(i) The minimum number of matches, and

(ii) The heat loads of each match.

The same type of information is also provided by the vertical MILP transshipment model discussed in section 8.3.3.3 which discriminates among equivalent number of matches using the assumption of vertical heat transfer.

The key question that arises at this point is how we can determine a HEN configuration that satisfies the criterion of minimum investment cost subject to the information provided by the targets of minimum utility cost and minimum number of matches applied in sequence.

Remark 1 Note that the minimum investment cost should be the desired criterion and not the criterion of minimum area since the minimum investment cost includes area calculations and it is not always true that the minimum area network corresponds to the minimum investment cost network.

8.4.1 Heat Exchanger Network Derivation

8.4.1.1 Problem Statement

Given the

(i) Minimum loads of hot and cold utilities,

(ii) Location of pinch point(s) and hence decomposed subnetworks,

(iii) Minimum number of matches in each subnetwork,

(iv) Heat loads of each match in each subnetwork,

determine a heat exchanger network configuration for each subnetwork that satisfies the criterion of minimum investment cost.

Remark 1 For (i) and (ii) we assumed a values of $HRAT$. For (iii) and (iv) we can have $HRAT = TIAT$ since we decompose into subnetworks based on the location of the pinch point(s). We can also have in (iii) and (iv) $EMAT = TIAT < HRAT$ for each subnetwork which may result in less units, less total area, and less investment cost. By relaxing $EMAT = TIAT$, that is, being strictly less than $HRAT$, more opportunities to make matches are introduced in the heat cascade. Note, however, that for the network derivation we may have $EMAT < HRAT$ of $EMAT = HRAT$ depending on which criterion of feasible heat exchange is considered. In principle, if $EMAT < HRAT$, $EMAT$ can take any small value ϵ close to zero. Also note that $EMAT$ is not a true optimization variable but simply a requirement for feasible heat exchange that can even be relaxed (i.e., may be ϵ from zero). If $EMAT - TIAT = \epsilon > 0$, then the only specification in the above problem statement is that of $HRAT$ based upon which (i) and (ii) are obtained. We will discuss later on how such a specification can be overcome.

In the following section we will discuss the approach proposed by Floudas *et al.* (1986) which allows for automatic derivation of minimum investment cost HEN configurations according to the above problem statement.

8.4.1.2 Basic Idea

The basic idea in the HEN derivation approach of Floudas *et al.* (1986) consists of the following steps:

(i) Showing that there exists a one-to-one correspondence between a match predicted by the MILP transshipment model and a feasible heat exchanger unit.

(ii) Postulating a representation of all desirable alternative HEN configurations (i.e., superstructure), which has as units the matches predicted by the MILP transshipment model, and heat loads calculated from this model.

(iii) Formulating the HEN superstructure as an optimization problem whose solution will provide a minimum investment cost heat exchanger network configuration, that is, it will determine the optimal:

(a) Stream interconnections,

(b) Stream flow rates and temperatures, and

(c) Areas of heat exchanger units.

Remark 1 Steps (i), (ii), and (iii) are applied to each decomposed subnetwork since for a given $HRAT$ we calculated first the minimum utility cost. The configurations of each subnetwork are then combined for the overall HEN configurations.

Step (i) was proved in Appendix A of Floudas *et al.* (1986), and hence the minimum number of matches predicted by the MILP transshipment model can always be represented with an equal number of heat exchanger units in a feasible HEN network.

In the next two section we will focus on steps (ii) and (iii).

8.4.1.3 Derivation of HEN Superstructure

The basic idea is deriving a HEN superstructure is to embed all alternative network structures using a graph-theoretical approach similar to the one described in the process synthesis chapter, in which each unit, input, and output is represented as a node in a graph with two–way arcs between each pair of units and one–way arcs from the inputs to the units/outputs and from the units to the outputs.

The key elements of the HEN superstructure are

(i) Heat exchanger units,

(ii) Mixers at the inlets of each exchanger,

(iii) Splitters at the outlets of each exchanger,

(iv) A mixer at each output stream, and

(v) A splitter at each input stream.

The heat exchanger units are denoted as large circles while the mixers and splitters are denoted as small circles. Note that each heat exchanger units has a mixer at its inlet and a splitter at its outlet.

The important feature of the HEN superstructure is that it consists of individual *stream superstructures* which can be derived independently. Then, the stream superstructures are combined into one overall superstructure which has embedded all configurations for each subnetwork.

Each stream superstructure has one input, one output, a number of heat exchangers equal to the matches that involve the particular stream, and all possible connections from the input to the units, between each pair of units, and from the units to the output. To illustrate such a representation, we will consider the following example that has

(i) One cold stream, $C1$;

(ii) Three hot streams, $H1$, $H2$, and $H3$; and

(iii) The matches $H1 - C1$, $H2 - C1$, $H3 - C1$ taking place in one subnetwork.

The superstructures of each stream $C1$, $H1$, $H2$, $H3$ are shown in Figure 8.15. Note that

(i) Each input stream has a splitter that features of a number of outlet streams equal to the number of heat exchangers that are associated with this stream;

(ii) Each outlet stream has a mixer that features of a number of inlet streams equal to the number of heat exchangers that are associated with this stream;

(iii) Each exchanger has a mixer at its inlet and a splitter at its outlet;

(iv) The mixer at the inlet of each exchanger is connected (i.e., there are inlet streams from) to the input splitter and the splitters of the other heat exchangers;

(v) The splitter at the outlet of each exchanger is connected to the output mixer and the mixers of the other heat exchangers.

Note that the two-way arcs between each pair of exchangers correspond to the arc between the splitter of exchanger 1 directed to the mixer of exchanger 2 and the arc between the splitter of exchanger 2 directed to the mixer of exchanger 1. Note also that the one–way arcs from the input to the units are the arcs from the input splitter directed to the mixers of the heat exchangers. Similarly, the one–way arcs from the units to the output are the arcs between the splitters of the exchangers directed to the output mixer.

By assigning flowrate heat capacities to all streams in the stream superstructures, and selectively setting some of them equal to zero, the incorporation of many interesting structures can be verified. These include the alternatives of

(i) Parallel structure,

(ii) Series structure,

(iii) Parallel-series,

(iv) Series-parallel, and

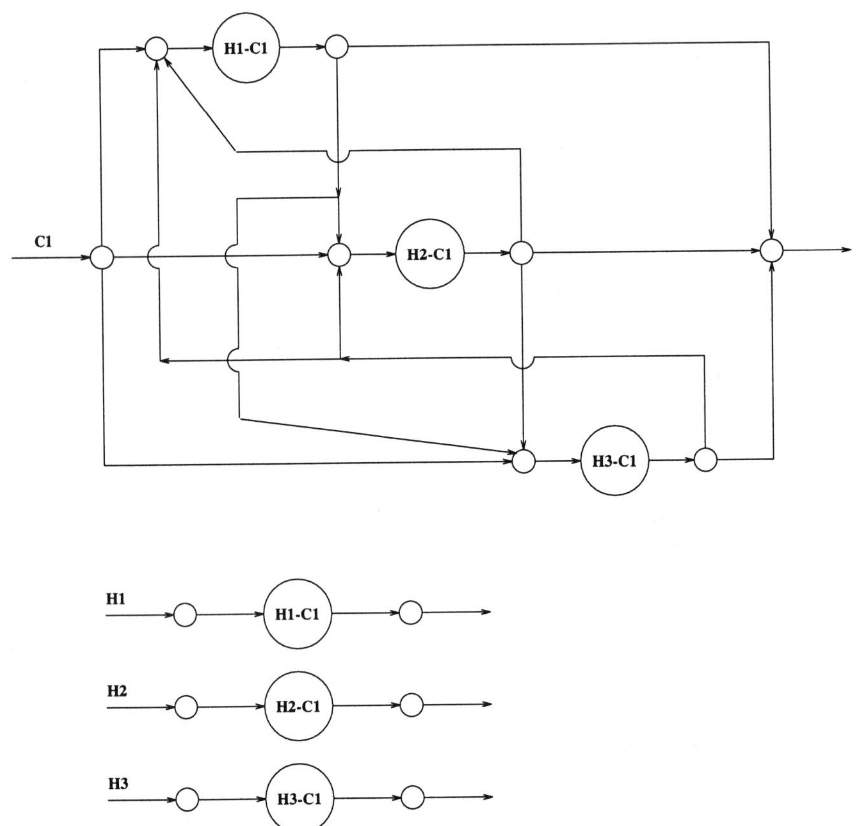

Figure 8.15: Stream superstructures of $C1$, $H1$, $H2$, $H3$.

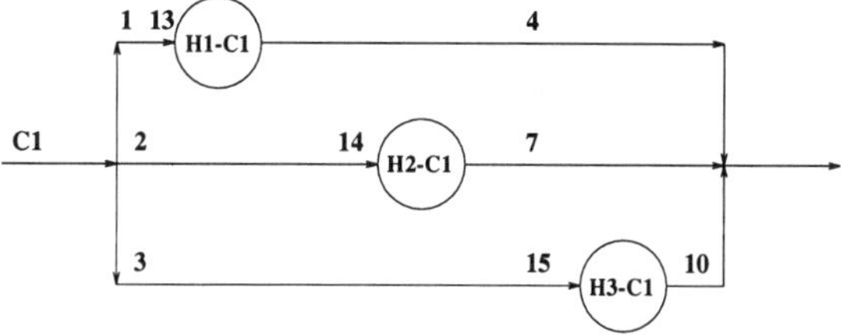

(a) Sequence in parallel

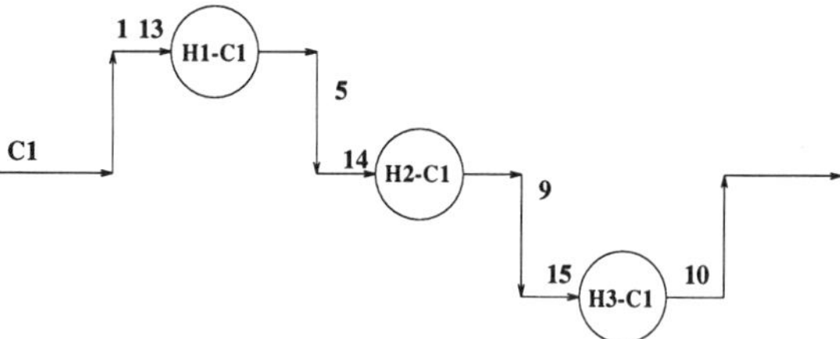

(b) Sequence in series

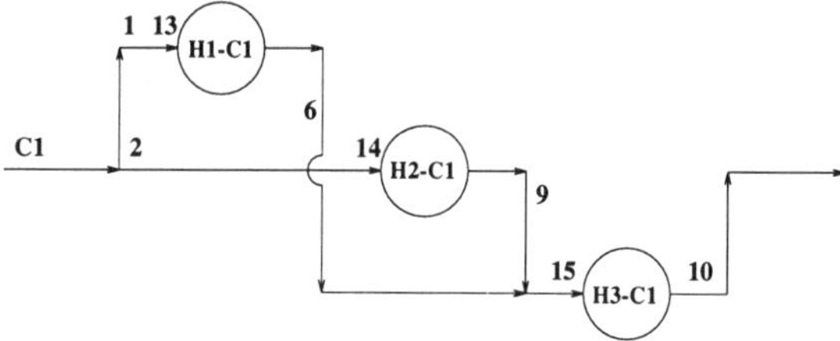

(c) Sequence in parallel-series

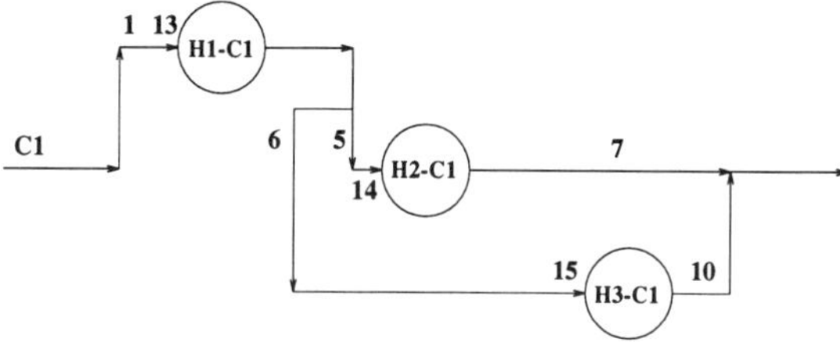

(d) Sequence in series-parallel

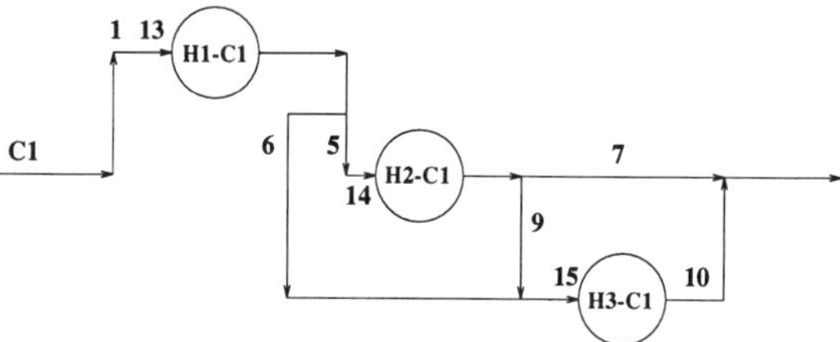

(e) Sequence in series-parallel with by-pass

Figure 8.16: Alternative structures embedded in stream $C1$ superstructure

(v) By-pass on (i), (ii), (iii), and (iv).

Figure 8.16 displays one instance of the above alternative cases.

The question that arises naturally now is how we can extract from the postulated superstructure a minimum investment cost configuration. This is addressed in the next section.

8.4.1.4 Mathematical Formulation of HEN Superstructure

To determine a minimum investment cost structure out of the many alternatives which are embedded in the HEN superstructure, we define variables for all streams of the superstructure to represent:

(i) The flowrate heat capacities of each stream,

(ii) The temperature of each stream, and

(iii) The areas of the heat exchangers.

Subsequently write the constraints of the superstructure which are

(a) Mass balances for the splitters,

(b) Mass balances for the mixers

(c) Energy balances for the mixers and exchangers,

(d) Feasibility constraints for heat exchange, and

(e) Nonnegativity constraints,

along with the objective function; that, is the investment cost of the heat exchangers that are postulated in the superstructure. The objective can be written directly as a function of temperatures, or as a function of the areas, which are defined as a function of temperatures. Note that constraints (a), (b), (d), and (e) are linear constraints, while the energy balance constraints (c) for the mixers and heat exchangers are nonlinear since they have bilinear products of unknown flow rates of the interconnecting streams times their unknown respective temperatures. The objective function is also a nonlinear function of the driving temperature forces expressed in the $(LMTD)$ form. As a result, the resulting mathematical model is a nonlinear programming NLP problem.

An optimal solution of such an optimization model will provide information on all unknown flow rates and temperatures of the streams in the superstructure (i.e., it will automatically determine the network configuration), as well as the optimal areas of the exchangers for a minimum investment cost network.

Instead of presenting the mathematical model for a general superstructure which can be found in Floudas *et al.* (1986), we will illustrate the derivation of such an optimization formulation through the following example:

Stream	T^{in} (K)	T^{in}(K)	FC_p (kW/K)	$h\left(\frac{\text{kW}}{m^2\text{K}}\right)$
$H1$	440	350	22	2
$C1$	349	430	20	2
$C2$	320	368	7.5	0.6667
S	500	500		
CW	300	320		
Cost of heat exchanger = $C_{ij} = 1300 \cdot A_{ij}^{0.6}$ \$/yr ; (A in m²)				

Table 8.6: Data for illustration of nonlinear model

Illustration 8.4.1 (Nonlinear Optimization Model)

This example is taken from Floudas *et al.* (1986) and features one hot, two cold process streams, one hot and one cold utility with data shown in Table 8.6. It is also given that $HRAT = TIAT = EMAT = 10$ K.

Applying the LP transshipment model for the minimum utility cost target yields:

(i) $QS = QW = 0$,

(ii) No pinch $\Rightarrow$ one subnetwork.

Applying the MILP transshipment model for the minimum number of matches target, we obtain:

$$\begin{array}{lclll} H1 - C1 & : & Q_{H1,C1} & = & 1620 \quad \text{kW}, \\ H1 - C2 & : & Q_{H1,C2} & = & 360 \quad \text{kW}. \end{array}$$

Based on the above information we can now postulate all stream superstructures which are shown in Figure 8.17. Note that we have also indicated in these stream superstructures the unknown variables of the interconnecting streams (i.e., flowrates, heat capacities, and temperatures) and the unknown areas of the two heat exchanger units.

We have introduced eight variables for the unknown flowrate heat capacities, (i.e., $F_1 - F_8$), four variables for the unknown intermediate temperatures (i.e., $T3$, $T56$, $T4$, $T78$) and two variables for the unknown areas (i.e., $A_{H1,C1}$, $A_{H1,C2}$).

Then, the resulting nonlinear programming NLP model for the minimum investment cost network is

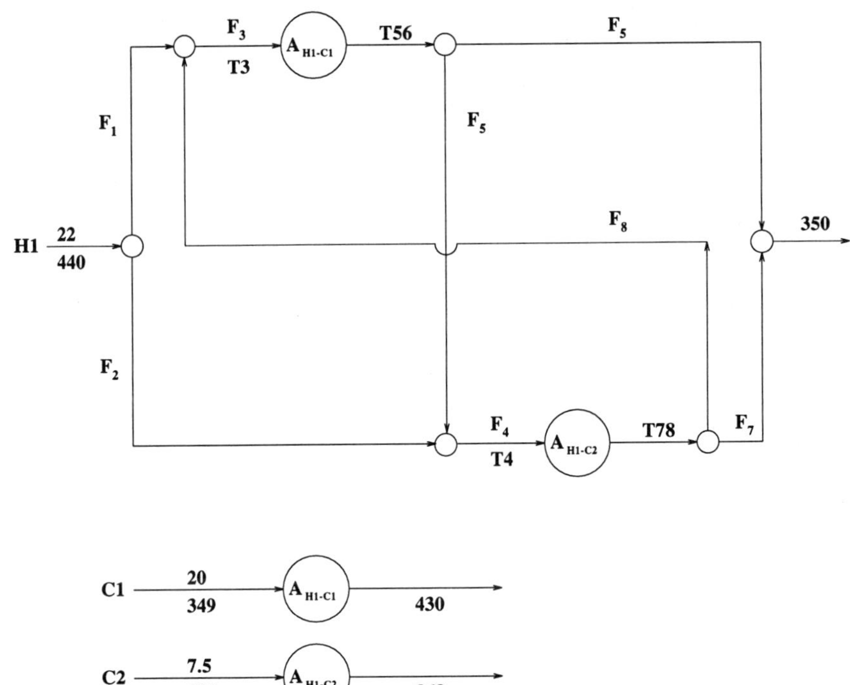

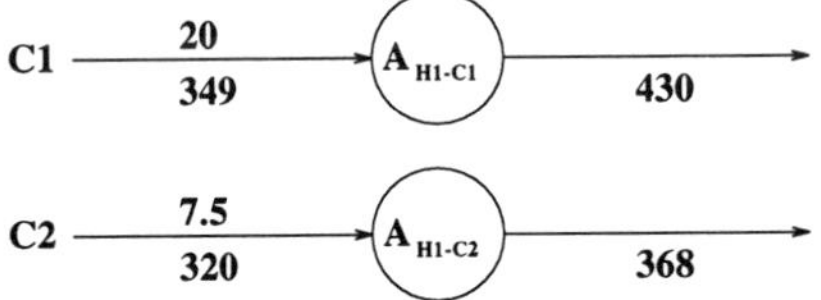

Figure 8.17: Stream superstructures for illustration 8.4.1

$$
\begin{array}{llll}
\min & 1300A^{0.6}_{H1,C1} + 1300A^{0.6}_{H1,C2} & & \\
s.t. & & & \\
& \left.\begin{array}{rcl} F_1 + F_2 & = & 22 \\ F_3 - F_5 - F_6 & = & 0 \\ F_4 - F_7 - F_8 & = & 0 \end{array}\right] & \text{Mass balances for splitters} & \\
& \left.\begin{array}{rcl} F_3 - F_1 - F_8 & = & 0 \\ F_4 - F_2 - F_6 & = & 0 \end{array}\right] & \text{Mass balances for mixers} & \\
& \left.\begin{array}{rcl} F_1 \cdot 440 - F_8 \cdot T78 - F_3 \cdot T3 & = & 0 \\ F_2 \cdot 440 - F_6 \cdot T56 - F_4 \cdot T4 & = & 0 \end{array}\right] & \text{Energy balances for mixers} & (\mathbf{P6}) \\
& \left.\begin{array}{rcl} F_3 \cdot (T3 - T56) & = & 1620 \\ F_4 \cdot (T4 - T78) & = & 360 \end{array}\right] & \text{Energy balances in exchangers} & \\
& \left.\begin{array}{rcl} T3 - 430 & \geq & 10 \\ T56 - 349 & \geq & 10 \\ T4 - 368 & \geq & 10 \\ T78 - 320 & \geq & 10 \end{array}\right] & \text{Feasibility constraints} & \\
& \left.\begin{array}{rcl} F_1, F_2, \ldots, F_8 & \geq & 0 \end{array}\right] & \text{Nonnegativity constraints} &
\end{array}
$$

where the areas in the objective function are defined as

$$
\begin{aligned}
A_{H1,C1} &= \frac{1620}{(1.0)(LMTD)_{H1,C1}}, \\
A_{H1,C2} &= \frac{360}{(0.5)(LMTD)_{H1,C2}}, \\
(LMTD)_{H1,C1} &= \frac{(T3 - 430) - (T56 - 349)}{\ln\left[(T3 - 430)/(T56 - 349)\right]}, \\
(LMTD)_{H1,C2} &= \frac{(T4 - 368) - (T78 - 320)}{\ln\left[(T4 - 368)/(T78 - 320)\right]}.
\end{aligned}
$$

Remark 1 Model **P6** corresponds to a nonlinear programming NLP problem since it has nonlinear objective and nonlinear constraints corresponding to the energy balances. It is important to note that **P6** always has a feasible solution because of the one-to-one correspondence property between matches and heat exchanger units (see Appendix A of Floudas *et al.* (1986)). This is an interesting feature on the grounds that we cannot know *a priori* whether any given NLP model has a feasible solution. Floudas *et al.* (1986) (see Appendix B) showed that increasing the flows of the recycle streams tends to result in an increase of the objective function. Since we minimize the objective function, then the model **P6** will try to set such flows to zero. This is also a desirable feature of **P6** since relatively simple structures will result from its solution that correspond to practically useful configurations.

Remark 2 Floudas and Ciric (1989) studied the mathematical structure of the objective and constraints of **P6** with target to attain the global solution of **P6**. As a result, they identified a number of key properties that are as follows:

Property 8.4.1
The objective function of **P6** *is convex (i.e., for a given set of matches and heat loads).*

Property 8.4.2
If the following heat capacities are fixed then the energy balances form a square system of linear equations. Furthermore, if the determinant of this system is not zero, then it has one unique solution with respect to the temperatures.

Floudas and Ciric (1989) present conditions under which the determinant is not zero.

Remark 3 The NLP model **P6** is nonconvex because of the bilinear equalities that correspond to the energy balances for mixers and exchangers. As a result, use of local NLP solvers (e.g., MINOS) yield local solutions. Floudas and Ciric (1989) showed that even for simple systems the global solution of **P6** may not be found and proposed the Global Optimal Search approach (Floudas *et al.*, 1989) for improving the chances of obtaining the global solution of **P6**. Note however that obtaining a global solution of **P6** can only be theoretically guaranteed via the use of global optimization algorithms (e.g., the **GOP** algorithm of Floudas and Visweswaran (1990; 1993), Visweswaran and Floudas (1990; 1992; 1993). Application of the **GOP** algorithm to heat exchanger network problems as well as a variety of design/control problems can be found in Visweswaran (1995).

Remark 4 In the considered example we have a fixed $EMAT = 10$ since we were given that $HRAT = TIAT = EMAT = 10$. Note, however, that EMAT participates linearly in the feasibility constraints. As a result, we can relax it in a straightforward fashion, which is to write the feasibility constraints as

$$
\begin{aligned}
T3 - 430 &\geq EMAT, \\
T56 - 349 &\geq EMAT, \\
T4 - 368 &\geq EMAT, \\
T78 - 320 &\geq EMAT, \\
EMAT &\geq \epsilon = 0.1.
\end{aligned}
$$

This way, we eliminate EMAT as a variable, and we simply use some lower bounds of $\epsilon = 0.1$.

Remark 5 We can also treat the heat loads of each match as variables since they participate linearly in the energy balances. The penalty that we pay, however, is that the objective function no longer satisfies the property of convexity, and hence we will have two possible sources of nonconvexities: the objective function and the bilinear equality constraints.

Remark 6 Note that when the driving forces at both ends of an exchanger are equal, then the $LMTD$ calculation may cause numerical difficulties arising because of division by zero. To avoid such numerical difficulties, a number of good approximations have been developed for the $LMTD$ calculation. These include:

(i) **The Paterson (1984) approximation:**

$$LMTD = \frac{2}{3}(\Delta T_1 \cdot \Delta T_2)^{\frac{1}{2}} + \frac{1}{3}\left(\frac{\Delta T_1 + \Delta T_2}{2}\right),$$

where

$$\begin{aligned} \Delta T_1 &= T^H_{in} - T^C_{out}, \\ \Delta T_2 &= T^H_{out} - T^C_{in}. \end{aligned}$$

(ii) **The Chen (1987) approximation:**

$$LMTD = \left[\Delta T_1 \cdot \Delta T_2 \cdot \left(\frac{\Delta T_1 + \Delta T_2}{2}\right)\right]^{\frac{1}{3}}.$$

The Chen approximation has the advantage that when either ΔT_1 or ΔT_2 equals zero, then the $LMTD$ is approximated to be zero, while the Paterson (1984) approximation yields a nonzero value. The Paterson (1984) approximation tends to slightly underestimate the area, while the Chen (1987) slightly overestimates the actual area.

Solving **P6** using GAMS/MINOS we determine a minimum investment cost network configuration automatically. The solution of **P6** is:

$$\begin{aligned} F_1 &= 20, \\ F_2 &= 2, \\ F_3 &= 20, \\ F_4 &= 8.125, \\ F_5 &= 13.875, \\ F_6 &= 6.125, \\ F_7 &= 8.125, \\ F_8 &= 0, \\ T3 &= 440\text{K}, \\ T4 &= 378.9\text{K}, \\ T56 &= 359\text{K}, \\ T78 &= 334.64\text{K}, \\ A_{H1,C1} &= 162\text{m}^2, \\ A_{H1,C2} &= 56.717\text{m}^2, \end{aligned}$$

which yields the HEN shown in Figure 8.18.

Note that the optimum network shown in Figure 8.18 features:

(i) An input splitter,

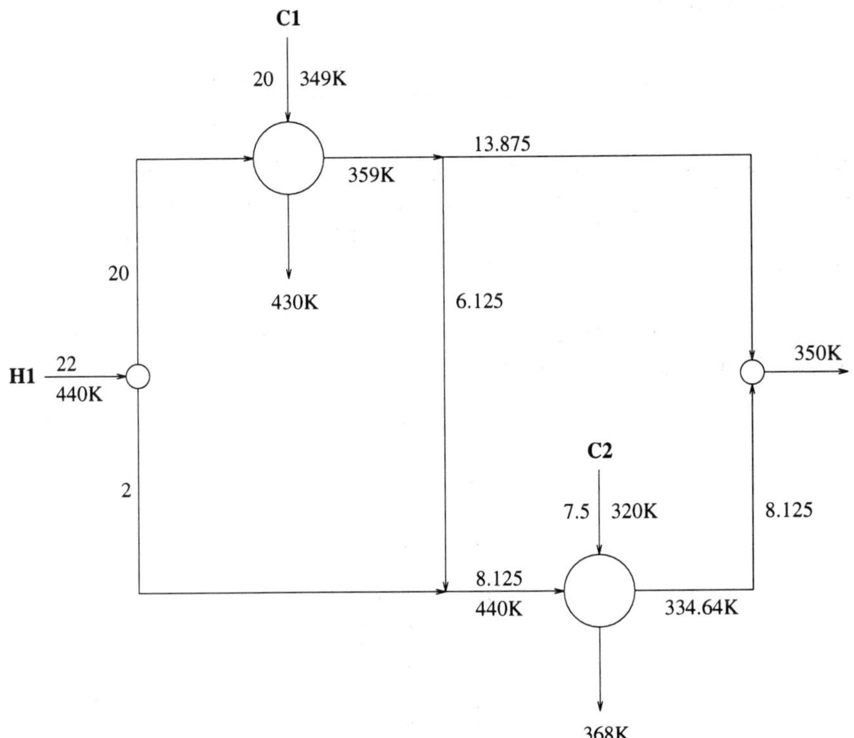

Figure 8.18: Minimum investment cost network

(ii) An exchanger $H1 - C1$ splitter,

(iii) An exchanger $H1 - C2$ mixer, and

(iv) A connecting stream form the splitter of $H1 - C1$ to the mixer of $H1 - C2$ exchanger.

Stream of flow rate $F2$ can be regarded as a by-pass of the $H1 - C1$ exchanger, while stream of flow rate $F5$ can be considered as a by-pass of the $H1 - C2$ exchanger.

Remark 7 Providing upper and lower bounds on all optimization variables is very important for avoiding numerical difficulties with any available nonlinear optimization solver. We can determine upper and lower bounds on the variables of **P6** through application of the constraints as follows:

<u>*Upper bounds on $T3$, $T4$*</u>

These are based on the input temperature of $H1$ and are

$$\begin{aligned} T3 &\leq 440, \\ T4 &\leq 440. \end{aligned}$$

<u>*Upper bounds on $T56$, $T78$*</u>

These are based on the minimum temperature drop between the inlet and outlet of each heat exchanger. Such a minimum temperature drop can be calculated if all input flow rate goes through the exchanger:

$$\begin{aligned} H1 - C1 &: T56 \leq 440 - \tfrac{1620}{22} = 366.3636, \\ H1 - C2 &: T78 \leq 440 - \tfrac{360}{22} = 423.6363. \end{aligned}$$

<u>*Lower bounds on $T3$, $T4$, $T56$, $T78$*</u>

These can be obtained from the feasibility constraints for this example and are

$$\begin{aligned} T3 &\leq 440, \\ T56 &\leq 359, \\ T4 &\leq 378, \\ T78 &\leq 330. \end{aligned}$$

Since $440 \leq T3 \leq 440$ we have

$$T3 = 440.$$

This implies that we cannot have a connecting stream from the splitter of exchanger $H1 - C2$ to the mixer of exchanger $H1 - C1$; that is, we will have

$$F8 = 0.$$

This simplifies the stream superstructure which is shown in Figure 8.19. This way, we eliminate the splitter of $H1 - C2$ exchanger. As a result, the model **P6** can be reduced by

(i) One variable F8 which is eliminated,

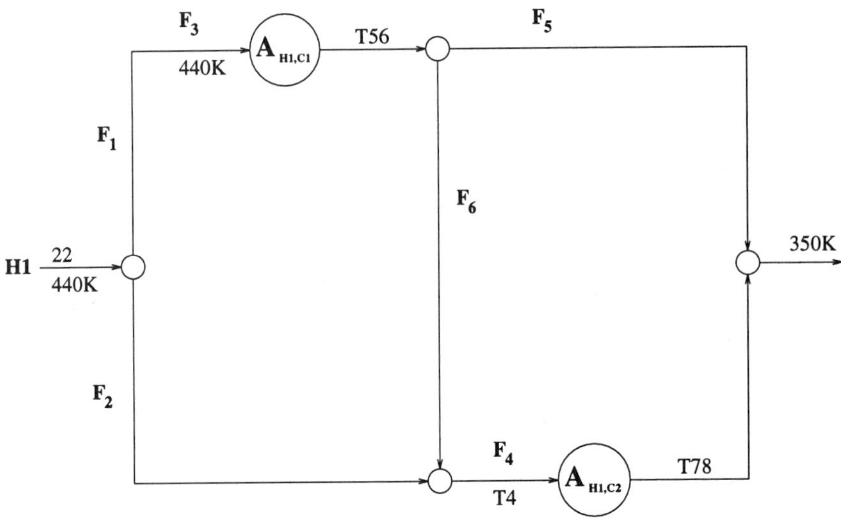

Figure 8.19: Simplified superstructure of $H1$ based on bound calculations

(ii) One mass balance for splitters of $H1 - C2$,

(iii) One mass balance for mixer of $H1 - C1$,

(iv) One energy balance for mixer of $H1 - C1$,

(v) One feasibility constraint of $H1 - C1$ which is automatically satisfied.

Note that the reduced model has one less bilinear constraint, which is very desirable since it represents a reduction in the constraints that are responsible for nonconvexities. Note also that from (iii) we can eliminate $F1$ since $F1 = F3$. Similarly, from (ii) we can eliminate $F7$ since $F4 = F7$. Therefore, the number of flowrate and heat capacity variables is reduced from 8 to 5, while the number of temperatures is reduced from 4 to 3.

Note that since the outlet temperature of $H1$ is 350 K, there are three alternative cases for the outlet temperatures of the two exchangers:

Case I: $T56 = T78 = 350\text{K}$,

Case II: $T56 < 350 < T78$,

Case III: $T78 < 350 < T56$.

Case I is denoted as *isothermal mixing*, and if it is assumed that it takes place from the very beginning, then it greatly simplifies the problem. In particular, the energy balances of the mixers and exchanger can be combined so as to eliminate completely the nonlinearities as follows:

$$F_3(T_3 - 350) = 1620 \quad \Rightarrow \quad F_3 \cdot T3 = 1620 + 350 \cdot F_3,$$
$$F_4(T_4 - 450) = 1620 \quad \Rightarrow \quad F_4 \cdot T4 = 360 + 350 \cdot F_4.$$

By substituting $F_3 \cdot T3$, $F_4 \cdot T4$ in the energy balances for the exchangers we have linear constraints:

$$\begin{aligned} F_1 \cdot 440 - F_3 \cdot 350 &= 1620, \\ F_2 \cdot 440 - F_6 \cdot 350 - F_4 \cdot 350 &= 360. \end{aligned}$$

Note, however, that cases I and II cannot take place because if $T56 \leq 350$, then the feasibility constraint

$$T56 - 349 \geq 10,$$

is violated. Hence only case III can take place.

Based on this analysis, we can reduce the feasible region by incorporating the constraint of case III

$$T78 < 350 < T56,$$

which in conjunction with the upper and lower bounds on the temperatures represent a very tight representation; that is,

$$330 \leq T78 \leq 350,$$

instead of

$$330 \leq T78 \leq 423.64.$$

After all the aforementioned analysis on obtaining tighter bounds on the temperatures, which resulted in eliminating streams from the HEN superstructure, the mathematical model **P6** becomes:

$$\begin{aligned} \min \quad & 1300 A_{H1,C1}^{0.6} + 1300 A_{H1,C2}^{0.6} \\ s.t. \quad & \\ & F_3 + F_2 = 22 \\ & F_3 - F_5 - F_6 = 0 \\ & F_4 - F_2 - F_6 = 0 \\ & F_2 \cdot 440 + F_6 \cdot T56 - F_4 \cdot T4 = 0 \\ & F_3 \cdot (440 - T56) = 1620 \\ & F_4 \cdot (T4 - T78) = 360 \\ & 378 \leq T4 \leq 440 \\ & 330 \leq T78 \leq 350 \\ & 359 \leq T56 \leq 366.3636 \\ & F_2, F_3, F_4, F_5, F_6 \geq 0 \end{aligned} \qquad \textbf{(P7)}$$

where $A_{H1,C1}$, $A_{H1,C2}$ are defined as in **P6**.

Note that the feasibility constraints are incorporated in the lower bound constraints of the temperatures.

Remark 8 Having obtained lower and upper bounds on the temperatures, we can now use the energy balances so as to obtain tighter lower and upper bounds on the flow rates F_3, F_4 through the exchanger as follows:

Upper Bounds on F_3, F_4

$$F_3^U = \frac{1620}{(440 - T56)^L} = \frac{1620}{(440 - T56^U)} = \frac{1620}{(440 - 366.36)} = 22,$$

which is already known since $F_3 + F_2 = 22$.

$$F_4^U = \frac{360}{(T4 - T78)^L} = \frac{360}{T4^L - T78^U} = \frac{360}{378 - 350} = 12.857.$$

Note that in the calculation of F_4^U we must maintain that:

$$T4^L - T78^U \geq 0,$$

since $T4 - T78 \geq 0$.

Lower Bounds on F_3, F_4

$$F_3^L = \frac{1620}{(440 - T56)^U} = \frac{1620}{(440 - T56^L)} = \frac{1620}{(440 - 359)} = 20,$$
$$F_4^L = \frac{360}{(T4 - T78)^U} = \frac{360}{T4^U - T78^L} = \frac{360}{440 - 330} = 3.2727.$$

Then the bounds on F_3, F_4 are

$$20 \leq F_3 \leq 22,$$
$$3.2727 \leq F_4 \leq 12.857.$$

Note that since the upper bound of F_4 is $12.857 < 22$, then we must have the stream of flow rate F_5 activated because the total flow rate of 22 cannot go through the $H1 - C2$ exchanger.

The incorporation of all such obtained bounds in the model is of crucial importance in obtaining good quality local solutions of the minimum investment cost HEN problem. The final NLP model takes the form:

$$
\begin{aligned}
& \min 1300 A_{H1,C1}^{0.6} + 1300 A_{H1,C2}^{0.6} \\
s.t.\quad & F_3 + F_2 = 22 \\
& F_3 - F_5 - F_6 = 0 \\
& F_4 - F_2 - F_6 = 0 \\
& F_2 \cdot 440 - F_6 \cdot T56 - F_4 \cdot T4 = 0 \\
& F_3 \cdot (440 - T56) = 1620 \\
& F_4 \cdot (T4 - T78) = 360 \qquad\qquad \textbf{(P8)} \\
& 378 \leq T4 \leq 440 \\
& 330 \leq T78 \leq 350 \\
& 359 \leq T56 \leq 366.3636 \\
& 20 \leq F_3 \leq 22 \\
& 3.2727 \leq F_4 \leq 12.857 \\
& F_2, F_5, F_6 \geq 0
\end{aligned}
$$

Note that bounds on F_2, F_5, F_6 can also be obtained by using the linear mass balances and the bounds on F_3 and F_4 as follows

$$
\begin{aligned}
0 \leq\; & F_2 \leq 2, \\
9.143 \leq\; & F_5 \leq 18.7273, \\
1.2727 \leq\; & F_6 \leq 10.857,
\end{aligned}
$$

which can be included in **P8**.

The important message from this analysis that is done prior to solving the NLP model is that we can eliminate part of the superstructure and obtain tight bounds on the flowrates and temperatures that help significantly the NLP solver to attain good quality solutions. All this analysis corresponds to monotonicity checks and can be performed automatically within the GAMS modeling system.

8.4.2 HEN Synthesis Strategy

The basic idea in the HEN synthesis strategy of Floudas *et al.* (1986) is to decompose the problem into the tasks of minimum utility first, minimum number of matches for the obtained utility target second, and network derivation for minimum investment cost that satisfies the first two tasks third. The rationale behind such a decomposition approach is that if we consider the total cost consisting of the utility and investment cost, then tasks 1 and 3 minimize the individual components of utility and investment cost, while the second task of minimum number of units exploits the economies of scale. The assumptions are that the energy cost is dominant, and that the minimum number of units solution that meets the minimum energy demend is close to the minimum investment solution. Figure 8.20 presents the steps of the HEN synthesis approach along with the optimization of the $HRAT$.

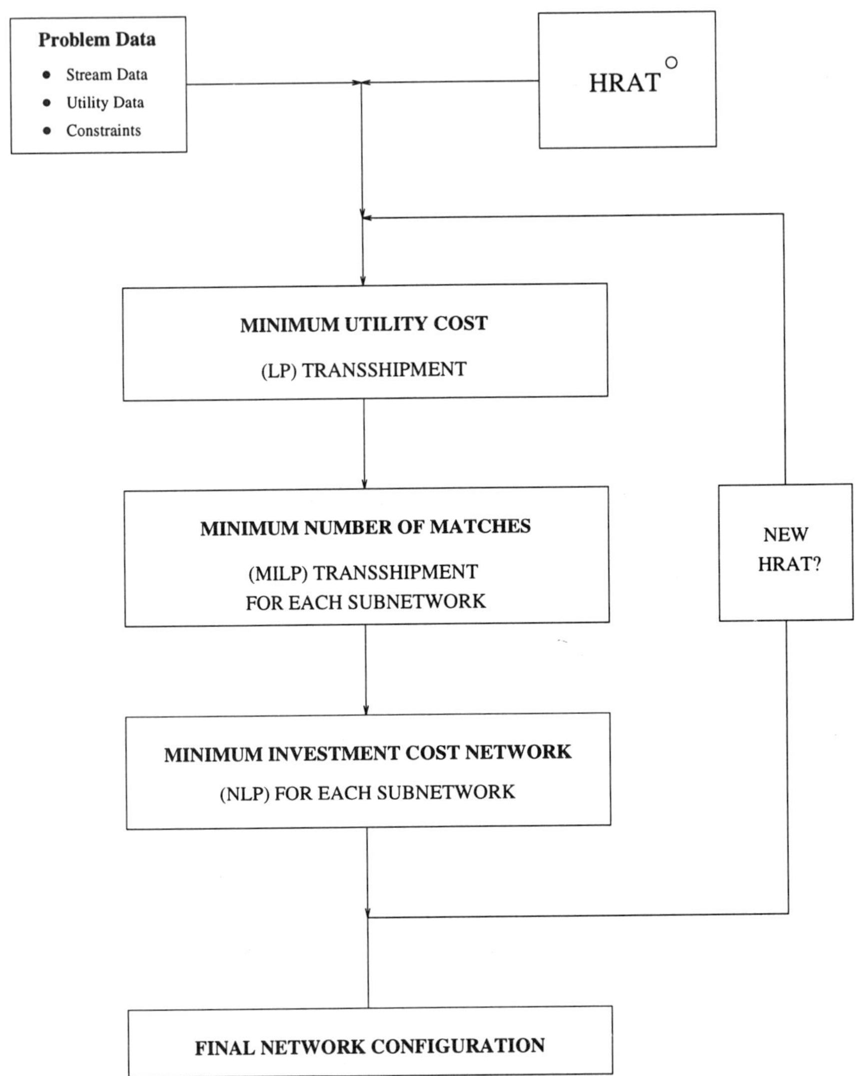

Figure 8.20: Steps of decomposition based HEN synthesis approach

For the given initial value of $HRAT = HRAT^o$ and the problem data (i.e., stream and utility data, constraints), the LP transshipment model is solved in the first step which provides information on the minimum loads of hot and cold utilities and the location of the pinch point(s). The pinch point(s) decompose the problem into subnetworks.

For each subnetwork, the MILP transshipment model is solved to determine the matches and the heat loads of each match. If more than one global solution exists then all global minimum number of matches solutions can be generated via the appropriate integer cuts. Note that in the modeling of the MILP we can relax $TIAT = EMAT = \epsilon$ where ϵ is a small positive number (e.g., $\epsilon = 0.1$).

For the global optimal solution(s) of the MILP transshipment model a HEN superstructure is postulated and formulated as an NLP problem whose solution provides a minimum investment cost network configuration. Note also that the $EMAT$ can be relaxed such that $EMAT = \epsilon$ if no other specific actions are imposed.

The only parameter that has been fixed in the above three sequential stages is the $HRAT$. We can subsequently update the $HRAT$ by performing a one-dimensional search using the golden section search method, which is shown as the outside loop in Figure 8.20.

The first implementation of this HEN synthesis strategy is reported as the program MAGNETS (Floudas *et al.*, 1986) in which the relaxation of $TIAT = EMAT = \epsilon$ for the MILP transshipment model and $EMAT = \epsilon$ for the NLP model were not included. Instead a fixed $HRAT$ was used. MAGNETS accepts input data either interactively or with free format data files. The user can study several networks since there are options for performing the calculations for different values of $HRAT$, options for imposing restricted matches, or options of generating automatically network configurations.

Remark 1 Note that in the minimum number of matches calculation, the vertical MILP transshipment could be used as a heuristic for ranking potential multiple global solutions of the MILP transshipment model of Papoulias and Grossmann (1983).

8.5 Simultaneous HEN Synthesis Approaches

In Section 8.4 we discussed decomposition-based HEN synthesis approaches that feature three separate tasks to be performed sequentially: (i) minimum utility cost, (ii) minimum number of matches, and (iii) minimum investment cost network configuration. Such a decomposition was motivated by the discovery of the pinch point on the one hand and by our inability in the 1980's to address the HEN synthesis problem as a single task problem. Application of such sequential synthesis approaches in many case studies resulted in good quality networks with respect to the total annualized cost which is a strong indication of the clever decomposition scheme.

The primary limitation however of sequential synthesis methods is that different costs associated with the design of HENs cannot be optimized simultaneously, and as a result the trade-offs are not taken into account appropriately. Early decisions in the selection of $HRAT$ and partitioning into subnetworks affect the number of units and areas of the units in the HEN configuration. Therefore, sequential synthesis methods can often lead to suboptimal networks.

Each task of the sequential approach introduces an element of difficulty toward attaining the minimum total annualized cost HEN configuration. The difficulty that arises in the minimum investment cost is due to the nonconvex NLP model which may exhibit several local solutions. Floudas and Ciric (1989) proposed an approach that searches for the global solution (even though it is not guaranteed) efficiently, and Visweswaran and Floudas (1995) applied the global optimization algorithm **GOP** that offers theoretical guarantee of global optimality.

The difficulty that arises in the selection of matches task is due to having multiple feasible combinations of matches which satisfy the minimum number of matches target. Discrimination among them can be achieved only via the vertical MILP model which however assumes the vertical heat transfer criterion.

The difficulty that arises in the utility targeting task is due to two sources. First, the requirement of a strict pinch assumes that energy is the dominant cost contribution and hence lower total cost solutions in which heat is allowed to be transferred across the pinch (i.e., pseudo-pinch case) may be excluded from consideration. Second, the specification of $HRAT$ requires iterating between several values of $HRAT$ to determine the optimal network and hence needs a significant amount of time and effort.

In this section we will focus on addressing the difficulties arising in the first two tasks of sequential HEN synthesis and we will discuss simultaneous approaches developed in the early 90s. More specifically, in section 8.5.1 we will discuss the simultaneous consideration of minimum number of matches and minimum investment cost network derivation. In section 8.5.2 we will discuss the *pseudo-pinch* concept and its associated simultaneous synthesis approach. In section 8.5.3, we will present an approach that involves no decomposition and treats $HRAT$ as an explicit optimization variable. Finally, in section 8.5.4, we will discuss the development of alternative simultaneous optimization models for heat integration which address the same single-task HEN problem as the approach of section 8.5.3.

8.5.1 Simultaneous Matches–Network Optimization

8.5.1.1 Problem Statement

Given the information provided by the minimum utility cost calculation (i.e., minimum loads of hot and cold utilities, location of pinch point(s)) determine the following:

(i) Matches that should take place,

(ii) Heat loads of each match, and

(iii) Heat exchanger network structure that satisfies the minimum investment cost criterion.

Remark 1 As can be seen in Figure 8.20, the above problem statement corresponds to the simultaneous consideration of the second and third rectangle inside the loop of the $HRAT$ optimization. The key feature is that we do not need to partition the HEN problem into subnetworks based on the location of its pinch point(s). Instead, we postulate all possible matches that may take place and optimize simultaneously for the selection of matches and their heat loads, as well as the minimum investment cost network structure.

Remark 2 Since the target of minimum number of matches is not used as a heuristic to determine the matches and heat loads with either the MILP transshipment model or the vertical MILP transshipment model, the above problem statement addresses correctly the simultaneous matches-network optimization.

Remark 3 Note that since we do not partition into subnetwork but treat the HEN problem as one network we may not have one-to-one correspondence between matches and heat exchangers units in the case of pinch point(s) taking place. In this case one match may have to take place in more than one (e.g., two) heat exchangers. This, however, can be remedied by simply postulating in the possible set of matches two heat exchangers for the matches that can take place across the pinch point(s) and let the optimization model identify whether there is a need for one, two or none of these units.

In the following section, we will discuss the approach proposed by Floudas and Ciric (1989) for simultaneous matches-network optimization.

8.5.1.2 Basic Idea

The basic idea in the simultaneous matches-network optimization approach of Floudas and Ciric (1989) consists of

(i) Postulating a representation of all possible matches and all possible alternative HEN configurations which is denoted as *hyperstructure*.

(ii) Formulating the matches HEN hyperstructure as a mixed-integer nonLinear programming MINLP problem whose solution will provide a minimum investment cost heat exchanger network configuration along with the:

   (a) Matches,

   (b) Heat loads of each match,

   (c) Stream interconnections,

   (d) Stream flow rates and temperatures, and

   (e) Areas of heat exchanger units.

Remark 1 Steps (i) and (ii) are applied to the overall HEN without decomposing it into subnetworks. It is assumed, however, that we have a fixed $HRAT$ for which we calculated the minimum utility cost. The $HRAT$ can be optimized by using the golden section search in the same way that we described it in Figure 8.20.

In the next two sections we will present in detail the hyperstructure generation (i.e., step (i)) and the optimization model (i.e., step (ii)).

8.5.1.3 Derivation of Matches – HEN Hyperstructure

The basic idea in postulating a Matches–HEN hyperstructure is to simultaneously embed:

Stream	T^{in} (K)	T^{out}(K)	FC_p (kW/K)
$H1$	450	350	12
$H2$	400	320	8
$C1$	340	420	10
$C2$	340	400	8
CW	300	320	22
$HRAT = 10 = TIAT = EMAT$			
$U = 0.8$kW/m^2K			
Cost of heat exchanger $= C_{ij} = 1300 \cdot A^{0.6}$ \$/yr ; (Ain m^2)			

Table 8.7: Data for illustrative example 8.5.1

(i) All alternative matches, and

(ii) All alternative network configurations.

Since we do not know, after the minimum utility cost calculation, the matches and their heat loads, we postulate all possible matches in (i) based upon the inlet and outlet temperatures of the streams. Then, having postulated the matches, we derive in (ii) all stream superstructures in the same way that we discussed in section 8.4.1.3 so as to include all potential flow patterns.

Remark 1 The primary difference between the hyperstructure and the superstructure of Floudas *et al.* (1986) is that the superstructure presented on section 8.4.1.3 involved only the matches selected by the MILP transshipment model which represents the minimum number of matches criterion. In addition to the matches, we also have information on the heat loads of each match which is not the case with the hyperstructure where we do not know the matches, as well as their heat loads.

Illustration 8.5.1 (Hyperstructure)

This example is a modified version of example 4 of Floudas and Ciric (1989) and features two hot, two cold streams, and one cold utility with data shown in Table 8.7.

For $HRAT = 10$ K, there is no pinch point and the heat load of the cold utility is $QW = 440$ kW. As a result in this case we have one-to-one correspondence between matches and heat exchanger units.

In Table 8.7 we have that $HRAT = TIAT = EMAT = 10$ K. Note, however, that the general case of $TIAT = EMAT = \epsilon < HRAT = 10$ K can be treated with the same methodology, as we will see in the next section.

From the inlet and outlet temperatures of the streams we have the following possible six matches:

$$\begin{array}{lcl} H1 & - & C1, \\ H1 & - & C2, \\ H1 & - & CW, \\ H2 & - & C1, \\ H2 & - & C2, \\ H2 & - & CW. \end{array}$$

As a result, the hyperstructure consists of stream superstructures in which $H1$ has three matches, $H2$ has three matches, $C1$ has two matches, $C2$ has two matches, and CW has two matches.

For simplicity of the presentation of the stream superstructures we will assume that the matches with the cold utility take place after the matches of the process streams. This way, the $H1$ and $H2$ superstructures will have two process matches with all possible interconnections and the cold utility match following in a series arrangement.

The hyperstructure for the illustrative example is shown in Figure 8.21 The variables also shown in Figure 8.21 are the unknown flow rates and temperatures of the streams. Note that the notation is based on capital letters for the hot streams (i.e., F, T) and lower case letters for the cold streams (i.e., f, t), while the superscripts denote the particular hot and cold streams.

As can be seen in Figure 8.21 we have six possible matches while the minimum number of units is four. In addition, we have all possible interconnections among the postulated six matches. Note that we do not need to include the superstructure for the cold utility since we can determine its individual flow rate from the heat loads to be calculated from the optimization model.

8.5.1.4 Mathematical Formulation of HEN Hyperstructure

To identify a minimum investment cost configuration out of the many embedded alternatives in the Matches-HEN hyperstructure, we define variables for two major components:

(i) The heat flow MILP transshipment model, and

(ii) The hyperstructure topology model.

In (i) we partition the temperature into intervals using $TIAT$ and introduce variables for the

(a) Existence of heat exchange in a pair of streams (i, j) (i.e., $y_{ij} = 0 - 1$);

(b) Heat exchange in a match (ij) at interval k (i.e., Q_{ijk});

(c) Heat loads of each match (i.e., Q_{ij}); and

(d) Heat residuals exiting each temperature interval k (i.e., R_k).

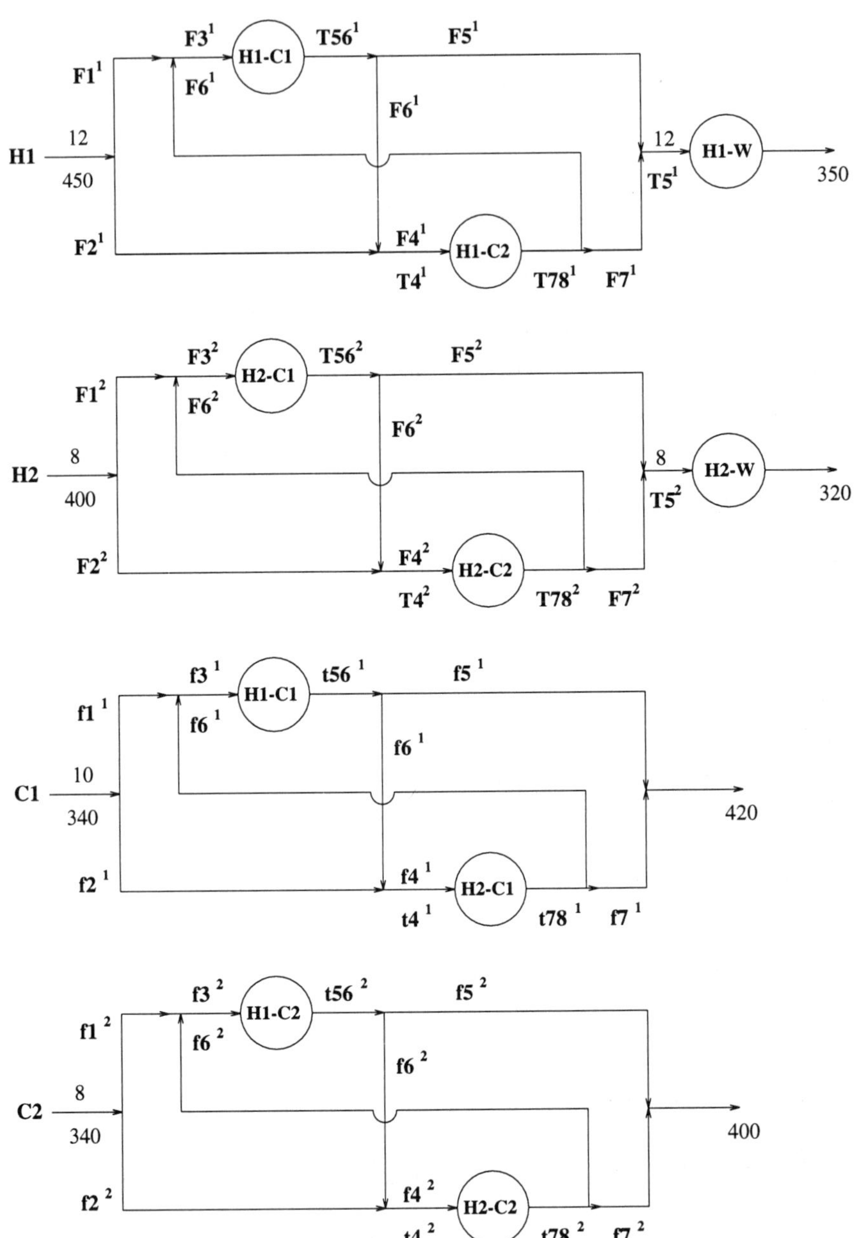

Figure 8.21: Hyperstructure for illustrative example 8.5.1

In (ii) we introduce variables for

(a) The flow rate heat capacities of each stream;

(b) The temperature of each stream; and

(c) The areas of the heat exchangers.

The set of constraints for the simultaneous Matches–HEN hyperstructure will then feature:

(A) The MILP transshipment model,

(B) The hyperstructure topology model plus the areas definitions, and

(C) The logical relationships between inlet flow rates and binary variables.

The model in (A) is identical to the MILP transshipment model described in section 8.3.2. The model in (B) consists of

(b.1) Mass balances for the splitters,

(b.2) Mass balances for the mixers,

(b.3) Energy balances for mixers and exchangers,

(b.4) Feasibility constraints for heat exchange,

(b.5) Utility load constraint imposed by the minimum utility cost calculation, and

(b.6) Nonnegativity and bound constraints on the flow rates.

We will next discuss (A), (B), and (C) via the illustrative example.

Illustration 8.5.2 (Optimization Model)

The complete optimization model for the illustrative example consists of parts (A), (B), and (C) which are presented in the following:

Part (A): The MILP transshipment model

To formulate the MILP transshipment model we first partition the whole temperature range into temperature intervals, which are shown in Figure 8.22, based on $TIAT = EMAT = 10\text{K} = HRAT$.

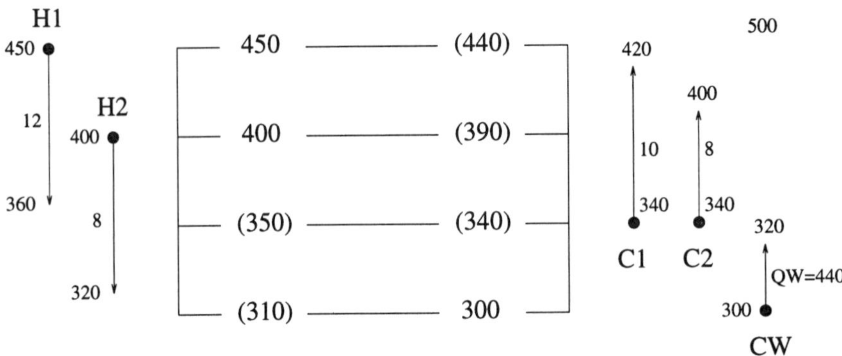

Figure 8.22: Temperature partitioning for illustration 8.5.2

Then the constraints of the MILP transshipment model take the following form:

$$\left.\begin{array}{rl} R_{H1,1} + Q_{H1,C1,1} + Q_{H1,C2,1} &= 600 \\ Q_{H1,C1,1} &= 300 \\ Q_{H1,C2,1} &= 80 \\ R_1 - R_{H1,1} &= 0 \end{array}\right] \quad TI-1,$$

$$\left.\begin{array}{rl} R_{H1,2} - R_{H1,1} + Q_{H1,C1,2} + Q_{H1,C2,2} &= 480 \\ R_{H2,2} + Q_{H2,C1,2} + Q_{H2,C2,2} &= 400 \\ Q_{H1,C1,2} + Q_{H2,C1,2} &= 500 \\ Q_{H1,C2,2} + Q_{H2,C2,2} &= 400 \\ R_2 - R_{H1,2} - R_{H2,2} &= 0 \end{array}\right] \quad TI-2,$$

$$\left.\begin{array}{rl} -R_{H1,2} + Q_{H1,W,3} &= 0 \\ -R_{H2,2} + Q_{H2,W,3} &= 240 \end{array}\right] \quad TI-3,$$

$$\left.\begin{aligned}
Q_{H1,C1} &= Q_{H1,C1,1} + Q_{H1,C1,2} \\
Q_{H1,C2} &= Q_{H1,C2,1} + Q_{H1,C2,2} \\
Q_{H1,W} &= Q_{H1,W,3} \\
Q_{H2,C1} &= Q_{H2,C1,2} \\
Q_{H2,C2} &= Q_{H2,C2,2} \\
Q_{H2,W} &= Q_{H2,W,3}
\end{aligned}\right] \quad \text{Definitions of heat loads,}$$

$$Q_{H1,W,3} + Q_{H2,W,3} = 440] \quad \text{Utility cost constraint,}$$

$$\left.\begin{aligned}
Q_{H1,C1} &\leq U_{H1,C1} \cdot y_{H1,C1} \\
Q_{H1,C2} &\leq U_{H1,C2} \cdot y_{H1,C2} \\
Q_{H1,W} &\leq U_{H1,W} \cdot y_{H1,W} \\
Q_{H2,C1} &\leq U_{H2,C1} \cdot y_{H2,C1} \\
Q_{H2,C2} &\leq U_{H2,C2} \cdot y_{H2,C2} \\
Q_{H2,W} &\leq U_{H2,W} \cdot y_{H2,W}
\end{aligned}\right] \quad \text{Logical constraints;}$$

$$Q_{H1,C1,1}, Q_{H1,C1,2}, Q_{H1,C2,1}, Q_{H1,C2,2}, Q_{H1,W,3} \geq 0,$$

$$Q_{H2,C1,2}, Q_{H2,C2,2}, Q_{H1,W,3} \geq 0,$$

$$R_{H1,1}, R_{H1,2}, R_{H2,2} \geq 0,$$

$$y_{H1,C1}, y_{H1,C2}, y_{H1,W}, y_{H2,C1}, y_{H2,C2}, y_{H2,W} = 0 - 1;$$

where

$$\begin{aligned}
U_{H1,C1} &= \min\{1080, 800\} &= 800, \\
U_{H1,C2} &= \min\{1080, 480\} &= 480, \\
U_{H1,W} &= \min\{1080, 440\} &= 440, \\
U_{H2,C1} &= \min\{400, 500\} &= 400, \\
U_{H2,C2} &= \min\{400, 400\} &= 400, \\
U_{H2,W} &= \min\{640, 440\} &= 440.
\end{aligned}$$

Remark 1 The temperature intervals could have been partitioned with

$$TIAT = EMAT = \epsilon < HRAT = 10,$$

and the part (A) model could be written in a similar way.

Part (B): Hyperstructure Topology Model and Area Definitions

Having defined variables for the flow rates and temperatures of the streams in the hyperstructure shown in Figure 8.21, we can write part (B) in a similar way to the superstructure model presented in section 8.4.1.4.

$$\left.\begin{array}{l}
F1^1 + F2^1 = 12 \\
F3^1 - F5^1 - F6^1 = 0 \\
F4^1 - F7^1 - F8^1 = 0 \\
F3^1 - F1^1 - F8^1 = 0 \\
F4^1 - F2^1 - F6^1 = 0 \\
\\
F1^1 \cdot 450 + F8^1 \cdot T78^1 - F3^1 \cdot T3^1 = 0 \\
F2^1 \cdot 450 + F6^1 \cdot T56^1 - F4^1 \cdot T4^1 = 0 \\
F5^1 \cdot T56^1 + F7^1 \cdot T78^1 - 12 \cdot T5^1 = 0 \\
\\
F3^1(T3^1 - T56^1) = Q_{H1,C1} \\
F4^1(T4^1 - T78^1) = Q_{H1,C2} \\
\\
12(T5^1 - 350) = Q_{H1,W} \\
T3^1 - t56^1 \geq 10 \\
T56^1 - t3^1 \geq 10 \\
T4^1 - t56^2 \geq 10 \\
T78^1 - t3^2 \geq 10 \\
\\
F1^1, \ldots, F8^1 \geq 0
\end{array}\right] \text{Stream } H1,$$

$$\left.\begin{array}{l}
F1^2 + F2^2 = 8 \\
F3^2 - F5^2 - F6^2 = 0 \\
F4^2 - F7^2 - F8^2 = 0 \\
F3^2 - F1^2 - F8^2 = 0 \\
F4^2 - F2^2 - F6^2 = 0 \\
\\
F1^2 \cdot 400 + F8^2 \cdot T78^2 - F3^2 \cdot T3^2 = 0 \\
F2^2 \cdot 400 + F6^2 \cdot T56^2 - F4^2 \cdot T4^2 = 0 \\
F5^2 \cdot T56^2 + F7^2 \cdot T78^2 - 8 \cdot T5^2 = 0 \\
\\
F3^2(T3^2 - T56^2) = Q_{H2,C1} \\
F4^2(T4^2 - T78^2) = Q_{H2,C2} \\
8(T5^2 - 320) = Q_{H2,W} \\
\\
T3^2 - t78^1 \geq 10 \\
T56^2 - t4^1 \geq 10 \\
T4^2 - t78^2 \geq 10 \\
T78^2 - t4^2 \geq 10 \\
\\
F1^2, \ldots, F8^2 \geq 0
\end{array}\right] \text{Stream } H2,$$

$$\left.\begin{array}{l}
f1^1 + f2^1 = 10 \\
f3^1 - f5^1 - f6^1 = 0 \\
f4^1 - f7^1 - f8^1 = 0 \\
f3^1 - f1^1 - f8^1 = 0 \\
f4^1 - f2^1 - f6^1 = 0 \\
\\
f1^1 \cdot 340 + f8^1 \cdot t78^1 - f3^1 \cdot t3^1 = 0 \\
f2^1 \cdot 340 + f6^1 \cdot t56^1 - f4^1 \cdot t4^1 = 0 \\
\\
f3^1(t3^1 - t56^1) = Q_{H1,C1} \\
f4^1(t4^1 - t78^1) = Q_{H2,C1} \\
\\
f1^1, \ldots, f8^1 \geq 0
\end{array}\right] \text{Stream } C1,$$

$$\left.\begin{array}{l}
f1^2 + f2^2 = 8 \\
f3^2 - f5^2 - f6^2 = 0 \\
f4^2 - f7^2 - f8^2 = 0 \\
f3^2 - f1^2 - f8^2 = 0 \\
f4^2 - f2^2 - f6^2 = 0 \\
\\
f1^2 \cdot 340 + f8^2 \cdot t78^2 - f3^2 \cdot t3^2 = 0 \\
f2^2 \cdot 400 + f6^2 \cdot t56^2 - f4^2 \cdot t4^2 = 0 \\
\\
f3^2(t3^2 - t56^2) = Q_{H1,C2} \\
f4^2(t4^2 - t78^2) = Q_{H2,C2} \\
\\
f1^2, \ldots, f8^2 \geq 0
\end{array}\right] \text{Stream } C2,$$

$$\begin{aligned}
A_{H1,C1} &= \frac{Q_{H1,C1}}{(0.8)(LMTD)_{H1,C1}}, \\
A_{H1,C2} &= \frac{Q_{H1,C2}}{(0.8)(LMTD)_{H1,C2}}, \\
A_{H1,W} &= \frac{Q_{H1,W}}{(0.8)(LMTD)_{H1,W}}, \\
A_{H2,C1} &= \frac{Q_{H2,C1}}{(0.8)(LMTD)_{H2,C1}}, \\
A_{H2,C2} &= \frac{Q_{H2,C2}}{(0.8)(LMTD)_{H2,C2}}, \\
A_{H2,W} &= \frac{Q_{H2,W}}{(0.8)(LMTD)_{H2,W}},
\end{aligned}$$

where the Paterson (1984) approximation is used for the $LMTD$s.

$$F3^1 - \frac{Q_{H1,C1}}{(\Delta T_{H1,C1,max})} \geq 0,$$

$$F4^1 - \frac{Q_{H1,C2}}{(\Delta T_{H1,C2,max})} \geq 0,$$
$$F3^2 - \frac{Q_{H2,C1}}{(\Delta T_{H2,C1,max})} \geq 0,$$
$$F4^2 - \frac{Q_{H2,C2}}{(\Delta T_{H2,C2,max})} \geq 0,$$
$$f3^1 - \frac{Q_{H1,C1}}{(\Delta T_{H1,C1,max})} \geq 0,$$
$$f4^1 - \frac{Q_{H1,C2}}{(\Delta T_{H1,C2,max})} \geq 0,$$
$$f3^2 - \frac{Q_{H2,C1}}{(\Delta T_{H2,C1,max})} \geq 0,$$
$$f4^2 - \frac{Q_{H2,C2}}{(\Delta T_{H2,C2,max})} \geq 0,$$

where $\Delta T_{H_i,C_j,max}$ are fixed to the largest possible temperature drops through each exchanger between H_i and C_j.

Part (C): Logical Constraints between Flow rates and Binary Variables

$$\begin{aligned}
F3^1 - 12 \cdot y_{H1,C1} &\leq 0 \\
F4^1 - 12 \cdot y_{H1,C2} &\leq 0 \\
F3^2 - 8 \cdot y_{H2,C1} &\leq 0 \\
F4^2 - 8 \cdot y_{H2,C2} &\leq 0 \\
f3^1 - 10 \cdot y_{H1,C1} &\leq 0 \\
f4^1 - 10 \cdot y_{H2,C1} &\leq 0 \\
f3^2 - 8 \cdot y_{H1,C2} &\leq 0 \\
f4^2 - 8 \cdot y_{H2,C2} &\leq 0
\end{aligned}$$

Remark 2 We can also write logical constraints for the areas A if we do not substitute the areas' expressions in the objective function. These would be

$$\begin{aligned}
A_{H1,C1} - \mathcal{U} \cdot y_{H1,C1} &\leq 0 \\
A_{H1,C2} - \mathcal{U} \cdot y_{H1,C2} &\leq 0 \\
A_{H1,W} - \mathcal{U} \cdot y_{H1,W} &\leq 0 \\
A_{H2,C1} - \mathcal{U} \cdot y_{H2,C1} &\leq 0 \\
A_{H2,C2} - \mathcal{U} \cdot y_{H2,C2} &\leq 0 \\
A_{H2,W} - \mathcal{U} \cdot y_{H2,W} &\leq 0
\end{aligned}$$

$$A_{H1,C1}, A_{H1,C2}, A_{H1,W}, A_{H2,C1}, A_{H2,C2}, A_{H2,W} \geq 0$$

Objective Function

The objective function of the simultaneous Matches–HEN hyperstructure problem can be written in two different ways depending on whether we substitute the expressions of the areas.

Case I: Substitution of Areas in the Objective

In this case we no longer need the logical constraints for the areas, and the objective takes the form:

$$\begin{aligned} \min \quad & 1300A_{H1,C1}^{0.6} \cdot y_{H1,C1} + 1300A_{H1,C2}^{0.6} \cdot y_{H1,C2} + 1300A_{H1,W}^{0.6} \cdot y_{H1,W} + \\ & 1300A_{H2,C1}^{0.6} \cdot y_{H2,C1} + 1300A_{H2,C2}^{0.6} \cdot y_{H2,C2} + 1300A_{H2,W}^{0.6} \cdot y_{H2,W}, \end{aligned}$$

where in the above objective the areas are replaced by their definitions.

Note that when $y_{H1,C1} = 0$, there is no contribution in the objective function.

Case II: Maintain the Area Definitions as Constraints

In this case we have the objective function as

$$\begin{aligned} \min \quad & 1300A_{H1,C1}^{0.6} + 1300A_{H1,C2}^{0.6} + 1300A_{H1,W}^{0.6} + \\ & 1300A_{H2,C1}^{0.6} + 1300A_{H2,C2}^{0.6} + 1300A_{H2,W}^{0.6}. \end{aligned}$$

Remark 3 In principle, we do not need to have the logical constraints for the areas since we have the logical constraints for the loads of the matches. Hence, if $y_{H1,C1} = 0$ for instance, then $Q_{H1,C1} = 0$ and therefore $A_{H1,C1} = 0$ from its definition constraint. To avoid, however, potential infeasibilities due to the nonlinear equality constraints that are the definitions of the areas we introduce the logical constraints for the areas.

Remark 4 The complete optimization model for the simultaneous matches-HEN problem consists of the objective function (note that Case I was used in Floudas and Ciric (1989) subject to the set of constraints presented in parts A,B and C). In this model we have binary variables y_{ij} denoting the potential existence of a match and continuous variables. As a result, the model is a MINLP problem. This MINLP model has a number of interesting features which are as follows:

(a) The binary variables participate linearly in the transshipment constraints and nonlinearly in the objective function. If the expression for the areas A_{ij} are held constant then the terms become linear in the binary variables.

(b) The heat loads Q_{ij} participate linearly in the set of energy balance constraints and nonlinearly only in the objective function. By fixing Q_{ij}, the objective function becomes the sum of convex functions (note that this is for case I).

(c) The energy balances are bilinear equality constraints in the flow rate heat capacities and temperatures and hence are nonconvex.

(d) The nonconvex nature of the MINLP model makes the determination of only local optimum solutions possible if local optimization solvers are utilized.

Match	Q (kW)	$A(m^2)$
$H1 - C1$	800	41.8
$H1 - C2$	800	41.8
$H2 - C2$	200	11.1
$H2 - CW$	440	15.2

Table 8.8: Matches, heat loads, areas of illustrative example

Solution Strategy Based on Decomposition

The solution strategy employed by Floudas and Ciric (1989) is based on v2-GBD discussed in the Fundamentals of Mixed-Integer Nonlinear Programming chapter. The interesting feature of this solution strategy is the selection of the sets x and y variables so as to take advantage of the underlying mathematical structure of the resulting MINLP model.

The standard selection of the y variables in a MINLP model is

$$y \equiv \{y_{ij}\};$$

that is, the binary variable y_{ij}. With such a selection, the primal subproblem is a nonconvex nonlinear optimization problem while the master subproblem is a $0-1$ programming problem with one scalar μ_B. Even though both the primal and the master subproblems are solvable (locally for primal, globally for master) with existing solvers, this selection scheme has the drawback that the master problem may generate solutions which are infeasible for the subsequent primal problem. This is due to having many integer combinations satisfying the minimum matches constraint but not necessarily satisfying the minimum utility constraint or the transshipment model. In other words, the master problem does not contain enough constraints so as to have feasibility in the transshipment model part of the primal problem. As a result, application of such a standard selection of the y variables can result in a large number of iterations.

The aforementioned difficulty can be overcome with the selection proposed by Floudas and Ciric (1989) in which the y variables consist of

$$y \equiv \{y_{ij}, Q_{ijk}, R_{ik}, Q_{ij}\};$$

that is, all variables involved in the MILP transshipment model (i.e., Part (A)). An immediate consequence of this selection is that the primal problem does not have any constraint related to the transshipment model (i.e., part (A)) but instead the linear set of constraints describing the transshipment model is moved directly to the master problem. As a result, any feasible solution of the master problem automatically satisfies the transshipment model constraints. The primal problem contains parts (B) and (C) constraints and is a nonconvex nonlinear programming problem, while the master problem is a MILP problem. Application of this strategy in the illustration example resulted in the HEN configuration shown in Figure 8.23. The solution data with respect to the matches selected their heat loads and the areas of the heat exchanger units are shown in Table 8.8.

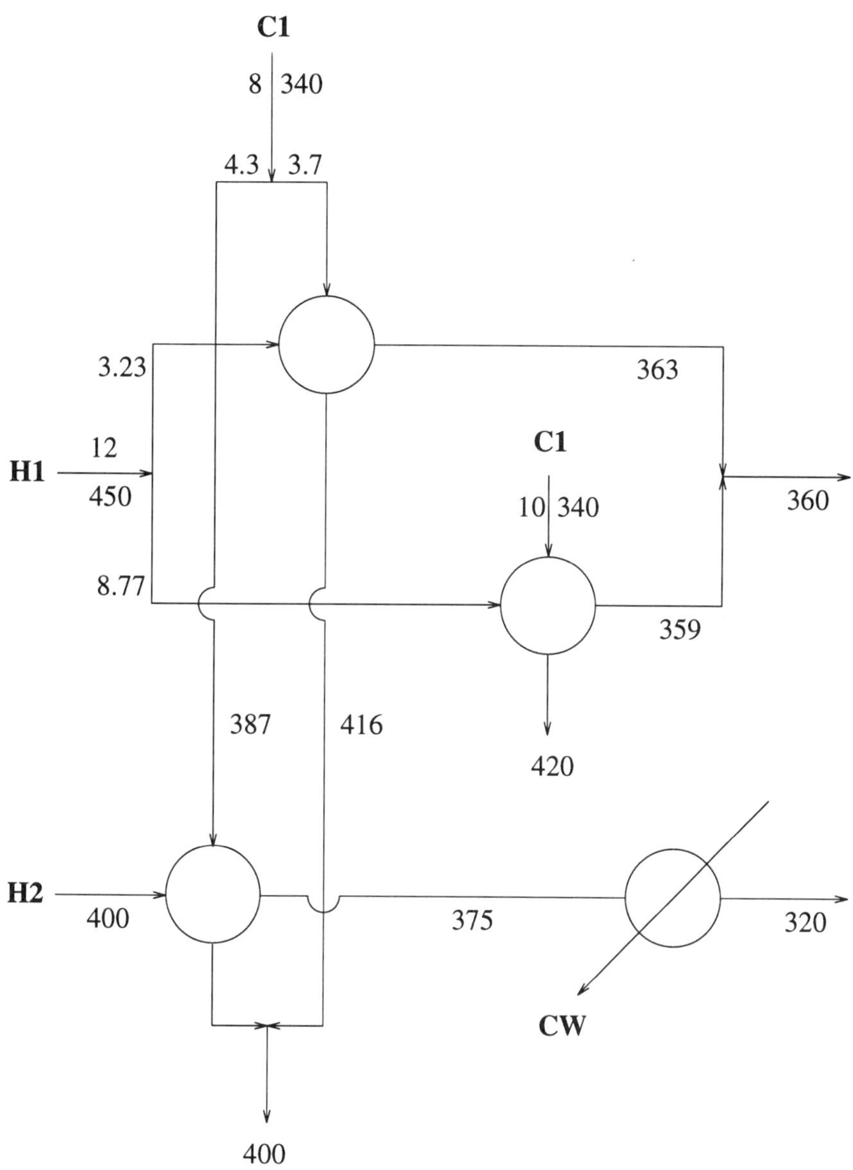

Figure 8.23: Optimal solution of illustrative example of simultaneous matches-HEN optimization

Note that the optimal network of Figure 8.23 has the minimum number of matches and involves splitting of streams $H1$ and $C2$. Note also that the minimum temperature constraints are not active (i.e., none of them is equal to $EMAT = 10$ K) which may imply that the splitting of streams occurs because of searching for the optimal solution. The illustrative example was solved using the library OASIS (Floudas, 1990).

8.5.2 Pseudo-Pinch

In Section 8.2.1 we defined three temperature approaches (i.e., $HRAT, TIAT$, and $EMAT$) that satisfy the relation:

$$HRAT \geq TIAT \geq EMAT.$$

In the presentation of the MILP transshipment models for minimum number of matches, we mentioned that $TIAT$ can be set equal to $EMAT$ which can be a small positive number $\epsilon > 0$:

$$TIAT = EMAT = \epsilon > 0.$$

In the *strict-pinch* case, in which we artificially do not allow for heat flow across the pinch and decompose the problem into subnetworks, we have that:

$$HRAT = TIAT = EMAT.$$

The $TIAT$ temperature approach, however, can in general be strictly less than $HRAT$:

$$HRAT > TIAT = EMAT$$

and this is denoted as the *pseudo-pinch* case. In the pseudo-pinch case, heat is allowed to flow across the pinch point if the trade-offs between the investment and operating cost suggest it.

In the next section we will illustrate graphically the effect of temperature partitioning with $TIAT = EMAT$ but strictly less than $HRAT$.

8.5.2.1 Partitioning with $EMAT = TIAT < HRAT$

We use the same illustrative example used in section 8.3.1.2 and 8.3.1.3 for $TIAT = HRAT$. Figure 8.24 shows the temperature intervals for $TIAT = HRAT$, and for $TIAT < HRAT$.

Strict-Pinch Case: $TIAT = HRAT$

Note that the maximum residual flow out of a temperature interval is given by the horizontal component of the intervals lower demarcation line (see the Appendix of Ciric and Floudas (1990)). For instance, in $TI - 1$ $(AIFG)$ we have R_1 equal to IA'. In $TI - 2$ (i.e., $AKEI$) we have

$$R_2 = EE'.$$

In $TI - 3$ (i.e., $KBHE$) we have

$$R_3 = 0,$$

since the horizontal component is simply the point H and we have a pinch at BH.

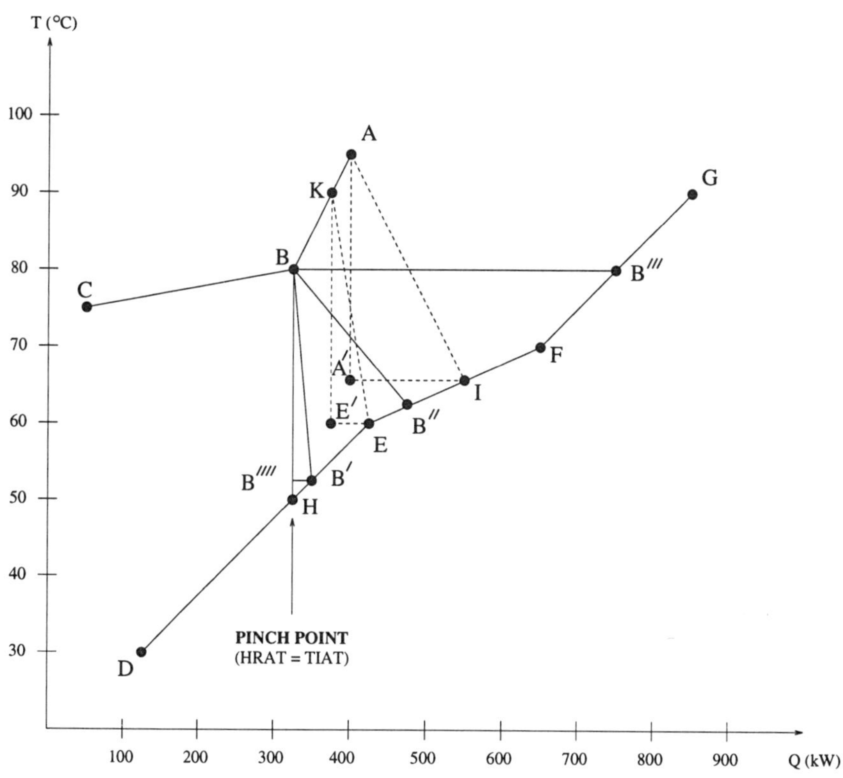

Figure 8.24: Temperature partitioning for $EMAT = TIAT < HRAT$

In $TI - 4$ (i.e., $CBHD$) we do not have any residual flow since it goes to the cold utility (i.e., last interval).

Pseudo-Pinch Case:

In the pseudo-pinch case we have

$$EMAT = TIAT < HRAT.$$

In this case, the perpendicular distances are strictly less than $HRAT$, and as $TIAT$ decreases the demarcation lines originating at hot stream inlets rotate upward toward the horizontal (see for instance point B', B'', B''' in Figure 8.24). The consequences of this rotation are that the horizontal components *increase*, which reflects an increase in the residual heat flow in each interval as well as the potential for heat flow across the pinch.

Note also that the maximum cross-pinch heat flow can be read directly from the **T–Q** of Figure 8.24. For instance if $EMAT = TIAT = BB''''$, then the maximum heat can cross the pinch is given by

$$R = B'B''''.$$

8.5.2.2 Analytical Calculation of Maximum Cross-Pinch Heat Flow

The maximum cross-pinch heat flow, R_p, can also be calculated analytically. In general, R_p is a piece-wise linear function of $TIAT$ and $HRAT$ and exhibits transition points for pinch points that take place in the following two cases:

Case I: The pinch point is located at a hot stream inlet whenever

$$T_p'' - TIAT$$

equals the inlet or outlet temperature of a cold stream where the slope of the cold composite curve changes.

Case II: The pinch point is located at a cold stream inlet whenever

$$T_p^c + TIAT$$

equals the inlet or outlet temperature of a cold stream where the slope of the hot composite curve changes.

For Case I, R_P is given by (see Ciric and Floudas (1990)):

$$\begin{aligned} R_p \;=\; & \sum_{j \in C_1} F_j^C \,(HRAT - TIAT) - \sum_{j \in C_2} F_j \,[T_j^{C,I} - (T_P^H - HRAT)] \\ & - \sum_{j \in C_3} F_j^C \,[(T_P^H - TIAT) - T_j^{C,O}]\,, \end{aligned}$$

where

F_j^C	is the flow rate heat capacity of cold stream j,
$T_j^{C,I}, T_j^{C,O}$	are the inlet and outlet temperatures of cold stream j,
C_1	is the set of cold streams in the range $(T_P^H - HRAT, T_P^H - TIAT)$,
C_2	is the subset of C_1 with inlet temperatures strictly greater than $(T_P^H - HRAT)$, and
C_3	is the subset of of C_1 with outlet temperatures strictly less than $(T_P^H - TIAT)$.

For Case II, R_P is provided by (see Ciric and Floudas (1990)):

$$\begin{aligned} R_p &= \sum_{i \in H_1} F_i^H (HRAT - TIAT) - \sum_{i \in H_2} F_i [(T_P^C + HRAT) - T_i^{H,I}] \\ &\quad - \sum_{i \in H_3} F_i^H [T_i^{H,O} - (T_P^C + TIAT)] . \end{aligned}$$

where

F_i^H	is the flow rate heat capacity of hot stream i,
$T_i^{H,I}, T_i^{H,O}$	are the inlet and outlet temperatures of hot stream i,
H_1	is the set of hot streams in the range $(T_P^H - HRAT, T_P^H - TIAT)$,
H_2	is the subset of H_1 with inlet temperatures strictly less than $(T_P^C - HRAT)$, and
H_3	is the subset of of H_1 with outlet temperatures strictly greater than $(T_P^C - TIAT)$.

The sets C_1, C_2, C_3, H_1, H_2, and H_3 can be obtained from the stream data.

We consider again the same illustrative example for which the temperature partitioning is depicted in Figure 8.24.

Note that the range of $TIAT$ can be partitioned into three regions:

Region I: BHE with $20 \leq TIAT \leq 30$.

Region II: BEF with $10 \leq TIAT \leq 20$.

Region III: BFB''' with $0 \leq TIAT \leq 10$.

Then, for each region we have:

Region I:
$$\left[\begin{aligned} C_1 &= \{C1\} \\ C_2 &= \{0\} \\ C_3 &= \{0\} \\ R_P &= 10 \cdot [30 - TIAT] \end{aligned} \right.$$

Region II:
$$\left[\begin{aligned} C_1 &= \{C1, C2\} \\ C_2 &= \{C2\} \\ C_3 &= \{0\} \\ R_P &= 10 \cdot [30 - TIAT] + 12.5 \cdot [30 - TIAT] \\ &\quad -12.5 \cdot [60 - (80 - 30)] \\ &= 22.5 \cdot [30 - TIAT] - 125 \end{aligned} \right.$$

Region III:
$$\left[\begin{aligned} C_1 &= \{C1, C2\} \\ C_2 &= \{C2\} \\ C_3 &= \{C2\} \\ R_P &= 10 \cdot [30 - TIAT] + 12.5 \cdot [30 - TIAT] \\ &\quad -12.5 \cdot [60 - (80 - 30)] - 22.5 \cdot [(80 - TIAT) - 70] \\ &= 10 \cdot [30 - TIAT] + 125 \end{aligned} \right.$$

If $TIAT = 25 < HRAT = 30$, then from region I we have

$$R_p = 50.$$

If $TIAT = 15$, then from region II we have

$$R_P = 212.5.$$

If $TIAT = 5$, then from region III we have

$$R_P = 375.$$

8.5.2.3 Pseudo-Pinch Synthesis Strategy

Ciric and Floudas (1990) proposed the following approach for the pseudo-pinch synthesis of HEN problem.

Step 1: Select $HRAT$. Select $TIAT = EMAT$ such that $TIAT < HRAT$.

Step 2: For the given $HRAT$, calculate the minimum utility loads needed.

Step 3: Apply the simultaneous Matches-HEN optimization model presented in section 8.5.1.

Return to step 1 and update $HRAT$ if on outer optimization loop over $HRAT$ is considered.

Remark 1 Ciric and Floudas (1990) through several examples demonstrated the effect of allowing heat across the pinch point on the simultaneous optimization of matches and network configurations. In particular, in their example 1, the optimal solution had one more unit than the minimum units criterion of one network is considered and one less unit if the HEN is decomposed into subnetworks based on the artificial concept of pinch point. Also, the effect of varying $TIAT$ was illustrated in their third example.

Remark 2 An important consequence of allowing heat flow across the pinch if the economic versus operating trade-offs suggest it, is that the optimal HEN structures that are obtained in Ciric and Floudas (1990) are simple and do not feature by-pass streams. As a result, these structures may be more interesting from the practical application point of view.

8.5.3 Synthesis of HENs Without Decomposition

8.5.3.1 Problem Statement

Given the information provided by the stream data (see also section 8.2), determine the minimum total annualized cost heat exchanger network by providing:

(i) The hot and cold utility loads,

(ii) The matches that take place,

(iii) The areas of heat exchanger unit, and

(iv) The topology (structure) of the heat exchanger network.

Remark 1 The above statement corresponds to the simultaneous consideration of all steps shown in Figure 8.20, including the optimization loop of the $HRAT$. We do not decompose based on the artificial *pinch-point* which provides the minimum utility loads required, but instead allow for the appropriate trade-offs between the operating cost (i.e., utility loads) and the investment cost (i.e., cost of heat exchangers) to be determined. Since the target of minimum utility cost is not used as heuristic to determine the utility loads with the LP transshipment model, but the utility loads are treated as unknown variables, then the above problem statement eliminates the last part of decomposition imposed in the simultaneous matches-network optimization presented in section 8.5.1.

Remark 2 In Figure 8.20 we have discussed the optimization loop of $HRAT$. Specifying a value of $HRAT$ allows the calculation of the minimum utility loads using the LP transshipment model, The optimization loop of $HRAT$ had to be introduced so as to determine the optimal value of $HRAT$ that gives the trade-off of operating and investment cost. Note, however, that in the approach of this section, in which we perform no decomposition at all, we do not specify the $HRAT$, but we treat the hot and cold utility loads as explicit unknown optimization variables. As a result, there is no need for the optimization loop of $HRAT$ since we will determine directly the utility loads.

Remark 3 For (ii)-(iv), the hyperstructure approach presented in the section of simultaneous matches-network optimization will be utilized.

Remark 4 The synthesis approach without decomposition to be presented in subsequent sections corresponds to the most general case of *pseudo-pinch* in which heat may be allowed to flow across the pinch if the trade-offs between the operating and investment cost suggest it. Hence, $TIAT = EMAT$ is strictly less than the $HRAT$ in the general case.

In the following section, we will discuss the approach proposed by Ciric and Floudas (1991) for the synthesis of heat exchanger networks without decomposition. Note that we will present the approach for the pseudo-pinch case (which is the most general). The approach for the strict-pinch case (which is a constrained scenario of the pseudo-pinch and as such features more structure) can be found in Ciric and Floudas (1991).

8.5.3.2 Basic Idea

The basic idea in the pseudo-pinch HEN synthesis without decomposition approach (Ciric and Floudas, 1991) consists of

(i) Treating the hot and cold utility loads as explicit optimization variables,

(ii) Allowing for heat flow across the pinch if needed,

(iii) Postulating a *hyperstructure* as discussed in section 8.5.1.3, and

(iv) Formulating a MINLP model with

(a) An objective function that minimizes the total annualized cost in which the areas of heat exchangers and the utility loads are explicit variables, and

(b) Constraints that feature the

(b.1) Pseudo-pinch MILP transshipment model, and

(b.2) Hyperstructure model presented in section 8.5.1.3.

Remark 1 The hot and cold utility loads participate linearly in the objective function (i.e., operating cost), and linearly in the pseudo-pinch MILP transshipment model. They also participate linearly in the energy balances of the utility exchangers postulated in the hyperstructure. This linear participation is very important in the MINLP mathematical model.

In the next section we will discuss and illustrate the mathematical model of the synthesis approach without decomposition.

8.5.3.3 Mathematical Model

We will discuss the mathematical model for the general case of *pseudo-pinch*. In this case, the $TIAT$ is set equal to $EMAT$ (i.e., $TIAT = EMAT$) which can be a small positive value, in general is strictly less than the $HRAT$. Hence, the temperature partitioning does not depend or vary with $HRAT$ (or the utility loads). The utility loads, however, are allowed to vary freely, since they are treated as explicit optimization variables.

Objective Function

The objective function of the mathematical model involves the minimization of the total annualized cost which consists of appropriately weighted investment and operating cost. It takes the form:

$$\min \quad \left[\sum_i \sum_j c_{ij} A_{ij}^{0.6} \cdot y_{ij} \right] + \left[\sum_{i \in HU} C_i Q_i^H + \sum_{j \in CU} C_j Q_j^C \right]$$

where

$Q_i^H \quad , \quad i \in HU$ is the unknown heat load of hot utility i,
$Q_j^C \quad , \quad j \in CU$ is the unknown heat load of cold utility j,
C_i, C_j, C_{ij} are the cost coefficients

Note that the binaries y_{ij} multiply A_{ij} in the objective function for the same reasons that we wrote the objective function of case I of the simultaneous matches-network optimization (see section 8.5.1.4).

Constraints

The set of constraints consists of the hyperstructure model presented in section 8.5.1.4 for the pseudo-pinch; that is,

(A) Pseudo-pinch MILP transshipment model,

(B) Hyperstructure topology model with area definition, and

(C) Logical relations between inlet flow rates and binaries.

Here, we will only present the pseudo-pinch MILP transshipment constraints since (B) and (C) were illustrated in section 8.5.1.4.

(A) Pseudo-Pinch MILP Transshipment Constraints

The pseudo-pinch MILP transshipment constraints are almost the same as the MILP transshipment constraints of Papoulias and Grossmann (1983) The key difference is that in the pseudo-pinch:

$$\begin{array}{lll} Q_{ik}^{H} & , \quad i \in HU_K & , \\ Q_{jk}^{C} & , \quad j \in CU_K & , \end{array}$$

are not fixed, but they are unknown optimization variables. Then, the heat load of each utility is given by

$$\begin{aligned} Q_i^H &= \sum_K Q_{ik}^H \quad , \quad i \in HU_k, \\ Q_j^C &= \sum_K Q_{jk}^C \quad , \quad j \in CU_k. \end{aligned}$$

Then, the pseudo-pinch MILP transshipment is of the following form:

$$\begin{aligned} R_{i,k} - R_{i,k-1} + \sum_{j \in CP_k \cup CU_k} Q_{ijk} &= Q_{ik}^H \quad , \quad i \in HP_k \\ R_{i,k} - R_{i,k-1} + \sum_{j \in CP_k} Q_{ijk} &= Q_{ik}^H \quad , \quad i \in HU_k \\ \sum_{i \in HP_k \cup HU_k} Q_{ijk} &= Q_{jk}^C \quad , \quad j \in CP_k \\ \sum_{j \in CP_k \cup CU_k} Q_{ijk} &= Q_{jk}^C \quad , \quad j \in CU_k \end{aligned}$$

$$\begin{aligned} R_k &= \sum_{i \in HP_k \cup HU_k} R_{ik} \\ Q_i^H &= \sum_{k \in TI} Q_{ik}^H \quad , \quad i \in HU_k \\ Q_j^C &= \sum_{k \in TI} Q_{jk}^H \quad , \quad j \in CU_k \\ Q_{ij} &= \sum_{k \in TI} Q_{ijk} \end{aligned}$$

$$\mathcal{L}_{ij} y_{ij} \leq Q_{ij} \leq \mathcal{U}_{ij} y_{ij}$$

$$Q_{ijk} \geq 0; Q_{ik}^H \geq 0, Q_{jk}^C \geq 0, R_{ik} \geq 0, k \in TI$$

$$R_0 = R_K = 0$$

$$y_{ij} = 0 - 1$$

Illustration 8.5.3 (MINLP Model)

This example is taken from the Ph.D. thesis of Ciric (1990) and was also used by Yee *et al.* (1990b). The stream data are shown in Table 8.9.

Stream	$T^{in}(^\circ F)$	$T^{out}(^\circ F)$	$FC_p(\text{kW}^\circ F^{-1})$
$H1$	500	320	6
$H2$	480	380	4
$H3$	460	360	6
$H4$	380	360	20
$H5$	380	320	12
$C1$	290	660	18
F	700	540	
W	300	180	
$U = 1.0\text{kW}/m^2\text{K}$			
Annual cost $= 1200A^{0.6}$ for all exchangers			
$C_F = 140$ \$/yr ; $C_W = 10$ \$/yr			

Table 8.9: Stream data for illustrative example 8.5.3

The $HRAT$ is allowed to vary freely between 1 K and 30 K, while $TIAT = EMAT = 1$ K. Since $TIAT = EMAT$ is in general strictly less than the $HRAT$, the problem was treated as a pseudo-pinch synthesis problem.

In the following, we will discuss the explicit mathematical formulation of this problem with the components (A), (B), and (C) of the constraints and the objective function.

(A) Pseudo-Pinch MILP Transshipment Constraints

To write the pseudo-pinch MILP transshipment constraints we have to perform the temperature partitioning first. Since $TIAT = EMAT = 1$ K, we have the partitioning shown in Figure 8.25. The pseudo-pinch MILP transshipment constraints are

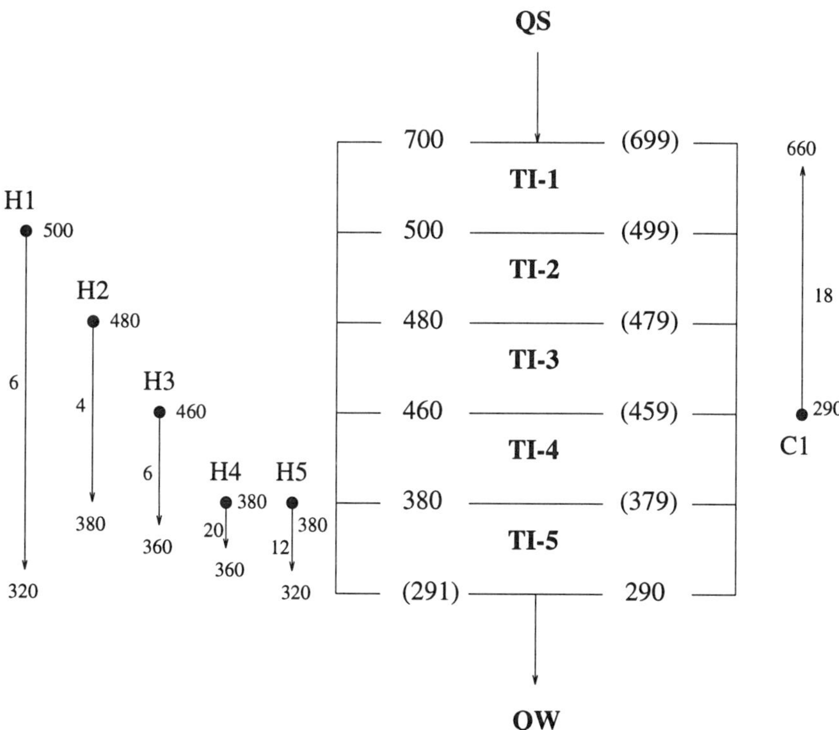

Figure 8.25: Temperature partitioning for illustrative example 8.5.3

$$\left.\begin{aligned} R_{S,1} + Q_{S,C1,1} &= QS \\ Q_{S,C1,1} &= 18(660-499) = 2898 \end{aligned}\right] \quad TI-1$$

$$\left.\begin{aligned} R_{S,2} - R_{S,1} + Q_{S,C1,2} &= 0 \\ R_{H1,2} + Q_{H1,C1,2} &= 6(500-480) = 120 \\ +Q_{S,C1,2} + Q_{H1,C1,2} &= 18(499-479) = 360 \end{aligned}\right] \quad TI-2$$

$$\left.\begin{aligned} R_{S,3} - R_{S,2} + Q_{S,C1,3} &= 0 \\ R_{H1,3} - R_{H1,2} + Q_{H1,C1,3} &= 6(480-460) = 120 \\ R_{H2,3} + Q_{H2,C1,3} &= 4(480-460) = 80 \\ Q_{S,C1,3} + Q_{H1,C1,3} + Q_{H2,C1,3} &= 18(479-459) = 360 \end{aligned}\right] \quad TI-3$$

$$\left.\begin{aligned} R_{S,4} - R_{S,3} + Q_{S,C1,4} &= 0 \\ R_{H1,4} - R_{H1,3} + Q_{H1,C1,4} &= 6(460-380) = 480 \\ R_{H2,4} - R_{H2,3} + Q_{H2,C1,4} &= 4(460-380) = 320 \\ R_{H3,4} + Q_{H3,C1,4} &= 6(460-380) = 480 \\ Q_{S,C1,4} + Q_{H1,C1,4} + & \\ +Q_{H2,C1,4} + Q_{H3,C1,4} &= 18(459-379) = 1440 \end{aligned}\right] \quad TI-4$$

$$\left.\begin{aligned} -R_{S,4} + Q_{S,C1,5} &= 0 \\ -R_{H1,4} + Q_{H1,C1,5} + Q_{H1,W,5} &= 6(380-320) = 360 \\ -R_{H2,4} + Q_{H2,C1,5} + Q_{H2,W,5} &= 0 \\ -R_{H3,4} + Q_{H3,C1,5} + Q_{H3,W,5} &= 6(380-360) = 120 \\ Q_{H4,C1,5} + Q_{H4,W,5} &= 20(380-360) = 400 \\ Q_{H5,C1,5} + Q_{H5,W,5} &= 12(380-320) = 720 \\ Q_{S,C1,5} + Q_{H1,C1,5} + Q_{H2,C1,5} + & \\ +Q_{H3,C1,5} + Q_{H4,C1,5} + Q_{H5,C1,5} &= 18(379-290) = 1602 \\ Q_{H1,W,5} + Q_{H2,W,5} + Q_{H3,W,5} + & \\ +Q_{H4,W,5} + Q_{H5,W,5} &= QW \end{aligned}\right] \quad TI-5$$

$$\left.\begin{aligned} R_1 &= R_{S,1} \\ R_2 &= R_{S,2} + R_{H1,2} \\ R_3 &= R_{S,3} + R_{H1,3} + R_{H2,3} \\ R_4 &= R_{S,4} + R_{H1,4} + R_{H2,4} + R_{H3,4} \end{aligned}\right] \quad \text{Definitions of residuals}$$

$$\left.\begin{aligned} Q_{S,C1} &= Q_{S,C1,1} + Q_{S,C1,2} + Q_{S,C1,3} + Q_{S,C1,4} + Q_{S,C1,5} \\ Q_{H1,C1} &= Q_{H1,C1,1} + Q_{H1,C1,2} + Q_{H1,C1,3} + Q_{H1,C1,4} + Q_{H1,C1,5} \\ Q_{H2,C1} &= Q_{H2,C1,3} + Q_{H2,C1,4} + Q_{H2,C1,5} \\ Q_{H3,C1} &= Q_{H3,C1,4} + Q_{H3,C1,5} \\ Q_{H4,C1} &= Q_{H4,C1,5} \\ Q_{H5,C1} &= Q_{H5,C1,5} \\ Q_{H1,W} &= Q_{H1,W,5} \\ Q_{H2,W} &= Q_{H2,W,5} \\ Q_{H3,W} &= Q_{H3,W,5} \\ Q_{H4,W} &= Q_{H4,W,5} \\ Q_{H5,W} &= Q_{H5,W,5} \end{aligned}\right] \quad \text{Definitions of heat loads}$$

$$\left.\begin{aligned} Q_{S,C1} - \mathcal{U}_{S,C1} \cdot y_{S,C1} &\leq 0 \\ Q_{H1,C1} - \mathcal{U}_{H1,C1} \cdot y_{H1,C1} &\leq 0 \\ Q_{H2,C1} - \mathcal{U}_{H2,C1} \cdot y_{H2,C1} &\leq 0 \\ Q_{H3,C1} - \mathcal{U}_{H3,C1} \cdot y_{H3,C1} &\leq 0 \\ Q_{H4,C1} - \mathcal{U}_{H4,C1} \cdot y_{H4,C1} &\leq 0 \\ Q_{H5,C1} - \mathcal{U}_{H5,C1} \cdot y_{H5,C1} &\leq 0 \\ Q_{H1,W} - \mathcal{U}_{H1,W} \cdot y_{H1,W} &\leq 0 \\ Q_{H2,W} - \mathcal{U}_{H2,W} \cdot y_{H2,W} &\leq 0 \\ Q_{H3,W} - \mathcal{U}_{H3,W} \cdot y_{H3,W} &\leq 0 \\ Q_{H4,W} - \mathcal{U}_{H4,W} \cdot y_{H4,W} &\leq 0 \\ Q_{H5,W} - \mathcal{U}_{H5,W} \cdot y_{H5,W} &\leq 0 \end{aligned}\right] \quad \text{Logical constraints}$$

where $\mathcal{U}_{S,C1}, \mathcal{U}_{H1,W}, \mathcal{U}_{H2,W}, \mathcal{U}_{H3,W}, \mathcal{U}_{H4,W}, \mathcal{U}_{H5,W}$ are some large numbers since we do not know the heat loads of the hot and cold utilities, and

$$\begin{aligned} \mathcal{U}_{H1,C1} &= 1080 \\ \mathcal{U}_{H1,C2} &= 240 \\ \mathcal{U}_{H1,W} &= 600 \\ \mathcal{U}_{H2,C1} &= 400 \\ \mathcal{U}_{H2,C2} &= 720 \end{aligned}$$

$$QS \geq 0, QW \geq 0$$

$$\left.\begin{array}{l}
Q_{S,C1,1}, Q_{S,C1,2}, Q_{S,C1,3}, Q_{S,C1,4}, Q_{S,C1,5} \geq 0 \\
Q_{H1,C1,2}, Q_{H1,C1,3}, Q_{H1,C1,4}, Q_{H1,C1,5}, Q_{H1,W,5} \geq 0 \\
Q_{H2,C1,3}, Q_{H2,C1,4}, Q_{H2,C1,5}, Q_{H2,W,5} \geq 0 \\
Q_{H3,C1,4}, Q_{H3,C1,5}, Q_{H3,W,5} \geq 0 \\
Q_{H4,C1,5}, Q_{H4,W,5} \geq 0 \\
Q_{H5,C1,5}, Q_{H5,W,5} \geq 0 \\
y_{S,C1}, y_{H1,C1}, y_{H2,C1}, y_{H3,C1}, y_{H4,C1}, y_{H5,C1} \\
y_{H1,W}, y_{H2,W}, y_{H3,W}, y_{H4,W}, y_{H5,W} = 0-1
\end{array}\right] \quad \text{Nonnegativity and integrality conditions}$$

Remark 1 Note that QS and QW are treated as unknown optimization variables which participate linearly in the set of constraints. Also, note that we must have

$$y_{S,C1} = 1$$

since there is no other hot process stream in $TI-1$. For simplicity at the presentation we will treat the match $S - C1$ as taking place after the other process-process matches in the hyperstructure representation to be presented next. To write the constraints for (B) and (C) we need to present first the hyperstructure representation of all possible alternatives for the illustrative example. From the stream data, the following potential matches have to be postulated:

$$\begin{array}{c}
S - C1 \\
H1 - C1 \\
H2 - C1 \\
H3 - C1 \\
H4 - C1 \\
H5 - C1 \\
H1 - W \\
H2 - W \\
H3 - W \\
H4 - W \\
H5 - W
\end{array}$$

out of which we know that $S - C1$ has to take place.

In postulating the hyperstructure representation we will make the assumption only for simplicity of presentation, that the utility matches take place after the process-process matches.

Remark 2 Note that each hot process stream has a postulated structure of a process-process match in series with the utility match. The cold stream $C1$, however, has a postulated structure with five process-process matches and then a hot utility match in series.

Remark 3 In the Ph.D. thesis of Ciric (1990), the full hyperstructure was derived and formulated as an MINLP problem. However, since the part of hyperstructure which corresponds to cold stream $C1$ involves five matches, then the mathematical model will be complex for presentation purposes. For this reason we will postulate the part of the hyperstructure of $C1$ by making the following simplification which eliminates a number of the interconnecting streams. This is only made for simplicity of the presentation, while the synthesis approach of Ciric and Floudas (1991) is for the general case.

Simplification of $C1$ part of hyperstructure

From Figure 8.25, we observe that

(i) $H4$ and $H5$ start at $TI - 5$,

(ii) $H3$ starts at $TI - 4$,

(iii) $H2$ starts at $TI - 3$,

(iv) $H1$ starts at $TI - 2$, and

(v) Hot utility is at $TI - 1$.

The simplification utilizes this information and can be stated as

> $C1$ will go through the potential matches with $H4$ and $H5$ (i.e. $H4 - C1$, $H5 - C1$) first, $H3 - C1$ second, $H2 - C1$ third, $H1 - C1$ fourth, and finally hot utility $C1$; after $C1$ goes through the level (i) matches it may be directed to any of the process matches of levels (ii), (iii), or (iv); after $C1$ goes through the level (ii) match it can only be directed to the matches of level (iii) and (iv) but not (i); after $C1$ goes through the level (iii) match then it can only be directed to the match of level (iv) but not to the matches of level (i) and (ii); after $C1$ goes through the match of level (iv) then it cannot be directed to the matches of levels (i), (ii), and (iii).

The above simplification can be depicted in the graph of Figure 8.26.

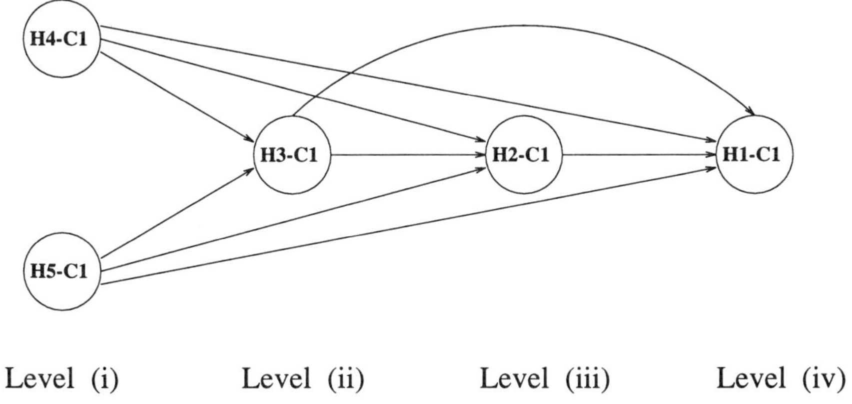

Figure 8.26: Simplification for hyperstructure

The hyperstructure of the illustration example with the above simplification for $C1$ is shown in Figure 8.27.

(B) Hyperstructure Topology and Area Definitions

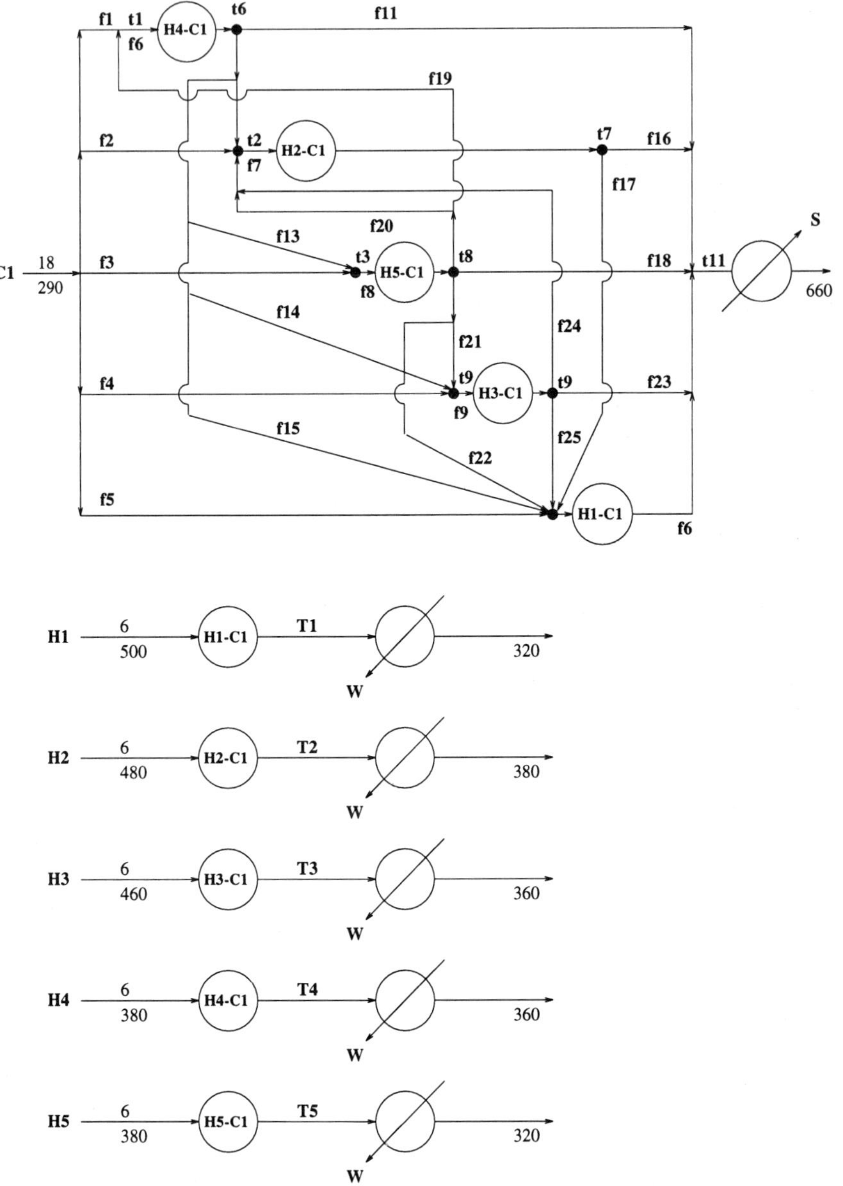

Figure 8.27: Hyperstructure for illustrative example

Based on the hyperstructure presented in Figure 8.27 and the variables for the interconnecting streams (i.e., flow rates and temperatures), constraints (B) are written as follows:

$$\left.\begin{array}{l}
f1 + f2 + f3 + f4 + f5 = 18 \\
f6 - f1 - f19 = 0 \\
f6 - f11 - f12 - f13 - f14 - f15 = 0 \\
f7 - f2 - f12 - f20 - f24 = 0 \\
f7 - f16 - f17 = 0 \\
f8 - f3 - f13 = 0 \\
f8 - f18 - f19 - f20 - f21 - f22 = 0 \\
f9 - f4 - f14 - f21 = 0 \\
f9 - f23 - f24 - f25 = 0 \\
f10 - f5 - f15 - f17 - f22 - f25 = 0
\end{array}\right] \text{Mass balances}$$

$$\left.\begin{array}{l}
f1 \cdot 290 + f19 \cdot t8 - f6 \cdot t1 = 0 \\
f2 \cdot 290 + f12 \cdot t6 + f20 \cdot t8 + f24 \cdot t9 - f7 \cdot t2 = 0 \\
f3 \cdot 290 + f13 \cdot t6 - f8 \cdot t3 = 0 \\
f4 \cdot 290 + f14 \cdot t6 + f21 \cdot t8 - f9 \cdot t4 = 0 \\
f5 \cdot 290 + f15 \cdot t6 + f17 \cdot t7 + f22 \cdot t8 + f25 \cdot t9 - f10 \cdot t5 = 0
\end{array}\right] \begin{array}{l}\text{Energy balances} \\ \text{in mixers}\end{array}$$

$$\left.\begin{array}{rcl}
f6 \cdot (t6 - t1) & = & Q_{H4,C1} \\
f7 \cdot (t7 - t2) & = & Q_{H2,C1} \\
f8 \cdot (t8 - t3) & = & Q_{H5,C1} \\
f9 \cdot (t9 - t4) & = & Q_{H3,C1} \\
f10 \cdot (t10 - t5) & = & Q_{H1,C1} \\
 & & \\
6 \cdot (500 - T1) & = & Q_{H1,C1} \\
4 \cdot (480 - T2) & = & Q_{H2,C1} \\
6 \cdot (460 - T3) & = & Q_{H3,C1} \\
20 \cdot (380 - T4) & = & Q_{H4,C1} \\
12 \cdot (380 - T5) & = & Q_{H5,C1} \\
 & & \\
18 \cdot (660 - t11) & = & Q_{S,C1} \\
 & & \\
6 \cdot (T1 - 320) & = & Q_{H1,W} \\
4 \cdot (T2 - 380) & = & Q_{H2,W} \\
6 \cdot (T3 - 360) & = & Q_{H3,W} \\
20 \cdot (T4 - 360) & = & Q_{H4,W} \\
12 \cdot (T5 - 320) & = & Q_{H5,W}
\end{array}\right] \begin{array}{l}\text{Energy balances} \\ \text{in heat exchangers}\end{array}$$

$$\left.\begin{array}{l} 500 - t10 \geq 1 \\ T1 - t5 \geq 1 \\ 480 - t7 \geq 1 \\ T2 - t2 \geq 1 \\ 460 - t9 \geq 1 \\ T3 - t4 \geq 1 \\ 380 - t6 \geq 1 \\ T4 - t1 \geq 1 \\ 380 - t8 \geq 1 \\ T5 - t3 \geq 1 \\ T1 \geq 320 \\ T2 \geq 380 \\ T3 \geq 360 \\ T4 \geq 360 \\ T5 \geq 320 \end{array}\right] \text{Feasibility constraints}$$

$$f1, f2, \ldots, f25 \geq 0] \text{ Nonnegativity constraints}$$

$$\left.\begin{array}{lcl} A_{H1,C1} & = & \frac{Q_{H1,C1}}{1 \cdot (LMTD)_{H1,C1}} \\ A_{H2,C1} & = & \frac{Q_{H2,C1}}{1 \cdot (LMTD)_{H2,C1}} \\ A_{H3,C1} & = & \frac{Q_{H3,C1}}{1 \cdot (LMTD)_{H3,C1}} \\ A_{H4,C1} & = & \frac{Q_{H4,C1}}{1 \cdot (LMTD)_{H4,C1}} \\ A_{H5,C1} & = & \frac{Q_{H5,C1}}{1 \cdot (LMTD)_{H5,C1}} \\ A_{S,C1} & = & \frac{QS}{1 \cdot (LMTD)_{S,C1}} \\ A_{H1,W} & = & \frac{Q_{H1,W}}{1 \cdot (LMTD)_{H1,W}} \\ A_{H2,W} & = & \frac{Q_{H2,W}}{1 \cdot (LMTD)_{H2,W}} \\ A_{H3,W} & = & \frac{Q_{H3,W}}{1 \cdot (LMTD)_{H3;W}} \\ A_{H4,W} & = & \frac{Q_{H4,W}}{1 \cdot (LMTD)_{H4,W}} \\ A_{H5,W} & = & \frac{Q_{H5,W}}{1 \cdot (LMTD)_{H5,W}} \end{array}\right] \text{Area definitions}$$

where the $LMTD$s are approximated using Paterson's formula.

$$\left.\begin{array}{ll} f6 - \frac{Q_{H4,C1}}{(\Delta T_{H4,C1,max})} & \geq 0 \\ f7 - \frac{Q_{H2,C1}}{(\Delta T_{H2,C1,max})} & \geq 0 \\ f8 - \frac{Q_{H5,C1}}{(\Delta T_{H5,C1,max})} & \geq 0 \\ f9 - \frac{Q_{H3,C1}}{(\Delta T_{H3,C1,max})} & \geq 0 \\ f10 - \frac{Q_{H1,C1}}{(\Delta T_{H1,C1,max})} & \geq 0 \end{array}\right] \begin{array}{l} \text{Linear bound} \\ \text{constraints on} \\ \text{inlet flow rates} \\ \text{of heat exchangers} \end{array}$$

where $\Delta T_{H_i,C_j,max}$ are fixed to the maximum possible temperature drops through each exchanger between H_i and $C1$.

(C) Logical Constraints between Flowrates and Binary Variables

$$
\begin{aligned}
f6 - 18 \cdot y_{H4,C1} &\leq 0 \\
f7 - 18 \cdot y_{H2,C1} &\leq 0 \\
f8 - 18 \cdot y_{H5,C1} &\leq 0 \\
f9 - 18 \cdot y_{H3,C1} &\leq 0 \\
f10 - 18 \cdot y_{H1,C1} &\leq 0
\end{aligned}
$$

Objective Function

The objective function represents the total annual cost and consists of the investment and operating cost properly weighted. In the following form of the objective function, we substitute the areas of the heat exchangers via the presented definitions in constraints (B):

$$
\begin{aligned}
\min \quad 1200 \quad & \{A_{H1,C1}^{0.6} \cdot y_{H1,C1} + A_{H2,C1}^{0.6} \cdot y_{H2,C1} + A_{H3,C1}^{0.6} \cdot y_{H3,C1} \\
& + A_{H4,C1}^{0.6} \cdot y_{H4,C1} + A_{H5,C1}^{0.6} \cdot y_{H5,C1} + A_{S,C1}^{0.6} \cdot y_{S,C1} \\
& + A_{H1,W}^{0.6} \cdot y_{H1,W} + A_{H2,W}^{0.6} \cdot y_{H2,W} + A_{H3,W}^{0.6} \cdot y_{H3,W} \\
& + A_{H4,W}^{0.6} \cdot y_{H4,W} + A_{H5,W}^{0.6} \cdot y_{H5,W}\}
\end{aligned}
$$

Remark 4 The presented optimization model is an MINLP problem. The binary variables select the process stream matches, while the continuous variables represent the utility loads, the heat loads of the heat exchangers, the heat residuals, the flow rates and temperatures of the interconnecting streams in the hyperstructure, and the area of each exchanger. Note that by substituting the areas from the constraints (B) into the objective function we eliminate them from the variable set. The nonlinearities in the in the proposed model arise because of the objective function and the energy balances in the mixers and heat exchangers. As a result we have nonconvexities present in both the objective function and constraints. The solution of the MINLP model will provide *simultaneously* the

(i) Hot and cold utility loads needed,

(ii) Matches that take place,

(iii) Heat loads of each match,

(iv) Areas of exchangers, and

(v) Topology (structure) of total annual cost HEN.

Due to the existence of nonconvexities, this solution is only a local optimum solution.

Solution Procedure

This example was solved using the solution procedure described in Ciric and Floudas (1991). Use of the **v2–GBD** is made via the library OASIS (Floudas, 1990). The partitioning of the variables was done such that

$$y \;:\; y_{H_i,C1}, y_{H_i,W} \quad , \quad i = 1,2,\ldots,5 \quad , \quad y_{S,C1}$$
$$x \;:\; \text{all continuous variables}$$

As a result, the primal problem has fixed the y variables that represent the selection of matches. Note that since the flowrates of the streams associated with unselected matches are zero, then we can delete from the primal problem the constraints of the unselected matches, which implies that we have a reduced size primal problem at each iteration.

The master problem formally consists of Lagrange functions and possible integer cuts. Note though that the transshipment model constraints (A) are included in the master problem, even though we project only on the binary variables. The transshipment model constraints in the master problem restrict the combinations of matches to the feasible ones based on the heat flow representation, and as such we avoid infeasible primal subproblems.

This solution approach was used to solve the illustrative example, and resulted in the HEN shown in Figure 8.28, with data on the matches heat loads and areas shown in Table 8.10.

Note that the HEN has a total annual cost of \$571,080 and consumes 3592 kW of fuel and 132 kW of cooling water. Note also that it features seven exchangers which is the minimum possible. Following the approach for Ciric and Floudas (1991) for calculating the heat loads of QS,QW as a function of $HRAT$ (linear in this example) we find that the optimal $HRAT$ is 8.42 K.

The topology of the HEN shown in Figure 8.28 features a split of $C1$ into three streams. In the top two branches there are two heat exchangers in series, while in the third branch there is one heat exchanger. Note that there is also a by-pass stream directed to the inlet of the $H1 - C1$ exchanger.

Note that the optimal HEN of Figure 8.28 cannot be obtained with the approach of Yee *et al.* (1990b) presented in Section 8.5.4. This is due to the isothermal mixing assumption that they have imposed so as to simplify their superstructure representation.

This example was also solved via the MAGNETS approach (Floudas *et al.*, 1986) in which a strict pinch was maintained (see Figure 8.20 and synthesis strategy of section 8.4.2), and a golden section search was applied for optimizing the loop of ΔT_{min}. The optimal solution found is shown in Figure 8.29 and features a total annual cost of \$572,900 with $\Delta T_{min,optimal} = 7.78$ K. Note also that this network contains eight units arranged in a series-parallel configuration.

8.5.4 Simultaneous Optimization Models for HEN Synthesis

8.5.4.1 Problem Statement

Given are a set of hot process streams, HP, to be cooled and a set of cold process streams, CP, to be heated with specified heat capacity flow rates. The inlet and outlet temperatures are specified as either exact values or via inequalities. A set of hot and cold utilities with specified inlet and outlet temperatures are also provided.

The objective is to determine the least annual cost heat exchanger network by providing:

Match	Q (kW)	$A(m^2)$
$H1 - C1$	948.454	79.391
$H2 - C1$	400	29.057
$H3 - C1$	600	57.488
$H4 - C1$	400	14.88
$H5 - C1$	720	25.504
$H1 - W$	131.546	6.28
$S - C1$	3591.546	32.113

Table 8.10: Matches, heat loads, areas for optimal solution

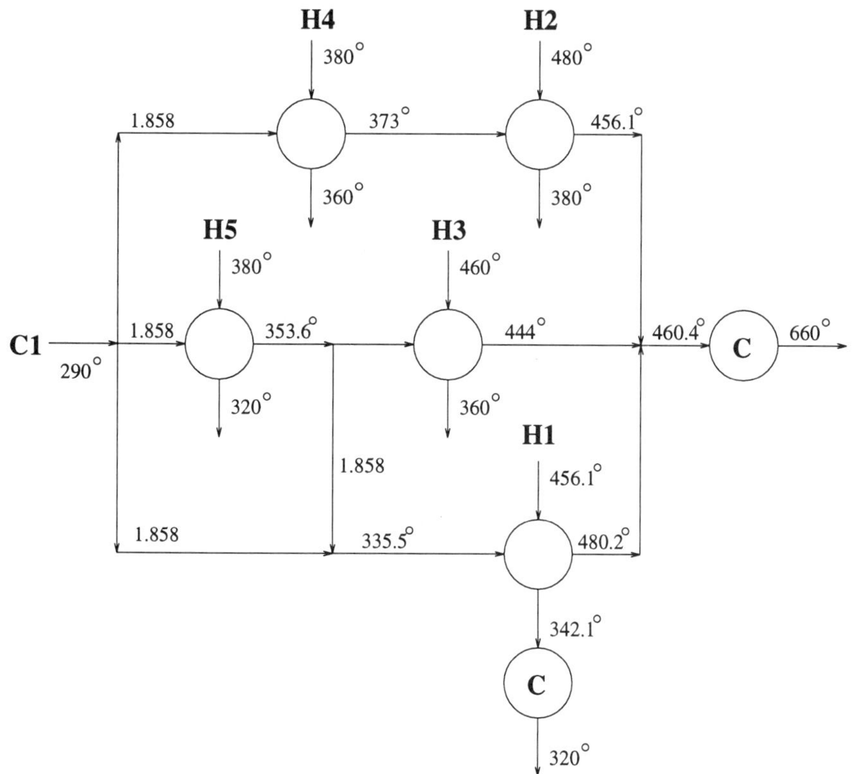

Figure 8.28: Optimal HEN structure for illustrative example 8.5.3

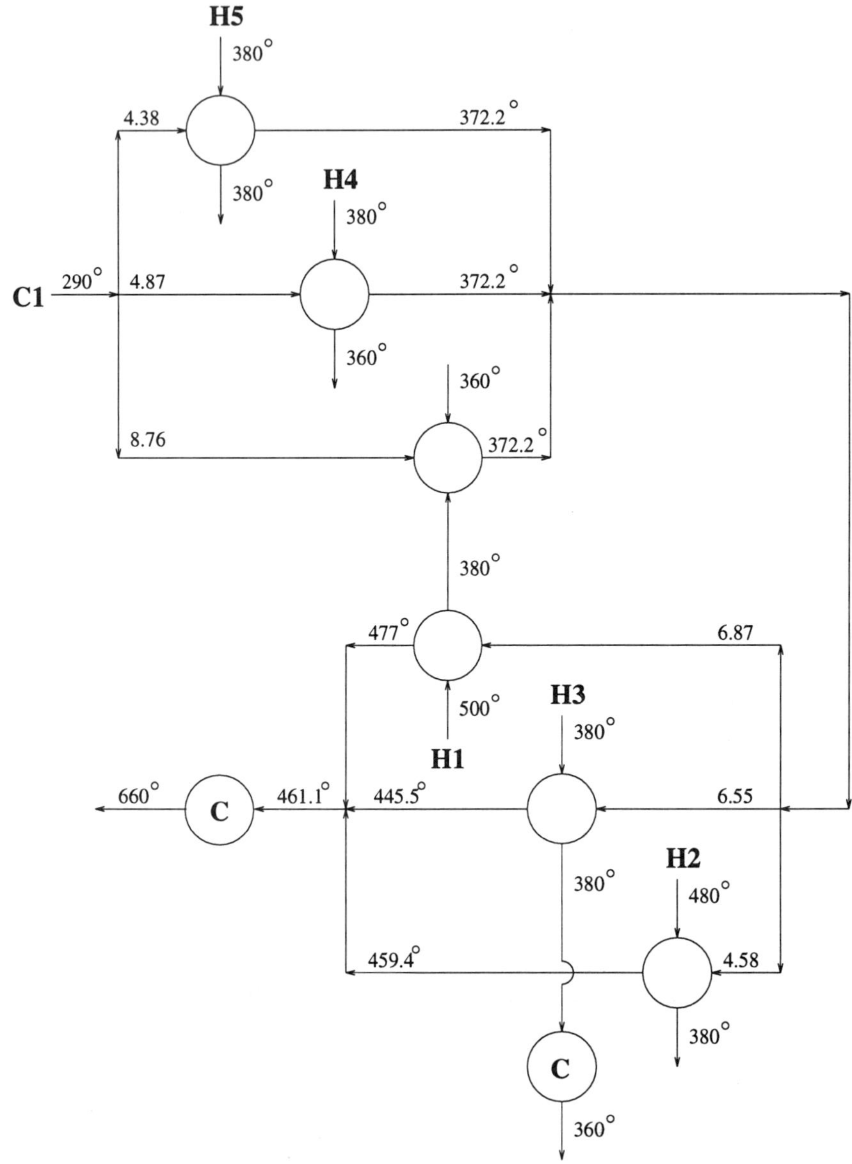

Figure 8.29: Optimal solution using MAGNETS for the illustrative example 8.5.3

(i) Hot and cold utilities required,

(ii) Matches that take place,

(iii) Heat loads of each match,

(iv) Areas of each exchanger, and

(v) Topology (structure) of the HEN.

Remark 1 The problem statement is identical to the problem statement of section 8.5.3.1 for the synthesis of HENs without decomposition (Ciric and Floudas, 1991). Note that as in section 8.5.3.1, there is no specification of any parameters so as to simplify or decompose the original problem into subproblems. In other words, the level of energy recovery (specified by fixing $HRAT$), the minimum approach temperature ($EMAT$), and the number of matches are not specified *a priori*. As a result, there is no decomposition into subnetworks based on the location of the pinch point(s), but instead the pinch point(s) are optimized simultaneously with the matches and the network topology. The approach presented for this problem is from Yee and Grossmann (1990) and is an alternative approach to the one of HEN synthesis without decomposition proposed by Ciric and Floudas (1991), and which was presented in section 8.5.3.

8.5.4.2 Basic Idea

The basic idea in the simultaneous optimization approach for HEN synthesis of Yee and Grossmann (1990) consists of

(i) Postulating a *simplified* superstructure based upon the problem data and the assumptions of (a) stage-wise representation and (b) isothermal mixing, and

(ii) Formulating the simplified superstructure as an MINLP model whose solution provides all the needed information stated in the problem definition.

Remark 1 The main motivation behind the development of the simplified superstructure was to end up with a mathematical model that features only linear constraints while the nonlinearities appear only in the objective function. Yee *et al.* (1990a) identified the assumption of isothermal mixing which eliminates the need for the energy balances, which are the nonconvex, nonlinear equality constraints, and which at the same time reduces the size of the mathematical model. These benefits of the isothermal mixing assumption are, however, accompanied by the drawback of eliminating from consideration a number of HEN structures. Nevertheless, as has been illustrated by Yee and Grossmann (1990), despite this simplification, good HEN structures can be obtained.

In the following two sections we will discuss the derivation of the simplified HEN superstructure and the resulting MINLP model.

8.5.4.3 Simplified HEN Superstructure

The basic idea of Yee *et al.* (1990a) in deriving a simplified HEN superstructure is based upon:

(i) A stage-wise representation, and

(ii) The isothermal mixing assumption.

In (i), the simplified superstructure consists of a number of stages, N_S, and in each stage many different possibilities for stream matching and sequencing are allowed to take place. In particular, at every stage, all possible matches are allowed to take place with a restriction imposed by (ii). The selection of the number of stages, N_S, is done such that

$$N_S \leq \max\{N_H, N_C\},$$

where N_H, N_C are the number of hot and cold stream, respectively. This selection can be regarded as arbitrary, but it is based on the observation that an optimal HEN does not usually involve a large number of heat exchangers, implying that a process stream does not exchange heat with many other streams. Nevertheless, the number of stages can be easily increased if a better representation is needed at the expense however of increasing the size of the model.

In (ii), the assumption of isothermal mixing implies that at each stage a process stream is split into a number of streams that are equal to the possible matches (i.e., parallel configuration), and the outlets of all matches feature the same temperature (i.e., isothermal mixing) which is the unknown outlet temperature of the considered stream from this stage. As a result of the isothermal mixing assumption, we can eliminate the energy balance constraints around each exchanger as well as at the mixing points and write only an overall balance for each stream within each stage, as well as an overall energy balance for each stream. The energy balances at each heat exchanger and mixing points can be eliminated because by assuming at each stage a parallel configuration for each stream with isothermal mixing of the outlet streams, then by determining the optimal temperatures of each stage we can backtrack and calculate the flow rates at each split stream. Note, however, that the isothermal mixing assumption excludes from consideration a number of structural alternatives.

The simplified HEN superstructure can be derived with the following procedure:

(1) Fix the number of stages, N_S, typically at

$$N_S = \max\{N_H, N_C\}.$$

(2) For each stage, each stream is split into a number of streams equal to the number of potential matches which are directed to the exchangers representing each potential match. The outlets of each exchanger feature the same temperature and are mixed at a mixing point where the temperature of the considered stream at the next stage is defined.

(3) The outlet temperature of each stream and at each stage are treated as unknown variables. Note that due to the isothermal mixing assumption these outlet temperatures are equal to the outlet temperatures of each heat exchanger in which this stream participates at the particular stage.

Illustration 8.5.4 (Simplified HEN Superstructure)

Stream	T^{in} (K)	T^{out} (K)	FC_p (kW/K)	Cost (\$/kW/yr)
$H1$	443	333	30	
$H2$	423	303	15	
$C1$	293	408	20	
$C2$	353	413	40	
S	450	450		80
W	293	313		20
$U = 0.8$kW/m^2K for all matches except ones involving steam				
$U = 1.2$kW/m^2K for matches involving steam				
Annual cost $= 1000A^{0.6}$ for all exchangers except heaters				
Annual cost $= 1200A^{0.6}$ for heaters, $A[=]m^2$				

Table 8.11: Data for illustrative example of simplified HEN superstructure

To illustrate the derivation of a simplified HEN superstructure of Yee *et al.* (1990a), we will consider the following example which is taken from Yee *et al.* (1990a) with data shown in Table 8.11 :

$$\left.\begin{array}{l} N_H = 2 \\ N_C = 2 \end{array}\right\} \Rightarrow N_S = 2.$$

From the data, it is possible to have the following matches at each stage:

$$H1 - C1$$
$$H1 - C2$$
$$H2 - C1$$
$$H2 - C2$$

while we assume for simplicity that the utility matches can be placed at the end of the stage representation.

For instance, in stage 1, $H1$ is split into two streams directed to exchangers $H1-C1$, $H1-C2$ (parallel configuration) with outlets having identical temperatures. The outlets are mixed, and this temperature, which is equal to the temperatures of the outlet streams of the $H1 - C1$, $H1 - C2$ exchangers, is the outlet temperature of stream $H1$ from stage 1. Similarly, the representation of hot stream $H2$ and cold streams $C1$ and $C2$ can be derived for stage 1.

In stage 2, we have an identical representation of matches for all streams to that of stage 1. At the right end of stage 2 we allow for coolers for $H1$ and $H2$, while in the left end of stage 1 we allow for heaters for $C1$ and $C2$. The simplified HEN superstructure for this example is shown in Figure 8.30.

Remark 1 In the simplified two-stage superstructure shown in Figure 8.30, we have four process matches at each stage, and four utility matches at the left and right end. Note however that if the number of stages becomes large (e.g., $N_S = 20$, which corresponds to having 20 hot or cold

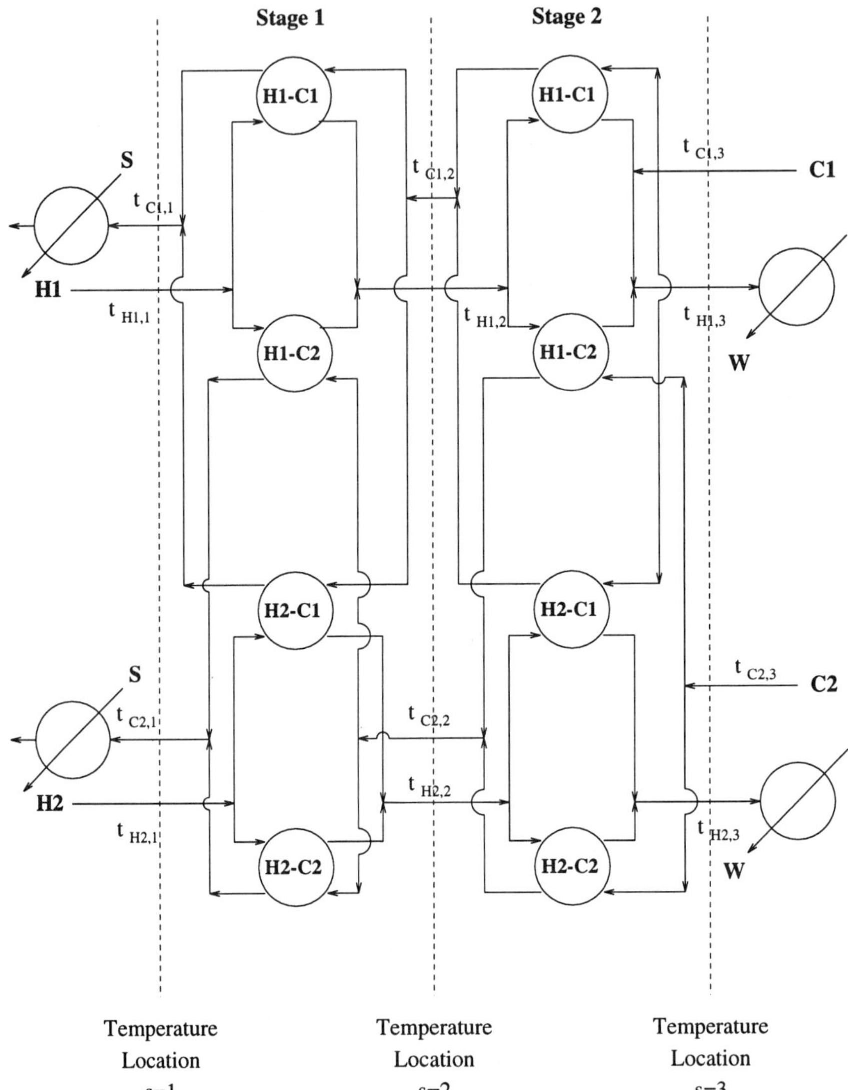

Figure 8.30: Simplified HEN superstructure for illustrative example 8.5.4

streams) then we postulate a multiple of all potential matches by the number of stages. As a result even though we may have the benefit of having linear constraints, as we will see in the next section, we may have a rather large model.

Remark 2 In Figure 8.30, the two stages are represented by eight heat exchangers with four possible matches at each stage and variable temperatures at each stage. Moreover, the structural alternatives of parallel and series configurations, as well as rematching of streams are embedded in the simplified superstructure.

Remark 3 What are not embedded in this simplified HEN superstructure are the structural alternatives that are excluded due to the isothermal mixing assumption. To understand better the meaning of the isothermal assumption, let us isolate from Figure 8.30 the part of the simplified superstructure for stream $H1$ which is shown in Figure 8.31. The isothermal mixing assumption states that the temperatures of stream $H1$ at the outlet of the exchanger, $H1 - C1$, $H1 - C2$ of the first stage are equal and hence the temperature of outlet stream from the mixer is also equal to them (i.e., $t_{H1,2}$). Similarly, we have for the outlet temperatures of the second stage. The combination of this assumption with the parallel configuration at each stage allows us to write only an overall heat balance for each stream at each stage. This is especially useful when the heat capacity flow rates are fixed in which case we no longer need to introduce the flow rate heat capacities of the solid streams in the model. The two major advantages of this are

(a) The dimensionality of the model is reduced, and

(b) The set of constraints become linear.

It is important to note, however, the cases for which the isothermal mixing assumption is rigorously valid. This is only when the resulting HEN structure does not have any splitting of the streams. For HEN configurations that have splits present, the isothermal mixing assumption may lead to an overestimation of the area cost and hence the investment cost. This may happen because it will attempt to restrict the trade-off of area among heat exchangers that are associated with split streams.

Furthermore, it is also possible that the resulting HEN structure may feature more heat exchangers than needed if splits are present. To overcome this limitation, resulting from the isothermal mixing assumption, Yee *et al.* (1990a) suggested the solution of a suboptimization problem in which the structure has been fixed but the flows and temperatures are being optimized so as to identify the optimal split flow and area distribution of the heat exchangers. This suboptimization has only continuous variables, linear constraints, and nonlinear objective and in certain cases may result in a reduction in the number of heat exchangers. Yee *et al.* (1990a) suggested an approach that combines the model of the simplified HEN superstructure with the solution of the NLP suboptimization model so as to arrive at an optimal HEN structure.

Despite however this suggested modification, the simplified HEN superstructure has a number of fundamental limitations

(i) It may exclude structures that are only feasible with nonisothermal mixing (e.g., by-pass streams), and

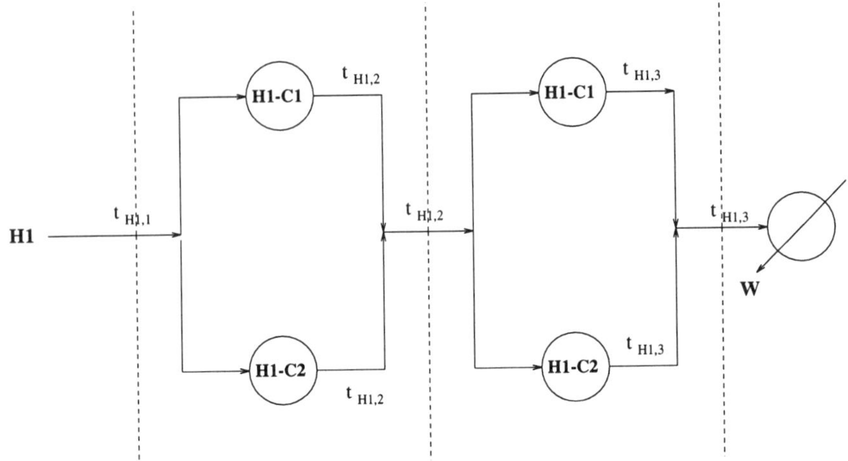

Figure 8.31: Restrictions imposed by the isothermal mixing assumption

(ii) It neglects from consideration the structures in which a split stream goes through two or more exchangers in series.

A number of such HEN structures which are excluded in the simplified HEN superstructure of Yee *et al.* (1990a) are shown in Figure 8.32.

Remark 4 From the HEN structures shown in Figure 8.32, it becomes apparent that the simplified HEN superstructure of Yee *et al.* (1990a) cannot identify the optimal solution of the illustrative example of section 8.4.1.1 which requires non-isothermal mixing with a by-pass, and it canno/t determine the optimal structure of the illustrative example of the HEN synthesis approach without decomposition of Ciric and Floudas (1991) since it involves the features of networks (a) and (c) o/f Figure 8.32.

Remark 5 The hyperstructure of Floudas and Ciric (1989) presented in section 8.5.1.3 has embedded all possible matches and all alternatives structures, while the simplified HEN superstructure of Yee *et al.* (1990a) has embedded all possible matches and a limited number of structures by the isothermal mixing assumption as many times as the number of stages. Note that both the hyperstructure and the simplified superstructure require as input the maximum number of times that two streams can exchange heat. The mathematical model of the hyperstructure has nonlinear constraints due to the inclusion of energy balances while the model of the modified HEN superstructure has only linear constraints which may make it more robust in the solution algorithm. Note, however, that the model complexity will undoubtedly face a trade-off in the case of many stages where the constraints may be linear but correspond to a large set, while the nonlinear constraints of the hyperstructure model do not depend on the number of stages and hence remain constant depending only on the potential matches.

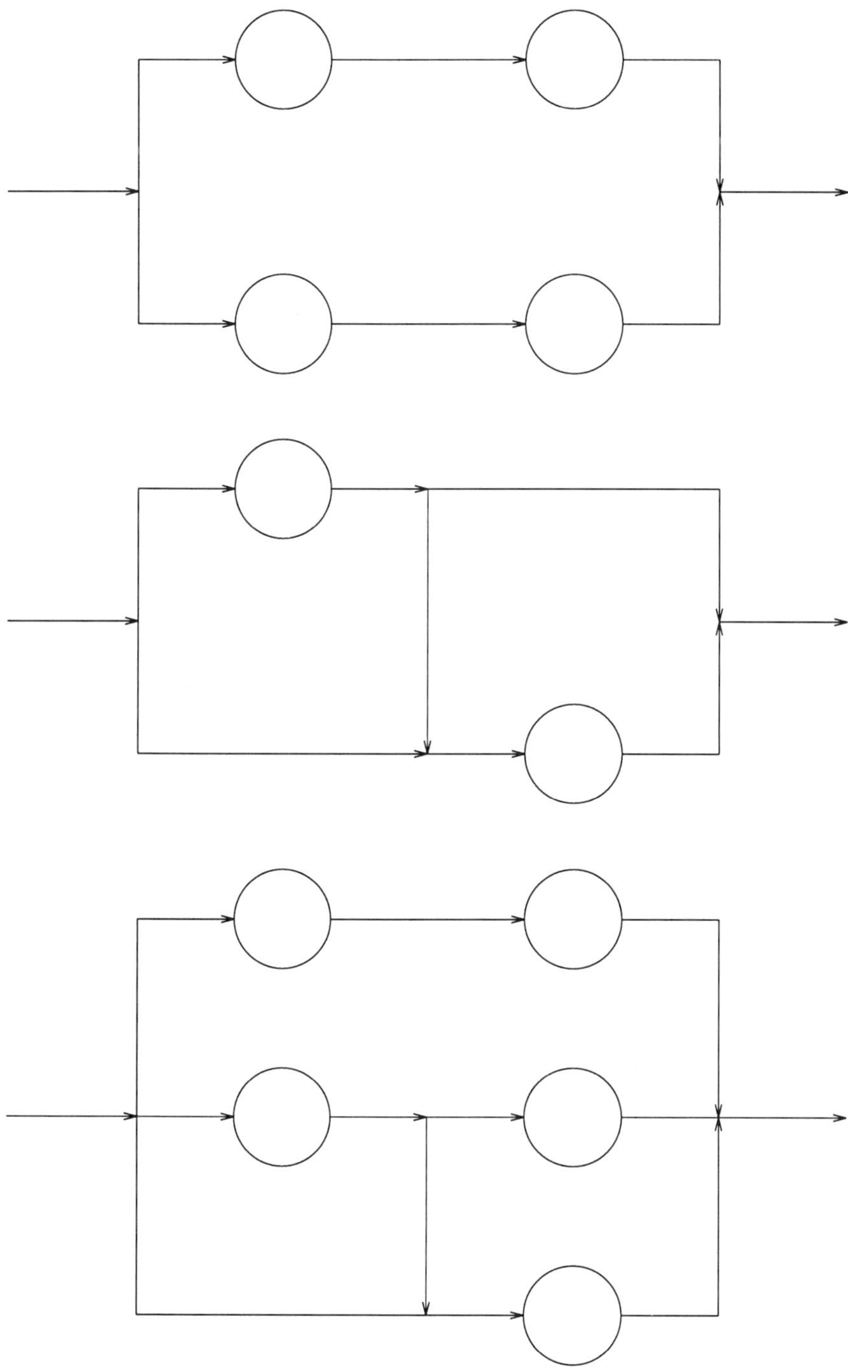

Figure 8.32: Structures excluded from simplified HEN superstructure

8.5.4.4 Mathematical Model of Simplified HEN Superstructure

To determine a minimum total annualized cost HEN structure out of the alternatives embedded in the simplified HEN superstructure presented in section 8.5.4.3, we define:

(i) Continuous variables for the heat flow, energy balances and feasibility constraints, and

(ii) Binary variables representing the existence of a process or utility match at each stage.

In (i), we introduce continuous variables for the

(a) Heat exchanged between hot and cold process streams, hot process stream and cold utility, and cold process stream and hot utility,

(b) Temperatures of hot and cold streams at the hot ends of each stage, and

(c) Temperature approaches for process matches and process-utility matches at each stage.

In (ii), we introduce binary variables for the

(a) Existence of process matches at each stage, and

(b) Existence of process-utility matches at each stage.

The set of constraints consists of

(A) Overall heat balance for each stream,

(B) Heat balance for each stream at each stage,

(C) Assignment of inlet temperatures,

(D) Feasibility of temperatures,

(E) Utility matches energy balances,

(F) Logical constraints, and

(G) Calculation of temperature approaches.

The objective function involves the minimization of the total annual cost.

We will discuss the set of constraints and the objective function for the illustrative example presented in section 8.5.4.3.

Illustration 8.5.5 (Optimization Model)

We will discuss the explicit formulation of the constraints and objective function for the simplified HEN superstructure of the illustrative example of section 8.5.4.3.

(A) – Overall heat balances for each stream :

$$
\begin{aligned}
Q_{H1,C1,1} + Q_{H1,C2,1} + Q_{H1,C1,2} + Q_{H1,C2,2} + Q_{H1,W} &= 30 \cdot (443 - 333) = 3300 \\
Q_{H2,C1,1} + Q_{H2,C2,1} + Q_{H2,C1,2} + Q_{H2,C2,2} + Q_{H2,W} &= 15 \cdot (423 - 303) = 1800 \\
Q_{H1,C1,1} + Q_{H2,C1,1} + Q_{H1,C1,2} + Q_{H2,C1,2} + Q_{C1,S} &= 30 \cdot (443 - 333) = 3300 \\
Q_{H2,C2,1} + Q_{H2,C2,1} + Q_{H2,C2,2} + Q_{H2,C2,2} + Q_{C2,S} &= 15 \cdot (423 - 303) = 1800
\end{aligned}
$$

The above overall heat balances state that the heat load of each process stream equals the sum of heat exchanged of that process stream with other process and utility streams. Note that the first subindex denotes the hot stream, the second the cold stream, and the third the stage.

(B) - Heat Balances for each stream at each stage :

$$
\left.
\begin{aligned}
Q_{H1,C1,1} + Q_{H1,C2,1} &= 30 \cdot (t_{H1,1} - t_{H1,2}) \\
Q_{H2,C1,1} + Q_{H2,C2,1} &= 15 \cdot (t_{H2,1} - t_{H2,2}) \\
Q_{H1,C1,1} + Q_{H2,C1,1} &= 20 \cdot (t_{C1,1} - t_{C1,2}) \\
Q_{H1,C2,1} + Q_{H2,C2,1} &= 40 \cdot (t_{C2,1} - t_{C2,2})
\end{aligned}
\right] \text{Stage 1}
$$

$$
\left.
\begin{aligned}
Q_{H1,C1,2} + Q_{H1,C2,2} &= 30 \cdot (t_{H1,2} - t_{H1,3}) \\
Q_{H2,C1,2} + Q_{H2,C2,2} &= 15 \cdot (t_{H2,2} - t_{H2,3}) \\
Q_{H1,C1,2} + Q_{H2,C1,2} &= 20 \cdot (t_{C1,2} - t_{C1,3}) \\
Q_{H1,C2,2} + Q_{H2,C2,2} &= 40 \cdot (t_{C2,2} - t_{C2,3})
\end{aligned}
\right] \text{Stage 2}
$$

where t's are the temperatures of hot and cold streams at the hot ends of each stage as shown in Figure 8.30.

The above heat balances take into account that in two adjacent stages the outlet temperature of a stream at stage 1 is its inlet temperature in stage 2.

(C) - Assignment of inlet temperatures :

$$
\begin{aligned}
t_{H1,1} &= 443 \\
t_{H2,1} &= 423 \\
t_{C1,3} &= 293 \\
t_{C2,3} &= 353
\end{aligned}
$$

We can substitute these values in the rest of the constraints and the objective function.

(D) - Feasibility of temperatures :

$$
\left.
\begin{aligned}
t_{H1,1} &\geq t_{H1,2} \\
t_{H1,2} &\geq t_{H1,3} \\
t_{H1,3} &\geq 333
\end{aligned}
\right] \text{Hot stream } H1
$$

$$\left.\begin{array}{rcl} t_{H2,1} & \geq & t_{H2,2} \\ t_{H2,2} & \geq & t_{H2,3} \\ t_{H2,3} & \geq & 303 \end{array}\right] \text{Hot stream } H2$$

$$\left.\begin{array}{rcl} t_{C1,3} & \leq & t_{C1,2} \\ t_{C1,2} & \leq & t_{C1,1} \\ t_{C1,1} & \leq & 408 \end{array}\right] \text{Cold stream } C1$$

$$\left.\begin{array}{rcl} t_{C2,3} & \leq & t_{C2,2} \\ t_{C2,2} & \leq & t_{C2,1} \\ t_{C2,1} & \leq & 413 \end{array}\right] \text{Cold stream } C2$$

These constraints need to be imposed so as to maintain a monotonic increase/decrease of the temperatures of each stage.

(E) - Energy balances for utility matches :

$$\begin{array}{rcl} Q_{H1,W} & = & 30 \cdot (t_{H1,3} - 333) \\ Q_{H2,W} & = & 15 \cdot (t_{H2,3} - 303) \\ Q_{C1,S} & = & 20 \cdot (408 - t_{C1,1}) \\ Q_{C2,S} & = & 40 \cdot (413 - t_{C2,1}) \end{array}$$

These constraints can be introduced directly in the overall balances (A) and the objective function, and hence can be eliminated (i.e., reduction of four variables and four constraints).

(F) - Logical constraints :

$$\left.\begin{array}{rcl} Q_{H1,C1,1} - \mathcal{U}_{H1,C1} \cdot y_{H1,C1,1} & \leq & 0 \\ Q_{H1,C2,1} - \mathcal{U}_{H1,C2} \cdot y_{H1,C2,1} & \leq & 0 \\ Q_{H2,C1,1} - \mathcal{U}_{H2,C1} \cdot y_{H2,C1,1} & \leq & 0 \\ Q_{H2,C2,1} - \mathcal{U}_{H2,C2} \cdot y_{H2,C2,1} & \leq & 0 \end{array}\right] \text{Stage 1}$$

$$\left.\begin{array}{rcl} Q_{H1,C1,2} - \mathcal{U}_{H1,C1} \cdot y_{H1,C1,2} & \leq & 0 \\ Q_{H1,C2,2} - \mathcal{U}_{H1,C2} \cdot y_{H1,C2,2} & \leq & 0 \\ Q_{H2,C1,2} - \mathcal{U}_{H2,C1} \cdot y_{H2,C1,2} & \leq & 0 \\ Q_{H2,C2,2} - \mathcal{U}_{H2,C2} \cdot y_{H2,C2,2} & \leq & 0 \end{array}\right] \text{Stage 2}$$

$$\left.\begin{array}{rcl} Q_{H1,W} - \mathcal{U}_{H1,C1} \cdot y_{H1,W} & \leq & 0 \\ Q_{H1,W} - \mathcal{U}_{H1,C2} \cdot y_{H1,W} & \leq & 0 \\ Q_{C1,S} - \mathcal{U}_{C1,S} \cdot y_{C1,S} & \leq & 0 \\ Q_{C2,S} - \mathcal{U}_{C2,S} \cdot y_{C2,S} & \leq & 0 \end{array}\right] \text{Utility matches}$$

where y's are $0-1$ variables and the $\mathcal{U}$'s are given as the smallest heat content of the two streams; that is,

$$\mathcal{U}_{H1,C1} = \min\{3300, 2300\} = 2300$$

$$\begin{aligned}\mathcal{U}_{H1,C2} &= \min\{3300, 2400\} = 2400\\ \mathcal{U}_{H2,C1} &= \min\{1800, 2300\} = 1800\\ \mathcal{U}_{H2,C2} &= \min\{1800, 2400\} = 1800\end{aligned}$$

and $\mathcal{U}_{H1,W}, \mathcal{U}_{H2,W}, \mathcal{U}_{C1,S}, \mathcal{U}_{C1,S}$ are some large upper bounds.

The logical constraints make certain that if a match does not take place (i.e., $y_{H1,C1,1} = 0$) then the corresponding heat loads become zero.

(G) - Calculation of temperature approaches :

$$\left.\begin{aligned}\Delta T_{H1,C1,1} &\leq t_{H1,1} - t_{C1,1} + \mathcal{U}(1 - y_{H1,C1,1})\\ \Delta T_{H1,C2,1} &\leq t_{H1,1} - t_{C2,1} + \mathcal{U}(1 - y_{H1,C2,1})\\ \Delta T_{H2,C1,1} &\leq t_{H2,1} - t_{C1,1} + \mathcal{U}(1 - y_{H2,C1,1})\\ \Delta T_{H2,C2,1} &\leq t_{H2,1} - t_{C2,1} + \mathcal{U}(1 - y_{H2,C2,1})\end{aligned}\right] \quad \begin{array}{l}\text{temperature}\\ \text{location } s = 1\end{array}$$

$$\left.\begin{aligned}\Delta T_{H1,C1,2} &\leq t_{H1,2} - t_{C1,2} + \mathcal{U}(1 - y_{H1,C1,2})\\ \Delta T_{H1,C2,2} &\leq t_{H1,2} - t_{C2,2} + \mathcal{U}(1 - y_{H1,C2,2})\\ \Delta T_{H2,C1,2} &\leq t_{H2,2} - t_{C1,2} + \mathcal{U}(1 - y_{H2,C1,2})\\ \Delta T_{H2,C2,2} &\leq t_{H2,2} - t_{C2,2} + \mathcal{U}(1 - y_{H2,C2,2})\end{aligned}\right] \quad \begin{array}{l}\text{temperature}\\ \text{location } s = 2\end{array}$$

$$\left.\begin{aligned}\Delta T_{H1,C1,3} &\leq t_{H1,3} - t_{C1,3} + \mathcal{U}(1 - y_{H1,C1,3})\\ \Delta T_{H1,C2,3} &\leq t_{H1,3} - t_{C2,3} + \mathcal{U}(1 - y_{H1,C2,3})\\ \Delta T_{H2,C1,3} &\leq t_{H2,3} - t_{C1,3} + \mathcal{U}(1 - y_{H2,C1,3})\\ \Delta T_{H2,C2,3} &\leq t_{H2,3} - t_{C2,3} + \mathcal{U}(1 - y_{H2,C2,3})\end{aligned}\right] \quad \begin{array}{l}\text{temperature}\\ \text{location } s = 3\end{array}$$

$$\left.\begin{aligned}\Delta T_{H1,W} &\leq t_{H1,3} - 313 + \mathcal{U}(1 - y_{H1,W})\\ \Delta T_{H1,W} &\leq t_{H2,3} - 313 + \mathcal{U}(1 - y_{H1,W})\\ \Delta T_{C2,S} &\leq 450 - t_{C1,1} + \mathcal{U}(1 - y_{C1,S})\\ \Delta T_{C2,S} &\leq 450 - t_{C2,1} + \mathcal{U}(1 - y_{C2,S})\end{aligned}\right] \quad \text{utility matches}$$

To incorporate the area of each match in the objective function we calculate the temperature approaches at all the locations of the simplified superstructure, as shown above. Note that we need to introduce the binary variables in these calculations so as to maintain feasible driving forces. To explain the need for using the binary variables, let us consider the calculation of $\Delta T_{H1,C1,1}$:

$$\text{If } y_{H1,C1,1} = 1, \text{then } \Delta T_{H1,C1,1} \leq t_{H1,1} - T_{C1,1};$$

that is, if the match takes place in stage 1, then the temperature approach is the one given by the difference of $t_{H1,1}$ and $t_{C1,1}$. If, however, $y_{H1,C1,1} = 0$, then the constraint becomes

$$\Delta T_{H1,C1,1} \leq t_{H1,1} - t_{C1,1} + \mathcal{U},$$

which corresponds to a relaxation of the constraint so as to avoid infeasibilities.

Note also that the above constraints are written as less than or equal (i.e., inequalities instead of equalities) since the cost of heat exchangers increases with higher values of the ΔT's. The

ΔT's are set greater than or equal to a small value ϵ; that is,

$$\Delta T\text{'s} \geq \epsilon,$$

so as to avoid infinite areas. In this case we have $EMAT = \epsilon$.

For the considered example we have

$$\begin{aligned} t_{H1,3} &\geq 333 > 313 \\ t_{C1,1} &\leq 408 < 450 \\ t_{C2,1} &\leq 413 < 450 \end{aligned}$$

As a result, we do not need to introduce the binary variables

$$\begin{gathered} y_{H1,W} \\ y_{C1,S} \\ y_{C2,S} \end{gathered}$$

for the (G) constraints since we always have feasibility of heat exchange. We need however to introduce the binary $y_{H2,W}$ since

$$t_{H2,3} \geq 303$$

As a result we eliminate the aforementioned three binary variables and have them as equal to 1 in the logical constraints (i.e., having upper bounds only on $Q_{H1,W}$, $Q_{C1,S}$, and $Q_{C2,S}$).

Note also that the set of constraints (A)-(G) exhibit the nice feature of linearity in the continuous and binary variables. Furthermore, the flow rates do not participate in the formulation at all and hence there is a reduction in the number of continuous variables by the number of flowrates. The penalty that we pay for this desirable feature is threefold: (i) we introduce more binary variables since we have possible process matches that are equal to the number of stages times the actual number of potential matches, (ii) we deal with a simplified set of alternatives that excludes a number of desirable structures, and (iii) we need to solve an NLP suboptimization problem to determine the flow rates of the split streams and possible reduce the number of heat exchangers if the resulting network exhibits splitting of the streams.

Objective function

The objective function consists of the total annual cost:

$$\begin{aligned} \min \quad & 1000\left(\frac{Q_{H1,C1,1}}{0.8(LMTD)_{H1,C1,1}}\right)^{0.6} + 1000\left(\frac{Q_{H1,C2,1}}{0.8(LMTD)_{H1,C2,1}}\right)^{0.6} \\ + \quad & 1000\left(\frac{Q_{H2,C1,1}}{0.8(LMTD)_{H2,C1,1}}\right)^{0.6} + 1000\left(\frac{Q_{H2,C2,1}}{0.8(LMTD)_{H2,C2,1}}\right)^{0.6} \\ + \quad & 1000\left(\frac{Q_{H1,C1,2}}{0.8(LMTD)_{H1,C1,2}}\right)^{0.6} + 1000\left(\frac{Q_{H1,C2,2}}{0.8(LMTD)_{H1,C2,2}}\right)^{0.6} \\ + \quad & 1000\left(\frac{Q_{H2,C1,2}}{0.8(LMTD)_{H2,C1,2}}\right)^{0.6} + 1000\left(\frac{Q_{H2,C2,2}}{0.8(LMTD)_{H2,C2,2}}\right)^{0.6} \\ + \quad & 1000\left(\frac{Q_{H1,W}}{0.8(LMTD)_{H1,W}}\right)^{0.6} + 1000\left(\frac{Q_{H2,W}}{0.8(LMTD)_{H2,W}}\right)^{0.6} \end{aligned}$$

Exchanger	Q (kW)	$A(m^2)$
1	628.8	22.8
1	271.2	19.3
1	2400	265.1
1	1400	179.0
1	400	38.3

Table 8.12: Heat loads and areas of optimal solution

$$\begin{aligned}
+ \quad & 1000\left(\frac{Q_{C1,S}}{0.8(LMTD)_{C1,S}}\right)^{0.6} + 1000\left(\frac{Q_{C2,S}}{0.8(LMTD)_{C2,S}}\right)^{0.6} \\
+ \quad & 80 \cdot [Q_{C1,S} + Q_{C2,S}] \\
+ \quad & 20 \cdot [Q_{H1,W} + Q_{H2,W}]
\end{aligned}$$

where Yee and Grossmann (1990) utilized the approximation of Chen for the $LMTD$s. For instance:

$$\begin{aligned}
(LMTD)_{H1,C1,1} &= \left[(\Delta T_{H1,C1,1}) \cdot (\Delta T_{H1,C1,2}) \cdot (\frac{\Delta T_{H1,C1,1} + \Delta T_{H1,C1,2}}{2})\right]^{\frac{1}{3}} \\
(LMTD)_{H1,W} &= \left[(\Delta T_{H1,W}) \cdot (333 - 293) \cdot (\frac{\Delta T_{H1,W} + (333 - 293)}{2})\right]^{\frac{1}{3}} \\
(LMTD)_{C1,S} &= \left[(\Delta T_{C1,S}) \cdot (450 - 408) \cdot (\frac{\Delta T_{C1,S} + (450 - 408)}{2})\right]^{\frac{1}{3}}
\end{aligned}$$

The objective function is nonlinear and nonconvex and hence despite the linear set of constraints the solution of the resulting optimization model is a local optimum. Note that the resulting model is of the MINLP type and can be solved with the algorithms described in the chapter of mixed-integer nonlinear optimization. Yee and Grossmann (1990) used the OA/ER/AP method to solve first the model and then they applied the NLP suboptimization problem for the fixed structure so as to determine the optimal flowrates of the split streams if these take place.

Despite the linear set of constraints, Yee and Grossmann (1990) proposed an initialization procedure for the solution of the relaxed NLP problem in the OA/ER/AP since this may improve the chances of obtaining the best relaxed solution (note that the relaxed NLP is nonconvex). This initialization procedure is presented in the appendix of their paper.

By setting $\epsilon = 0.1$, the solution of the MINLP model resulted in a network with split streams. As a result, the NLP suboptimization problem had to be solved to determine the flowrates of the split streams. The final network found by Yee and Grossmann (1990) is shown in Figure 8.31. Note that only cooling water is needed with a utility cost of 8,000 \$/yr and the network features a total annual cost of \$80,274. The matches and areas of the heat exchangers are shown in Table 8.12. Note also that exchangers 1 and 2 involve rematching of the streams $H1$, $C1$, and that the network features the minimum number of heat exchangers.

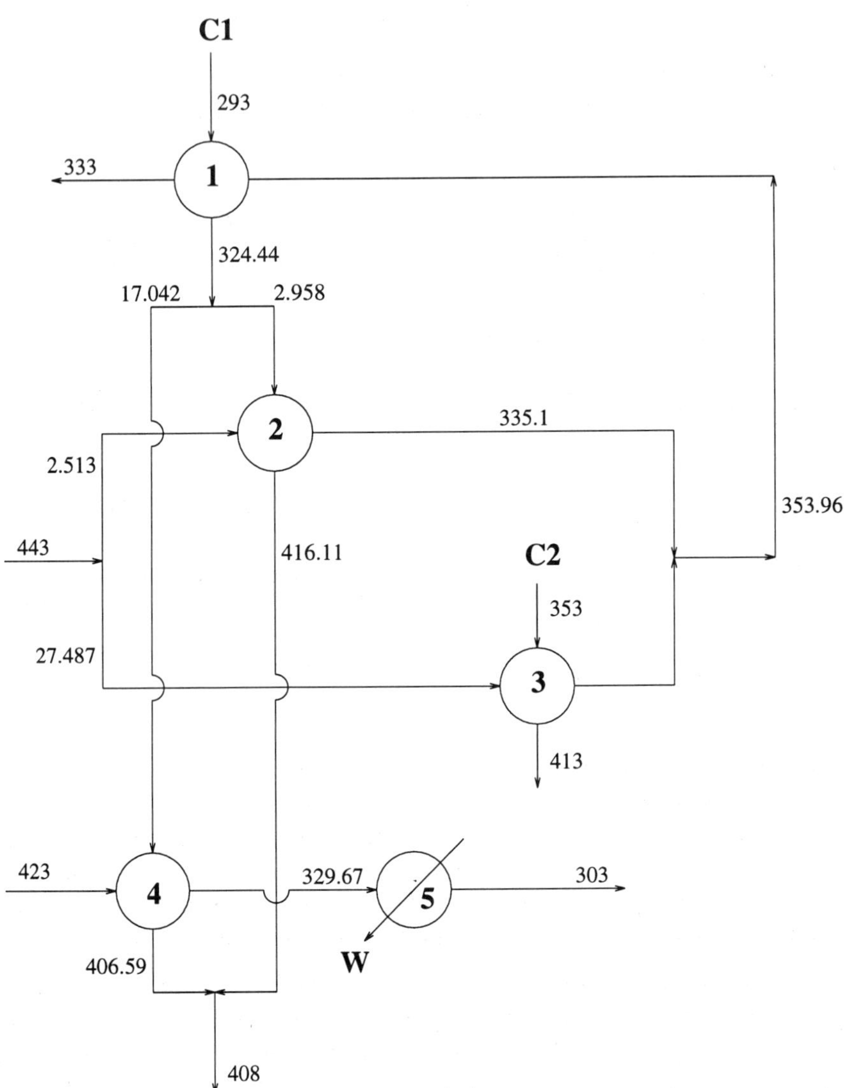

Figure 8.33: Optimal HEN structure for illustrative example

Summary and Further Reading

This chapter presents optimization-based approaches for the synthesis of heat exchanger networks. Sections 8.1 and 8.2 introduce the reader to the overall problem definition, key temperature approaches, and outline the different types of approaches proposed in the last three decades. For further reading, refer to the review paper of Gundersen and Naess (1988) and the suggested references.

Section 8.3 presents the targets for HEN synthesis. More specifically, section 8.3.1 focuses on the minimum utility cost target and its formulation as a linear programming LP transshipment model; Section 8.3.2 presents the minimum number of matches target which is formulated as a mixed-integer linear programming MILP problem; Section 8.3.2 discusses the vertical heat transfer criterion and presents the vertical MILP transshipment model that can be used to distinguish among solutions with the same minimum number of matches. For further reading on optimization models for target in HEN synthesis, refer to the suggested references and to the recent approaches proposed by Colberg and Morari (1990) and Yee *et al.* (1990a; 1990b).

Section 8.4 presents a decomposition-based approach for the HEN synthesis. Section 8.4.1 discusses an approach for the automatic generation of minimum investment cost HENs that satisfy the targets of minimum utility cost and minimum matches, and section 8.4.2 outlines the decomposition-based synthesis strategy. Further reading in this subject can be found in the suggested references.

Section 8.5 discusses the recently proposed simultaneous HEN synthesis approaches which do not rely on postulating targets that decompose and simplify the problem. Section 8.5.1 focuses on the simultaneous search for matches and network configurations, and discusses the HEN hyperstructure representation and its formulation as a mixed-integer nonlinear programming MINLP problem. Section 8.5.2 presents the pseudo-pinch concept, a calculation of the maximum cross-pinch flow and the pseudo-pinch synthesis strategy. Section 8.5.3 presents an approach for HEN synthesis that does not require decomposition, and discusses the resulting MINLP model, along with the proposed algorithmic procedure. Section 8.5.4 presents an alternative optimization approach for synthesis of HENs without decomposition, which aims at creating linear structure in the set of constraints of the MINLP model. For further reading in this subject refer to the suggested references.

Problems on Heat Exchanger Network Synthesis

1. Given is a process that consists of the following set of hot and cold process streams, hot and cold utilities:

Stream	FC_p (kW/K)	T^{IN} (K)	T^{OUT} (K)
$H1$	111.844	502.65	347.41
$H2$	367.577	347.41	320.00
$H3$	29.7341	405.48	310.00
$C1$	9.236	320.00	670.00
$C2$	112.994	368.72	450.00
$C3$	107.698	320.00	402.76
F		700.00	700.00
CW		295.00	325.00

$\Delta T_{min} = 15$K

Cost of heat exchangers $= 1300A^{0.6}$ \$/yr , A = m^2

Cost of furnace $= 0.45754Q^{0.7}$ \$/yr

Fuel cost $=$ 174.022 \$/kWyr

CW cost $=$ 5.149 \$/kWyr

(i) Formulate the LP transshipment model, and solve it to determine the minimum utility cost.

(ii) Formulate and solve the MILP transshipment model for the minimum number of matches at each subnetwork.

(iii) Postulate a superstructure for each subnetwork, formulate it as an NLP model, and solve it to determine a minimum investment cost HEN structure. For the overall heat transfer coefficient use $U = 0.2$ kW/(m^2 K) for all matches.

2. Given is the following set of hot and cold streams, hot and cold utilities, ΔT_{min} and cost data:

Stream	FC_p (kW/K)	T^{IN} (K)	T^{OUT} (K)
$C1$	24.795	288.888	650.000
$H1$	7.913	630.555	338.888
$H2$	5.803	583.333	505.555
$H3$	2.374	555.555	319.444
$H4$	31.652	494.444	447.222
$H5$	6.3305	477.777	311.111
$H6$	65.943	422.222	383.333
F		700.000	700.000
CW		300.000	333.333

$\Delta T_{min} = 7\text{K}$

Cost of heat exchangers $= 1300A^{0.6}$ \$/yr , A = m^2

Cost of furnace $= 0.45754Q^{0.7}$ \$/yr

Fuel cost $= 174.022$ \$/kWyr

CW cost $= 4.634$ \$/kWyr

$U = 0.3$ kW/(m^2 K) for all matches

(a) Address (i), (ii), and (iii) of problem 1.

(b) Repeat (a) for $\Delta T_{min} = 6\text{K}$, and $\Delta T_{min} = 10\text{K}$.

3. Given is the data of problem 2, and the minimum utility consumption for $\Delta T_{min} = 7\text{K}$,

(i) Postulate a hyperstructure without decomposing into subnetworks,

(ii) Formulate the MINLP model of the postulated hyperstructure for the simulatneous matches-HEN optimization, and

(iii) Solve the MINLP using v2–GBD and the algorithmic procedure described in Section 8.5.1.4.

4. Given is the following set of hot and cold process streams, utilities, and cost data:

Stream	FC_p (kW/K)	T^{IN} (K)	T^{OUT} (K)
$H1$	30	280	60
$H2$	45	180	20
$C1$	40	20	160
$C2$	60	120	260
S		300	300
CW		10	20

$HRAT = 32\text{K}$

$TIAT = \Delta T_{min} = 20\text{K}$

$U = 0.8\ \text{kW/(m}^2\ \text{K)}$

Cost of heat exchangers $= 1300A^{0.6}$ \$/yr , A = m^2

(i) Determine the minimum utility consumption.

(ii) Determine the maximum heat flow across the pinch point.

(iii) Formulate and solve the match-HEN optimization model that corresponds to the hyperstructure representation.

(iv) Does the resulting network structure satisfy the minimum number of matches criterion? Why?

5. Given is the data of problem 2 with $TIAT = 5°F$. Treat the problem as a pseudo-pinch and consider the synthesis approach without decomposition described in Section 8.5.3.

   (i) Find the utility consumption levels as a function of the unknown $HRAT$,

   (ii) Formulate and solve the MINLP model for the HEN synthesis without decomposition approach to determine the optimal value of $HRAT$, matches, heat loads of matches, heat loads of utilities, and HEN structure.

   (iii) Compare the solution with the one resulting from the decomposition approach in problem 2.

6. Given is the following set of hot and cold process streams, utilities, and cost data:

Stream	FC_p (kW/K)	$T^{IN}(^\circ C)$	$T^{OUT}(^\circ C)$
$H1$	20	150	60
$H2$	80	90	60
$C1$	25	20	125
$C2$	30	25	100
S		180	180
CW		10	15

$HRAT = 20^\circ C$

Cost of heat exchangers $= 8600 + 670A^{0.83}$ \$/yr , A = m^2

$U = 0.05$ kW/(m^2 $^\circ C$)

(i) Postulate the simplified superstructure according to the discussion in section 8.5.4.

(ii) Formulate the MINLP simulatneous optimization model that corresponds to this simplified superstructure.

(iii) Solve the resulting MINLP model using OA/ER/AP and v2–GBD.

Chapter 9 Distillation-based Separation Systems Synthesis

This chapter presents two applications of MINLP methods in the area of separations. Section 9.1 provides an overall introduction to the synthesis of separation systems. Section 9.2 focuses on sharp heat–integrated distillation sequencing. Section 9.3 presents an application of nonsharp separation synthesis.

9.1 Introduction

The synthesis of distillation-based separation sequences has been one of the most important subjects of investigation in the area of process synthesis. This is attributed to the significant contribution of separation processes to the total capital investment and operating expenses of a chemical plant. As a result, a lot of interest has been generated in the development of systematic approaches that select optimal sequences of distillation columns. Westerberg (1985) provided a comprehensive review of the distillation-based separation synthesis approaches, as well as presented a classification of different types of separation problems along with their associated challenges. Nishida *et al.* (1981) and Smith and Linnhoff (1988) reviewed the general separation synthesis problem (i.e., not only distillation-based) and presented the progress made.

To illustrate the nature of the distillation-based separation system synthesis problem, let us consider its generic definition shown in Figure 9.1, which is as follows:

> Given a number of input multicomponent streams which have specified amounts for each component, create a cost-optimal configuration of distillation columns, mixers, and splitters that produces a number of multicomponent products with specified composition of their components.

The products feature components that exist in the input streams and can be obtained by redistributing the components existing in the input streams, while the cost-optimal configuration corresponds to the least total annual cost one.

Most of distillation columns or sequences can be classified as

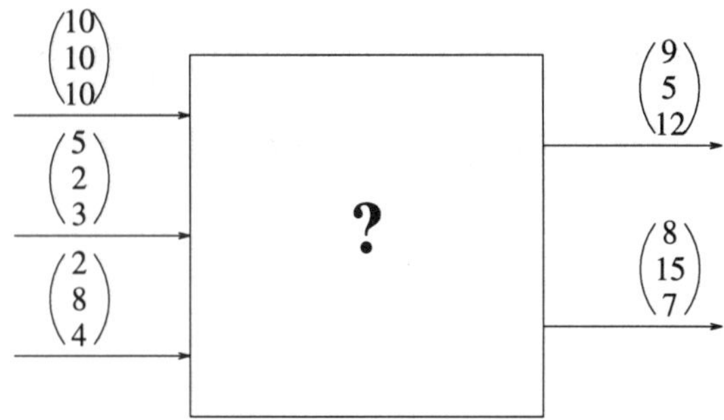

Figure 9.1: Generic distillation based separation problem

(i) Sharp,

(ii) Nonsharp,

(iii) Simple,

(iv) Complex,

(v) Heat-integrated, and

(vi) Thermally coupled.

In (i), a column separates its feed into products without overlap in the components. An example is the separation of a stream consisting of four components A, B, C, and D via a distillation column, into one product consisting of only A and another product featuring B, C, and D. If all columns are sharp, then the separation sequence is termed as sharp sequence.

In (ii), distribution of the light key, heavy key, and/or intermediate components is allowed in both the distillate and bottoms of a distillation column. For example, an input stream with four components A, B, C, and D in which the light key is component B and the heavy key component is C may have at the top of the column A, B, and C and B, C, and D at the bottoms of the non-sharp distillation column.

In (iii), a distillation column features one feed stream and two product streams (i.e., a distillate and a bottoms stream).

In (iv), a distillation column may have several feeds and several products which may include side streams.

In (v), by operating the columns at different pressures alternatives of heat integration between reboilers and condensers take place so as to reduce the utility consumption.

In (vi), a combination of complex columns with heat integration is allowed to take place.

It should be noted that the distillation-based separation system synthesis problem corresponds to a combinatorial problem, since the number of sequences and configurations increase with the number of components, and becomes even more difficult when consideration of heat integration alternatives takes place.

In the next three sections, we will discuss MINLP optimization approaches for the synthesis of:

(a) Heat-integrated sharp and simple distillation sequences, and

(b) Nonsharp and simple distillation sequences.

9.2 Synthesis of Heat-integrated Sharp Distillation Sequences

The significant contribution of the utilities to the total cost of a distillation-based separation sequence provides an incentive to synthesize the sequences of distillation columns which feature heat integration alternatives.

The first systematic studies of heat-integrated distillation sequences were performed by Rathore *et al.* (1974b, 1974b) who presented an algorithm based on dynamic programming for fixed and variable pressures of the distillation columns. Sophos *et al.* (1978) and Morari and Faith (1980) developed branch and bound approaches based upon Lagrangian theory. Umeda *et al.* (1979) presented an evolutionary approach that utilizes heat availability diagrams to improve the heat integration of a given distillation sequence. Naka *et al.* (1982) developed an approach that synthesizes heat-integrated distillation sequences based upon a minimization of the loss of available energy. Linnhoff *et al.* (1983) considered the design of distillation sequences in conjunction with the heat integration of the overall process and developed insights based on the notion of the pinch point. Andrecovich and Westerberg (1985) developed a simple synthesis method that is based upon utility bounding, assuming that $Q\Delta T$, the product of the condenser or reboiler heat duty and the temperature difference between the reboiler and condenser, is constant for each distillation column. Andrecovich and Westerberg (1985) formulated the synthesis problem of heat-integrated distillation sequences as a mixed-integer linear programming MILP problem. This formulation is based upon a superstructure that consists of distillation columns at discretized pressure levels. Westerberg and Andrecovich (1985) removed the assumption of constancy on $Q\Delta T$ and showed that the minimum utility target can be calculated by formulating it as a small linear programming problem. They also showed that even though the calculated utility target may increase, the synthesized sequences remain the same as the ones for constant $Q\Delta T$. Gomez and Seader (1985) used the thermodynamic principles of (a) minimum reversible work of separation and (b) second-law analysis for irreversibility, to synthesize distillation sequences. Isla and Cerda (1987) chose the least utility cost as the design target and formulated the synthesis problem of heat-integrated distillation trains as a mixed-integer programming problem that can identify the optimal heat-integrated sequence from the utility point of view. Meszarus and Fonyo (1986) developed a two-level predictor-based search method that utilizes the heat-cascade theory and sets lower bounds for all feasible heat-integrated separation sequences. They ordered the structures of distillation sequences according to the lower bounds and optimized the heat matches starting with the structure of the lowest lower-bounding value.

Floudas and Paules (1988) proposed a systematic framework by considering the column pressures as continuous variables in a MINLP model of a superstructure that incorporates all desirable heat integration alternatives along with the different sequences of distillation columns.

9.2.1 Problem Statement

Given

(i) A single multicomponent feed stream of known composition and flow rate to be separated into pure component products, and

(ii) A set of hot and cold utilities at different temperature levels,

determine the minimum total annual cost heat–integrated sequence of distillation columns.

The distillation columns are assumed to be sharp and simple. The thermodynamic state (e.g., saturated vapor, liquid) of the feed, distillate, and bottoms streams of each column are assumed known. The condenser type of each column is known (e.g., total condenser).

The heat integration alternatives include the potential heat exchange of the reboiler of a column with the condensers of the other columns postulated in the representation of alternatives, as well as possibilities of heat exchange with utilities available. Hence, we have the following categories of matches that may take place:

(i) Process-process matches which take place between condensers and reboilers, and

(ii) Process-utility matches which take place between condensers and cold utilities and/or reboilers and hot utilities.

Note that the pressure (temperature) of each distillation column is not fixed, but it is treated as explicit optimization variable. As a result, structural optimization takes into account simultaneously the effect of operating the distillation columns at different pressures (temperatures) which provides the trade-off between the investment and operating cost.

In the following sections, we will discuss the approach of Floudas and Paules (1988) for the synthesis of heat-integrated sharp and simple distillation sequences.

9.2.2 Basic Idea

The basic idea in the synthesis approach of heat-integrated distillation sequences proposed by Floudas and Paules (1988) consists of

(i) Treating the pressure (temperature) of each distillation column as an unknown continuous variable,

(ii) Quantifying the effect of these unknown continuous variables (i.e., temperatures or pressures) on the cost of each column, the heat duties of the reboilers and condensers, and the feasibility constraints,

(iii) Postulating a representation of all possible separation sequences as well as all possible heat integration alternatives, and

(iv) Formulating the postulated superstructure of heat-integrated distillation sequences alternatives as a mixed-integer nonlinear programming MINLP problem whose solution will provide a least total annual cost configuration along with

   (a) Distillation columns selected,
   (b) Heat integration matches between condensers-reboilers, condensers-cold utilities, and reboilers-hot utilities,
   (c) Heat loads of each match,
   (d) Temperatures of each reboiler and condenser, and
   (e) Areas of heat exchangers.

9.2.3 Derivation of Superstructure

In postulating a superstructure of alternatives for the heat-integrated distillation sequencing problem, we simultaneously embed:

(i) All different sequences of distillation columns, and

(ii) All alternative matches of heat exchange between condensers-reboilers, condensers-cold utilities, and reboilers-hot utilities.

The heat integration alternatives consist of only the different matches that may take place and do not include structural alternatives of the heat exchanger network topology presented in the hyperstructure or superstructure sections of HEN synthesis.

Illustration 9.2.1 (Superstructure)
This example is taken from Floudas and Paules (1988) and consists of a three-component feed stream to be separated into pure component products. Two types of hot utilities are available along with one cold utility. The data for this example are shown in Table 9.1.

The superstructure for the illustrative example is shown in Figure 9.2. Note that we have two sequences of distillation columns (i.e., column 1 followed by column 2, and column 3 followed by column 4) since the feed stream has three components. In Figure 9.2, the potential process-process matches and the process-utility matches are indicated. The variables shown in Figure 9.2 are the unknown flow rates for the column sequences (i.e., F_1, F_2, F_3, F_4) and existence variables for the columns (i.e., y_1, y_2, y_3, y_4) and the process-process matches (i.e., $y_{C1,R2}, y_{C2,R1}, y_{C3,R4}, y_{C4,R3}$). The temperatures of the condensers and reboilers of each column, as well as the existence variables of the process-utility matches are not shown in Figure 9.2.

Remark 1 It is important to emphasize the relation between the condensers and reboiler streams to the hot and cold streams, which is

(i) A condenser is a hot stream to be cooled, and

(ii) A reboiler is a cold stream to be heated.

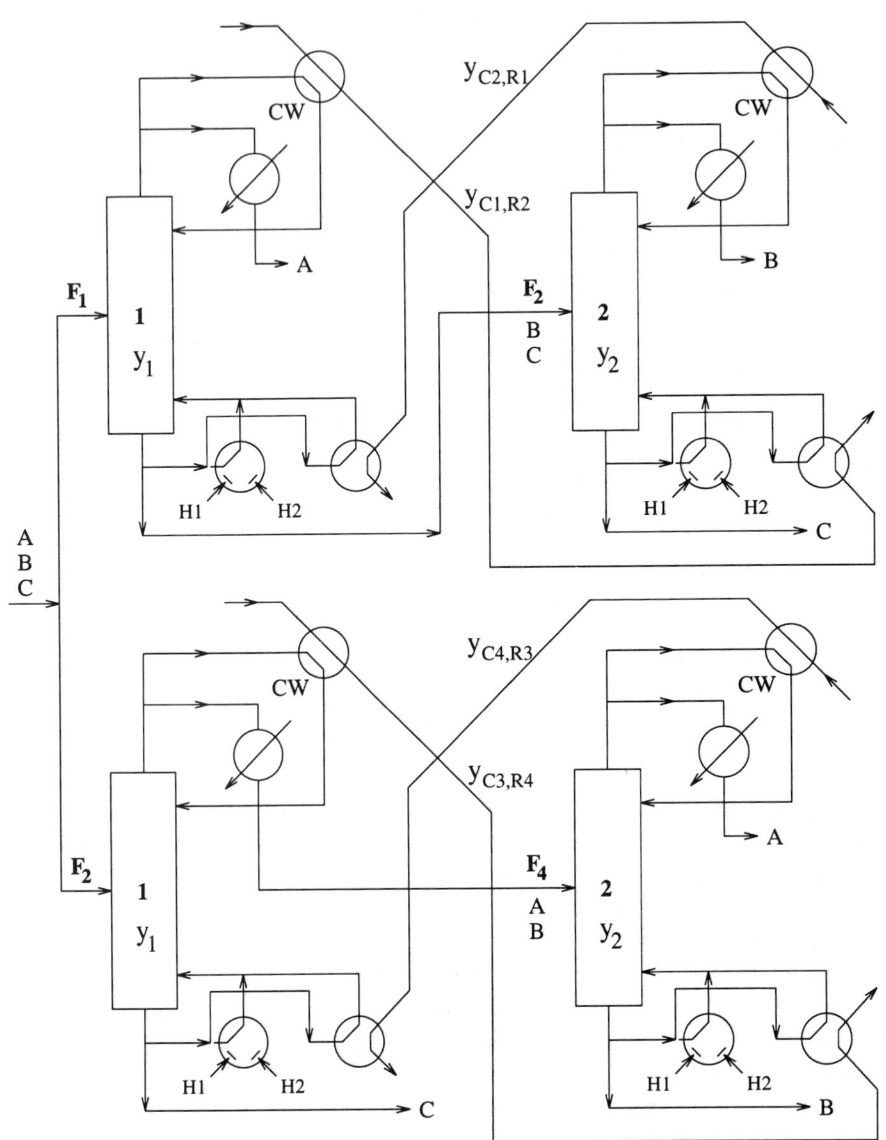

Figure 9.2: Superstructure for illustrative example

Component	Feed composition
Hexane (A)	0.80
Benzene (B)	0.10
Heptane (C)	0.10
Feed flow rate = 396 kgmol/hr	
Low pressure steam	at 421 K; Cost = $3.110 \cdot 10^6/kJ$
Exhaust steam	at 373 K; Cost = $1.140 \cdot 10^6/kJ$
Cooling water (20K rise)	at 305 K; Cost = $0.159 \cdot 10^6/kJ$
$\Delta T_{min} = 10K$	

Table 9.1: Data for illustrative example 9.2.1

9.2.4 Mathematical Formulation of Superstructure

To determine a least total annual cost heat-integrated sequence of distillation columns from the postulated superstructure, we define variables for

(i) The distillation sequencing, and

(ii) The heat integration alternatives.

In (i) we introduce variables for

(a) The existence of a distillation column (i.e., $y_1, y_2, y_3, y_4 = 0 - 1$),

(b) The feed flowrates in the sequencing (i.e., F_1, F_2, F_3, F_4),

(c) The temperatures of the condensers and reboilers of each column (i.e., $T_1^C, T_2^C, T_3^C, T_4^C, T_1^R, T_2^R, T_3^R, T_4^R$), and

(d) The heat loads of the condensers and reboilers of each column (i.e., $Q_1^C, Q_2^C, Q_3^C, Q_4^C, Q_1^R, Q_2^R, Q_3^R, Q_4^R$).

In (ii), we introduce variables for

(a) The existence of heat exchange between condensers-reboilers (i.e., $y_{C1,R2}, y_{C2,R1}, y_{C3,R4}, y_{C4,R3} = 0-1$), condensers-cold utility (i.e., $y_{C1,W}, y_{C2,W}, y_{C3,W}, y_{C4,W} = 0 - 1$), and hot utilities-reboilers (i.e., $y_{S1,R1}, y_{S2,R1}, y_{S1,R2}, y_{S2,R2}, y_{S1,R3}, y_{S2,R3}, y_{S1,R4}, y_{S2,R4} = 0 - 1$), and

(b) The heat loads of each potential match (i.e., $Q_{C1,R2}, Q_{C2,R1}, Q_{C3,R4}, Q_{C4,R3}, Q_{C1,W}, Q_{C2,W}, Q_{C3,W}, Q_{C4,W}, Q_{S1,R1}, Q_{S2,R1}, Q_{S1,R2}, Q_{S2,R2}, Q_{S1,R3}, Q_{S2,R3}, Q_{S1,R4}, Q_{S2,R4}$).

The set of constraints for the synthesis of heat-integrated distillation sequences problem will consist of

(i) Mass balances,

(ii) Energy balances,

(iii) Heat duty definitions,

(iv) Temperature definitions,

(v) Critical or imposed bounds on temperatures,

(vi) Minimum temperature approach constraints,

(vii) Logical constraints,

(viii) Pure integer constraints, and

(ix) Nonnegativity and integrality constraints.

The objective function consists of

(i) The investment cost for the distillation columns and the heat exchangers, and

(ii) The operating cost for the hot and cold utilities consumption.

Note that the investment cost of each distillation column can be expressed as a nonlinear fixed charge relaxation in which the nonlinearities are introduced due to the cost dependence on both the temperature and the column and the feed flow rate.

We will discuss the development of such a model for the illustrative example.

Illustration 9.2.2 (Optimization Model)
The complete set of constraints for the superstructure of the illustrative example is as follows:

(i) <u>Mass balances</u>

$$
\begin{aligned}
F_1 + F_3 &= 396 \\
F_2 - 0.2F_1 &= 0 \\
F_4 - 0.9F_2 &= 0
\end{aligned}
$$

(ii) <u>Energy balances</u>

$$
\begin{aligned}
Q_1^C &= Q_{C1,R2} + Q_{C1,W} \\
Q_2^C &= Q_{C2,R1} + Q_{C2,W} \\
Q_3^C &= Q_{C3,R4} + Q_{C3,W} \\
Q_4^C &= Q_{C4,R3} + Q_{C4,W} \\
Q_1^R &= Q_{C2,R1} + Q_{S1,R1} + Q_{S2,R1} \\
Q_2^R &= Q_{C1,R2} + Q_{S1,R2} + Q_{S2,R2} \\
Q_3^R &= Q_{C4,R3} + Q_{S1,R3} + Q_{S2,R3} \\
Q_4^R &= Q_{C3,R4} + Q_{S1,R4} + Q_{S2,R4}
\end{aligned}
$$

The mass and energy balances constitute a linear set of constraints. The energy balances for the condensers state that all cooling required by a condenser of a particular column must be transferred either to a reboiler of another column in the same sequence or to a cold utility. Similarly, the energy balances for the reboilers state that all heating required by a condenser must be provided by either the condenser of another column in the same sequence or by the two available hot utilities.

(iii) Heat duty definitions

$$
\begin{aligned}
Q_1^C &= 32.4 + 0.0225 \cdot T_1^C + S1_1 - S2_1 \\
Q_2^C &= 25.0 + 0.0130 \cdot T_2^C + S1_2 - S2_2 \\
Q_3^C &= 3.75 + 0.0043 \cdot T_3^C + S1_3 - S2_3 \\
Q_4^C &= 35.1 + 0.0156 \cdot T_4^C + S1_4 - S2_4 \\
Q_1^R &= Q_1^C \\
Q_2^R &= Q_2^C \\
Q_3^R &= Q_3^C \\
Q_4^R &= Q_4^C
\end{aligned}
$$

The coefficients of the above linear expressions are obtained via regression analysis of the simulation data taken at a variety of pressure levels Floudas and Paules (1988). Note that in the above definitions we have introduced a set of slack variables. These are introduced so as to prevent infeasibilities from arising from the equality constraints whenever a column does not participate in the activated sequence. These slack variables participate in the set of logical constraints and are both set to zero if the corresponding column exists, while they are activated to nonzero value if the column does not exist, so as to relax the associated equality constraints.

(iv) Temperature definitions

$$
\begin{aligned}
T_1^R &= 9.541 + 1.028 \cdot T_1^C + S5_1 - S6_1 \\
T_2^R &= 12.24 + 1.050 \cdot T_2^C + S5_2 - S6_2 \\
T_3^R &= 8.756 + 1.029 \cdot T_3^C + S5_3 - S6_3 \\
T_4^R &= 9.181 + 1.005 \cdot T_4^C + S5_4 - S6_4
\end{aligned}
$$

Slack variables need to be introduced in the linear constraints between the reboiler and condenser temperatures so as to avoid potential infeasibilities. Note that the expressions in (iii) and (iv) are linear because this is the result of the regression analysis of the simulation data. In the general case, however, they may be nonlinear.

(v) Critical or imposed bounds on temperatures

$$
\begin{aligned}
T_1^{C,MIN} - T_1^C - \mathrm{U}(1 - y_1) &\leq 0 \\
T_2^{C,MIN} - T_2^C - \mathrm{U}(1 - y_2) &\leq 0 \\
T_3^{C,MIN} - T_3^C - \mathrm{U}(1 - y_3) &\leq 0
\end{aligned}
$$

$$
\begin{aligned}
T_4^{C,MIN} - T_4^C - \mathrm{U}(1-y_4) &\le 0 \\
T_1^R - T_1^{R,MAX} &\le 0 \\
T_2^R - T_2^{R,MAX} &\le 0 \\
T_3^R - T_3^{R,MAX} &\le 0 \\
T_4^R - T_4^{R,MAX} &\le 0
\end{aligned}
$$

where $T_i^{C,MIN}, T_i^{R,MAX}$, $i = 1,2,3,4$ are the minimum condenser temperatures and maximum reboiler temperatures. The minimum condenser temperatures are introduced so as to disallow operating a condenser at a temperature below the one that corresponds to a bubble pressure of 1 atm. The maximum reboiler temperatures are imposed so as to avoid operation close to the critical points.

If the normal boiling point of the overhead composition is below the coldest cold utility, then the minimum condenser temperature in a column will be equal to the temperature of the coldest cold utility plus ΔT_{min}. They take the values of $341.92, 343.01, 353.54$ and 341.92 for tasks 1, 2, 3 and 4 respectively. The maximum reboiler temperature, if critical conditions are not approached by the hottest hot utility, will be the temperature of the hottest hot utility minus ΔT_{min}. Based on the data, the maximum reboiler temperature is 411 K.

Note that if for instance $y_1 = 0$, then the first constraint is relaxed because of the less or equal form, while it is maintained as bound when $y_1 = 1$. Note that we do not need to introduce binary variables for the reboiler constraints due to their less or equal form.

(vi) Minimum temperature approach constraints

$$
\left.
\begin{aligned}
T_1^C - T_2^R + \mathrm{U}(1-y_{C1,R2}) &\ge 10 \\
T_2^C - T_1^R + \mathrm{U}(1-y_{C2,R1}) &\ge 10 \\
T_3^C - T_4^R + \mathrm{U}(1-y_{C3,R4}) &\ge 10 \\
T_4^C - T_3^R + \mathrm{U}(1-y_{C4,R3}) &\ge 10
\end{aligned}
\right] \quad Process - process\ matches
$$

$$
\left.
\begin{aligned}
421 - T_1^R + \mathrm{U}(1-y_{S1,R1}) &\ge 10 \\
421 - T_2^R + \mathrm{U}(1-y_{S1,R2}) &\ge 10 \\
421 - T_3^R + \mathrm{U}(1-y_{S1,R3}) &\ge 10 \\
421 - T_4^R + \mathrm{U}(1-y_{S1,R4}) &\ge 10 \\
373 - T_1^R + \mathrm{U}(1-y_{S2,R1}) &\ge 10 \\
373 - T_2^R + \mathrm{U}(1-y_{S2,R2}) &\ge 10 \\
373 - T_3^R + \mathrm{U}(1-y_{S2,R3}) &\ge 10 \\
373 - T_4^R + \mathrm{U}(1-y_{S2,R4}) &\ge 10
\end{aligned}
\right] \quad Hot\ utilities - reboiler\ matches
$$

Note that we do not need to include the ΔT_{min} constraints for the condenser-cold utility matches since they have already been included in (v). Note also that if, for instance, $y_{C1,R2} = 1$, then we have

$$
T_1^C - T_2^R \ge 10,
$$

which is the desirable feasibility constraint. If, however, $y_{C1,R2} = 0$, then we have

$$T_1^C - T_2^R + \mathrm{U} \geq 10,$$

which is a relaxation of the ΔT_{min} constraint.

(vii) Logical constraints

$$\left.\begin{aligned} F_1 - \mathrm{U}y_1 &\leq 0 \\ F_2 - \mathrm{U}y_2 &\leq 0 \\ F_3 - \mathrm{U}y_3 &\leq 0 \\ F_4 - \mathrm{U}y_4 &\leq 0 \end{aligned}\right] \quad \textit{Mass balances}$$

$$\left.\begin{aligned} Q_{C1,R2} - \mathrm{U}y_{C1,R2} &\leq 0 \\ Q_{C2,R1} - \mathrm{U}y_{C2,R1} &\leq 0 \\ Q_{C3,R4} - \mathrm{U}y_{C3,R4} &\leq 0 \\ Q_{C4,R3} - \mathrm{U}y_{C4,R3} &\leq 0 \end{aligned}\right] \quad \textit{Heat integration loads}$$

$$\left.\begin{aligned} Q_{C1,W} - \mathrm{U}y_{C1,W} &\leq 0 \\ Q_{C2,W} - \mathrm{U}y_{C2,W} &\leq 0 \\ Q_{C3,W} - \mathrm{U}y_{C3,W} &\leq 0 \\ Q_{C4,W} - \mathrm{U}y_{C4,W} &\leq 0 \end{aligned}\right] \quad \textit{Cold utility loads}$$

$$\left.\begin{aligned} Q_{S1,R1} - \mathrm{U}y_{S1,R1} &\leq 0 \\ Q_{S1,R2} - \mathrm{U}y_{S1,R2} &\leq 0 \\ Q_{S1,R3} - \mathrm{U}y_{S1,R3} &\leq 0 \\ Q_{S1,R4} - \mathrm{U}y_{S1,R4} &\leq 0 \\ Q_{S2,R1} - \mathrm{U}y_{S2,R1} &\leq 0 \\ Q_{S2,R2} - \mathrm{U}y_{S2,R2} &\leq 0 \\ Q_{S2,R3} - \mathrm{U}y_{S2,R3} &\leq 0 \\ Q_{S2,R4} - \mathrm{U}y_{S2,R4} &\leq 0 \end{aligned}\right] \quad \textit{Hot utility loads}$$

$$\left.\begin{aligned} Q_1^C + Q_1^R - \mathrm{U}y_1 &\leq 0 \\ Q_2^C + Q_2^R - \mathrm{U}y_2 &\leq 0 \\ Q_3^C + Q_3^R - \mathrm{U}y_3 &\leq 0 \\ Q_4^C + Q_4^R - \mathrm{U}y_4 &\leq 0 \end{aligned}\right] \quad \textit{Reboiler and condenser loads}$$

$$\left.\begin{aligned} S1_1 + S2_1 - \mathrm{U}(1 - y_1) &\leq 0 \\ S1_2 + S2_2 - \mathrm{U}(1 - y_2) &\leq 0 \\ S1_3 + S2_3 - \mathrm{U}(1 - y_3) &\leq 0 \\ S1_4 + S2_4 - \mathrm{U}(1 - y_4) &\leq 0 \\ \\ S3_1 + S4_1 - \mathrm{U}(1 - y_1) &\leq 0 \\ S3_2 + S4_2 - \mathrm{U}(1 - y_2) &\leq 0 \\ S3_3 + S4_3 - \mathrm{U}(1 - y_3) &\leq 0 \\ S3_4 + S4_4 - \mathrm{U}(1 - y_4) &\leq 0 \end{aligned}\right] \quad \textit{Slack variables}$$

The logical constraints for the slack variables coupled with the nonnegativity of the slacks imply that if for instance, $y_1 = 1$, then

$$\begin{aligned} S1_1 + S2_1 &= 0 \Rightarrow S1_1 = S2_1 = 0, \\ S3_1 + S4_1 &= 0 \Rightarrow S3_1 = S4_1 = 0. \end{aligned}$$

If, however, $y_1 = 0$, then the slacks can take any value that relaxes the associated constraints.

(viii) Pure integer constraints

$$\left.\begin{aligned} y_2 - y_1 &\leq 0 \\ y_4 - y_3 &\leq 0 \end{aligned}\right] \textit{Sequence constraints}$$

$$\left.\begin{aligned} y_{S1,R1} + y_{S2,R1} &\leq 1 \\ y_{S1,R2} + y_{S2,R2} &\leq 1 \\ y_{S1,R3} + y_{S2,R3} &\leq 1 \\ y_{S1,R4} + y_{S2,R4} &\leq 1 \end{aligned}\right] \textit{Utility limits}$$

$$\left.\begin{aligned} y_{C1,R2} + y_{C2,R1} &\leq 1 \\ y_{C3,R4} + y_{C4,R3} &\leq 1 \end{aligned}\right] \textit{Single direction of heat integration}$$

The sequence constraints state, for instance, that if $y_1 = 0$, then $y_2 = 0$ (i.e., column 2 does not exist if column 1 does not exist). The utility limits constraints state that not more than one hot utility is allowed to be used at each reboiler.

The single direction of heat integration constraints disallows mutual heat integration of two columns in both directions. The temperature approach constraints will make this infeasible, but the above constraints help in speeding the enumerative procedure of the combinations.

For example, setting

$$y_{C1,R2} = 1 \quad \text{and} \quad y_{C2,R1} = 0,$$

results in the following for (vi):

$$\begin{aligned} T_1^C - T_2^R &\geq 10, \\ T_2^C - T_1^R + \mathcal{U} &\geq 10, \end{aligned}$$

which relaxes the second direction of heat integration.

(ix) Nonnegativity and integrality constraints

$$\begin{aligned} F_1, F_2, F_3, F_4 &\geq 0 \\ Q_i^C, Q_i^R &\geq 0, \quad i = 1,2,3,4 \\ Q_{C1,R2}, Q_{C2,R1}, Q_{C3,R4}, Q_{C4,R3} &\geq 0 \\ Q_{C1,W}, Q_{C2,W}, Q_{C3,W}, Q_{C4,W} &\geq 0 \\ Q_{S1,R1}, Q_{S1,R2}, Q_{S1,R3}, Q_{S1,R4} &\geq 0 \end{aligned}$$

$$
\begin{aligned}
Q_{S2,R1}, Q_{S2,R2}, Q_{S2,R3}, Q_{S2,R4} &\geq 0 \\
S1_i, S2_i, S3_i, S4_i, &\geq 0, \quad i = 1,2,3,4 \\
y_1, y_2, y_3, y_4 &= 0-1 \\
y_{C1,R2}, y_{C2,R1}, y_{C3,R4}, y_{C4,R3} &= 0-1 \\
y_{C1,W}, y_{C2,W}, y_{C3,W}, y_{C4,W} &= 0-1 \\
y_{S1,R1}, y_{S1,R2}, y_{S1,R3}, y_{S1,R4} &= 0-1 \\
y_{S2,R1}, y_{S2,R2}, y_{S2,R3}, y_{S2,R4} &= 0-1
\end{aligned}
$$

Note that the set of constraints are linear in the continuous and binary variables, which is an attractive feature for the algorithms to be used for the solution of the optimization model.

Objective function

The objective function represents the total annual cost which consists of (i) the investment cost for the distillation columns and heat exchangers and (ii) the operating cost for the hot and cold utilities.

(i) Investment cost

The investment cost of each distillation column is expressed as a nonlinear fixed charge cost and the following expressions result from performing regression analysis on the simulation data taken at a number of operating pressures Paules (1990).

$$
\begin{aligned}
C_{C1} &= 151.125 \cdot y_1 + 0.003375(T_1^C - T_1^{C,MIN}) \cdot F_1, \\
C_{C2} &= 180.003 \cdot y_2 + 0.000893(T_2^C - T_2^{C,MIN}) \cdot F_2, \\
C_{C3} &= 4.2286 \cdot y_3 + 0.004458(T_3^C - T_3^{C,MIN}) \cdot F_3, \\
C_{C4} &= 213.42 \cdot y_4 + 0.003176(T_4^C - T_4^{C,MIN}) \cdot F_4.
\end{aligned}
$$

Note that the above expressions are nonlinear due to the bilinear terms between the temperatures and the flowrates. They are also scaled by 1000.

The investment cost of each heat exchanger is also expressed as a nonlinear fixed charge cost (linear though in the areas) with the following expressions:

$$
\left.
\begin{aligned}
C^E_{C1,R2} &= 3.392 \cdot y_{C1,R2} + 0.0893 \cdot \left[\frac{Q_{C1,R2}}{U(T_1^C - T_2^R)}\right] \\
C^E_{C2,R1} &= 3.392 \cdot y_{C2,R1} + 0.0893 \cdot \left[\frac{Q_{C2,R1}}{U(T_2^C - T_1^R)}\right] \\
C^E_{C3,R4} &= 3.392 \cdot y_{C3,R4} + 0.0893 \cdot \left[\frac{Q_{C3,R4}}{U(T_3^C - T_4^R)}\right] \\
C^E_{C4,R3} &= 3.392 \cdot y_{C4,R3} + 0.0893 \cdot \left[\frac{Q_{C4,R3}}{U(T_4^C - T_3^R)}\right]
\end{aligned}
\right] \quad \text{Process–process matches}
$$

Hot utility – process matches:

$$
\begin{aligned}
C^E_{S1,R1} &= 3.392 \cdot y_{S1,R1} + 0.0893 \cdot \left[\frac{Q_{S1,R1}}{\mathcal{U}(421 - T_1^R)}\right] \\
C^E_{S1,R2} &= 3.392 \cdot y_{S1,R2} + 0.0893 \cdot \left[\frac{Q_{S1,R2}}{\mathcal{U}(421 - T_2^R)}\right] \\
C^E_{S1,R3} &= 3.392 \cdot y_{S1,R3} + 0.0893 \cdot \left[\frac{Q_{S1,R3}}{\mathcal{U}(421 - T_3^R)}\right] \\
C^E_{S1,R4} &= 3.392 \cdot y_{S1,R4} + 0.0893 \cdot \left[\frac{Q_{S1,R4}}{\mathcal{U}(421 - T_4^R)}\right] \\
C^E_{S2,R1} &= 3.392 \cdot y_{S2,R1} + 0.0893 \cdot \left[\frac{Q_{S2,R1}}{\mathcal{U}(373 - T_1^R)}\right] \\
C^E_{S2,R2} &= 3.392 \cdot y_{S2,R2} + 0.0893 \cdot \left[\frac{Q_{S2,R2}}{\mathcal{U}(373 - T_2^R)}\right] \\
C^E_{S2,R3} &= 3.392 \cdot y_{S2,R3} + 0.0893 \cdot \left[\frac{Q_{S2,R3}}{\mathcal{U}(373 - T_3^R)}\right] \\
C^E_{S2,R4} &= 3.392 \cdot y_{S2,R4} + 0.0893 \cdot \left[\frac{Q_{S2,R4}}{\mathcal{U}(373 - T_4^R)}\right]
\end{aligned}
$$

Cold utility – condenser matches:

$$
\begin{aligned}
C^E_{C1,W} &= 3.392 \cdot y_{C1,W} + 0.0893 \cdot \left[\frac{Q_{C1,W}}{\mathcal{U}(LMTD)_{C1,W}}\right] \\
C^E_{C2,W} &= 3.392 \cdot y_{C2,W} + 0.0893 \cdot \left[\frac{Q_{C2,W}}{\mathcal{U}(LMTD)_{C2,W}}\right] \\
C^E_{C3,W} &= 3.392 \cdot y_{C3,W} + 0.0893 \cdot \left[\frac{Q_{C3,W}}{\mathcal{U}(LMTD)_{C3,W}}\right] \\
C^E_{C4,W} &= 3.392 \cdot y_{C4,W} + 0.0893 \cdot \left[\frac{Q_{C4,W}}{\mathcal{U}(LMTD)_{C4,W}}\right]
\end{aligned}
$$

where

$$
\begin{aligned}
(LMTD)_{C1,W} &= \frac{20}{\ln(\frac{T_1^C - 305}{T_1^C - 325})} \\
(LMTD)_{C2,W} &= \frac{20}{\ln(\frac{T_2^C - 305}{T_2^C - 325})} \\
(LMTD)_{C3,W} &= \frac{20}{\ln(\frac{T_3^C - 305}{T_3^C - 325})} \\
(LMTD)_{C4,W} &= \frac{20}{\ln(\frac{T_4^C - 305}{T_4^C - 325})}
\end{aligned}
$$

and the overall heat transfer coefficient $\mathcal{U}$ is 0.0028.

(ii) Operating Cost

The operating cost consists of the hot and cold utility cost which takes the form:

$$\begin{aligned} & 3.110\,[Q_{S1,R1} + Q_{S1,R2} + Q_{S1,R3} + Q_{S1,R4}] \\ + \; & 1.140\,[Q_{S2,R1} + Q_{S2,R2} + Q_{S2,R3} + Q_{S2,R4}] \\ + \; & 0.159\,[Q_{C1,W} + Q_{C2,W} + Q_{C3,W} + Q_{C4,W}], \end{aligned}$$

where the cost coefficients of the investment cost are in 10^6 \$ and Q's are in kJ.

Then the objective function is

$$OBJ = \frac{1}{\tau} \cdot (investment\ cost) + \beta \cdot (operating\ Cost),$$

where τ is the payout time ($\tau = 0.4$), and β is the correction factor for income taxes ($\beta = 0.52$).

The resulting model is a MINLP with linear constraints and nonlinear objective function. The objective function terms can be convexified using the exponential transformation with the exception of the condenser-cold utility expressions. If we replace the LMTDs with 2-3 times the ΔT_{min}, then, the whole objective can be convexified Floudas and Paules (1988). This implies that its global solution can be attained with the OA/ER or the v2-GBD algorithms.

The optimization model for the illustrative example was solved with the OA/ER algorithm using the library OASIS (Floudas, 1990) and resulted in the optimal sequence shown in Figure 9.3. This solution features a heat-integrated scheme involving the upper sequence with columns 1 and 2. The condenser temperature (and hence corresponding column pressure) of column 2 is raised to a level high enough so as to have heat transfer from the condenser of column 2 to the reboiler of column 1. Low-pressure steam is selected for use in column 2 since the column temperature is high enough to disallow the use of exhaust steam. Exhaust steam is used for heating the reboiler of the column 1. There is no need for cold utility in column 2, while cooling water is used for the condenser of column 1. The total annual cost for this heat-integrated sequence is \$1,096,770.

9.3 Synthesis of Nonsharp Distillation Sequences

During the last three decades a lot of attention has been on the separation of a single multicomponent feed stream into pure substances using sharp separations. Relative to this effort, there has been less work for nonsharp separation synthesis, as well as the separation of several multicomponent feed streams into several multicomponent streams.

Nath (1977) investigated a subproblem of the general separation problem that involves the separation of a single multicomponent feed stream into multicomponent products and considered systematically the introduction of nonsharp splits. Muraki and Hayakawa (1984) developed a procedure for the synthesis of sharp split distillation sequences with bypass to produce two multicomponent products. In the first stage, they used a heuristic procedure to identify a column ordering. In the second stage, advantage is taken of the graphical representation of the separation

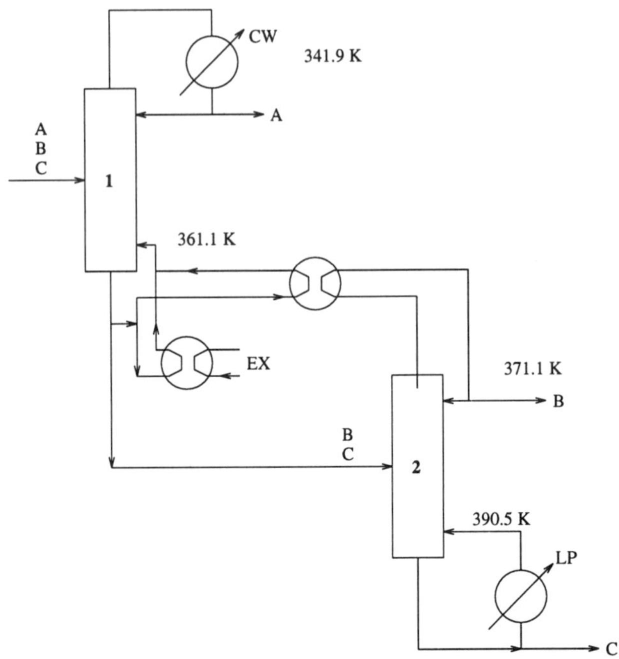

Figure 9.3: Optimal solution for illustrative example

process through the use of the *material allocation diagram* (MAD) proposed by Nath (1977) so as to optimize the separation process by the introduction of division and blending. A MAD representation of a nonsharp separation product is not always unique. Muraki *et al.* (1986) extended this work to the synthesis of sequences with one feed and multiple multicomponent products. They proposed a method for constructing a *modified material allocation diagram* (MMAD). It should be noted though, that the above approaches have addressed subproblems of the general separation problem.

Floudas (1987) addressed the general separation problem. The basic idea was to derive a superstructure, based upon simple and sharp columns, which has embedded in it all possible separation sequences and all possible options of splitting, mixing, and bypassing of the streams. Simple columns are defined as columns with a single feed and one top and one bottom product (no side streams). The formulation was a nonlinear nonconvex programming (NLP) problem and the solution of the nonconvex NLP provided an optimal separation sequence that corresponds to a local optimum. Floudas and Anastasiadis (1988) proposed a mixed-integer linear programming (MILP) formulation for the general distillation-based separation problem. They illustrated that this formulation can be utilized as an initialization procedure for the determination of a good starting point for the nonconvex NLP formulation of the general separation problem. Wehe and Westerberg (1987) described an algorithmic procedure for the synthesis of distillation sequences with bypasses, using sharp separators. Their procedure involved examining various structurally different flowsheets and determining lower and upper bounds for each of these flowsheets. The lower bound is established by using a relaxation of the corresponding nonlinear program and the best solution of the nonlinear program provides the upper bound. Upper and lower bounds very close to each other were considered as implying optimally.

Bamopoulos *et al.* (1988) studied separation sequences with nonsharp separations. They represented the problem in terms of a *component recovery matrix* (R-matrix) which is an algebraic extension of the *material allocation diagram* (MAD) proposed by Nath (1977). The (R-matrix) representation of a nonsharp product is unique, as opposed to the MAD representation. Furthermore, a MAD can be typically constructed for the restricted case of a single feed and every component being assigned to at most two products. Manipulation of the component recovery matrix generates several plausible flowsheets which include the operations of distillation, stream splitting, and stream mixing. They proposed a two-stage approach. In the first step, a heuristic ordering of options, coupled with a depth-first technique, is proposed. In the second step, a best-first technique is proposed for searching for the few better schemes. Heuristic rules are employed to optimize the partially developed sequences. This work provides interesting insights into the complex domain of nonsharp separations but does not guarantee that all possible sequences are studied. Cheng and Liu (1988) proposed a simple heuristic method for the systematic synthesis of initial sequences for nonsharp multicomponent separations. They introduced the *component assignment diagram* (CAD) for representation of the problem and the *separation specification table* (SST) for feasibility analysis. Their procedure is useful for generating good initial sequences but it may generate more columns than required. Muraki and Hayakawa (1987) considered the degrees of separation sharpness and finding optimal degrees of separation sharpness and the ratios of stream divisions for a given separation sequence. Muraki and Hayakawa (1988) coupled this with their earlier two-stage strategy Muraki and Hayakawa (1984). The first stage involves searching for the separation sequence. In the second stage a search is made for optimum values of the degrees of separation sharpness and ratios of stream divisions. These two stages are repeated until an optimum process is synthesized.

Wehe and Westerberg (1990) provided a bounding procedure for finding the minimum number of columns required for separating a single multicomponent feed into multicomponent products using a nonsharp separation sequence. Aggarwal and Floudas (1990) proposed a systematic approach for the synthesis of distillation sequences involving nonsharp splits of components by considering the key component recoveries explicitly as optimization variables. A superstructure was developed that allows for nonsharp splits and involves all possible sequences, as well as options for splitting, mixing, and bypassing. This superstructure is then formulated as a mixed-integer nonlinear programming MINLP problem whose solution provides an optimal distillation sequence. It was also demonstrated that this work can be extended so as to handle nonsharp splits for the case of nonadjacent key components.

Aggarwal and Floudas (1992) extended the synthesis approach of nonsharp separations so as to allow for heat integration alternatives. The pressure of each column and the key component recoveries are treated explicitly as optimization variables and a two-level decomposition approach was proposed for the solution of the resulting MINLP model.

9.3.1 Problem Statement

> Given a single multicomponent feed stream of known conditions (i.e. flowrate, composition, temperature and pressure) to be separated into a number of multicomponent products of specified compositions, determine an optimal distillation configuration that performs the desired separation task by allowing the use of nonsharp columns and which satisfies the criterion of minimum total annual cost.

Each column performs a simple split (i.e., we have simple distillation columns). The thermodynamic state of the feed streams, distillates and bottoms streams, and the type of condenser used in each column are assumed to be known.

No heat integration takes place, and hence hot and cold utilities provide the required loads in the reboilers and condensers, respectively. Also, it is assumed that all columns operate at fixed pressures (i.e., the pressure or temperature of each column is fixed).

In each column, we allow only the light and heavy key components to be distributed in the distillate and bottom products, and the key components are considered to be adjacent components after they have been arranged in order of relative volatility.

Note that the feed composition of each column (with the exception of the first column) is now variable since the recoveries of the light and heavy key components in each column are variables themselves.

In the following sections, we will discuss the approach of Aggarwal and Floudas (1990) for the synthesis of nonsharp distillation configurations.

9.3.2 Basic Idea

The basic idea in the synthesis of nonsharp distillation sequences approach of Aggarwal and Floudas (1990) consists of

(i) Treating the recoveries of the light and heavy key components as variables in each distillation column,

(ii) Postulating a non-sharp separation superstructure which allows for nonsharp splits and involves all alternative configurations as well as options for mixing, splitting, and by-passing, and

(iii) Formulating the nonsharp separation superstructure as a MINLP model whose solution provides a minimum total annual cost separation sequence with information on

   (a) The columns selected,
   (b) The optimal recoveries of the light and heavy key components, and
   (c) The topology with stream interconnections, mixing, splitting, and potential by-passing.

Remark 1 Since the light and heavy key recoveries of each column are treated explicitly as unknown optimization variables, then the cost of each nonsharp distillation column should be a function of its feed flow rate, feed composition, as well as the recoveries of the key components.

9.3.3 Nonsharp Separation Superstructure

The basic idea in the derivation of a nonsharp separation superstructure is to embed:

(i) A number of nonsharp distillation columns that is equal to the number of separation breakpoints between the components; that is, to have one column for each pair of light and heavy key components,

(ii) All possible sequences for series, parallel or series-parallel arrangements, and

(iii) Options for splitting, mixing, and by-passing.

Since each column performs the separation between adjacent key components, the superstructure would have a maximum of $(n - 1)$ columns for a feed stream of n components. Having one column for each pair of adjacent light and heavy key components does not exclude configurations with prefractionators since a component which is a light key in one column can be the heavy key for a subsequent column. In each column, the nonkey components go entirely to the top or bottom product depending on whether they are lighter or heavier than the corresponding key components. In each column, the light and heavy key components are allowed to distribute in the top and bottom products.

Illustration 9.3.1 (Nonsharp Separation Superstructure)
The nonsharp separation superstructure for a four-component feed stream to be separated into two products P_1 and P_2 is shown in Figure 9.4. Since the feed stream consists of four components (i.e., A, B, C, D) then there are three separation breakpoints (i.e., $A/BCD, AB/CD$, and ABC/D) and hence we postulate in the superstructure shown in Figure 9.4 the following three separation tasks:

Task I : A/BCD Light Key = A, Heavy key = B ;

Task II : AB/CD Light Key = B, Heavy key = C ;

Task III : ABC/D Light Key = C, Heavy key = D .

In Figure 9.4, letters within parentheses stand for the components that will be present if nonsharp separation takes place. For example, stream 7 will consist of component B only if the recovery of this component in column I is incomplete (which means that column I is a nonsharp separator. The bottom product of column I (stream 10) can, in principle, contain all four components. If the feed to column I, stream 6, contains all four components, then stream 10 would contain all four components for non-sharp separation and only B, C, and D for sharp separation. Sharp separation is taken to mean recoveries greater than 99%. If stream 6 consists of only A and B, then stream 10 would contain A and B or B only, depending on whether the separation is nonsharp or sharp. The operation of columns II and III can be described in a similar fashion. This shows that the superstructure consists of both options for each column – sharp or nonsharp separation. The main steps in the derivation of the superstructure for a four-component feed stream are as follows:

1. The feed stream is split into five streams, three of which go to three columns (streams 1, 2, and 3) and the other two are overall bypasses to the two products (streams 4 and 5).

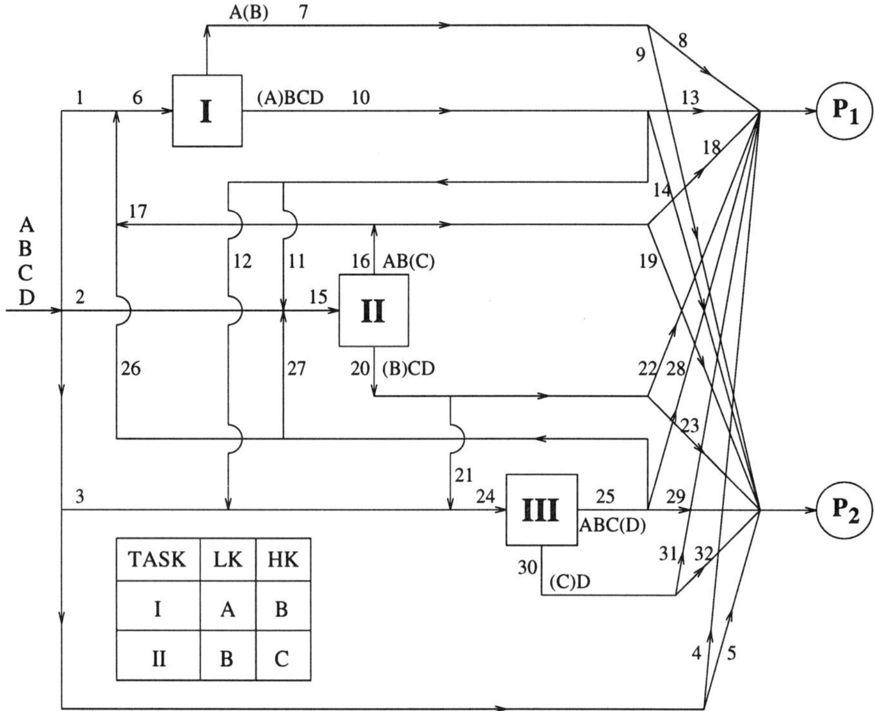

Figure 9.4: Illustration of nonsharp separation superstructure

2. The outlet streams from each column are split into

   (a) Two streams that are directed to the two multicomponent products (e.g., split stream 10 into streams 13 and 14); and

   (b) Two streams to be sent to the other columns. There are some exceptions to this step. If the outlet stream from a column does not contain the key components for some of the other columns, then no split stream goes to those columns (e.g., stream 16, which is the top product of column II, can have components A, B, and C only. Therefore, a split stream (stream 17) goes to column I which has A and B as key components, but there is no split stream going to column III since this column has C and D as key components, whereas stream 16 does not contain component D).

The creation of such a superstructure can be generalized for any number of components. The number of columns is determined from the number of components – one column for each separation breakpoint. This superstructure consists of:

(i) An initial splitter for the feed stream;

(ii) A splitter at each product stream of each column;

(iii) A mixer at the inlet of each column; and

(iv) A mixer prior to each desired multicomponent product.

The proposed superstructure is the most general one under the assumptions stated earlier. If for a particular problem, practical limitations make certain configurations undesirable or infeasible, they can be eliminated by deleting some streams from the superstructure or imposing restrictions in the mathematical formulation. From the superstructure shown in Figure 9.4, many configurations with sharp and/or nonsharp separators can be obtained by setting the flow rates of some of the streams to zero and fixing some of the recoveries of the key components. A few representative examples are

- A sharp sequence with I, II, and III in series results from setting streams 2, 3, 12, 17, 26, and 27 to zero and fixing the recoveries of key components for all columns at 0.99 (which corresponds to sharp separation).

- Nonsharp sequence with I, II, and III in parallel by setting streams 11, 12, 17, 21, 26, and 27 to zero and leaving the recoveries as variables.

- Sequence with nonsharp columns I and II in parallel and then in series with sharp column III by setting streams 3, 11, 17, 26, and 27 to zero and fixing the key component recoveries for column III at the upper bound (0.99).

- Sequence with only columns II and III by setting stream 6 to zero.

9.3.4 Mathematical Formulation of Nonsharp Separation Superstructure

To determine a minimum total annual cost configuration of non-sharp distillation columns out of the many alternatives embedded in the postulated superstructure, we define variables for the

(i) Flow rates of all streams,

(ii) Compositions of all interconnecting streams,

(iii) Component flow rates for the feed streams to columns, and

(iv) Recoveries of light and heavy key components.

Then, the set of constraints consists of

(a) Mass balances for the splitters,

(b) Definitions of component flow rates for feed streams to the columns,

(c) Key component balances for each column,

(d) Component balances for each column,

(e) Component balances for mixers at the inlet of each column,

(f) Mass balances for each component in the final mixers,

(g) Summation of mole fractions,

(h) Logical constraints,

(i) Bounds on recoveries, and

(j) Nonnegativity and integrality constraints.

The objective function represents the total annual cost consisting of the investment cost of the columns and heat exchangers and the operating cost of the hot and cold utilities.

We will discuss the mathematical formulation through the following illustrative example.

Illustration 9.3.2 (Mathematical Model)
The example considered here is taken from Aggarwal and Floudas (1990) and consists of a three-component feed stream to be separated into two products, P_1 and P_2, which are also three-component streams. The data for this problem is shown in Table 9.2. Since the feed stream features three components, there are two separation breakpoints (i.e., A/BC and AB/C) and hence the superstructure has two columns. The nonsharp superstructure for this example is shown in Figure 9.5. Note that all streams are numbered so as to define the required variables which are

(i) Flow rates: $F_1, F_2, \ldots, F_{20}$.

Components	Feed (kgmol h^{-1})	Products (kgmol h^{-1})	
		P_1	P_2
Propane	100	30	70
iso-Butane	100	50	50
n-Butane	100	30	70
Utilities available:			
Cooling water @ 305 K (20K rise), cost = \$0.159/10^6 kJ			
Steam @ 343 kPa, cost = \$2.45/10^6 kJ			
Payback period = 2.5 yr.			
Tax correction factor = 0.52			

Table 9.2: Data for illustrative example

(ii) Compositions of all interconnecting feed and product streams:

$$\left.\begin{array}{ll} \text{Stream 5} : & x_{A,5}, x_{B,5}, x_{C,5} \\ \text{Stream 13} : & x_{A,13}, x_{B,13}, x_{C,13} \end{array}\right\} \textit{Feed streams}$$

$$\left.\begin{array}{ll} \text{Stream 6} : & x_{A,6}, x_{B,6} \\ \text{Stream 9} : & x_{A,9}, x_{B,9}, x_{C,9} \end{array}\right\} \textit{Products of task I}$$

$$\left.\begin{array}{ll} \text{Stream 14} : & x_{A,14}, x_{B,14}, x_{C,14} \\ \text{Stream 18} : & x_{B,18}, x_{C,18} \end{array}\right\} \textit{Products of task II}$$

(iii) Component flowrates for feeds to columns

$$\begin{array}{ll} f_{A,5}, f_{B,5}, f_{C,5} & : \text{Task I} \\ f_{A,13}, f_{B,13}, f_{C,13} & : \text{Task II} \end{array}$$

(iv) Recoveries of key components

$$\begin{array}{ll} r^{lk}_{A,1}, r^{hk}_{B,1} & : \text{Task I} \\ r^{lk}_{B,1}, r^{hk}_{C,1} & : \text{Task II} \end{array}$$

The set of constraints for the illustrative example is

(a) Mass balances for splitters

$$\begin{aligned} F_1 + F_2 + F_3 + F_4 &= 300 \\ F_6 - F_7 - F_8 &= 0 \\ F_9 - F_{10} - F_{11} - F_{12} &= 0 \\ F_{14} - F_{15} - F_{16} - F_{17} &= 0 \\ F_{18} - F_{19} - F_{20} &= 0 \end{aligned}$$

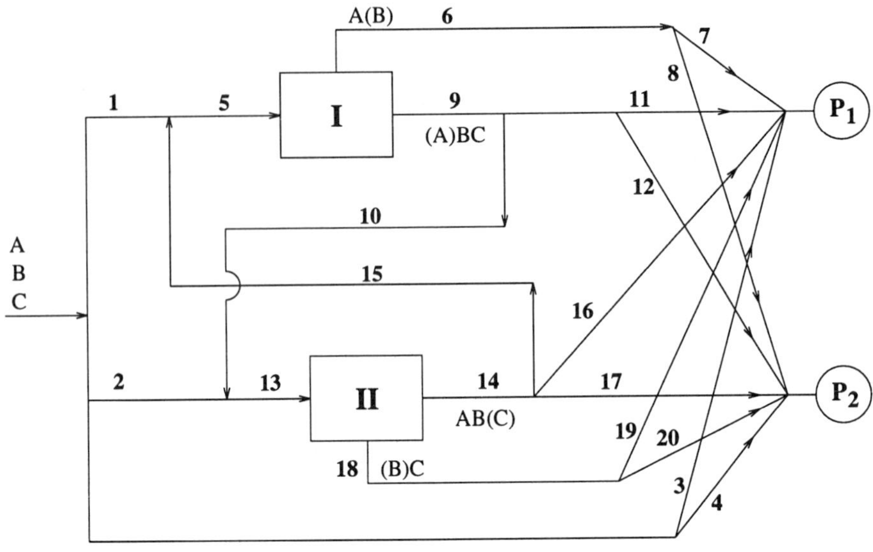

TASK	LK	HK
I	A	B
II	B	C

Figure 9.5: Superstructure for illustrative example 9.3.2

(b) Definitions of component flow rates for column feeds

$$\left.\begin{aligned} f_{A,5} - F_5 \cdot x_{A,5} &= 0 \\ f_{B,5} - F_5 \cdot x_{B,5} &= 0 \\ f_{C,5} - F_5 \cdot x_{C,5} &= 0 \end{aligned}\right] \quad Task\ I$$

$$\left.\begin{aligned} f_{A,13} - F_{13} \cdot x_{A,13} &= 0 \\ f_{B,13} - F_{13} \cdot x_{B,13} &= 0 \\ f_{C,13} - F_{13} \cdot x_{C,13} &= 0 \end{aligned}\right] \quad Task\ II$$

(c) Key component balances

$$\left.\begin{aligned} F_6 \cdot x_{A,6} - r^{lk}_{A,1} \cdot f_{A,5} &= 0 \\ F_9 \cdot x_{B,9} - r^{hk}_{B,1} \cdot f_{B,5} &= 0 \end{aligned}\right] \quad Task\ I$$

$$\left.\begin{aligned} F_{14} \cdot x_{B,14} - r^{lk}_{B,2} \cdot f_{B,13} &= 0 \\ F_{18} \cdot x_{C,18} - r^{hk}_{C,2} \cdot f_{C,13} &= 0 \end{aligned}\right] \quad Task\ II$$

(d) Component balances

$$\left.\begin{aligned} f_{A,5} - F_6 \cdot x_{A,6} - F_9 \cdot x_{A,9} &= 0 \\ f_{B,5} - F_6 \cdot x_{B,6} - F_9 \cdot x_{B,9} &= 0 \\ f_{C,5} - F_9 \cdot x_{C,9} &= 0 \end{aligned}\right] \quad Task\ I$$

$$\left.\begin{aligned} f_{A,13} - F_{14} \cdot x_{A,14} &= 0 \\ f_{B,13} - F_{14} \cdot x_{B,14} - F_{18} \cdot x_{B,18} &= 0 \\ f_{C,13} - F_{14} \cdot x_{C,14} - F_{18} \cdot x_{C,18} &= 0 \end{aligned}\right] \quad Task\ II$$

(e) Mass balances for each component in mixers prior to columns

$$\left.\begin{aligned} f_{A,5} - (0.333)F_1 - F_{15} \cdot x_{A,14} &= 0 \\ f_{B,5} - (0.333)F_1 - F_{15} \cdot x_{B,14} &= 0 \\ f_{C,5} - (0.333)F_1 - F_{15} \cdot x_{C,14} &= 0 \end{aligned}\right] \quad Task\ I$$

$$\left.\begin{aligned} f_{A,13} - (0.333)F_2 - F_{10} \cdot x_{A,9} &= 0 \\ f_{B,13} - (0.333)F_2 - F_{10} \cdot x_{B,9} &= 0 \\ f_{C,13} - (0.333)F_2 - F_{10} \cdot x_{C,9} &= 0 \end{aligned}\right] \quad Task\ II$$

(f) Mass balances for each component in final mixers

$$\begin{aligned} (0.333)F_3 + F_7 \cdot x_{A,6} + F_{11} \cdot x_{A,9} + F_{16} \cdot x_{A,14} &= 30 \\ (0.333)F_3 + F_7 \cdot x_{B,6} + F_{11} \cdot x_{B,9} + F_{16} \cdot x_{B,14} + F_{19} \cdot x_{B,19} &= 50 \\ (0.333)F_3 + F_{11} \cdot x_{C,9} + F_{16} \cdot x_{C,14} + F_{19} \cdot x_{C,19} &= 30 \end{aligned}$$

Note that we only need to write the mass balances for each component for product P_1. From the other balances, we automatically satisfy the component balances for product P_2.

(g) Summation of mole fractions

$$\left.\begin{aligned} x_{A,5} + x_{B,5} + x_{C,5} &= 1 \\ x_{A,13} + x_{B,13} + x_{C,13} &= 1 \end{aligned}\right] \quad Feeds\ to\ columns$$

$$\left.\begin{aligned} x_{A,6} + x_{B,6} &= 1 \\ x_{A,9} + x_{B,9} + x_{C,9} &= 1 \end{aligned}\right] \quad Products\ of\ task\ I$$

$$\left.\begin{aligned} x_{A,14} + x_{B,14} + x_{C,14} &= 1 \\ x_{B,18} + x_{C,18} &= 1 \end{aligned}\right] \quad Products\ of\ task\ II$$

where y_1, y_2 are 0-1 variables denoting the existence of columns I and II, respectively.

(h) Logical Constraints

$$\begin{aligned} F_5 - 300 \cdot y_1 &\leq 0 \\ F_{13} - 300 \cdot y_2 &\leq 0 \end{aligned}$$

(i) Bounds on recoveries

$$\left.\begin{array}{rcl} 0.85 \leq & r^{lk}_{A,1} & \leq 1 \\ 0.85 \leq & r^{hk}_{B,1} & \leq 1 \end{array}\right] \quad Task\ I$$

$$\left.\begin{array}{rcl} 0.85 \leq & r^{lk}_{B,2} & \leq 1 \\ 0.85 \leq & r^{hk}_{C,2} & \leq 1 \end{array}\right] \quad Task\ II$$

The lower bounds on the recoveries were set to 0.85 on the grounds that a number of simulations performed showed that to avoid the distribution of nonkey components it was necessary to keep the recoveries of the key components greater than or equal to 0.85.

(j) Nonnegativity and integrality conditions

$$\begin{array}{rcl} F_1, F_2, \ldots, F_{20} & \geq & 0 \\ f_{A,5}, f_{B,5}, f_{C,5}, f_{A,13}, f_{B,13}, f_{C,13} & \geq & 0 \\ x_{A,5}, x_{B,5}, x_{C,5}, x_{A,13}, x_{B,13}, x_{C,13} & \geq & 0 \\ x_{A,6}, x_{B,6}, x_{A,9}, x_{B,9}, x_{C,9} & \geq & 0 \\ x_{A,14}, x_{B,14}, x_{C,14}, x_{B,18}, x_{C,18} & \geq & 0 \\ y_1, y_2 & = & 0-1 \end{array}$$

Remark 1 Note that the set of constraints is nonlinear due to constraints (b), (c), (d), (e), and (f). These constraints feature bilinear terms of

(i) Stream flow rates and component compositions, and

(ii) Recoveries of key components and component flow rates.

As a result, the above constraints are nonconvex.

The objective function represents the total annual cost and takes into account the investment cost of each column and its associated exchangers as well as the operating cost, and it is derived using the approach presented in Aggarwal and Floudas (1990). It is of the following form:

$$\begin{aligned} MIN \quad & 23947 \cdot y_1 + 75835 \cdot y_2 \\ & + \left[-1399.04 + 935.14 \cdot r^{lk}_{A,1} + 773.08 r^{hk}_{B,1} - 57.19 \cdot x_{A,5} + 426.56 \cdot x_{B,5}\right] \cdot F_5 \\ & + \left[-6615.88 + 3381.47 \cdot r^{lk}_{B,2} + 3733.49 r^{hk}_{C,2} + 163.71 \cdot x_{A,13} + 2899.96 \cdot x_{B,13}\right] \cdot F_{13}, \end{aligned}$$

where the coefficients are obtained via regression analysis of shortcut simulation data taken at different flow rates and compositions.

Remark 2 Note that the objective function is also nonconvex due to the bilinear products of

(i) column feed flow rates and recoveries, and

(ii) column feed flow rates and component compositions.

Algorithmic Strategy:

The resulting optimization model is a nonconvex MINLP and the use of OA/ER/AP or v2-GBD can only provide a local optimum solution. In fact, application of v2-GBD or OA/ER/AP by projecting only on the binary variables showed that there exist several local solutions and the solution found depends heavily on the starting point. As a result, Aggarwal and Floudas (1990) proposed another projection scheme, which is

$$\begin{aligned} y\text{-variables} \quad &: \quad y_1, y_2, \\ & \quad\;\; F_1, F_2, \ldots, F_{20}, \\ & \quad\;\; f_{A,5}, f_{B,5}, f_{C,5}, f_{A,13}, f_{B,13}, f_{C,13} \\ x\text{-variables} \quad &: \quad \text{remaining variables.} \end{aligned}$$

and made use of the v3-GBD (Floudas *et al.*, 1989) which provided consistent performance with respect to locating a least total annual cost sequence from all starting points. Note, though, that the use of v3-GBD does not provide theoretical guarantee of obtaining the best possible solution.

It should also be noted that the selection of the above y-variables (i) makes the primal problem in v3-GBD a linear programming LP problem, (ii) allows the linear constraints to be directly transferred into the relaxed master problem, and (iii) makes the master problem a mixed-integer linear programming MILP problem.

To reduce potential infeasible primal problems, Aggarwal and Floudas (1990) proposed in Appendix B of this paper a set of linear constraints that can be incorporated in the relaxed master problem.

Applying the above algorithmic strategy, the optimal solution found, shown in Figure 9.6, has a total annual cost of \$156,710. It features (i) task II first and then task I, (ii) a total by-pass of the feed to the two products, and (iii) the bottoms of task II is directed to product P_2.

Note that the recoveries in the optimal solution shown in Figure 9.6 are at their lower bounds for $r^{lk}_{A,1}$, $r^{lk}_{B,2}$ and $r^{hk}_{C,2}$, and at the upper bound for $r^{hk}_{B,1}$. A comparison between sharp and nonsharp sequences Aggarwal and Floudas (1990) shows that significant savings can result from optimizing the degree of nonsharp separation. In particular, in example 2 of their paper, which corresponds to the same optimization model presented in the illustrative example but with different product compositions, the optimal solution consists of a single column with a by-pass stream. This corresponds to 70% cost reduction versus the sharp separation case.

In Appendix A of Aggarwal and Floudas (1990), a procedure is presented that calculates upper bounds on the flow rates of the overall by-pass streams. These bounds are useful in restricting the feasible domain of the remaining flow rates.

The illustrative example was solved using the v3-GBD with the APROS (Paules and Floudas, 1989) methodology, and is included in the library OASIS (Floudas, 1990). This example has also been solved using the MINOPT (Rojnuckarin and Floudas, 1994) in which the special features of radial search are automatically incorporated.

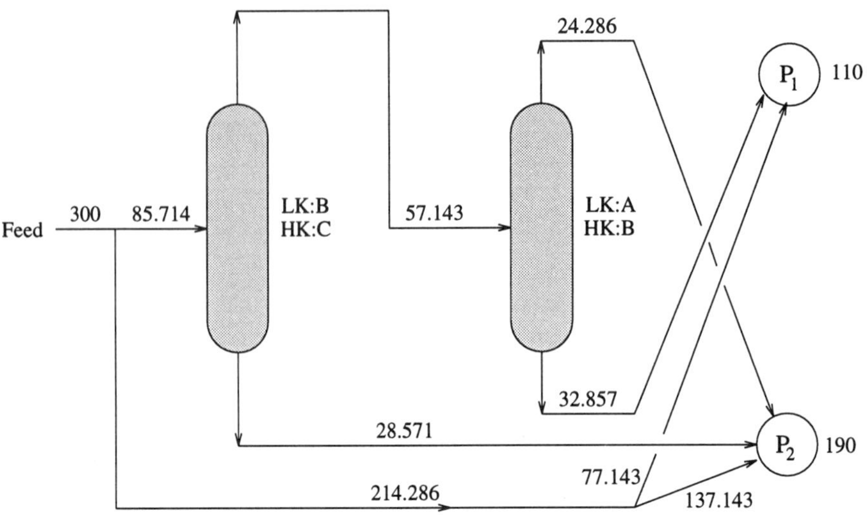

Figure 9.6: Optimal Nonsharp separation sequence for illustrative example

Summary and Further Reading

This chapter discusses mixed-integer nonlinear optimization applications in the area of distillation-based separation system synthesis. Section 9.1 introduces the reader to the generic definition of separation system synthesis and provides a classification of the different types of separations based on distillation column sequencing. For further reading on separation system synthesis the reader is referred to the reviews by Westerberg (1985) and Nishida *et al.* (1981)

Section 9.2 presents key issues that arise in the synthesis of heat-integrated sharp distillation columns and presents a mixed-integer nonlinear optimization model which allows for treating the pressure of each column explicitly as a variable while at the same time heat integration alternatives are incorporated. For further reading in this subject the reader is referred to Floudas and Paules (1988) and the other suggested references.

Section 9.3 describes the synthesis of nonsharp distillation sequences without considering the heat integration alternatives. The mixed-integer nonlinear optimization approach of Aggarwal and Floudas (1990) is discussed and illustrated via an example that involves a three-component feed to be separated into two three component products. Further reading material in this subject can be found in Aggarwal and Floudas (1990). Reading material on the synthesis of heat-integrated nonsharp separation sequences can be found in Aggarwal and Floudas (1992).

Chapter 10 Synthesis of Reactor Networks and Reactor-Separator-Recycle Systems

This chapter discusses the application of MINLP methods in the synthesis of reactor networks with complex reactions and in the synthesis of reactor–separator–recycle systems.

10.1 Introduction

Despite the importance of reactor systems in chemical engineering processes, very few systematic procedures for the optimal synthesis of reactor networks have been proposed. The main reason for the scarcity of optimization strategies for reactor networks is the difficulty of the problem itself. The large number of alternatives along with the highly nonlinear equations that describe these systems have led to the development of a series of heuristic and intuitive rules that provide solutions only for simple cases of reaction mechanisms. Most of the studies considered single reactors with a specified mixing pattern and focused on investigating the effect of temperature distribution, residence time distribution, or catalyst dilution profile on its performance.

In the sequel, we will briefly review the approaches developed based on their classification: (i) isothermal operation and (ii) nonisothermal operation.

Isothermal Operation

Trambouze and Piret (1959) proposed graphical and analytical criteria for selecting the type of reactor. Levenspiel (1962) reported heuristic rules for optimal yield and selectivity in stirred tank and tubular reactors. Aris (1964, 1969) applied dynamic programming to determine the optimal amounts of by-passes and cold streams in a multistage reaction system within a fixed structure. Gillespie and Carberry (1966) studied the Van der Vusse reaction with an intermediate level of mixing and demonstrated the potential advantages of recycle reactors for such a complex reaction. Horn and Tsai (1967) studied the effects of global and local mixing using the adjoint variables of optimization theory.

Jackson (1968) proposed an algebraic structure for the reactor representation consisting of parallel ideal tubular reactors that were interconnected with side streams at various sink and source points. Different flow configurations and mixing patterns could be obtained by varying the number and the positions of the sink and source points, as well as the levels of the sidestreams. By deliberate manipulation of the flow configuration, potential improvements in the reactor performance could be investigated. Ravimohan (1971) modified Jackson's model so as to handle cases of local mixing. In addition to the source and sink points, he introduced a third type of points in which he placed ideal continuous stirred tank reactors of known volume. Glasser and Jackson (1984) showed how a discrete residence time distribution (macromixing) with micromixing only between streams having equal remaining life can model an arbitrary system of flow mixing and splitting with plug sections in between. Jackson and Glasser (1985) proposed a mathematical model for the reactor performance (general mixing model) which contains the macromixing and micromixing as explicit functions. The model is valid for steady flow reactors in which no volume change on mixing or reaction takes place and includes the segregated and the Zwietering (1959) maximum mixed model as special cases. Glasser *et al.* (1987) studied geometrical approaches for determining the attainable region in concentration space of an isothermal reacting system. Once the region is known, the optimization problem is greatly facilitated and the efficiency of the method is demonstrated by studying a number of two-dimensional examples.

Paynter and Haskins (1970) represented the reactor network by an axial dispersion model (ADR). In this model the reactor network consisted of a plug flow reactor with variable dispersion coefficient, and the synthesis problem was formulated as an optimal control nonlinear programming problem. Waghmare and Lim (1981) dealt with the optimal configuration of isothermal reactors and noted that optimal feeding strategies have spatial duals with PFR and CSTR configurations. Since optimal feeding strategies have duals with batch reactors, they concluded that batch reactors can be represented by CSTR and PFR combinations. Chitra and Govind (1981, 1985) classified the various reaction mechanisms into different categories. According to the mechanism, they applied either heuristic rules or a direct search procedure that determines the maximum yield of a serial network of CSTRs.

Achenie and Biegler (1986) used a modified Jackson model with ADRs instead of PFRs. They assumed constant density systems and a piecewise constant dispersion coefficient. The synthesis problem, based on this model, was formulated as an optimal control nonlinear programming problem. The solution of the problem consisted of successively solving a set of nonlinear two-point boundary value ordinary differential equations and a quadratic programming problem. Achenie and Biegler (1990) reformulated the problem with a superstructure consisting of recycled PFRs in series. The density of the reaction mixture was assumed constant, and the reactor type was determined by the value of the recycle ratio which ranged from 0 (PFR) to ∞ (CSTR). The mathematical formulation of the synthesis problem resulted in an optimal control problem and was solved by successively solving a system of ODEs and a quadratic programming problem. From a different point of view, Achenie and Biegler (1988) developed a targeting technique in which they searched for upper bounds (targets) of the objective function. Using a modified Ng and Rippin model which is based on isothermal homogeneous systems and does not cover all mixing states, they formulated the synthesis problem with decision variables the macromixing and micromixing functions. Although information about the optimal mixing pattern could be obtained, this approach was not able to explicitly provide the optimal reactor network configuration.

Kokossis and Floudas (1990) proposed a systematic approach for the optimization problem of reactor networks that involve complex reaction schemes under isothermal operation. A reactor representation is proposed that utilizes as a basis the continuous stirred tank reactor and approximates the plug flow reactor as a cascade of CSTRs. A reactor network superstructure is presented in which all possible structural configurations of interconnected reactors are embedded. Different types of recycle reactors are also incorporated that can handle different feeding, recycling, and bypassing strategies. The reactor network synthesis problem that is formulated based upon this superstructure is a large-scale mixed integer nonlinear programming problem, and its solution provides information about the optimal reactor network configuration.

Nonisothermal Operation

The synthesis of nonisothermal reactor networks has been studied traditionally through fixed structures of reactor units in which potential improvements can be identified from intuitive rearrangements of the assumed reactor system. Dynamic programming techniques were applied to the reactor design problem associated with a single exothermic reaction processed by a cascade of either adiabatically operated beds (Aris;1960a, 1960b) or continuous stirred tank reactors (Aris, 1960c). The approach assumed a linear profit function for the system, and the analysis provided valuable graphical interpretation of the method and justified the use of adiabatic stages for studying the nonisothermal problem. From a similar perspective, Aris (1960d) made use of analytical expressions of the maximum reaction rate and the Denbigh's notion of maximization of rectangles in order to obtain optimal holding times and temperatures for the single exothermic reaction. The potential effect of cold shots along a reactor system of three adiabatic beds was next investigated by Lee and Aris (1963), who studied the catalytic oxidation of sulfur dioxide. The same reaction was also studied by King (1965), who assumed a reactor system made up of a continuous stirred tank and an ideal plug flow reactor. Both reactors operated adiabatically and dynamic programming tools were used for the optimal search.

The study of single exothermic reaction mechanisms offers favorable features (i.e., optimal disjoint policy and existence of an optimal reaction rate with respect to the temperature) which, as Mah and Aris (1964) pointed out, are not generally obtained in more complicated cases. These authors studied the optimal policy for a stirred tank reactor processing reversible consecutive reactions and discussed the formidable computational effort associated with the determination of an overall optimal policy. Wang and Fan (1964) applied the single recursion relation of Katz's algorithm for the optimization of multistage chemical processes and provided an alternative tool for searching the optimal temperature profile and the holding times of a battery of stirred tank reactors in which a single reversible reaction takes place. In a later work, Fan *et al.* (1965) extended this work and studied the effect of a total recycle in the performance of the cascade.

Dyson and Horn (1967) studied the performance of an exothermic reactor by using the maximization of the outlet conversion as the objective function. The system featured a preheater for the inlet stream, and the analysis provided the limiting performance of the system in terms of the allocation of a cold shot along the reactor. This pioneering work has exemplified the structural alternatives of a nonisothermal reactor network and, despite its reference to a single reversible reaction, it defined a new perspective for the synthesis problem. Dyson *et al.* (1967) properly classified the structural alternatives and the existing options for controlling the temperature of an exothermic reactor (e.g., indirect and direct intercooling and interheating, perfect control) and they discussed alternative approaches for handling cases of more complicated reactions. In a later work, Dyson and Horn (1969) determined the minimum mass of catalyst required by an adiabatic tubular reactor with feed preheater and direct intercooling by cold feed bypass for the single exothermic reaction. The inlet and outlet conversions and the outlet flow rates were fixed, and the minimization was performed over the inlet temperature and the catalyst and cold shot distributions.

Russell and Buzelli (1969) studied a class of competitive-consecutive reactions and presented comparisons among different alternative structures (CSTRs and PFRs). In an effort to study the minimization of catalyst problem for the water gas reaction, Caha *et al.* (1973) proposed a structure consisting of three serial adiabatic beds featuring options for indirect cooling. The optimization of multistage catalytic reactors was also studied by Brusset *et al.* (1978), who used dynamic programming methods. No options for cold shots were considered, and the optimal search has been based upon an economic objective function. On the other side, Burghardt and Patzek (1977) discarded any options for indirect cooling (heat exchangers) and studied the problem only in terms of potential cold shots.

The optimization of nonisothermal reactor systems, that process complex reaction mechanisms, was addressed by Chitra and Govind (1985), who proposed a serial structure of recycled PFRs with indirect cooling. By classifying the reaction mechanisms according to a set of proposed criteria (i.e., type I, II, and III), these authors suggested graphical or analytical procedures according to the type of the considered mechanism. The same reactor unit (recycled PFR) was also used by Achenie and Biegler (1990), and the resulting configuration was optimized using successive quadratic programming techniques. Both of the above research groups assumed only serial combinations of units and resulted in a nonlinear programming formulation unable to handle cases where different types of reactors have to be considered. In these cases, as Achenie and Biegler (1990) pointed out, only a mixed integer nonlinear programming MINLP formulation would be the appropriate alternative.

From a completely different standpoint, Glasser *et al.* (1987), Hildebrandt and Glasser (1990), and Hildebrandt *et al.* (1990) proposed that the synthesis strategy be focused on the construction of the attainable region of the particular reaction, a concept first introduced by Horn (1964). For the isothermal problems the strategy provided an elegant tool for the graphical interpretation of the synthesis problem, and the extension of the method to more complicated mechanisms and nonisothermal problems unavoidably goes through the proposition of a series of curves and three-dimensional graphical simulations. While geometric concepts lead to tools for visualizing and constructing an attainable region in the concentration space, obtaining the attainable region in higher dimensions can become much more difficult. Toward addressing the higher dimensionality problems, Balakrishna and Biegler (1992a) proposed convex two-dimensional projections of the multidimensional problems for isothermal reactor networks and derived general sufficient conditions for the construction of the attainable region. Balakrishna and Biegler (1992b) extended this approach to nonisothermal systems and systems with variable feed compositions, while Balakrishna and Biegler (1993) proposed an MINLP based unified targeting approach for the coupling of reactors, separators, and heat integration options. A recent review of the synthesis of chemical reactor networks is provided by Hildebrandt and Biegler (1994).

Kokossis and Floudas (1994) proposed a superstructure of alternatives for the nonisothermal reactor system and subsequently formulate and solve the synthesis problem according to the proposed general structure. The presented approach is based upon a generalization of the isothermal reactor superstructure by Kokossis and Floudas (1990). The nonisothermal case features alternatives related to the type of temperature control for the reactors and includes, apart from the pure adiabatic operation, options for perfectly controlled units and directly or indirectly intercooled or interheated reactors. Once these options are embedded into a nonisothermal reactor unit, a general reactor superstructure is proposed which includes the structural alternatives for the reactor system.

The approach is applicable to homogeneous exothermic or endothermic complex reactions and the synthesis problem formulated as a mixed integer nonlinear programming MINLP problem can handle both thermodynamic and economic objective functions. The design variables of the problem include the concentrations and temperatures of the participating units, the inlet, outlet and interconnecting streams among the reactors, the amount of heat added or removed from each unit, the sizes of the reactors, as well as economic functions associated with the fixed and operating costs or the profit of the process. The solution of the MINLP problem will provide information about the optimal type of temperature control, the optimal temperature profile, the feeding, recycling, and bypassing strategy, as well as the optimal type and size of the reactor units (CSTR, PFR) to be used.

In the following sections, we will discuss the synthesis of isothermal and nonisothermal reactor networks with complex reaction mechanisms.

10.2 Synthesis of Isothermal Reactor Networks

10.2.1 Problem Statement

Given

(i) A complex reaction mechanism and its kinetics (e.g., arbitrary nonlinear kinetics),

(ii) Different types of reactors (e.g., CSTRs, PFRs), and

(iii) A performance index which is a function of outlet stream compositions and the reactor volumes (e.g., yield, selectivity),

determine a reactor network that optimizes the prescribed performance index and provides information on

(a) The number and type of reactors,

(b) The volumes of each reactor,

(c) The feeding, recycling and by-passing, and

(d) The topology of the reactor network and the optimal values of the flowrates and compositions of each stream.

Remark 1 The operating conditions for each reactor (i.e., temperature, pressure) are assumed fixed at their nominal values and hence we have isothermal operation.

Remark 2 The reacting mixture is assumed to be an ideal solution (i.e., zero excess solution property). As a result of this assumption we do not need data of the excess molar volume as a function of the reacting components.

Remark 3 Homogeneous types of reactions are only considered, and no catalytic action takes place in any of the reactors.

Remark 4 The representation of the reactor units is based on CSTRs and PFRs which represent the two different extremes in reactor behavior. Note, however, that reactors with various distribution functions can be incorporated.

In the following we will discuss the approach of Kokossis and Floudas (1990) for the synthesis of isothermal complex reactor networks.

10.2.2 Basic Idea

The basic idea in the synthesis approach of isothermal reactor networks with complex reaction mechanisms proposed by Kokossis and Floudas (1990) consists of

(i) Representing each reactor unit by a single or several stirred tank reactors in series,

(ii) Postulating a representation of all possible structural alternatives which include side-streams, recycles, by-passes as well as reactors in series, parallel, series-parallel and parallel-series configurations,

(iii) Formulating the postulated representation as a mixed-integer nonlinear programming MINLP problem that optimizes the performance index subject to the constraints that describe the representation, and

(iv) Solving the resulting MINLP model providing information on:

  (a) The reactors selected,

  (b) The interconnections among the reactors,

  (c) The values of the stream flow rates, compositions and reactor volumes, and

  (d) The topology of the reactor network configuration.

Remark 1 The resulting model does not involve differential equations, but it is instead algebraic due to the representation of each reactor unit of (i) which is discussed in the following section.

10.2.3 Reactor Unit Representation

The representation of different types of reactor units in the approach proposed by Kokossis and Floudas (1990) is based on the ideal CSTR model, which is an algebraic model, and on the approximation of plug flow reactor, PFR units by a series of equal volume CSTRs. The main advantage of such a representation is that the resulting mathematical model consists of only algebraic constraints. At the same time, however, we need to introduce binary variables to denote the existence or not of the CSTR units either as single units or as a cascade approximating PFR units. As a result, the mathematical model will consist of both continuous and binary variables.

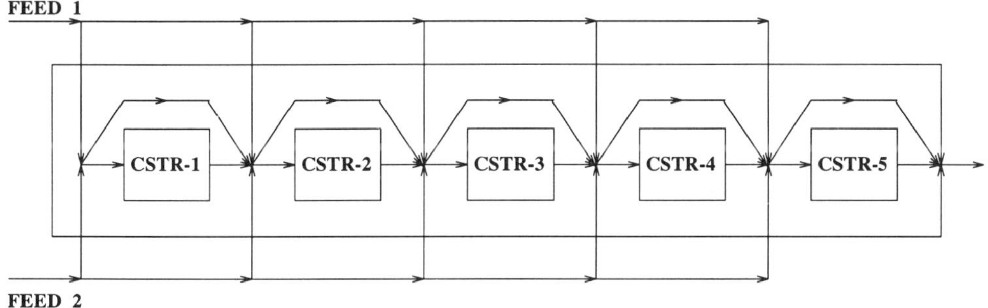

Figure 10.1: Representation of a PFR unit with five CSTRs and two feeds

Remark 1 If no approximation is introduced in the PFR model, then the mathematical model will consist of both algebraic and differential equations with their related boundary conditions (Horn and Tsai, 1967; Jackson, 1968). If in addition local mixing effects are considered, then binary variables need to be introduced (Ravimohan, 1971), and as a result the mathematical model will be a mixed-integer optimization problem with both algebraic and differential equations. Note, however, that there do not exist at present algorithmic procedures for solving this class of problems.

Remark 2 Note that the approximation of PFRs with a cascade of equal volume CSTRs is improved as we increase the number of CSTRs. By doing so however we increase the number of binary variables that denote their existence, and hence the combinatorial problem becomes more difficult to handle. Usually approximations of a PFR with 5-10 CSTRs are adequate. Kokossis and Floudas (1990) studied the effect of this approximation and their proposed approach could deal with approximations of PFRs with 200 equal volume CSTRs in series.

Using the approximation of a PFR with a cascade of equal volume CSTRs in series, we allow for all feed streams (i.e., fresh feeds, streams from the outlets of other reactors) to be distributed to any of the CSTRs that approximate the PFR unit. In addition, potential by-passes around each CSTR that approximates the PFR unit are introduced so as to include cases in which full utilization of these CSTRs is not desirable.

To illustrate the PFR unit representation, let us consider the representation of a PFR with five CSTRs shown in Figure 10.1. Figure 10.1 illustrates a PFR with five CSTRs and two feed streams. These feed streams can represent fresh feed (e.g., feed 1) and the outlet stream of another reactor (e.g., feed 2). Note that the feed streams are allowed to distribute to any of the CSTRs. Also note the incorporation of by-passes around each CSTR unit.

Remark 3 Batch and semibatch reactors can also be studied by considering their space equivalent PFRs. The space equivalent of a single batch reactor can be regarded as a PFR with no side streams. The optimal holding time for the batch operation can be determined by the optimal PFR length and the assumed linear velocity of the fluid in the PFR.

Remark 4 Semibatch reactors can be represented via the proposed approach since their performance can be approximated by a sequence of batch operations. In this case, the space equivalent of a semibatch reactor consists of a series arrangement of PFRs each featuring a separate fresh feed stream. This representation can be obtained from Figure 10.1 if we replace each CSTR by a cascade of 5-10 CSTRs (i.e., sub-PFR representation) and having each side stream denoting additions of fresh reactants at different times of operations. In this case, each sub-PFR represents a different batch operation of the overall semibatch process. The optimal holding time at the inlet of the third sub-PFR can be obtained by the optimal lengths of the first two sub-PFRs and the assumed fluid velocity in each sub-PFR.

Remark 5 Note also that reactors of various distribution functions can be treated with this approach, on the grounds that different distribution functions are usually approximated via cascades of CSTRs. In this case, we can treat the number of CSTRs as a variable or provide a variety of alternative reactors each featuring different numbers of CSTRs. Kokossis and Floudas (1990), present examples for batch, semibatch reactors and different distribution functions.

10.2.4 Reactor Network Superstructure

The basic reactor units used in the derivation of the reactor network superstructure are

(i) Ideal CSTRs, and

(ii) Ideal PFRs approximated as a cascade of equal volume CSTRs.

Postulating a reactor network representation of alternatives requires that we embed:

(a) Different types of reactors,

(b) All possible interconnections among them which include reactors in series, parallel, series-parallel, parallel-series, and

(c) Options for feed stream splitting, stream mixing, and by-passing.

Illustration 10.2.1 (Reactor Network Superstructure)
This example is taken from Kokossis and Floudas (1990) and features the Van der Vusse reaction:

$$A \xrightarrow{k_1} B \xrightarrow{k_2} C$$
$$2A \xrightarrow{k_3} D$$

with

$$\begin{aligned}
k_1 &= 10\ s^{-1}\ (first\ order\ reaction\ ) \\
k_2 &= 1\ s^{-1}\ (first\ order\ reaction\ ) \\
k_3 &= 1\ lt/gmol\ s\ \ (first\ order\ reaction\ ) \\
Feed\ flow\ rate &=\ 100\ lt/s\ of\ pure\ A \\
Feed\ concentration &=\ 5.8\ gmol/lt
\end{aligned}$$

The objective is to maximize the yield of B.

A superstructure is postulated that consists of one CSTR and one PFR unit. The PFR unit is approximated via a cascade of five equal volume CSTR units. The superstructure is depicted in Figure 10.2.

Remark 1 Note that the CSTR unit, denoted as R-1, and the PFR unit, denoted as R-2 and consisting of a cascade of five CSTRs of equal volume are interconnected in all possible ways. Also note that R-1 features a by-pass while in R-2 the feed can be distributed to the inlets of any of the five CSTRs that approximate the PFR. The outlet stream of the CSTR is directed to the final destination as well as to the inlets of the CSTRs that approximate the PFR unit. At the outlet of the PFR unit there is a recycle stream for the PFR which can be distributed to any inlet of the CSTRs.

Remark 2 Note that at the inlet of each unit (i.e., CSTR, PFR, CSTR approximating PFR) there is a mixer while at the outlet of each unit there is a splitter. There is also a splitter of the feed and a final mixer. As a result, in addition to the unknown flow rates of each stream of the superstructure we have also as unknowns the compositions of outlets of each unit. Finally, the volumes of each reactor unit are treated as unknown variables.

Remark 3 The superstructure shown in Figure 10.2 contains the discrete alternatives of existence or not of the reactor units, as well as the continuous alternatives of components, and volumes of the reactor units. Such a superstructure contains many structural alternatives of interest which can be obtained by setting a selected number of flow rates to zero values. Kokossis and Floudas (1990) provide a few such illustrations for superstructures of two CSTRs and two PFRs. In Figure 10.2, the variables: (i) flow rates for all streams except the by-passes and (ii) component compositions ($i = A, B, C, D$) are also indicated.

10.2.5 Mathematical Formulation of Reactor Superstructure

To determine a reactor network that optimizes the performance index (e.g., maximizes yield) out of the many embedded alternatives in the superstructure of Figure 10.2, we define variables for

(i) The existence of each reactor unit (i.e., $y_1 = 0 - 1$ for $R - 1$, $y_2 = 0 - 1$ for $R - 2$, and z_1, z_2, z_3, z_4, z_5 for units $C - 1$ to $C - 5$); due to the approximation of $R - 2$ with five CSTRs, there is a relation between y_2 and $z_1, \ldots, z_5$.

(ii) The stream flow rates (i.e. $F_1, F_2, \ldots F_{29}$).

(iii) The component compositions (molar fractions) at the inlets and outlets of each reactor unit (i.e., $x_{7,i}, x_{14,i}, x_{16,i}, x_{18,i}, x_{20,i}, x_{22,i}$ for the inlets and $x_{8,i}, x_{15,i}, x_{17,i}, x_{19,i}, x_{21,i}, x_{23,i}$ for the outlets.

(iv) The component compositions (molar fractions) at the final destination (i.e., x_A, x_B, x_C, x_D), and

(v) The volumes of the reactor units (i.e., V_1 for $R - 1$, V_2 for $R - 2$, and V for each of $C - 1, C - 2, C - 3, C - 4$, and $C - 5$).

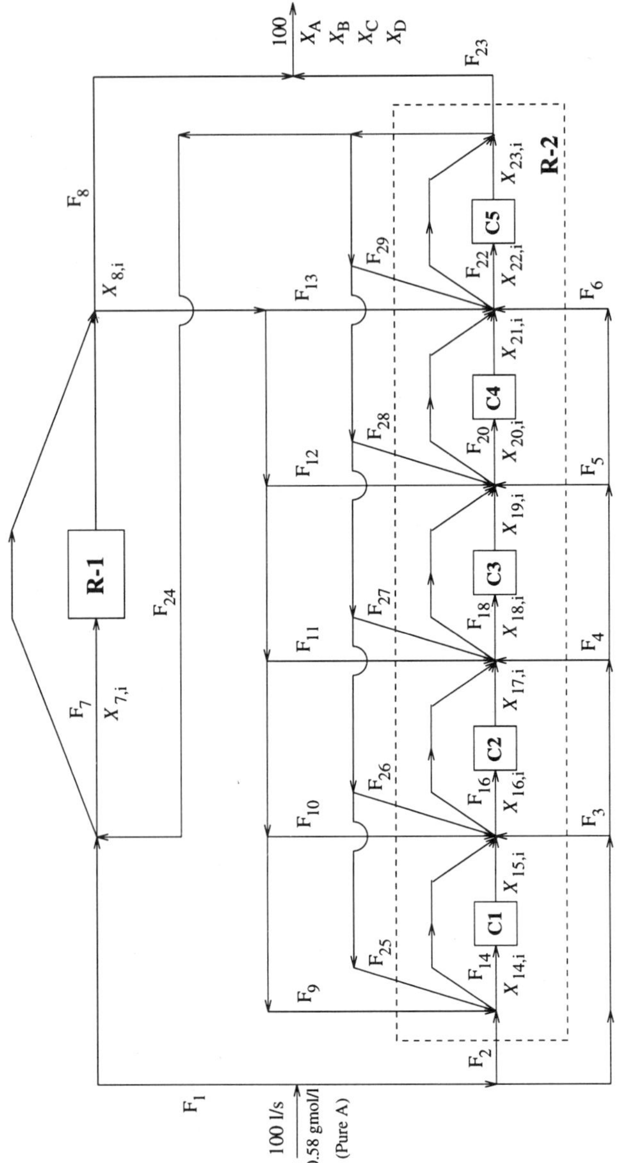

Figure 10.2: Reactor network superstructure for illustrative example

The set of constraints will consist of

(i) Mass balances for splitters of feed and reactor units,

(ii) Component balances for mixers at the inlets of reactor units,

(iii) Component balances for final mixer,

(iv) Component balances around each reactor,

(v) Summation of mole fractions,

(vi) Logical constraints and volume constraints, and

(vii) Nonnegativity and integrality constraints.

The objective function will optimize the considered performance measure (e.g., maximize yield, selectivity).

For simplicity of the presentation we will not consider the by-passes around each reactor unit. In the sequel, we will discuss the optimization model for the illustrative example.

Illustration 10.2.2 (Optimization Model)
The set of constraints for the superstructure of Figure 10.2 of the illustrative example is as follows (note that we omit the by-pass streams around the reactor units for simplicity of the presentation):

(i) Mass balances for splitters

$$\begin{aligned} F_1 + F_2 + F_3 + F_4 + F_5 + F_6 &= 100 \\ F_7 - F_8 - F_9 - F_{10} - F_{11} - F_{12} - F_{13} &= 0 \\ F_{22} - F_{23} - F_{24} - F_{25} - F_{26} - F_{27} - F_{28} - F_{29} &= 0 \end{aligned}$$

(ii) Component balances for mixers at inlets of reactors

$$\left.\begin{aligned} F_1 \cdot 0.58 + F_{24} \cdot x_{23,A} - F_7 \cdot x_{7,A} &= 0 \\ F_{24} \cdot x_{23,B} - F_7 \cdot x_{7,B} &= 0 \\ F_{24} \cdot x_{23,C} - F_7 \cdot x_{7,C} &= 0 \\ F_{24} \cdot x_{23,D} - F_7 \cdot x_{7,D} &= 0 \end{aligned}\right] \quad \mathrm{R-1}$$

$$\left.\begin{aligned} F_2 \cdot 0.58 + F_9 \cdot x_{8,A} + F_{25} \cdot x_{23,A} - F_{14} \cdot x_{14,A} &= 0 \\ F_9 \cdot x_{8,B} + F_{25} \cdot x_{23,B} - F_{14} \cdot x_{14,B} &= 0 \\ F_9 \cdot x_{8,C} + F_{25} \cdot x_{23,C} - F_{14} \cdot x_{14,C} &= 0 \\ F_9 \cdot x_{8,D} + F_{25} \cdot x_{23,D} - F_{14} \cdot x_{14,D} &= 0 \end{aligned}\right] \quad \mathrm{C-1}$$

$$\left.\begin{aligned} F_{14} \cdot x_{15,A} + F_{10} \cdot x_{8,A} + F_{26} \cdot x_{23,A} - F_{16} \cdot x_{16,A} &= 0 \\ F_{14} \cdot x_{15,B} + F_{10} \cdot x_{8,B} + F_{26} \cdot x_{23,B} - F_{16} \cdot x_{16,B} &= 0 \\ F_{14} \cdot x_{15,C} + F_{10} \cdot x_{8,C} + F_{26} \cdot x_{23,C} - F_{16} \cdot x_{16,C} &= 0 \\ F_{14} \cdot x_{15,D} + F_{10} \cdot x_{8,D} + F_{26} \cdot x_{23,D} - F_{16} \cdot x_{16,D} &= 0 \end{aligned}\right] \quad \mathrm{C-2}$$

$$\left.\begin{aligned}
F_{16}\cdot x_{17,A}+F_{11}\cdot x_{8,A}+F_{27}\cdot x_{23,A}-F_{18}\cdot x_{18,A} &= 0\\
F_{16}\cdot x_{17,B}+F_{11}\cdot x_{8,B}+F_{27}\cdot x_{23,B}-F_{18}\cdot x_{18,B} &= 0\\
F_{16}\cdot x_{17,C}+F_{11}\cdot x_{8,C}+F_{27}\cdot x_{23,C}-F_{18}\cdot x_{18,C} &= 0\\
F_{16}\cdot x_{17,D}+F_{11}\cdot x_{8,D}+F_{27}\cdot x_{23,D}-F_{18}\cdot x_{18,D} &= 0
\end{aligned}\right]\quad C-3$$

$$\left.\begin{aligned}
F_{18}\cdot x_{19,A}+F_{12}\cdot x_{8,A}+F_{28}\cdot x_{23,A}-F_{20}\cdot x_{20,A} &= 0\\
F_{18}\cdot x_{19,B}+F_{12}\cdot x_{8,B}+F_{28}\cdot x_{23,B}-F_{20}\cdot x_{20,B} &= 0\\
F_{18}\cdot x_{19,C}+F_{12}\cdot x_{8,C}+F_{28}\cdot x_{23,C}-F_{20}\cdot x_{20,C} &= 0\\
F_{18}\cdot x_{19,D}+F_{12}\cdot x_{8,D}+F_{28}\cdot x_{23,D}-F_{20}\cdot x_{20,D} &= 0
\end{aligned}\right]\quad C-4$$

$$\left.\begin{aligned}
F_{20}\cdot x_{21,A}+F_{13}\cdot x_{8,A}+F_{29}\cdot x_{23,A}-F_{22}\cdot x_{22,A} &= 0\\
F_{20}\cdot x_{21,B}+F_{13}\cdot x_{8,B}+F_{29}\cdot x_{23,B}-F_{22}\cdot x_{22,B} &= 0\\
F_{20}\cdot x_{21,C}+F_{13}\cdot x_{8,C}+F_{29}\cdot x_{23,C}-F_{22}\cdot x_{22,C} &= 0\\
F_{20}\cdot x_{21,D}+F_{13}\cdot x_{8,D}+F_{29}\cdot x_{23,D}-F_{22}\cdot x_{22,D} &= 0
\end{aligned}\right]\quad C-5$$

(iii) Component balances for final mixer

$$\begin{aligned}
F_8\cdot x_{8,A}+F_{23}\cdot x_{23,A}-100\cdot x_A &= 0\\
F_8\cdot x_{8,B}+F_{23}\cdot x_{23,B}-100\cdot x_B &= 0\\
F_8\cdot x_{8,C}+F_{23}\cdot x_{23,C}-100\cdot x_C &= 0\\
F_8\cdot x_{8,D}+F_{23}\cdot x_{23,D}-100\cdot x_D &= 0
\end{aligned}$$

(iv) Component balances around each reactor

$$\left.\begin{aligned}
F_7\cdot x_{7,A}-F_7\cdot x_{8,A} &+ V_1\cdot(-A_{R_1}-C_{R_1}) = 0\\
F_7\cdot x_{7,B}-F_7\cdot x_{8,B} &+ V_1\cdot(A_{R_1}-B_{R_1}) = 0\\
F_7\cdot x_{7,C}-F_7\cdot x_{8,C} &+ V_1\cdot(B_{R_1}) = 0\\
F_7\cdot x_{7,D}-F_7\cdot x_{8,D} &+ V_1\cdot(C_{R_1}) = 0
\end{aligned}\right]\quad R-1$$

where

$$\begin{aligned}
A_{R_1}-10\cdot x_{8,A} &= 0\\
B_{R_1}-1\cdot x_{8,B} &= 0\\
C_{R_1}-1\cdot(x_{8,A})^2 &= 0
\end{aligned}$$

are the reaction rates.

$$\left.\begin{aligned}
F_{14}\cdot x_{14,A}-F_{14}\cdot x_{15,A} &+ V_{C_1}\cdot(-A_{C_1}-C_{C_1}) = 0\\
F_{14}\cdot x_{14,B}-F_{14}\cdot x_{15,B} &+ V_{C_1}\cdot(A_{C_1}-B_{C_1}) = 0\\
F_{14}\cdot x_{14,C}-F_{14}\cdot x_{15,C} &+ V_{C_1}\cdot(B_{C_1}) = 0\\
F_{14}\cdot x_{14,D}-F_{14}\cdot x_{15,D} &+ V_{C_1}\cdot(C_{C_1}) = 0
\end{aligned}\right]\quad C-1$$

where

$$\begin{aligned}
A_{C_1}-10\cdot x_{15,A} &= 0\\
B_{C_1}-1\cdot x_{15,B} &= 0\\
C_{C_1}-1\cdot(x_{15,A})^2 &= 0
\end{aligned}$$

$$\left.\begin{array}{rcll}
F_{16}\cdot x_{16,A}-F_{16}\cdot x_{17,A} & + & V_{C_2}\cdot(-A_{C_2}-C_{C_2}) & =0\\
F_{16}\cdot x_{16,B}-F_{16}\cdot x_{17,B} & + & V_{C_2}\cdot(A_{C_2}-B_{C_2}) & =0\\
F_{16}\cdot x_{16,C}-F_{16}\cdot x_{17,C} & + & V_{C_2}\cdot(B_{C_2}) & =0\\
F_{16}\cdot x_{16,D}-F_{16}\cdot x_{17,D} & + & V_{C_2}\cdot(C_{C_2}) & =0\\
\text{where}\quad A_{C_2}-10\cdot x_{17,A} & = & 0 & \\
B_{C_2}-1\cdot x_{17,B} & = & 0 & \\
C_{C_2}-1\cdot(x_{17,A})^2 & = & 0 &
\end{array}\right]\ \mathrm{C-2}$$

$$\left.\begin{array}{rcll}
F_{18}\cdot x_{18,A}-F_{18}\cdot x_{19,A} & + & V_{C_3}\cdot(-A_{C_3}-C_{C_3}) & =0\\
F_{18}\cdot x_{18,B}-F_{18}\cdot x_{19,B} & + & V_{C_3}\cdot(A_{C_3}-B_{C_3}) & =0\\
F_{18}\cdot x_{18,C}-F_{18}\cdot x_{19,C} & + & V_{C_3}\cdot(B_{C_3}) & =0\\
F_{18}\cdot x_{18,D}-F_{18}\cdot x_{19,D} & + & V_{C_3}\cdot(C_{C_3}) & =0\\
\text{where}\quad A_{C_3}-10\cdot x_{19,A} & = & 0 & \\
B_{C_3}-1\cdot x_{19,B} & = & 0 & \\
C_{C_3}-1\cdot(x_{19,A})^2 & = & 0 &
\end{array}\right]\ \mathrm{C-3}$$

$$\left.\begin{array}{rcll}
F_{20}\cdot x_{20,A}-F_{20}\cdot x_{21,A} & + & V_{C_4}\cdot(-A_{C_4}-C_{C_4}) & =0\\
F_{20}\cdot x_{20,B}-F_{20}\cdot x_{21,B} & + & V_{C_4}\cdot(A_{C_4}-B_{C_4}) & =0\\
F_{20}\cdot x_{20,C}-F_{20}\cdot x_{21,C} & + & V_{C_4}\cdot(B_{C_4}) & =0\\
F_{20}\cdot x_{20,D}-F_{20}\cdot x_{21,D} & + & V_{C_4}\cdot(C_{C_4}) & =0\\
\text{where}\quad A_{C_4}-10\cdot x_{21,A} & = & 0 & \\
B_{C_4}-1\cdot x_{21,B} & = & 0 & \\
C_{C_4}-1\cdot(x_{21,A})^2 & = & 0 &
\end{array}\right]\ \mathrm{C-4}$$

$$\left.\begin{array}{rcll}
F_{22}\cdot x_{22,A}-F_{22}\cdot x_{23,A} & + & V_{C_5}\cdot(-A_{C_5}-C_{C_5}) & =0\\
F_{22}\cdot x_{22,B}-F_{22}\cdot x_{23,B} & + & V_{C_5}\cdot(A_{C_5}-B_{C_5}) & =0\\
F_{22}\cdot x_{22,C}-F_{22}\cdot x_{23,C} & + & V_{C_5}\cdot(B_{C_5}) & =0\\
F_{22}\cdot x_{22,D}-F_{22}\cdot x_{23,D} & + & V_{C_5}\cdot(C_{C_5}) & =0\\
\text{where}\quad A_{C_5}-10\cdot x_{23,A} & = & 0 & \\
B_{C_5}-1\cdot x_{23,B} & = & 0 & \\
C_{C_5}-1\cdot(x_{23,A})^2 & = & 0 &
\end{array}\right]\ \mathrm{C-5}$$

(v) Summation of mole fractions

$$\begin{aligned}
x_{7,A}+x_{7,B}+x_{7,C}+x_{7,D} &= 1\\
x_{8,A}+x_{8,B}+x_{8,C}+x_{8,D} &= 1\\
x_{14,A}+x_{14,B}+x_{14,C}+x_{14,D} &= 1\\
x_{15,A}+x_{15,B}+x_{15,C}+x_{15,D} &= 1\\
x_{16,A}+x_{16,B}+x_{16,C}+x_{16,D} &= 1\\
x_{17,A}+x_{17,B}+x_{17,C}+x_{17,D} &= 1\\
x_{18,A}+x_{18,B}+x_{18,C}+x_{18,D} &= 1\\
x_{19,A}+x_{19,B}+x_{19,C}+x_{19,D} &= 1\\
x_{20,A}+x_{20,B}+x_{20,C}+x_{20,D} &= 1
\end{aligned}$$

$$
\begin{aligned}
x_{21,A} + x_{21,B} + x_{21,C} + x_{21,D} &= 1 \\
x_{22,A} + x_{22,B} + x_{22,C} + x_{22,D} &= 1 \\
x_{23,A} + x_{23,B} + x_{23,C} + x_{23,D} &= 1 \\
x_A + x_B + x_C + x_D &= 1
\end{aligned}
$$

(vi) Volume constraints and logical constraints

$$
\begin{aligned}
V_2 - V_{C_1} - V_{C_2} - V_{C_3} - V_{C_4} - V_{C_5} &= 0 \\
V_{C_1} = V_{C_2} = V_{C_3} = V_{C_4} = V_{C_5}
\end{aligned}
$$

$$
\begin{aligned}
V_1 - \mathrm{U} \cdot y_1 &\leq 0 \\
V_2 - \mathrm{U} \cdot y_2 &\leq 0 \\
F_7 - \mathrm{U} \cdot y_1 &\leq 0 \\
F_{14} - \mathrm{U} \cdot z_1 &\leq 0 \\
F_{16} - \mathrm{U} \cdot z_2 &\leq 0 \\
F_{18} - \mathrm{U} \cdot z_3 &\leq 0 \\
F_{20} - \mathrm{U} \cdot z_4 &\leq 0 \\
F_{22} - \mathrm{U} \cdot z_5 &\leq 0 \\
y_1 + y_2 &\geq 1 \\
V_{C_1} - \mathrm{U} \cdot z_1 &\leq 0 \\
V_{C_2} - \mathrm{U} \cdot z_2 &\leq 0 \\
V_{C_3} - \mathrm{U} \cdot z_3 &\leq 0 \\
V_{C_4} - \mathrm{U} \cdot z_4 &\leq 0 \\
V_{C_5} - \mathrm{U} \cdot z_5 &\leq 0
\end{aligned}
$$

where U is a specified upper bound.

(vii) Nonnegativity and integrality constraints

$$
\begin{aligned}
F_1, F_2, \ldots, F_{29} &\geq 0 \\
V_1, V_2, V_{C_1}, V_{C_2}, V_{C_3}, V_{C_4}, V_{C_5}, &\geq 0 \\
x_{7,i}, x_{8,i}, x_{14,i}, x_{15,i}, x_{16,i}, x_{17,i}, x_{18,i}, & \\
x_{19,i}, x_{20,i}, x_{21,i}, x_{22,i}, x_{23,i} &\geq 0 \quad i = A, B, C, D \\
x_A, x_B, x_C, x_D &\geq 0 \\
y_1, y_2, z_1, z_2, z_3, z_4, z_5 &= 0 - 1
\end{aligned}
$$

The objective function

The objective function represent the total profit and takes the form:

$$
MAX \; x_B
$$

Remark 1 The resulting optimization model is an MINLP problem. The objective function is linear for this illustrative example (note that it can be nonlinear in the general case) and does not involve any binary variables. Constraints (i), (v), and (vi) are linear in the continuous variables and the binary variables participate separably and linearly in (vi). Constraints (ii), (iii), and (iv) are nonlinear and take the form of bilinear equalities for (ii) and (iii), while (iv) can take any nonlinear form dictated by the reaction rates. If we have first-order reaction, then (iv) has bilinear terms. Trilinear terms will appear for second-order kinetics. Due to this type of nonlinear equality constraints, the feasible domain is nonconvex, and hence the solution of the above formulation will be regarded as a local optimum.

Remark 2 (Algorithmic approach) The MINLP model for the illustrative example can be solved using the v2-GBD by projecting on the binary variables $y_1, y_2, z_1, z_2, z_3, z_4, z_5$. Kokossis and Floudas (1990) proposed an effective initialization procedure that makes the solution of the MINLP model robust. This initialization strategy consists of the following steps:

Step 1 Solution of a pseudo-primal problem

For each primal problem, a pseudo-primal is formulated and solved. The pseudo-primal has only the currently active units of the superstructure and hence its objective function and constraints considers only the variables which are associated to the active units.

Step 2 Solution of a relaxed primal problem

Having obtained the solution vector $\bar{x}$ of the pseudo-primal problem, we relax the set of equality constraints of the primal problem $\boldsymbol{h}(\boldsymbol{x}) = 0$ to the form

$$-\alpha \leq \boldsymbol{h}(\boldsymbol{x}) \leq \alpha$$

and solve the relaxed-primal problem (with all constraints) by minimizing α in a small region around the optimal solution $\bar{x}$ of the pseudo-primal.

Step 3 Solution of the primal problem

If $\alpha = 0$ is obtained in step 2 for the relaxed primal problem, then we have a feasible starting point for the primal problem which is used to initialize the levels of its variables.

If $\alpha \neq 0$, then the primal is regarded infeasible at this iteration, and we proceed to the master problem with the Lagrange multipliers obtained from the relaxed primal problem at step 2.

Step 2 can be applied in an iterative fashion by increasing gradually the region around $\bar{x}$ until a feasible point is obtained. The above algorithmic procedure has been implemented in the computer program OREAN using the algorithmic methodology APROS (Paules and Floudas, 1989) and it is part of the OASIS library (Floudas, 1990).

Solving the MINLP model for the illustrative example with OREAN using 200 CSTRs as an approximation of the PFR, we obtain as optimal solution a single PFR unit. The starting point for the algorithm was a single CSTR unit. The optimal yield is 0.4364, and the solution is shown in Figure 10.3.

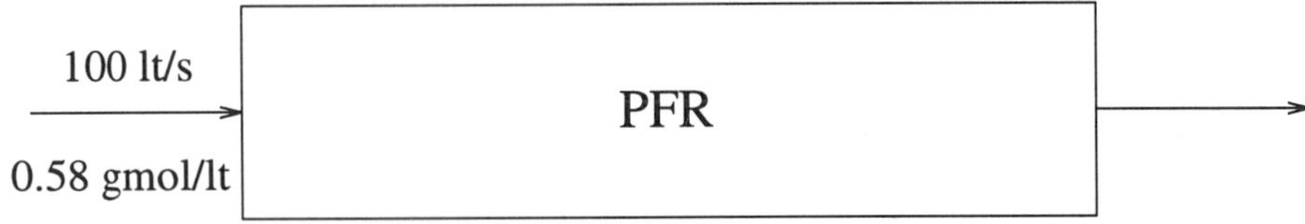

Figure 10.3: Optimal solution for illustrative example

The same example was solved using MINOPT (Rojnuckarin and Floudas, 1994) by treating the PFR model as a differential model. The required input files are shown in the MINOPT manual. Kokossis and Floudas (1990) applied the presented approach for large-scale systems in which the reactor network superstructure consisted of four CSTRs and four PFR units interconnected in all possible ways. Each PFR unit was approximated by a cascade of equal volume CSTRs (up to 200-300 CSTRs in testing the approximation). Complex reactions taking place in continuous and semibatch reactors were studied. It is important to emphasize that despite the complexity of the postulated superstructure, relatively simple structure solutions were obtained with the proposed algorithmic strategy.

Kokossis and Floudas (1994) extended the MINLP approach so as to handle nonisothermal operation. The nonisothermal superstructure includes alternatives of temperature control for the reactors as well as options for directly or indirectly intercooled or interheated reactors. This approach can be applied to any homogeneous exothermic or endothermic reaction and the solution of the resulting MINLP model provides information about the optimal temperature profile, the type of temperature control, the feeding, recycling, and by-passing strategy, and the optimal type and size of the reactor units.

10.3 Synthesis of Reactor-Separator-Recycle Systems

10.3.1 Introduction

In most chemical processes reactors are sequenced by systems that separate the desired products out of their outlet reactor streams and recycle the unconverted reactants back to the reactor system. Despite the fact that process synthesis has been developed into a very active research area, very few systematic procedures have been proposed for the synthesis of reactor/separator/recycle systems. The proposed evolutionary approaches are always based upon a large number of heuristic rules to eliminate the wide variety of choices. Many of these heuristics are actually extensions of results obtained by separately studying the synthesis problem of reactor networks or separator systems, and therefore the potential trade-offs resulting from the coupling of the reactors with the separators have not been investigated.

The delay in the development of a general synthesis strategy that will set the basis for a rigorous and systematic search for the optimal reactor/separator/recycle configuration is mainly due to the difficulties arising from the large number of structural alternatives and the nonlinear design equations of the participating units. Instead of focusing on an optimal search procedure, the various proposed methods have restricted the synthesis problem to a limited search around a feasible operation point. Thus, although it is often possible to identify directions of potential improvements, the finally obtained solution can never be claimed to, structurally and/or operationally, be even a local optimal point.

In the optimization of reactor/separator/recycle systems a very limited number of publications exist. Conti and Paterson (1985) proposed a heuristic evolutionary technique to solve the synthesis problem. First a hierarchy of heuristics is adopted that target to (i) minimize process complexity, (ii) maximize process yield, and (iii) select the distillation column which minimizes total heat load. According to the proposed hierarchy, a base case design is quickly established where everything is specified. Incremental changes are then introduced to the system by successively relaxing the heuristics so that a low-cost process topology is obtained at each stage. It is evident that since there is no unique way of relaxing any of the above heuristics, arbitrary decisions should often be made to provide directions of potential changes in the system. Performing a case study for the Van der Vusse reaction, the authors were able to compare their results with reported optimal solutions for the isolated reactor system. The comparison made clear that a design based upon the maximization of reactor yield is much inferior to the design based upon the integrated flowsheet.

Floquet *et al.* (1985) proposed a tree searching algorithm in order to synthesize chemical processes involving reactor/separator/recycle systems interlinked with recycle streams. The reactor network of this approach is restricted to a single isothermal CSTR or PFR unit, and the separation units are considered to be simple distillation columns. The conversion of reactants into products, the temperature of the reactor, as well as the reflux ratio of the distillation columns were treated as parameters. Once the values of the parameters have been specified, the composition of the outlet stream of the reactor can be estimated and application of the tree searching algorithm on the alternative separation tasks provides the less costly distillation sequence. The problem is solved for several values of the parameters and conclusions are drawn for different regions of operation.

Pibouleau *et al.* (1988) provided a more flexible representation for the synthesis problem by replacing the single reactor unit by a cascade of CSTRs. They also introduced parameters for defining the recovery rates of intermediate components into the distillate, the split fractions of top and bottom components that are recycled toward the reactor sequence, as well as parameters for the split fractions of the reactor outlet streams. A benzene chlorination process was studied as an example problem for this synthesis approach. In this example, the number of CSTRs in the cascade was treated as a parameter that ranged from one up to a maximum of four reactors. By repeatedly solving the synthesis problem, an optimum number of CSTRs was determined.

Kokossis and Floudas (1991) present a general approach based upon mathematical programming techniques for the synthesis of reactor/separator/recycle systems. A superstructure is postulated with all the different alternatives for the reactor and separator network, as well as for their possible interconnections. Different separation tasks, different types of reactors, different reactor configurations and different feeding, recycling, and bypassing strategies are considered. The synthesis problem is formulated as a mixed integer nonlinear programming problem MINLP. The continuous variables include the stream flow rates and compositions of the reactors and separators while the integer variables describe the existence of the reactors and the distillation columns. The solution of the MINLP formulation will provide an optimal configuration of the reactor/separator/recycle system.

In the following section, we will present the synthesis approach of reactor-separator-recycle systems proposed by Kokossis and Floudas (1991).

10.3.2 Problem Statement

Given

(i) A complex reaction mechanism of known kinetics (e.g., linear or nonlinear),

(ii) Different types of reactors (e.g., CSTRs, PFRs),

(iii) Distillation columns which can be used to separate the desired products from the output of the reactors and recycle the unconverted reactants to the reactors, and

(iv) A performance criterion (e.g., total cost, profit, yield, selectivity of desired products, conversion of reactants),

determine a reactor-separator-recycle system that optimizes the selected performance criterion and provides information on

(a) The reactor network (i.e., number, type and sizes of reactors; feeding, recycling, and bypassing strategy; and the topology of the reactor network),

(b) The separator network (i.e., appropriate distillation columns and sequences; sizes of distillation columns; duties of reboilers and condensers), and

(c) The interconnection between the two networks through the allocation of the outlet reactor streams and the allocation of the recycle(s) from the distillation columns to the inlet reactor streams.

Remark 1 The reactor network consists of ideal CSTRs and PFRs interconnected in all possible ways (see superstructure of reactor network). The PFRs are approximated as a cascade of equal volume CSTRs. The reactors operate under isothermal conditions.

Remark 2 The separators are sharp and simple distillation columns (i.e., sharp splits of light and heavy key components without distribution of component in both the distillate and bottoms; one feed and two products). The operating conditions of the distillation columns (i.e., pressure, temperature, reflux ratio) are fixed at nominal values. Hence, heat integration options are not considered, and the hot and cold utilities are directly used for heating and cooling requirements, respectively.

10.3.3 Basic Idea

The basic idea in the reactor-separator-recycle synthesis approach of Kokossis and Floudas (1991) consists of

(i) Combining the reactor network representation with the separation network representation to create a reactor-separator-recycle superstructure which includes all structural alternatives of interest; this superstructure should account for all possible interconnection between the reactor network and the separation network, include all options of reactor configurations and all distillation sequences,

(ii) Modeling the reactor-separator-recycle superstructure as a mixed-integer nonlinear programming MINLP problem, and

(iii) Solving the MINLP model providing information on

  (a) The reactor network topology and volumes of reactors,

  (b) The distillation network topology and the sizes of the units, and

  (c) The allocation of recycles.

In the following sections, we will discuss the above steps (i), (ii), and (iii).

10.3.4 Reactor-Separator-Recycle Superstructure

The derivation of a superstructure for a reactor-separator-recycle system is based on combining the individual representations of the reactor network and the separation network. The new additional element is the allocation of the potential recycle streams from the separation system to the inlets of the reactors in the reactor system.

Illustration 10.3.1
This example is taken from Kokossis and Floudas (1991) and features the following reaction mechanism:

$$A \xrightarrow{k_1} B \xrightarrow{k_2} C$$

where B is the desired product, A is the fresh feed, and C is a by-product.

The reactor network representation corresponds to a superstructure of three CSTRs. In the separation network, the components A, B, and C are to be separated via sharp and simple distillation columns according to their relative volatilities which are in the order of A, B, and C.

The potential separation tasks are $A/BC, AB/C, B/C$, and A/B. The component which should be considered for recycling is A and flows as distillate from columns A/BC and A/B. The generated superstructure, shown in Figure 10.4, features all the possible interconnections among the reactor units (streams 7, 8, 10, 11, 15, and 16), among the reactor and the separation units (streams 5, 6, 12, 13, 17, and 18) as well as all the potential recycles from the separation network to the reactors (streams 25, 26, and 27).

The different configurations for the reactor/separator/recycle system can be obtained by eliminating the appropriate streams of the proposed superstructure. Thus, elimination of all but streams 1, 7, 11, 14, 18, 22, 24, and 27 results in the configuration shown in Figure 10.5(a) where the three CSTRs are connected in series, the separator system includes columns A/BC and B/C and a total recycle of A is fed into the first CSTR. Should all but streams 1, 2, 3, 4, 5, 9, 12, 14, 17, 21, 23, and 26 be eliminated from the superstructure, the configuration of Figure 10.5(b) is obtained where the CSTRs are connected in parallel, the separator network consists of columns AB/C and A/B and the recycle stream from A/B feeds the second CSTR. A different configuration

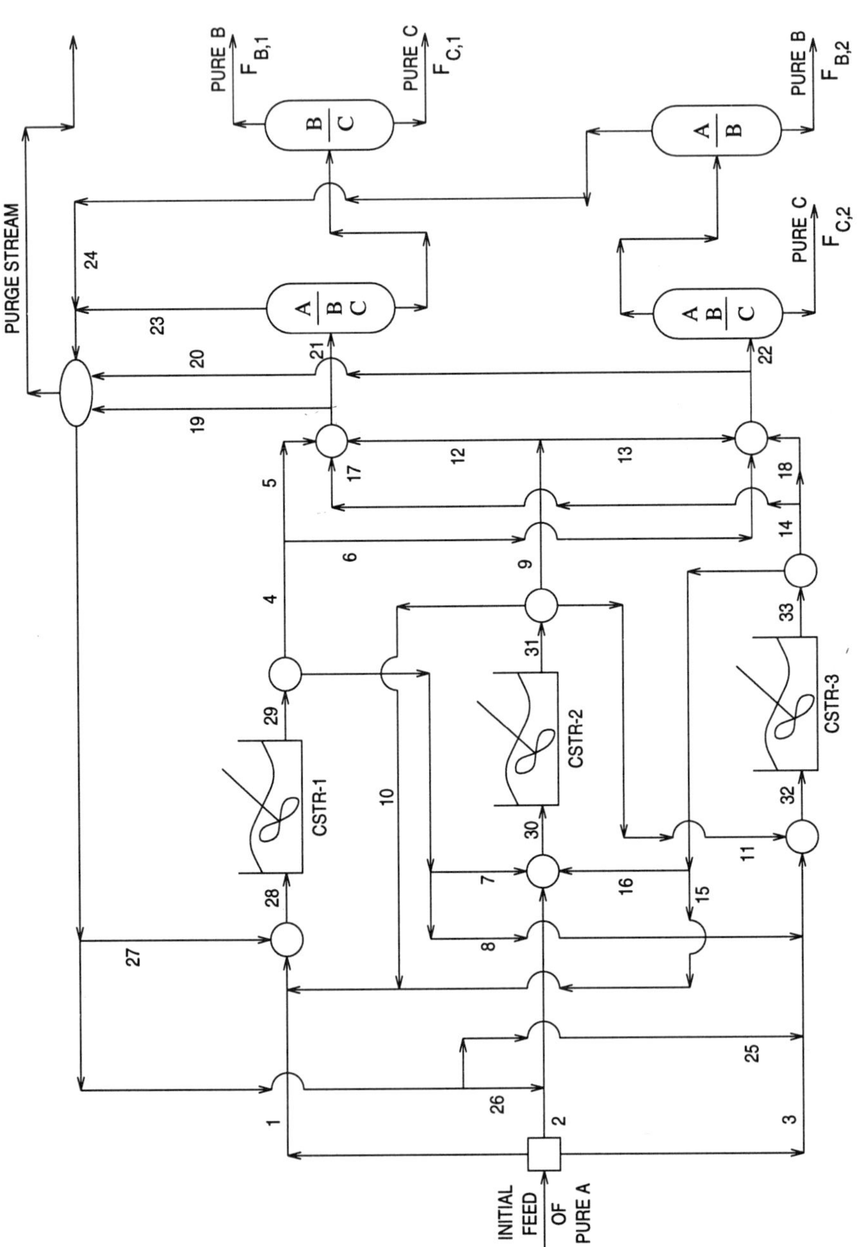

Figure 10.4: Superstructure of illustrative example

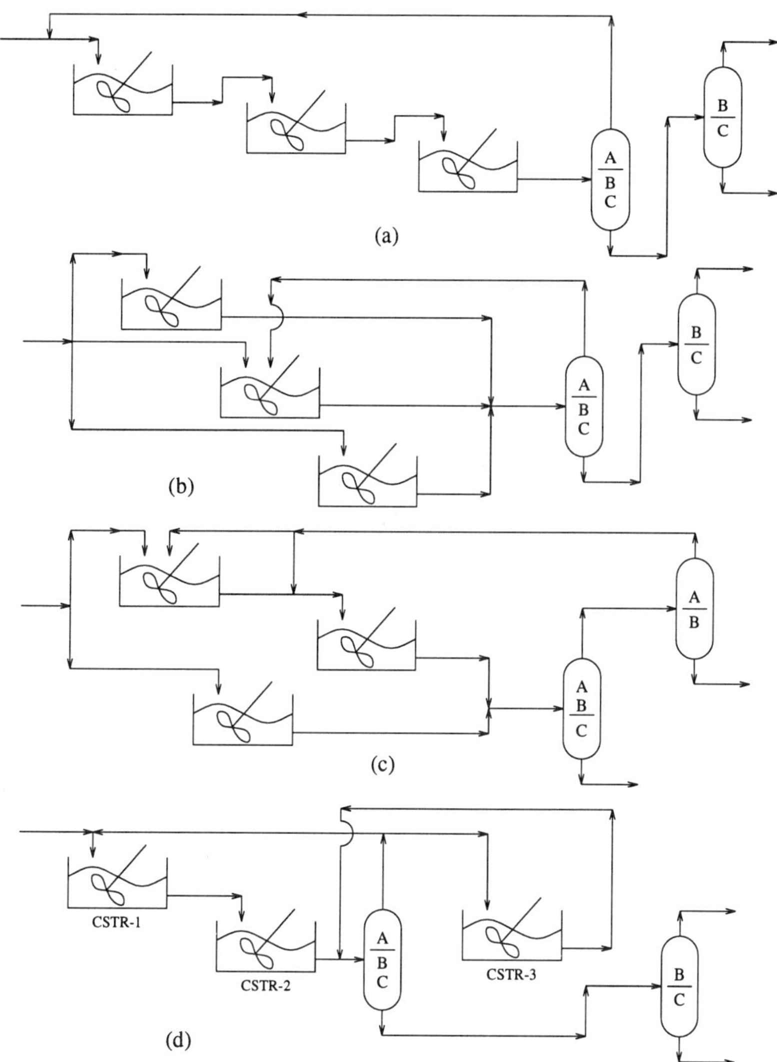

Figure 10.5: Four structural alternatives embedded in the superstructure of Figure 10.4

is illustrated in Figure 10.5(c) in case that streams $1, 2, 8, 9, 14, 17, 12, 21, 23, 25$, and 27 are only activated. In the reactor network, the first and third CSTR are connected in series and the second reactor is in parallel with the serial arrangement of the other two CSTRs. The separator system consists of columns AB/C and A/B while the recycle of A is fed into the first and third CSTR. As a final example, the reactor/separator/recycle configuration of Figure 10.5(d) is shown. The configuration results by considering only streams $1, 7, 9, 12, 14, 17, 21, 23, 25$, and 27 from the superstructure of Figure 3.1 and consists of reactors CSTR-1 and CSTR-2 in series with CSTR-2 feeding column A/BC the distillate of which feeds CSTR-3 and CSTR-1.

Remark 1 In the reactor-separator-recycle superstructure shown in Figure 10.4 for the $A \xrightarrow{k_1} B \xrightarrow{k_2} C$ reaction mechanism, we have recycle options for the fresh feed A only. Note, however, that if the reaction mechanism were of reversible reactions:

$$A \rightleftarrows B \rightleftarrows C$$

where B is the desired product, A is the feed, and C is the by-product, then we would have incentive to recycle both A and C. In this case we would have two recycle streams and the superstructure should include all such recycle options. The above case is studied in Kokossis and Floudas (1991) for the production of ethylbenzene.

Remark 2 Note that we no longer have known compositions of A, B, and C in the feeds to the distillation columns. As a result of this, the cost of each column depends on the variable feed composition.

10.3.5 Mathematical Formulation

To determine a reactor-separator-recycle system that optimizes the performance criterion (e.g., maximum profit) from the rich set of alternatives postulated in the superstructure, we define variables for

(i) The existence of each reactor unit (e.g., $y_1, y_2, y_3 = 0 - 1$ for the CSTRs of Figure 10.4),

(ii) The existence of the distillation columns (e.g., $z_1 = 0 - 1$ for A/BC, $z_3 = 0 - 1$ for B/C, $z_2 = 0 - 1$ for AB/C, and $z_4 = 0 - 1$ for A/B),

(iii) The stream flowrates (i.e., $F_1, F_2, \ldots, F_{32}$),

(iv) The component compositions (molar fractions) at the inlets of each reactor unit (i.e., $x_{28,i}, x_{30,i}, x_{32,i}, i = A, B, C$) and at the outlets of each reactor unit (i.e., $x_{29,i}, x_{31,i}, x_{33,i}, i = A, B, C$),

(v) The component compositions (molar fractions) at the inlets of the first distillation columns (i.e., $x_{21,i}, x_{22,i}, i = A, B, C$),

(vi) The volumes of the reactor units (i.e., V_1, V_2, V_3),

(vii) The equal heat duties of the reboilers and condensers at each column (i.e., Q_1, Q_2, Q_3, Q_4), and the inlets of the B/C and A/B column (i.e., $x_{2,i}$ for $i = B, C$ and $x_{4,i}$ for $i = A, B$), and

(viii) The feed flow rate of pure A (i.e., F_A) and the final flowrates of pure B and pure C (i.e., $F_{B,1}, F_{C,1}, F_{B,2}, F_{C,2}$).

The set of constraints consists of

(i) Mass balances for splitters of feed and reactor units,

(ii) Component balances for mixers at the inlets of the reactor units,

(iii) Component balances for mixers at the inlets of the first distillation columns of the different sequences,

(iv) Component balances at the mixers prior to the recycle alternatives,

(v) Component balances around each reactor,

(vi) Component balances for each distillation sequence,

(vii) Summation of mole fractions,

(viii) Logical constraints and volume constraints, and

(ix) Nonnegativity and integrality constraints.

The objective function will optimize the selected performance criterion (e.g., profit).

In the remaining part of this section, we will present the optimization model for the illustrative example $A \longrightarrow B \longrightarrow C$.

Illustration 10.3.2 (Optimization Model)
This example is taken from Kokossis and Floudas (1991). The chemical reactions of this liquid-phase process are given as

$$\begin{aligned} C_6H_6 + Cl_2 &\xrightarrow{k_1} C_6H_5Cl + HCl \\ C_6H_5Cl + Cl_2 &\xrightarrow{k_2} C_6H_4Cl_2 + HCl \end{aligned}$$

Further chlorination reactions can also take place, but since they involve insignificant amounts of reactants they have been considered to be negligible. The kinetics of the process were studied by McMullin (1948), who showed that the chlorination of benzene (A), monochlorobenzene (B) and dichlorobenzene (C) is in all cases first-order and irreversible.

In the reaction level, pure A reacts to the desired product B, waste product C and hydrochloric acid. The kinetic constants are $k_1 = 0.412\ h^{-1}$ and $k_2 = 0.055\ h^{-1}$. The produced hydrochloric acid is eliminated at the reaction level output by a stripping operation whose cost is not taken into account. Although all the reactions are exothermic, internal coils are used in the reactors to remove the evolved heat and, therefore, keep the temperature in the reactors constant.

In the separation level, unreacted A is separated and recycled toward the reactor network, valuable product B, of which the demand is assumed to be 50 $kmol/hr$ and product C. The volatility ranking of these components is $\alpha_A > \alpha_B > \alpha_C$. Thus, the possible separation tasks are A/BC (column 1), AB/C (column 2), B/C (column 3), and A/B (column 4).

The cost parameters for this example problem are

- Cost of steam = \$21.67/10^3 kJyr
- Cost of cold water = \$4.65/10^3 kJyr
- Purchase price of benzene = \$27.98/kmol
- Purchase price of chlorine = \$19.88/kmol
- Sale price of monochlorobenzene = \$92.67/kmol
- Payout time = 2.5 yr
- Income tax rate = 0.52
- Operating hours/year = 8,000

The set of constraints for the superstructure shown in Figure 10.4 (omitting the purge stream and streams 19 and 20, which correspond to recycles prior to separation, for simplicity of the presentation) is as follows:

(i) Mass balances for splitters of feed and reactor units

$$\begin{aligned} F_1 + F_2 + F_3 - F_A &= 0 \\ F_{28} - F_5 - F_6 - F_7 - F_8 &= 0 \\ F_{30} - F_{12} - F_{13} - F_{10} - F_{11} &= 0 \\ F_{32} - F_{17} - F_{18} - F_{15} - F_{16} &= 0 \end{aligned}$$

(ii) Component balances for mixers at inlets of reactors

$$\left.\begin{aligned} F_1 + F_{10} \cdot x_{31,A} + F_{15} \cdot x_{33,A} + F_{27} - F_{28} \cdot x_{28,A} &= 0 \\ F_{10} \cdot x_{31,B} + F_{15} \cdot x_{33,B} - F_{28} \cdot x_{28,B} &= 0 \\ F_{10} \cdot x_{31,C} + F_{15} \cdot x_{33,C} - F_{28} \cdot x_{28,C} &= 0 \end{aligned}\right] \quad CSTR-1$$

$$\left.\begin{aligned} F_2 + F_7 \cdot x_{29,A} + F_{16} \cdot x_{33,A} + F_{26} - F_{30} \cdot x_{30,A} &= 0 \\ F_7 \cdot x_{29,B} + F_{16} \cdot x_{33,B} - F_{30} \cdot x_{30,B} &= 0 \\ F_7 \cdot x_{29,C} + F_{16} \cdot x_{33,C} - F_{30} \cdot x_{30,C} &= 0 \end{aligned}\right] \quad CSTR-2$$

$$\left.\begin{aligned} F_3 + F_8 \cdot x_{29,A} + F_{11} \cdot x_{31,A} + F_{25} - F_{32} \cdot x_{32,A} &= 0 \\ F_8 \cdot x_{29,B} + F_{11} \cdot x_{31,B} - F_{32} \cdot x_{32,B} &= 0 \\ F_8 \cdot x_{29,C} + F_{11} \cdot x_{31,C} - F_{32} \cdot x_{32,C} &= 0 \end{aligned}\right] \quad CSTR-3$$

(iii) Component balances for mixers at inlets of distillation columns

$$\left.\begin{aligned} F_5 \cdot x_{29,A} + F_{12} \cdot x_{31,A} + F_{17} \cdot x_{33,A} - F_{21} \cdot x_{21,A} &= 0 \\ F_5 \cdot x_{29,B} + F_{12} \cdot x_{31,B} + F_{17} \cdot x_{33,B} - F_{21} \cdot x_{21,B} &= 0 \\ F_5 \cdot x_{29,C} + F_{12} \cdot x_{31,C} + F_{17} \cdot x_{33,C} - F_{21} \cdot x_{21,C} &= 0 \end{aligned}\right] \; Column\ 1$$

$$\left.\begin{aligned} F_6 \cdot x_{29,A} + F_{13} \cdot x_{31,A} + F_{18} \cdot x_{33,A} - F_{22} \cdot x_{22,A} &= 0 \\ F_6 \cdot x_{29,B} + F_{13} \cdot x_{31,B} + F_{18} \cdot x_{33,B} - F_{22} \cdot x_{22,B} &= 0 \\ F_6 \cdot x_{29,C} + F_{13} \cdot x_{31,C} + F_{18} \cdot x_{33,C} - F_{22} \cdot x_{22,C} &= 0 \end{aligned}\right] \; Column\ 2$$

(iv) Component balances at mixer prior to recycles

$$F_{23} + F_{24} - F_{25} - F_{26} - F_{27} = 0$$

Due to the assumption of not having streams 19 and 20 and no purge stream, the component balances become a total mass balance at the mixer.

(v) Component balances around each reactor

$$\left.\begin{aligned} F_{28} \cdot x_{28,A} - F_{28} \cdot x_{29,A} &+ V_1 \cdot (-A_1) = 0 \\ F_{28} \cdot x_{28,B} - F_{28} \cdot x_{29,B} &+ V_1 \cdot (A_1 - B_1) = 0 \\ F_{28} \cdot x_{28,C} - F_{28} \cdot x_{29,C} &+ V_1 \cdot (B_1) = 0 \\ \text{where} \quad A_1 - 0.412 \cdot x_{29,A} &= 0 \\ B_1 - 0.055 \cdot x_{29,B} &= 0 \end{aligned}\right] \; CSTR-1$$

$$\left.\begin{aligned} F_{30} \cdot x_{30,A} - F_{30} \cdot x_{31,A} &+ V_2 \cdot (-A_2) = 0 \\ F_{30} \cdot x_{30,B} - F_{30} \cdot x_{31,B} &+ V_2 \cdot (A_2 - B_2) = 0 \\ F_{30} \cdot x_{30,C} - F_{30} \cdot x_{31,C} &+ V_2 \cdot (B_2) = 0 \\ \text{where} \quad A_2 - 0.412 \cdot x_{31,A} &= 0 \\ B_2 - 0.055 \cdot x_{31,B} &= 0 \end{aligned}\right] \; CSTR-2$$

$$\left.\begin{aligned} F_{32} \cdot x_{32,A} - F_{32} \cdot x_{33,A} &+ V_3 \cdot (-A_3) = 0 \\ F_{32} \cdot x_{32,B} - F_{32} \cdot x_{33,B} &+ V_3 \cdot (A_3 - B_3) = 0 \\ F_{32} \cdot x_{32,C} - F_{32} \cdot x_{33,C} &+ V_3 \cdot (B_3) = 0 \\ \text{where} \quad A_3 - 0.412 \cdot x_{33,A} &= 0 \\ B_3 - 0.055 \cdot x_{33,B} &= 0 \end{aligned}\right] \; CSTR-3$$

(vi) Component balances for each distillation sequence

$$\left.\begin{aligned} F_{23} - F_{21} \cdot x_{21,A} &= 0 \\ F_{B,1} - F_{21} \cdot x_{21,B} &= 0 \\ F_{C,1} - F_{21} \cdot x_{21,B} &= 0 \end{aligned}\right] \; Columns\ 1\ and\ 3$$

$$\left.\begin{aligned} F_{24} - F_{22} \cdot x_{22,A} &= 0 \\ F_{B,2} - F_{22} \cdot x_{22,B} &= 0 \\ F_{C,2} - F_{22} \cdot x_{22,B} &= 0 \end{aligned}\right] \; Columns\ 2\ and\ 4$$

(vii) Summation of mole fractions

$$\begin{aligned}
x_{28,A} + x_{28,B} + x_{28,C} &= 1 \\
x_{29,A} + x_{29,B} + x_{29,C} &= 1 \\
x_{30,A} + x_{30,B} + x_{30,C} &= 1 \\
x_{31,A} + x_{31,B} + x_{31,C} &= 1 \\
x_{32,A} + x_{32,B} + x_{32,C} &= 1 \\
x_{33,A} + x_{33,B} + x_{33,C} &= 1 \\
x_{21,A} + x_{21,B} + x_{21,C} &= 1 \\
x_{22,A} + x_{22,B} + x_{22,C} &= 1
\end{aligned}$$

(viii) Logical constraints

$$\left.\begin{aligned}
F_{28} - \mathcal{U} y_1 &\leq 0 \\
F_{30} - \mathcal{U} y_2 &\leq 0 \\
F_{32} - \mathcal{U} y_3 &\leq 0 \\
V_1 - \mathcal{U} y_1 &\leq 0 \\
V_2 - \mathcal{U} y_2 &\leq 0 \\
V_3 - \mathcal{U} y_3 &\leq 0 \\
F_{21} - \mathcal{U} z_1 &\leq 0 \\
F_{22} - \mathcal{U} z_2 &\leq 0 \\
z_3 - z_1 &\leq 0 \\
z_4 - z_2 &\leq 0
\end{aligned}\right] \quad \textit{Reactor network}$$

(ix) Nonnegativity and integrality constraints

$$\begin{aligned}
F_1, F_2, \ldots, F_{32} &\geq 0 \\
V_1, F_2, V_3 &\geq 0 \\
x_{28,i}, x_{29,i}, x_{30,i}, x_{31,i}, x_{32,i}, x_{33,i} &\geq 0 \quad i = A, B, C \\
x_{21,i}, x_{22,i} &\geq 0 \quad i = A, B, C \\
F_A, F_{B,1}, F_{C,1}, F_{B,2}, F_{C,2} &\geq 0 \\
y_1, y_2, y_3, z_1, z_2, z_3, z_4 &= 0 - 1
\end{aligned}$$

The objective function

The objective function represent the total profit and takes the form:

$$\begin{aligned}
MAX \quad & (92.67) \cdot (8,000) \cdot (F_{B,1} + F_{B,2}) - (27.98) \cdot (8,000) \cdot (F_A) \\
& -(19.88) \cdot (8,000) \cdot (F_A) + \frac{1}{2.5}\{C_R^{cap} + C_{Column}^{cap}\} + 0.52 \cdot \{C^{oper}\}
\end{aligned}$$

where

$$
\begin{aligned}
C_R^{cap} &= C_{R1} + C_{R2} + C_{R3} \\
C_{Column}^{cap} &= C_{C1} + C_{C2} + C_{C3} + C_{C4} \\
C^{oper} &= 21.67(Q_1 + Q_2 + Q_3 + Q_4) + 4.65(Q_1 + Q_2 + Q_3 + Q_4) \\
C_{R1} &= 25,795 \cdot y_1 + 8,178 \cdot V_1 \\
C_{R2} &= 25,795 \cdot y_2 + 8,178 \cdot V_2 \\
C_{R3} &= 25,795 \cdot y_3 + 8,178 \cdot V_3 \\
C_{C1} &= 132,718 \cdot z_1 + F_{21} \cdot (369 \cdot x_{21,A} - 1,114 \cdot x_{21,B}) \\
C_{C2} &= 211,547 \cdot z_2 - F_{22} \cdot (1,010 \cdot x_{22,A} - 479 \cdot x_{22,B}) \\
C_{C3} &= 25,000 \cdot z_3 + F_3 \cdot (6,985 \cdot x_{3,B} - 3,870 \cdot x_{3,B}^2) \\
C_{C4} &= 86,944 \cdot z_4 + F_4 \cdot (1,136 \cdot x_{4,A}) \\
F_3 &= F_{21}(1 - x_{21,A}) \\
F_4 &= F_{22}(1 - x_{22,C}) \\
x_{3,B} &= \frac{x_{21,B}}{(x_{21,B} + x_{21,C})}, \quad x_{4,A} = \frac{x_{22,A}}{(x_{22,A} + x_{22,B})} \\
Q_1 &= F_{21} \cdot (3 + 36.1 \cdot x_{21,A} + 7.7 \cdot x_{21,B}) \\
Q_2 &= F_{22} \cdot (16.18 + 16.83 \cdot x_{22,A} + 42.14 \cdot x_{22,B}) \\
Q_3 &= F_3 \cdot (26.212 + 29.45 \cdot x_{3,B}) \\
Q_4 &= F_4 \cdot (10.70 + 28.41 \cdot x_{4,A})
\end{aligned}
$$

Remark 1 The mathematical model is an MINLP problem since it has both continuous and binary variables and nonlinear objective function and constraints. The binary variables participate linearly in the objective and logical constraints. Constraints (i), (iv), (vii), and (viii) are linear while the remaining constraints are nonlinear. The nonlinearities in (ii), (iii), and (vi) are of the bilinear type and so are the nonlinearities in (v) due to having first-order reactions. The objective function also features bilinear and trilinear terms. As a result of these nonlinearities, the model is nonconvex and hence its solution will be regarded as a local optimum unless a global optimization algorithm is utilized.

Kokossis and Floudas (1991) proposed a solution strategy for such mathematical models that is based on v2-GBD with an initialization scheme similar to the one presented in section 10.2.5. The algorithmic procedure was implemented using the APROS methodology in the library OASIS (Floudas, 1990).

Solving the MINLP model by projecting on the binary variables resulted in the optimal configuration shown in Figure 10.6.

Note that there exists a series sequence of the three reactors with the top distillation sequence selected (i.e., $A/BC, B/C$) with a recycle to the first reactor.

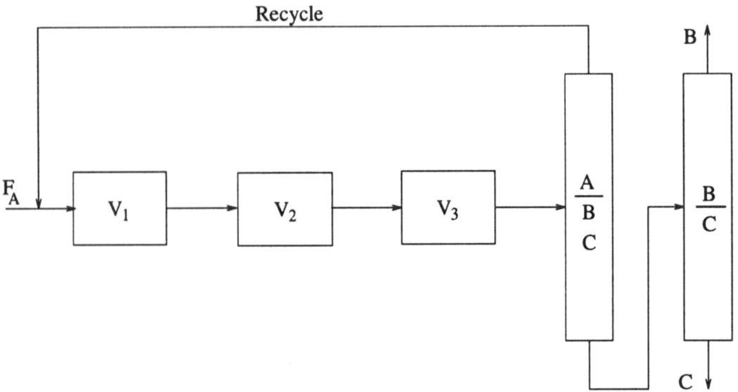

Figure 10.6: Optimal solution of MINLP model

Summary and Further Reading

This chapter presents an introduction to the key issues of reactor-based and reactor-separator-recycle systems from the mixed-integer nonlinear optimization perspective. Section 10.1 introduces the reader to the synthesis problems of reactor-based systems and provides an outline of the research work for isothermal and nonisothermal operation. Further reading on this subject can be found in the suggested references and the recent review by Hildebrandt and Biegler (1994).

Section 10.2 describes the MINLP approach of Kokossis and Floudas (1990) for the synthesis of isothermal reactor networks that may exhibit complex reaction mechanisms. Section 10.3 discusses the synthesis of reactor-separator-recycle systems through a mixed-integer nonlinear optimization approach proposed by Kokossis and Floudas (1991). The problem representations are presented and shown to include a very rich set of alternatives, and the mathematical models are presented for two illustrative examples. Further reading material in these topics can be found in the suggested references, while the work of Kokossis and Floudas (1994) presents a mixed-integer optimization approach for nonisothermal reactor networks.

Bibliography

L. K. E. Achenie and L. T. Biegler. Algorithmic synthesis of chemical reactor networks using mathematical programming. *Ind. Eng. Chem. Fundam.*, 25:621, 1986.

L. K. E. Achenie and L. T. Biegler. Developing targets for the performance index of a chemical reactor network: Isothermal systems. *Ind. Eng. Chem.*, 27:1811, 1988.

L. K. E. Achenie and L. T. Biegler. A Superstructure based approach to chemical reactor network synthesis. *Comp. & Chem. Eng.*, 14:23, 1990.

W. P. Adams and H. D. Sherali. Mixed integer bilinear programming problems. *Math. Progr.*, 59:279, 1993.

A. Aggarwal and C. A. Floudas. Synthesis of general distillation sequences - nonsharp separations. *Comp. & Chem. Eng.*, 14(6):631, 1990.

A. Aggarwal and C. A. Floudas. A decomposition approach for global optimum search in QP, NLP and MINLP problems. *Ann. of Oper. Res.*, 25:119, 1990b.

A. Aggarwal and C. A. Floudas. Synthesis of heat integrated nonsharp distillation sequences. *Comp. & Chem. Eng.*, 16:89, 1992.

M. J. Andrecovich and A. W. Westerberg. An MILP formulation for heat-integrated distillation sequence synthesis. *AIChE J.*, 31(9):1461, 1985.

M. J. Andrecovich and A. W. Westerberg. A simple synthesis method based on utility bounding for heat-integrated distillation sequences. *AIChE J.*, 31(3):363, 1985.

R. Aris. Studies in optimization - I: The optimum design of adiabatic reactors with several beds. *Chem. Eng. Sci.*, 12:243, 1960a.

R. Aris. Studies in optimization - II: Optimum temperature gradients in tubular reactors. *Chem. Eng. Sci.*, 13:75, 1960b.

R. Aris. Studies in optimization - III: The optimum operating conditions in sequences of stirred tank reactors. *Chem. Eng. Sci.*, 13:75, 1960c.

R. Aris. Studies in optimization - IV: The optimum conditions for a single reaction. *Chem. Eng. Sci.*, 13:197, 1960d.

R. Aris. *Discrete Dynamic Programming*. Blaisdell, New York, 1964.

R. Aris. *Elementary Chemical Reactor Analysis*. Prentice Hall, Englewood Cliffs, NJ, 1969.

M. Avriel. *Nonlinear Programming : Analysis and Methods*. Prentice-Hall Inc., Englewood Cliffs, New Jersey, 1976.

M. Avriel, W. E. Dewart, S. Schaible, and I. Zang. *Generalized Concavity*. Plenum Press, New York, 1988.

M. J. Bagajewicz and V. Manousiouthakis. On the generalized benders decomposition. *Comp. & Chem. Eng.*, 15(10):691, 1991.

S. Balakrishna and L. T. Biegler. A constructive targetting approach for the synthesis of isothermal reactor networks. *Ind. Eng. Chem. Res.*, 31:300, 1992a.

S. Balakrishna and L. T. Biegler. Targetting strategies for the synthesis and heat integration of nonisothermal reactor networks. *Ind. Eng. Chem. Res.*, 31:2152, 1992b.

S. Balakrishna and L. T. Biegler. A unified approach for the simultaneous synthesis of reaction, energy and separation systems. *Ind. Eng. Chem. Res.*, 3:1372, 1993.

E. Balas. Nonconvex quadratic programming via generalized polars. *SIAM J. Appl. Math.*, 28:335, 1975.

E. Balas. Disjunctive programming. *Ann. of Discrete Math.*, 5:3, 1979.

G. Bamopoulos, R. Nath, and R. L. Motard. Heuristic synthesis of nonsharp separation sequences. *AIChE J.*, 34(5):763, 1988.

M. S. Bazaraa, H. D. Sherali, and C. M. Shetty. *Nonlinear Programming: Theory and Algorithms,* 2nd *Edition*. J. Wiley, New York, 1993.

E. M. L. Beale. *Mathematical programming in practice*. Wiley, 1968.

E. M. L. Beale. *Integer programming*, page 409. The state of the art in numerical analysis, D. Jacobs Ed. Academic Press, 1977.

E. M. L. Beale. Branch and bound methods for mathematical programming systems. *Ann. of Discrete Math.*, 5:201, 1979.

E. M. L. Beale and J. J. H. Forrest. Global optimization using special ordered sets. *Math. Progr.*, 10:52, 1976.

E. M. L. Beale and J. A. Tomlin. Special facilities in a general mathematical programming system for nonconvex problems using ordered sets of variables. In J. Lawrence, editor, *The fifth international conference on Oper. Res.*, page 447, London, 1970. Tavistock Publications.

J. F. Benders. Partitioning procedures for solving mixed-variables programming problems. *Numer. Math.*, 4:238, 1962.

D. L. Bertsekas, G. S. Lower, N. R. Sandell, and T. A. Posbergh. Optimal short term scheduling of large-scale power systems. *IEEE Trans. Automatic Control*, AC–28:1, 1983.

D. B. Birewar and I. E. Grossmann. Incorporating scheduling in the optimal design of multiproduct batch plants. *Comp. & Chem. Eng.*, 13:141, 1989.

D. B. Birewar and I. E. Grossmann. Simultaneous synthesis, sizing and scheduling of multiproduct batch plants. *Comp. & Chem. Eng.*, 29(11):2242, 1990.

J.A. Bloom. Solving an electricity generating capacity expansion planning problem by generalized benders' decomposition. *Oper. Res.*, 31(5):84, 1983.

B. Borchers and J. E. Mitchell. An improved branch and bound algorithm for mixed integer nonlinear programs. Technical Report RPI Math Report No. 200, Rensselaer Polytechnic Institute, 1991.

E. A. Boyd. Fenchel cutting planes for integer programs. *Oper. Res.*, 42(1):53, 1994.

A. Brooke, D. Kendrick, and A. Meeraus. *GAMS: A User's Guide*. Scientific Press, Palo Alto, CA., 1988.

H. Brusset, D. Depeyre, C. Richard, and P. Richard. Use of dynamic programming in the optimization of a multistage reactor. *I & EC Proc. Des. and Dev.*, 17:355, 1978.

A. Burghardt and T. Patzek. Constrained optimization of cold-shot converters. *Chem. Eng. Jnl.*, 16:153, 1977.

J. Caha, V. Hlavacek, M. Kubicek, and M. Marek. Study of the optimization of chemical engineering equipment. Numerical solution of the optimization of an adiabatic reactor. *Inter. Chem. Eng.*, 13:466, 1973.

J. Cerda and A. W. Westerberg. Synthesizing heat exchanger networks having restricted stream/stream match using transportation problem formulations. *Chem. Eng. Sci.*, 38:1723, 1983.

J. Cerda, A. W. Westerberg, D. Mason, and B. Linnhoff. Minimum utility usage in heat exchanger network synthesis - a transportation problem. *Chem. Eng. Sci.*, 38:373, 1983.

J. J. J. Chen. Comments on improvements on a replacement for the logarithmic mean. *Chem. Eng. Sci.*, 42:2488, 1987.

S. H. Cheng and Y. A. Liu. Studies in chemical process design and synthesis : 8. A simple heuristic method for systematic synthesis of initial sequences for sloppy multicomponent separations. *I&EC Res.*, 27:2304, 1988.

S. P. Chitra and R. Govind. Yield optimization for complex reactor systems. *Chem. Eng. Sci.*, 36:1219, 1981.

S. P. Chitra and R. Govind. Synthesis of optimal serial reactor structures for homogeneous reactions. *AIChE J.*, 31:185, 1985.

A. R. Ciric. *Global optimum and retrofit issues in heat exchanger network and utility system synthesis*. PhD thesis, Princeton University, 1990.

A. R. Ciric and C. A. Floudas. A retrofit approach of heat exchanger networks. *Comp. & Chem. Eng.*, 13(6):703, 1989.

A. R. Ciric and C. A. Floudas. Application of the simultaneous match-network optimization approach to the pseudo-pinch problem. *Comp. & Chem. Eng.*, 14:241, 1990.

A. R. Ciric and C. A. Floudas. A mixed-integer nonlinear programming model for retrofitting heat exchanger networks. *I&EC Res.*, 29:239, 1990.

A. R. Ciric and C. A. Floudas. Heat exchanger network synthesis without decomposition. *Comp. & Chem. Eng.*, 15:385, 1991.

W. F. Clocksin and C. S. Mellish. *Programming in prolog*. Springer–Verlag, New York, 1981.

R. D. Colberg and M. Morari. Area and capital cost targets for heat exchanger network synthesis with constrained matches and unequal heat transfer coefficients. *Comp. & Chem. Eng.*, 14:1, 1990.

G. A. P. Conti and W. R. Paterson. Chemical reactors in process synthesis. *Process System Engineering, Symp. Series*, 92:391, 1985.

H. P. Crowder, E. L. Johnson, and M. W. Padberg. Solving large–scale zero–one linear programming problems. *Oper. Res.*, 31:803, 1983.

R. J. Dakin. A tree search algorithm for mixed integer programming problems. *Computer J.*, 8:250, 1965.

J. E. Dennis and R. B. Schnabel. *Numerical Methods for Unconstrained Optimization and Nonlinear Equations*. Prentice Hall, 1983.

W. B. Dolan, P. T. Cummings, and M. D. LeVan. Process optimization via simulated annealing: application to network design. *AIChE J.*, 35:725, 1989.

W. B. Dolan, P. T. Cummings, and M. D. LeVan. Algorithmic efficiency of simulated annealing for heat exchanger network design. *Comp. & Chem. Eng.*, 14:1039, 1990.

N. J. Driebeek. An algorithm for the solution of mixed integer programming problems. *Management Sci.*, 12:67, 1966.

M. A. Duran and I. E. Grossmann. An outer approximation algorithm for a class of mixed-integer nonlinear programs. *Math. Prog.*, 36:307, 1986a.

M. A. Duran and I. E. Grossmann. A mixed-integer nonlinear programming algorithm for process systems synthesis. *AIChE J.*, 32(4):592, 1986b.

D. C. Dyson and F. J. M. Horn. Optimum distributed feed reactors for exothermic reversible reactions. *J. Optim. Theory and Appl.*, 1:40, 1967.

D. C. Dyson and F. J. M. Horn. Optimum adiabatic cascade reactor with direct intercooling. *I & EC Fund.*, 8:49, 1969.

D. C. Dyson, F. J. M. Horn, R. Jackson, and C. B. Schlesinger. Reactor optimization problems for reversible exothermic reactions. *Canadian J. of Chem. Eng.*, 45:310, 1967.

T. F. Edgar and D. M. Himmelblau. *Optimization of Chemical Processes.* McGraw Hill, New York, 1988.

L. T. Fan, L. E. Erickson, R. W. Sucher, and G. S. Mathad. Optimal design of a sequence of continuous-flow stirred-tank reactors with product recycle. *I & EC*, 4:431, 1965.

N. M. Faqir and I. A. Karimi. Design of multipurpose batch plants with multiple production routes. In *Proceedings FOCAPD '89, Snowmass, Colorado*, page 451, 1990.

A. V. Fiacco and G. P. McCormick. *Nonlinear Programming: Sequential Unconstrained Minimization Techniques.* Society for Industrial and Applied Mathematics, 1968.

M. L. Fisher. The Lagrangian relaxation method for solving integer programming problems. *Management Sci.*, 27:1, 1981.

R. Fletcher. *Practical Methods of Optimization.* J. Wiley, New York, 2nd edition, 1987.

R. Fletcher, J. A. J. Hall, and W. R. Johns. Flexible retrofit design of multiproduct batch plants. *Comp. & Chem. Eng.*, 15(12):843, 1991.

R. Fletcher and S. Leyffer. Solving mixed integer nonlinear programs by outer approximation. *Math. Progr.*, 66(3):327, 1994.

O. E. Flippo. *Stability, Duality and Decomposition in General Mathematical Programming.* PhD thesis, Erasmus University, Rotterdam, The Netherlands, 1990.

O. E. Flippo and A. H. G. Rinnoy Kan. A note on benders decomposition in mixed integer quadratic programming. *Oper. Res. Lett.*, 9:81, 1990.

O. E. Flippo and A. H. G. Rinnoy Kan. Decomposition in general mathematical programming. *Math. Progr.*, 60:361, 1993.

P. Floquet, L. Pibouleau, and S. Domenech. Reactor separator sequences synthesis by a tree searching algorithm. *Process System Engineering, Symp. Series*, 92:415, 1985.

P. Floquet, L. Pibouleau, and S. Domenech. Mathematical programming tools for chemical engineering process design synthesis. *Chem. Eng. and Proc.*, 23(2):99, 1988.

C. A. Floudas. Separation synthesis of multicomponent feed streams into multicomponent product streams. *AIChE J.*, 33(4):540, 1987.

C. A. Floudas. *OASIS: Discrete-Continuous Optimization ApproacheS In Process Systems.* Computer-Aided Systems Laboratory, Dept. of Chemical Engineering, Princeton University, 1990. Part I: Documentation, Part II: Input Files.

C. A. Floudas, A. Aggarwal, and A. R. Ciric. Global optimum search for nonconvex NLP and MINLP problems. *Comp. & Chem. Eng.*, 13(10):1117, 1989.

C. A. Floudas and S. H. Anastasiadis. Synthesis of distillation sequences with several multicomponent feed and product streams. *Chem. Eng. Sci.*, 43(9):2407, 1988.

C. A. Floudas and A. R. Ciric. Strategies for overcoming uncertainties in heat exchanger network synthesis. *Comp. & Chem. Eng.*, 13(10):1133, 1989.

C. A. Floudas, A. R. Ciric, and I. E. Grossmann. Automatic synthesis of optimum heat exchanger network configurations. *AIChE J.*, 32(2):276, 1986.

C. A. Floudas and I. E. Grossmann. Synthesis of flexible heat exchanger networks for multiperiod operation. *Comp. & Chem. Eng.*, 10(2):153, 1986.

C. A. Floudas and I. E. Grossmann. Algorithmic approaches to process synthesis: logic and global optimization. In *Proceedings of Foundations of Computer-Aided Design, FOCAPD '94*, Snowmass, Colorado, 1994.

C. A. Floudas and P. M. Pardalos. *A Collection of Test Problems for Constrained Global Optimization Algorithms*, volume 455 of *Lecture Notes in Computer Science*. Springer-Verlag, Berlin, Germany, 1990.

C. A. Floudas and P. M Pardalos. *Recent Advances in Global Optimization*. Princeton Series in Computer Science. Princeton University Press, Princeton, New Jersey, 1992.

C. A. Floudas and P. M Pardalos. *State of the Art in Global Optimization : Computational Methods and Applications*. Nonconvex Optimization and Its Applications. Kluwer Academic Publishers, 1995.

C. A. Floudas and G. E. Paules, IV. A mixed-integer nonlinear programming formulation for the synthesis of heat integrated distillation sequences. *Comp. & Chem. Eng.*, 12(6):531, 1988.

C. A. Floudas and V. Visweswaran. A global optimization algorithm (GOP) for certain classes of nonconvex NLPs: I. theory. *Comp. & Chem. Eng.*, 14:1397, 1990.

C. A. Floudas and V. Visweswaran. A primal-relaxed dual global optimization approach. *J. Optim. Theory and Appl.*, 78(2):187, 1993.

B. Ganish, D. Horsky, and K. Srikanth. An approach to optimal positioning of a new product. *Management Sci.*, 29:1277, 1983.

A. M. Geoffrion. Duality in nonlinear programming: A simplified applications-oriented development. *SIAM Rev.*, 13(1):1, 1971.

A. M. Geoffrion. Generalized benders decomposition. *J. Optim. Theory and Appl.*, 10(4):237, 1972.

A. M. Geoffrion. *Perspectives on Optimization*, chapter -Duality in nonlinear programming: A simplified applications-oriented development, page 65. Addison–Wesley, Philipines, 1972b.

A. M. Geoffrion and R. McBride. Lagrangian relaxation applied to capacitated facility location problems. *AIIE Trans.*, 10:40, 1978.

G. M. Geoffrion and G. W. Graves. Multicommodity distribution system design by Benders decmoposition. *Management Sci.*, 20(5):822, 1974.

A. Georgiou and C. A. Floudas. An optimization model for the generic rank determination of structural matrices. *Int. J. Control*, 49(5):1633, 1989.

A. Georgiou and C. A. Floudas. Structural analysis and synthesis of feasible control systems. Theory and Applications. *Chem. Eng. Res. and Dev.*, 67:600, 1990.

J. C. Geromel and M. R. Belloni. Nonlinear programs with complicating variables: theoretical analysis and numerical experience. *IEEE Trans. Syst. Man. Cynernetics*, SMC–16:231, 1986.

P. E. Gill, W. Murray, and M. H. Wright. *Practical Optimization*. Academic Press, New York, 1981.

B. Gillespie and J. J. Carberry. Influence of mixing on isothermal reactor yield and adiabatic reactor conversion. *I & EC Fund.*, 5:164, 1966.

D. Glasser, D. Hildebrandt, and C. M. Crowe. A Geometric approach to steady flow reactors: The attainable region and optimization in concentration space. *ACS*, 26:1803, 1987.

D. Glasser and R. Jackson. A Generalized residence time distribution model for a chemical reactor. In *8th International Symposium on Chemical Reaction Engineering, I.Ch.E. Symposium Series No. 87*, page 535, 1984.

F. Glover. Improved linear integer programming formulations of nonlinear integer problems. *Mgmt. Sci.*, 22(4):445, 1975.

M. A. Gomez and J. D. Seader. Synthesis of distillation trains by thermodynamic analysis. *Comp. & Chem. Eng.*, 9:311, 1985.

R. E. Gomory. Outline of an algorithm for integer solutions to linear programs. *Bull. American Math. Soc.*, 64:275, 1958.

R. E. Gomory. An algorithm for the mixed integer problem. Technical Report RM–2597, The Rand Corporation, 1960.

L. E. Grimes, M. D. Rychener, and A. W. Westerberg. The synthesis and evolution of networks of heat exchange that feature the minimum number of units. *Chem. Eng. Comm.*, 14:339, 1982.

I. E. Grossmann. Mixed-integer programming approach for the synthesis of integrated process flowsheets. *Comp. & Chem. Eng.*, 9(5):463, 1985.

I. E. Grossmann. MINLP optimization strategies and algorithms for process synthesis. In J. J. Siirola, I. E. Grossmann, and G. Stephanopoulos, editors, *Foundations of computer aided process design, '89*, page 105, Snowmass, Colorado, 1989.

I. E. Grossmann. MINLP optimization strategies and algorithms for process synthesis. In *Proc. 3rd. Int. Conf. on Foundations of Computer-Aided Process Design*, page 105, 1990.

I. E. Grossmann. Mixed-integer nonlinear programming techniques for the synthesis of engineering systems. *Res. Eng. Des.*, 1:205, 1990.

I. E. Grossmann, L. T. Biegler, and A. W. Westerberg. Retrofit design of processes. In G. V. Reklaitis and H. D. Spriggs, editors, *Foundations of computer aided process operations, '87*, page 403, Park City, Utah, 1987.

I. E. Grossmann and C. A. Floudas. Active constraint strategy for flexibility analysis in chemical processes. *Comp. & Chem. Eng.*, 11(6):675, 1987.

I. E. Grossmann, I. Quesada, R. Ramon, and V. T. Voudouris. Mixed-integer optimization techniques for the design and scheduling of batch processes. In *Proc. of NATO Advanced Study Institute on Batch Process Systems Engineering*, 1992.

I. E. Grossmann and R. W. H. Sargent. Optimal design of multipurpose chemical plants. *I&EC Proc. Des. Dev.*, 18(2):343, 1979.

T. Gundersen. Achievements and future challenges in industrial design applications of process systems engineering. In *Process Systems Engineering, '91*, page I.1.1, Montebello, Canada, 1991.

T. Gundersen and I. E. Grossmann. Improved optimization strategies for automated heat exchanger network synthesis through physical insights. *Comp. & Chem. Eng.*, 14(9):925, 1990.

T. Gundersen and L. Naess. The Synthesis of cost optimal heat exchanger networks, an industrial review of the state of the art. *Comp. & Chem. Eng.*, 12(6):503, 1988.

O. K. Gupta. *Branch and bound experiments in nonlinear integer programming*. PhD thesis, Purdue University, 1980.

E. Hansen. *Global optimization using interval analysis*. Marcel Dekker Inc., New York, 1992.

J. E. Hendry, D. F. Rudd, and J. D. Seader. Synthesis in the design of chemical processes. *AIChE J.*, 19(1), 1973.

M. R. Hestenes. *Optimization Theory: The Finite Dimensional Case*. J. Wiley, New York, 1975.

D. Hildebrandt and L. T. Biegler. Synthesis of chemical reactor networks. In *Proceedings of Foundations of Computer Aided Process Design, FOCAPD'94*, 1994.

D. Hildebrandt and D. Glasser. The attainable region and optimal reactor structures. *Chem. Eng. Sci.*, 45:2161, 1990.

D. Hildebrandt, D. Glasser, and C. M. Crowe. Geometry of the attainable region generated by reaction and mixing: with and without constraints. *I & EC Res.*, 29:49, 1990.

V. Hlavacek. Synthesis in the design of chemical processes. *Comp. & Chem. Eng.*, 2:67, 1978.

H. H. Hoang. Topological optimization of networks: A nonlinear mixed integer model employing generalized Benders decomposition. *IEEE Trans. Automatic Control*, AC–27:164, 1982.

E. C. Hohmann. *Optimum networks for heat exchange*. PhD thesis, University of Southern California, 1971.

K. Holmberg. On the convergence of the cross decomposition. *Math. Progr.*, 47:269, 1990.

K. Holmberg. Variable and constrained duplication techniques in primal–dual decomposition methods. Technical Report Report Lith-MAT-R-1991-23, Linkoping University, 1991.

K. Holmberg. Generalized cross decomposition applied to nonlinear integer programming problems: duality gaps and convexification in parts. *Optimization*, 23:341, 1992.

J. N. Hooker. Resolution vs. cutting plane solution of inference problems. *Oper. Res. Lett.*, 7(1), 1988.

F. Horn. Attainable and non-attainable regions in chemical reaction technique. In *3rd European Symposium on Chemical Reaction Engineering*, page 1. Pergamon Press, New York, 1964.

F. J. M. Horn and M. J. Tsai. The use of the adjoint variables in the development of improvement criteria for chemical reactors. *J. Optim. Theory and Appl.*, 1:131, 1967.

R. Horst and P. M. Pardalos. *Handbbok of global optimization*. Nonconvex Optimization and Its Applications. Kluwer Academic Publishers, 1995.

R. Horst and H. Tuy. *Global Optimization: Deterministic Approaches*. Springer-Verlag, Berlin, Germany, 1990.

C. S. Hwa. Mathematical formulation and optimization of heat exchange networks using separable programming. *AIChE-I. Chem. Eng. Symp. Series*, No. 4:101, 1965.

M. A. Isla and J. Cerda. A general algorithmic approach to the optimal synthesis of energy-efficient distillation train designs. *Chem. Eng. Comm.*, 54:353, 1987.

R. Jackson. Optimization of chemical reactors with respect to flow configuration. *J. Optim. Theory and Appl.*, 2:240, 1968.

R. Jackson and D. Glasser. A General mixing model for steady flow chemical reactors. *Chem. Eng. Commun.*, 42:17, 1985.

R. B. Jarvis and C. C. Pantelides. *DASOLV, A Differential-Algebraic Equation Solver*. Center for Process Engineering, Dept. of Chemical Engineering, Imperial College, 1992.

R. E. Jeroslow and J. Wang. Solving propositional satisfiability problems. *Ann. of Math. and AI*, 1:167, 1990.

R. G. Jeroslow. Cutting plane theory: Disjunctive methods. *Ann. of Discrete Math.*, 1:293, 1977.

R. G. Jeroslow and J. K. Lowe. Modeling with integer variables. *Math. Progr. Study*, 22:167, 1984.

R. G. Jeroslow and J. K. Lowe. Experimental results on the new techniques for integer programming formulations. *J. Oper. Res. Soc.*, 36(5):393, 1985.

B. Kalitventzeff and F. Marechal. The management of a utility network. In *Process Systems Engineering. PSE '88, Sydney, Australia*, page 223, 1988.

R. P. King. Calculation of the optimal conditions for chemical reactors of the combined type. *Chem. Eng. Sci.*, 20:537, 1965.

G. R. Kocis and I. E. Grossmann. Relaxation strategy for the structural optimization of process flow sheets. *I&EC Res.*, 26(9):1869, 1987.

G. R. Kocis and I. E. Grossmann. Global optimization of nonconvex MINLP problems in process synthesis. *I&EC Res.*, 27(8):1407, 1988.

G. R. Kocis and I. E. Grossmann. Computational experience with DICOPT solving MINLP problems in process systems engineering. *Comp. & Chem. Eng.*, 13:307, 1989a.

A. C. Kokossis and C. A. Floudas. Optimization of complex reactor networks-I. isothermal operation. *Chem. Eng. Sci.*, 45(3):595, 1990.

A. C. Kokossis and C. A. Floudas. Optimal synthesis of isothermal reactor-separator-recycle systems. *Chem. Eng. Sci.*, 46:1361, 1991.

A. C. Kokossis and C. A. Floudas. Optimization of complex reactor networks - II: non-isothermal operation. *Chem. Eng. Sci.*, 49:1037, 1994.

A. C. Kokossis and C. A. Floudas. Optimization of complex reactor networks - II. Non-isothermal Operation. *Chem. Eng. Sci.*, 49(7):1037, 1994.

E. Kondili, C. C. Pantelides, and R. W. H. Sargent. A general algorithm for short term scheduling of batch operations – I. MILP formulation. *Comp. & Chem. Eng.*, 17(2):211, 1993.

Z. Kravanja and I. E. Grossmann. PROSYN - An MINLP process synthesizer. *Comp. & Chem. Eng.*, 14:1363, 1990.

A. H. Land and A. G. Doig. An automatic method for solving discrete programming problems. *Econometrica*, 28:497, 1960.

R. Lazimy. Mixed-integer quadratic programming. *Math. Progr.*, 22:332, 1982.

R. Lazimy. Improved algorithm for mixed-integer quadratic programs and a computational Study. *Math. Progr.*, 32:100, 1985.

K. Lee and R. Aris. Optimal adiabatic bed reactors for sulfur dioxide with cold shot cooling. *I & EC*, 2:300, 1963.

O. Levenspiel. *Chemical Reactor Engineering, An Introduction to the Design of Chemical Reactors*. John Wiley, NY, 1962.

B. Linnhoff, H. Dunford, and R. Smith. Heat integration of distillation columns into overall processes. *Chem. Eng. Sci.*, 38(8):1175, 1983.

B. Linnhoff and J. R. Flower. Synthesis of heat exchanger networks, Part I. Systematic generation of energy optimal networks. *AIChE J.*, 24:633, 1978a.

B. Linnhoff and J. R. Flower. Synthesis of heat exchanger networks, Part II. Evolutionary generation of networks with various criteria of optimality. *AIChE J.*, 24:642, 1978b.

D. L. Luenberger. *Linear and Nonlinear Programming*. Addison-Wesley, Reading, MA, 2nd edition, 1984.

M. L. Luyben and C. A. Floudas. Analyzing the interaction of design and control, Part 1: A multiobjective framework and application to binary distillation synthesis. *Comp. & Chem. Eng.*, 18(10):933, 1994a.

M. L. Luyben and C. A. Floudas. Analyzing the interaction of design and control, Part 2: Reactor–separator–recycle system. *Comp. & Chem. Eng.*, 18(10):971, 1994b.

T. L. Magnanti and R. T. Wong. Accelerating benders decomposition: algorithmic enhancement and model selection criterion. *Oper. Res.*, 29(3):464, 1981.

R. S. H. Mah and R. Aris. Optimal policies for first-order consecutive reversible reactions. *Chem. Eng. Sci.*, 19:541, 1964.

O. L. Mangasarian. *Nonlinear Programming*. McGraw Hill, New York, 1969.

R. K. Martin and L. Schrage. Subset coefficient reduction cuts for $0-1$ mixed integer programming. *Oper. Res.*, 33:505, 1985.

H. Mawengkang and B. A. Murtagh. Solving nonlinear integer programs with large scale optimization software. *Ann. of Oper. Res.*, 5:425, 1986.

G. P. McCormick. *Nonlinear Programming: Theory, Algorithms and Applications*. J. Wiley, New York, New York, 1983.

D. McDaniel and M. Devine. A modified Benders partitioning algorithm for mixed integer programming. *Management Sci.*, 24(3):312, 1977.

R.B. McMullin. *Chem. Eng. Progr.*, 44:183, 1948.

I. Meszarus and Z. Fonyo. A new bounding strategy for synthesizing distillation schemes with energy integration. *Comp. & Chem. Eng.*, 10:545, 1986.

M. Minoux. *Mathematical Programming: Theory and Algorithms.* J. Wiley, New York, 1986.

M. Morari and C. D. Faith. The synthesis of distillation trains with heat integration. *AIChE J.*, 26(6):916, 1980.

M. Muraki and T. Hayakawa. Separation processes synthesis for multicomponent products. *J. Chem. Eng. Japan*, 17(5):533, 1984.

M. Muraki and T. Hayakawa. Multicomponent separation process synthesis with separation sharpness. *J. Chem. Eng. Japan*, 20(2):195, 1987.

M. Muraki and T. Hayakawa. Synthesis of a multicomponent multiproduct separation process with nonsharp separators. *Chem. Eng. Sci.*, 43(2):259, 1988.

M. Muraki, K. Kataoka, and T. Hayakawa. Evolutionary synthesis of multicomponent multiproduct separation processes. *Chem. Eng. Sci.*, 41:1843, 1986.

K. G. Murty and S. N. Kabadi. Some NP-complete problems in quadratic and nonlinear programming. *Math. Progr.*, 39:117, 1987.

Y. Naka, M. Terashita, and T. Takamatsu. A thermodynamic approach to multicomponent distillation systems synthesis. *AIChE J.*, 28:812, 1982.

R. Nath. *Studies in the Synthesis of Separation Processes.* PhD thesis, University of Houston, Texas, 1977.

G. L. Nemhauser and L. A. Wolsey. *Integer and Combinatorial Optimization.* J. Wiley, New York, 1988.

A. Neumaier. *Interval methods for systems of equations.* Cambridge University Press, Cambridge, England, 1990.

N. Nishida, G. Stephanopoulos, and A. W. Westerberg. A review of process synthesis. *AIChE J.*, 27(3):321, 1981.

O. Odele and S. Macchietto. Computer aided molecular design: A novel method for optimal solvent selection. *Fluid Phase Equilibria*, 82:47, 1993.

G. M. Ostrovsky, M. G. Ostrovsky, and G. W. Mikhailow. Discrete optimization of chemical processes. *Comp. & Chem. Eng.*, 14(1):111, 1990.

M. Padberg and G. Rinaldi. A branch–and–cut algorithm for the resolution of large scale symmetric travelling salesman problems. *SIAM Rev.*, 33:60, 1991.

S. Papageorgaki and G. V. Reklaitis. Optimal design of multipurpose batch plants: 1. Problem formulation. *I&EC Res.*, 29(10):2054, 1990.

S. Papageorgaki and G. V. Reklaitis. Optimal design of multipurpose batch plants: 2. A decomposition solution strategy. *I&EC Res.*, 29(10):2062, 1990.

K. Papalexandri and E. N. Pistikopoulos. An MINLP retrofit approach for improving the flexibility of heat exchanger networks. *Ann. Oper. Res.*, 42:119, 1993.

S. A. Papoulias and I. E. Grossmann. A structural optimization approach in process synthesis - II. Heat recovery networks. *Comp. & Chem. Eng.*, 7:707, 1983.

P. M. Pardalos and J. B. Rosen. *Constrained global optimization: Algorithms and applications*, volume 268 of *Lecture Notes in Computer Science*. Springer Verlag, Berlin, Germany, 1987.

P. M. Pardalos and G. Schnitger. Checking local optimality in constrained quadratic programming is NP-Hard. *Oper. Res. Lett.*, 7(1):33, 1988.

P. M. Pardalos and S. A. Vavasis. Quadratic programming with one negative eigenvalue is NP-hard. *J. Global Optim.*, 1:15, 1991.

R. G. Parker and R. L. Rardin. *Discrete optimization*. Academic Press, 1988.

W. R. Paterson. A replacement for the logarithmic mean. *Chem. Eng. Sci.*, 39:1635, 1984.

G. E. Paules, IV and C. A. Floudas. Synthesis of flexible distillation sequences for multiperiod operation. *Comp. & Chem. Eng.*, 12(4):267, 1988.

G. E. Paules, IV and C. A. Floudas. APROS: Algorithmic development methodology for discrete-continuous optimization problems. *Oper. Res.*, 37(6):902, 1989.

G.E Paules, IV. *Synthesis and analysis of flexible and complex heat-integrated distillation sequences*. PhD thesis, Princeton University, Princeton, NJ, 1990.

G.E. Paules, IV and C.A. Floudas. Stochastic programming in processcsynthesis: A two-stage model with MINLP recourse for multiperiod heat-integrated distillation sequences. *Comp. & Chem. Eng.*, 16(3):189, 1992.

J. D. Paynter and D. E. Haskins. Determination of optimal reactor type. *Chem. Eng. Sci.*, 25:1415, 1970.

C.C. Petersen. A note on transforming the product of variables to linear form in linear programs. Working paper, Purdue University, 1971.

T.K. Pho and L. Lapidus. Topics in computer-aided design: Part II. Synthesis of optimal heat exchanger networks by tree searching algorithms. *AIChE J.*, 19:1182, 1973.

L. Pibouleau, P. Floquet, and S. Domenech. Optimal synthesis of reactor separator systems by nonlinear programming method. *AIChE J.*, 34:163, 1988.

P. Psarris and C. A. Floudas. Improving dynamic operability in MIMO systems with time delays. *Chem. Eng. Sci.*, 45(12):3505, 1990.

R. Raman and I. E. Grossmann. Relation between MILP modeling and logical inference for chemical process synthesis. *Comp. & Chem. Eng.*, 15:73, 1991.

R. Raman and I. E. Grossmann. Integration of logic and heuristic knowledge in MINLP optimization for process synthesis. *Comp. & Chem. Eng.*, 16:155, 1992.

R. L. Rardin and V. E. Unger. Surrogate constraints and the strength of bounds derived from $0-1$ Benders partitioning procedures. *Oper. Res.*, 24(6):1169, 1976.

R.N.S. Rathore, K.A. Van Wormer, and G.J. Powers. Synthesis strategies for multicomponent separation systems with energy integration. *AIChE J.*, 20(3):491, 1974a.

R.N.S. Rathore, K.A. Van Wormer, and G.J. Powers. Synthesis of distillation systems with energy integration. *AIChE J.*, 20(5):940, 1974b.

A. L. Ravimohan. Optimization of chemical reactor networks. *J. Optim. Theory and Appl.*, 8:204, 1971.

G. V. Reklaitis. Perspectives on scheduling and planning process operations. In *Proc. 4th. Int. Symp. on Process Systems Engineering, Montreal, Canada*, page III.2.1, 1991.

G. V. Reklaitis, A. Ravindran, and K. M. Ragsdel. *Engineering Optimization: Methods and Applications*. J. Wiley, New York, 1983.

S. H. Rich and G. J. Prokapakis. Multiple routings and reaction paths in project scheduling. *Ind. Eng. Chem. Res.*, 26(9):1940, 1986.

S. H. Rich and G. J. Prokapakis. Scheduling and sequencing of batch operations in a multipurpose plant. *Ind. Eng. Chem. Res.*, 25(4):979, 1986.

R. T. Rockefellar. *Convex Analysis*. Princeton University Press, Princeton, N.J., 1970.

A. Rojnuckarin and C.A. Floudas. *MINOPT, A Mixed Integer Nonlinear Optimizer*. Computer Aided Systems Laboratory, Dept. of Chemical Engineering, Princeton University, N.J., 1994.

R. Rouhani, W. Lasdon, L. Lebow, and A. D. Warren. A generalized benders decomposition approach to reactive source planning in power systems. *Math. Progr. Study*, 25:62, 1985.

T. W. F. Russell and D. T. Buzelli. Reactor analysis and process synthesis for a class of complex reactions. *I & EC Proc. Des. and Dev.*, 8:2, 1969.

N. Sahinidis and I. E. Grossmann. Reformulation of multiperiod MILP models for planning and scheduling of chemical processes. *Comp. & Chem. Eng.*, 15:255, 1991.

N. Sahinidis, I. E. Grossmann, R. Fornari, and M. Chathrathi. Long range planning model for the chemical process industries. *Comp. & Chem. Eng.*, 13:1049, 1989.

N. V. Sahinidis and I. E. Grossmann. Convergence properties of generalized Benders decomposition. *Comp. & Chem. Eng.*, 15(7):481, 1991.

A. Schrijver. *Theory of linear and integer programming*. Wiley–Interscience series in discrete mathematics and optimization. J. Wiley, 1986.

N. Shah and C. C. Pantelides. Optimal long-term campaign planning and design of batch operations. *Ind. Eng. Chem. Res.*, 30:2309, 1991.

N. Shah, C. C. Pantelides, and R. W. H. Sargent. A general algorithm for short term scheduling of batch operations – II. Computational issues. *Comp. & Chem. Eng.*, 17(2):229, 1993.

R. Smith and B. Linnhoff. The design of separators in the context of overall processes. *Chem. Eng. Res. Dev.*, 66:195, 1988.

A. Sophos, G. Stephanopoulos, and M. Morari. Synthesis of optimum distillation sequences with heat integration schemes. paper 42d 71st Annual AIChE Meeting, Miami, FL, 1978.

G. Stephanopoulos. Synthesis of process flowsheets: An adventure in heuristic design or a utopia of mathematical programming? In R. S. H. Mah and W. D. Seader, editors, *Foundations of computer aided process design*, volume 2, page 439, New York, 1981. Engineering Foundation.

J. A. Tomlin. An improved branch and bound method for integer programming. *Oper. Res.*, 19:1070, 1971.

F. E. Torres. Linearization of mixed integer products. *Math. Progr.*, 49:427, 1991.

P.J. Trambouze and L. E. Piret. Continuous stirred tank reactors: Designs for maximum conversions of raw material to desired product. *AIChE J.*, 5:384, 1959.

T. Umeda, T. Harada, and K. Shiroko. A thermodynamic approach to the synthesis of heat integration systems in chemical processes. *Comp. & Chem. Eng.*, 3:273, 1979.

T. Umeda, K. Niida, and K. Shiroko. A thermodynamic approach to heat integration in distillation systems. *AIChE J.*, 25:423, 1979.

T. J. Van Roy. Cross decomposition for mixed integer programming. *Math. Progr.*, 25:46, 1983.

T. J. Van Roy. A cross decomposition algorithm for capacitated facility location. *Oper. Res.*, 34(1):145, 1986.

T. J. Van Roy and L. A. Wolsey. Valid inequalities for mixed $0-1$ programs. *Discrete Appl. Math.*, 14:199, 1986.

T. J. Van Roy and L. A. Wolsey. Solving mixed $0-1$ programs by automatic reformulation. *Oper. Res.*, 35:45, 1987.

J. Vaselenak, I. E. Grossmann, and A. W. Westerberg. An embedding formulation for the optimal scheduling and design of multiproduct batch plants. *I&EC Res.*, 26(1):139, 1987.

J. Vaselenak, I. E. Grossmann, and A. W. Westerberg. Optimal retrofit design of multipurpose batch plants. *I&EC Res.*, 26(4):718, 1987.

S. Vavasis. *Nonlinear Optimization: Complexity Issues.* Oxford University Press, New York, N.Y., 1991.

J. Viswanathan and I. E. Grossmann. A combined penalty function and outer approximation for MINLP optimization. *Comp. & Chem. Eng.*, 14(7):769, 1990.

M. Viswanathan and L. B. Evans. Studies in the heat integration of chemical process plants. *AIChE J.*, 33:1781, 1987.

V. Visweswaran. *Global optimization of nonconvex, nonlinear problems.* PhD thesis, Princeton University, 1995.

V. Visweswaran and C. A. Floudas. A Global optimization algorithm (GOP) for certain classes of nonconvex NLPs: II. Application of theory and test problems. *Comp. & Chem. Eng.*, 14:1419, 1990.

V. Visweswaran and C. A. Floudas. Unconstrained and constrained global optimization of polynomial functions in one variable. *J. Global Optim.*, 2:73, 1992.

V. Visweswaran and C. A. Floudas. New properties and computational improvement of the GOP algorithm for problems with quadratic objective function and onstraints. *J. Global Optim.*, 3(3):439, 1993.

V. Visweswaran and C. A. Floudas. Application of the GOP algorithm to process synthesis problems. *in preparation*, 1995.

K. Vlahos. Generalized cross decomposition: Application to electricity capacity planning. Technical Report 200, London Business School, 1991.

V. T. Voudouris and I. E. Grossmann. Mixed integer linear programming reformulations for batch process design with discrete equipment sizes. *Ind. Eng. Chem. Res.*, 31:1315, 1992.

V. T. Voudouris and I. E. Grossmann. Optimal synthesis of multiproduct batch plants with cyclic scheduling and inventory considerations. *Ind. Eng. Chem. Res.*, 32:1962, 1993.

R. S. Waghmare and H. C. Lim. Optimal operation of isothermal reactors. *Ind. Eng. Chem. Fundam.*, 20:361, 1981.

M. Walk. *Theory of Duality in Mathematical Programming.* Springer–Verlag, Wien-New York, 1989.

C. S. Wang and L. T. Fan. Optimization of some multistage chemical processes. *I & EC Fund.*, 3:38, 1964.

R. R. Wehe and A. W. Westerberg. An algorithmic procedure for the synthesis of distillation sequences with bypass. *Comp. & Chem. Eng.*, 11(6):619, 1987.

R. R. Wehe and A. W. Westerberg. A bounding procedure for the minimum nmumber of columns in nonsharp distillation sequences. *Chem. Eng. Sci.*, 45(1):1, 1990.

M. C. Wellons and G. V. Reklaitis. Scheduling of multipurpose batch chemical plants: 1. Formulation of single-product campaigns. *I&EC Res.*, 30(4):671, 1991.

M. C. Wellons and G. V. Reklaitis. Scheduling of multipurpose batch chemical plants: 1. Multiple product campaign formulation and production planning. *I&EC Res.*, 30(4):688, 1991.

A. W. Westerberg. A review of process synthesis. In R. G. Squires and G. V. Reklaitis, editors, *ACM Symposium Series*, Washington, D.C., 1980. The Am. Chem. Soc.

A. W. Westerberg. The synthesis of distillation based separation systems. *Comp. & Chem. Eng.*, 9(5):421, 1985.

A. W. Westerberg. Synthesis in engineering design. *Comp. & Chem. Eng.*, 13(415):365, 1989.

A. W. Westerberg and M. J. Andrecovich. Utility bounds for non-constant $Q\Delta T$ for heat integrated distillation sequence synthesis. *AIChE J.*, 31(9):1475, 1985.

P. Williams. *Model building in mathematical programming*. Wiley, Chichester, 1988.

L. A. Wolsey. A resource decomposition algorithm for general mathematical programs. *Math. Progr. Study*, 14:244, 1981.

T. F. Yee and I. E. Grossmann. Simultaneous optimization models for heat integration - II. Heat exchanger network synthesis. *Comp. & Chem. Eng.*, 14(10):1165, 1990.

T. F. Yee, I. E. Grossmann, and Z. Kravanja. Simultaneous optimization models for heat integration - I. Area and energy targeting and modeling of multi-stream exchangers. *Comp. & Chem. Eng.*, 14(10):1151, 1990a.

T. F. Yee, I. E. Grossmann, and Z. Kravanja. Simultaneous optimization models for heat integration - III. Area and energy targeting and modeling of multi-stream exchangers. *Comp. & Chem. Eng.*, 14(11):1185, 1990b.

Th. N. Zwietering. The degree of mixing in continuous flow systems. *Chem. Eng. Sci.*, 11:1, 1959.

Index

In this index, page numbers are shown in **bold** type to indicate a major source of information about the item being indexed. Page numbers in *italics* indicate a definition.

C

D

E

H

I

J

P

Q

R

S